DESIGN-BUILD FOR THE DESIGN PROFESSIONAL

DESIGN-BUILD FOR THE DESIGN PROFESSIONAL

G. William Quatman II, FAIA, Esq.

ASPEN LAW & BUSINESS
A Division of Aspen Publishers, Inc.
Gaithersburg New York

Printed in the United States of America

2 3 4 5 6 7 8 9 0

Library of Congress Cataloging-in-Publication Data

Quatman, G. William, 1958-
Design-build for the design professional / G. William Quatman
p. cm.
Includes bibliographical references and index.
ISBN 0-7355-1727-4
1. Construction contracts—United States. 2. Architectural contracts—United States. I. Title.

KF902 .Q38 2001
343.73′07869—dc21 00-063955

About Aspen Law & Business

Aspen Law & Business—comprising the former Harcourt Professional Publishing, Prentice Hall Law & Business, Little, Brown and Company's Professional Division, and Wiley Law Publications—is a leading publisher of authoritative treatises, practice manuals, services, and journals for accountants, auditors, attorneys, financial and tax advisors, corporate and bank directors, and other business professionals. Our mission is to provide practical solution-based how-to information keyed to the latest legislative, judicial, and regulatory developments.

We offer publications in the areas of banking and finance; bankruptcy; business and commercial law; corporate law; pensions, benefits, and labor; insurance law; securities; taxation; intellectual property; government and administrative law; matrimonial and family law; environmental and health law; international law; legal practice and litigation; and criminal law.

Other Aspen products treating construction law issues include:

Alternative Clauses to Standard Construction Contracts (Second Edition)
Architect and Engineer Liability: Claims Against Design Professionals (Second Edition)
Calculating Construction Damages
Calculating Lost Labor Productivity in Construction Claims
California Construction Law (Sixteenth Edition)
Construction Change Order Claims
Construction Delay Claims (Third Edition)
Construction Defect Claims and Litigation
Construction Industry Forms (Second Edition)
Construction Joint Venture Forms and Practice Guide
Construction Law Handbook
Construction Litigation: Practice Guide with Forms
Construction Litigation: Representing the Contractor (Second Edition)
Construction Litigation: Representing the Owner (Second Edition)
Construction Scheduling: Preparation, Liability and Claims
Construction Subcontracting Manual: Practice Guide with Forms
Construction Surety and Bonding Law
Design-Build Contracting Claims
Design-Build Contracting Formbook
Design-Build Contracting Handbook
Differing Site Condition Claims
Fifty State Construction Lien & Bond Law (Second Edition)
Handling Construction Defect Claims: Western States (Third Edition)
Legal Guide to AIA Documents (Fourth Edition)
Practical Guide to Construction Contract Surety Claims
Proving and Pricing Construction Claims (Second Edition)
State-by-State Guide to Architect, Engineer, and Contractor Licensing

Sweet on Construction Industry Contracts: Major AIA Documents (Fourth Edition)
2001 Construction Law Update

ASPEN LAW & BUSINESS
A Division of Aspen Publishers, Inc.
A Wolters Kluwer Company
www.aspenpublishers.com

SUBSCRIPTION NOTICE

This Aspen Law & Business product is updated on a periodic basis with supplements to reflect important changes in the subject matter. If you purchased this product directly from Aspen Law & Business, we have already recorded your subscription for the update service.

If, however, you purchased this product from a bookstore and wish to receive future updates and revised or related volumes billed separately with a 30-day examination review, please contact our Customer Service Department at 1-800-234-1660 or send your name, company name (if applicable), address, and title of the product to:

ASPEN LAW & BUSINESS
A Division of Aspen Publishers, Inc.
7201 McKinney Circle
Frederick, MD 21704

ABOUT THE AUTHOR

G. William Quatman, FAIA is a licensed architect and an attorney with the law firm of Shughart Thomson & Kilroy, P.C., in Kansas City, Missouri. The law firm has offices in Missouri, Kansas, and Colorado. Mr. Quatman has two degrees in architecture from the University of Kansas and his J.D. from the University of Missouri–Kansas City. He worked for general contractors and architects prior to obtaining his law degree. His law practice is devoted exclusively to the representation of design professionals, contractors, and owners in the construction industry. He is legal counsel not only to the Kansas City chapter of the American Institute of Architects but for the Builders Association of Missouri and the Mid-America Chapter of the Design-Build Institute of America. He is a faculty member and teaches courses on construction law and design-build for the University of Kansas School of Architecture. He has written many articles and spoken to industry groups on the topic of design-build, including two National AIA Conventions. He is the past president of AIA/Missouri and he has written design-build ordinances and legislation in his home state of Missouri. In May 2000, Mr. Quatman was made a Fellow of the American Institute of Architects in recognition of his leadership and service to the profession. He can be reached at (816) 421-3355 or wquatman@kc.stklaw.com on the Internet.

The views expressed in this book are those of the author and do not necessarily reflect the views of the American Institute of Architects, the Design-Build Institute of America, or any other organization of which the author is a member.

To my wife, Denise, and my family, with all my love.

SUMMARY OF CONTENTS

CONTENTS

Chapter 12
CONTRACTOR AND PROFESSIONAL LICENSING LAWS . . . 167

Chapter 13
INSURANCE AND BONDING FOR DESIGN-BUILD 195

Chapter 14

PUBLIC SECTOR DESIGN-BUILD 221

Chapter 17
DESIGN-BUILD PROJECT TYPES AND PROJECT STUDIES 273

Chapter 18
DESIGN-BUILD CONTRACTS 293

Chapter 19
COST ESTIMATING FOR DESIGN-BUILD FIRMS 309

Chapter 20
PROJECT MANAGEMENT 319

PREFACE

While writing this book, I had the honor of being named a Fellow of the American Institute of Architects, of which I am both humbled and proud. As I nearly completed my research and writing, I came upon a law book published in 1894 titled *Architect, Owner and Builder Before the Law.* The author, an architect named T. M. Clark, did not list his business accomplishments, his degrees, or even his company.[1] The only professional accomplishment he listed was seven words, in small print, just below his name; those words, "Fellow of the American Institute of Architects," and his writing told me much about the author. I would have liked to have known him, as we appear to share the same interest in law and architecture. Mr. Clark stated in his own preface, that his goal was to prepare through his legal learning and research a book that would give both the lawyer and the layman a convenient reference, stating that, "he hopes that in this way, if in no other, his work may be found of use to a profession in which he counts many kind friends."

As I read Mr. Clark's book, which no doubt served as an excellent reference in its day, I was struck by the similarities in the legal issues then and today. Most of what he wrote could be republished as an accurate description of the law 100 years later as it relates to architects. Design-build was an issue for design professionals in the 1890s and architects were concerned about ethics, contracts, fees, liability and their leadership role in construction. While we share similar thoughts, I am sure that the world was quite a different place in 1894. As I worked on this book, typing my thoughts on a laptop computer, updating my publisher through e-mail, I wondered who might pick up this book in 100 years (if we even have bound books then). How different will construction be at the next turn of the century? Will architects take back their leadership role? Will design-build be a passing phase that marked a trend at the start of the new millennium, a mere footnote in a book written by another architect/lawyer in the year 2100? Only time will tell. But it is my hope that my research and this book is of some benefit to architects and lawyers and, like Mr. Clark, that this work may be found of use to two professions in which I am fortunate to count many kind friends.

G. WILLIAM QUATMAN
Fellow of the American Institute of Architects
Kansas City, Missouri

[1] According to the Henry Withey, AIA's *Biographical Dictionary of American Architects* (Deceased), New Age Publishing Co. (1956), Theodore Minot Clark, FAIA was a Boston architect who lived from 1845 to 1909. After graduation from Harvard at the age of 20, he worked in the office of Henry Hobson Richardson. He had his own practice until 1880, when he became a Professor of Architecture at M.I.T., where he remained on the faculty for eight years. He wrote several books and was editor of American Architect and Building News from 1888 until 1909. He was active in the Boston Society of Architects and the AIA.

DESIGN-BUILD FOR THE DESIGN PROFESSIONAL

CHAPTER 1

THE GROWTH, OPPORTUNITIES, AND APPEAL OF DESIGN-BUILD

§ 1.01 INTRODUCTION

Design-build is growing at a record pace. Everyone is talking about it and some design firms are doing more than talking—they are forming their own design-build firms to take advantage of this exploding market. But before jumping on the design-build train, architects and engineers need to step back and look at the practical, legal, and financial implications of becoming "contractors." Design professionals have been trained for years to limit liability and avoid risk. Yet, the very things that design professionals have been taught to avoid (*i.e.*, warranties, site safety, indemnity, cost and time guarantees) are the same things that owners want design-build contractors to deliver.

Why haven't architects and engineers led the way to date? There are many theories: architects and engineers are great at design, but lack the ability to manage a construction project; design professionals are not risk takers; or simply that architects and engineers lack the financial resources to get bonds and buy needed equipment. Whatever the reason, some design firms have seen the growing trend of design-build, have become frustrated in their role as a subcontractor to contractors who are leading design-build jobs, and have decided to stop watching the parade and are now setting themselves up as prime design-builders.

This book looks at the business and legal concerns that an architect or engineer should consider when offering design-build services, whether as a prime contractor or a subcontractor.

§ 1.02 WHAT IS DESIGN-BUILD?

Under the traditional method of construction contracting, the owner hires a design professional (architect or engineer) under one contract, and then hires a general contractor under a separate contract. The owner, as shown in Figure 1-1, is literally caught in between these two entities. If there is a construction problem, often the designer blames the contractor and vice versa, with a lot

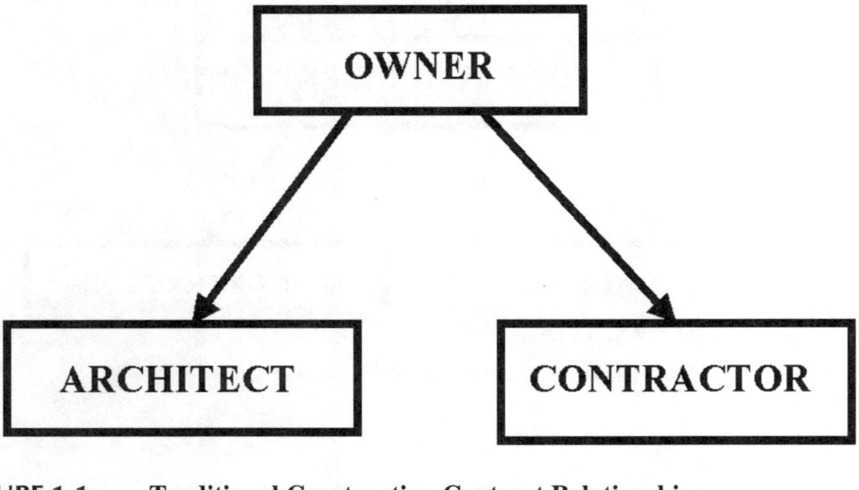

FIGURE 1-1. Traditional Construction Contract Relationships.

of unproductive time spent on finger-pointing rather than problem solving. Design-build, by contrast, is the type of construction contracting where one entity provides the owner with both the design services and the construction needed to meet the owner's program. The owner gets both the design and construction under one contract from a single source rather than the traditional method of separate contracts with the architect/engineer and with a general contractor. (*See* Figure 1-2). Its advantages over traditional contracting methods include early involvement of the contractor for input on construction methods and cost control, speed of construction and, most importantly—for the owner—single-point responsibility, as compared to "finger-point" responsibility. Design-build is growing in use nationally. However, the courts have rendered only a few decisions dealing with the topic and the law on design-build is still developing nationwide.

The American Society of Civil Engineers (ASCE) says that design-build is "coherent, contemporary and controversial."[1] It is certainly all of those things and more, with $73.7 billion worth of design-build work in the United States projected for 2000 and increases predicted each year after that. In a 1999 survey by the American Institute of Architects (AIA), a full one-third of the firms reported that design-build was a major initiative in their office. Architects and engineers should make plans for this rapidly expanding market.

§ 1.03 SUDDEN RESURRECTION OF DESIGN-BUILD

Design-build has a history as long as there has been construction in the world. Dating back to the earliest civilizations, builders of shelter also designed what they

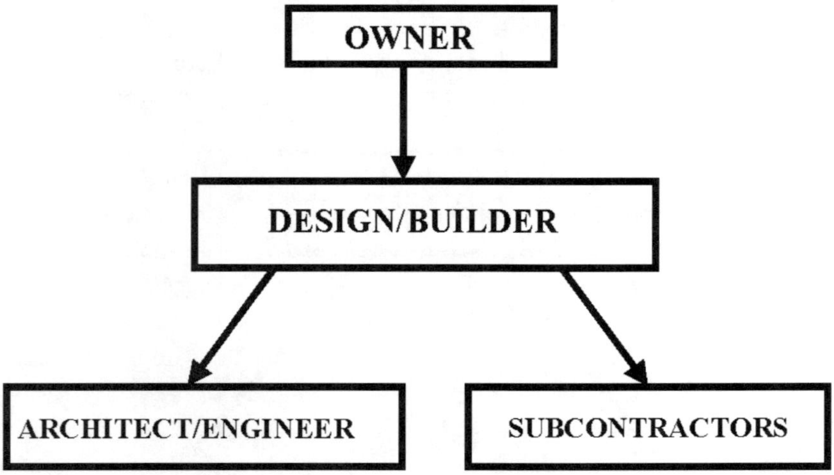

FIGURE 1-2. Design-Build Contracting Relationships.

[1] ASCE Policy Statement on Design-Build in the Federal Sector (1992).

built. The great civilizations of the past were built by Master Builders who both designed and directed the construction of the temples and other buildings. It is in more recent history that the role of designer has become separated from the act of construction. As explained in Chapter 2, from 1909 to 1978, the AIA Code of Ethics prohibited its members from engaging in building contracting, and the Institute frowned on design-build as a method of project delivery. In 1978, the AIA Board lifted the ban and in 1985, the AIA started publishing design-build contract forms. Congress permitted a limited use of design-build by the Department of Defense from 1985 to 1991 and in 1992 permitted broader use. In 1993, the Design-Build Institute of America (DBIA) was formed to promote design-build, growing from a handful of members to over 650 members in just six years. In early 1995, the AIA and the Associated General Contractors of America (AGC) published guidelines for procurement of design-build services on government projects. In 1996, Congress enacted laws to authorize all agencies of the federal government to utilize design-build. Over 30 bills were introduced in state legislatures nationwide in 1999 to permit use of design-build by various governmental agencies, and many local governments have started using design-build on everything from fire stations to city halls and prisons to wastewater treatment plants. In the United States design-build has grown from $6 billion per year in 1982 to over $73 billion predicted for 2000.

As of September 2000, over two-thirds of the states have passed some form of legislation to permit design-build on public and private projects and statistics show that design-build is taking over larger segments of the construction market each year. (*See* Appendix E to this book). Industry forecasters predict that design-build will account for 50 percent of all non-residential construction by the year 2005. According to Robert Ivy, "No one, however, can afford to ignore design-build today."[2]

§ 1.04 WHAT'S DRIVING THIS?

Most industry observers attribute the increase in design-build to two factors. The first is simplicity. One contract, as opposed to two or more, with an early fixed cost is attractive to owners. The second factor is that there are less disputes. Owners have grown tired of legal disputes, change orders, and bid conflicts. They find that design-build offers them less lawyers, less money and less time. Dr. James Broaddus says that, "Many of the disagreements between the architect and contractor are resolved by the owner of the project. The owner often ends up paying. Design-build clarifies the risks for the owner with single-source accountability and has been shown to have the least amount of claims and litigation."[3] Ellerbe Becket's president and CEO Bob Degenhardt says the traditional method of building is fragmented, as opposed to design-build which is integrated: "Currently, the construction industry operates much like it did 50 years ago: A bunch of fragments join together to construct a project, but they often have competing and conflicting

---

[2]Robert Ivy, FAIA, *Not to Be Ignored*, Architectural Record, Oct. 1998, p. 15.

[3]Comments of Dr. Broaddus, Former President of DBIA, at design-build conference in Kansas City, Missouri on January 11, 2000.

goals that lead to cost overruns, delays and litigation."[4] Design-build streamlines the process, with one point of accountability, an aspect that owners find seductive.

One design-build contractor says that while some clients like design-build due to single point responsibility, quite simply: "We find that a lot of our clients choose design-build because they're in a hurry! Design-build is a faster way to get things done." While some skeptics claim that design-build is "contractor driven" and being pushed by the DBIA, not owners, studies would show otherwise, particularly in the public sector. Design-build's popularity is owner driven, largely due to lower cost, shortened project delivery times, and single-point responsibility for project design and construction. "Faster-cheaper-better" is what owners are after, and they seem to be finding that through design-build. In its 1995 report on Design-build, the Building Futures Council found that, "Public owners that have been permitted to deviate from the traditional system have uniformly concluded that design-build and/or negotiated contracts generally produce more value for the dollar."[5] The report stated that by consolidating separate responsibilities for design and construction, areas for potential disputes are minimized. More importantly, the report found that, "In many instances, agency officials interviewed by the Committee indicated that when public agencies have utilized design-build, projects have come in 15 to 20 percent below the engineers' estimate."

A survey of over 500 construction customers in the United Kingdom in 1995 by the University of Reading found that customers wanted five things. According to the report:

1. Customers want greater value from their buildings, achieved through a clearer focus on meeting the functional business needs of the customer.
2. Customers' immediate priorities are to reduce capital costs and improve the quality of new buildings. More important in the longer term is the reduction of occupancy costs.
3. Customers want single-point responsibility, the key reason to use design-build, plus a guaranteed maximum price and the avoidance of design and construction risks.
4. Customers believe that significant value improvement and cost reduction can be gained by the integration of design and construction, and that they can achieve greater certainty of time and cost by using design-build.
5. Finally, customers consider that contractors can add significantly to the design process, but they do not believe that contractors have the experience and understanding of the creative design process and how it can add value to their buildings.[6]

---

[4] *Ellerbe Becket Develops New Strategy*, Kansas City Star, June 12, 1998.

[5] *Report on Design-Build as an Alternative Construction Delivery Method for Public Owners*, Building Futures Council (1995).

[6] *Designing and Building a World-Class Industry*, Project Delivery Report, AIA Design-Build PIA, Sept. 1998.

The report concluded that design-build can meet these customer needs, but that to be successful, design-build "needs to become design-led."

§ 1.05 GROWTH OF DESIGN-BUILD

Owners have grown tired of the disputes, egos, and lawyers that seem to drive many construction projects. So they have turned to design-build in increasing numbers, seeking a better way to streamline construction, with fewer delays and disputes. In 1982, design-build accounted for about $6 billion in construction dollars. By 1985, design-build was only 10 percent to 15 percent of the building market.[7] Then, from 1986 to 1992, design-build grew from $18 billion to $49 billion, a jump of 172 percent.[8] The number of design-build contracts then doubled between 1995 and 1996.[9] In 1997, revenues among the nation's top 100 design-build firms topped $39 billion with design-build increasing to 33 percent of the construction market in 1998. Design-build is growing in every market sector. Statistics reported in 1998 show increased use of design-build in every major construction group.[10] Here are the reported numbers showing increases from April 1995 to April 1996:

- Surface Transportation up 81%
- Public Buildings up 96%
- Commercial Facilities up 135%
- Transportation Buildings up 149%
- Sports Facilities up 151%
- Industrial/Manufacturing up 319%

These figures, especially in surface transportation (*i.e.*, roads, highways, light rail), are probably higher since these numbers were reported now that the whole country seems to be embracing design-build in greater numbers each year. Statistics compiled by DBIA and F.W. Dodge show that the total number of design-build contracts is increasing by more than 100 percent annually. The use of design-build in large civil projects is growing by as much as 150 percent each year, depending on the category of construction.[11] According to the U.S. Department of Commerce, construction had its eighth consecutive year of overall growth in 1999, increasing by 6 percent to $705 billion. Residential construction accounts for 45 percent of the total, with public and non-residential construction

[7] *The Dollars Behind Design Rights*, Engineering News Record, May 15, 1995, p. 9.

[8] DBIA's Introduction to Design-Build (1994), quoting Engineering News Record in 1993.

[9] *It Takes Cooperation*, Engineering News Record, August 18, 1997, p. 114.

[10] *Design-Build Survey*, Zweig White & Assoc., Inc., 1998, p. 21.

[11] *Designer-Led Design-Build: Why it Works for Contractors*, Design Build Intelligence, Oct. 15, 1999.

building accounting for $342 billion.[12] With an estimated 33 percent market share of design-build work, this represents more than $112 billion of construction work that will be passed up if a firm is not involved in design-build.

Worldwide construction is reportedly over $3.6 trillion in 1999,[13] with much of the rest of the world already using design-build, turnkey, and Engineer-Procure-Construct (EPC) contracts on a regular basis. Opportunities for design professions in design-build are not limited to the United States.

§ 1.06 MORE DESIGN PROFESSIONALS ARE GOING INTO DESIGN-BUILD

Architects and engineers have some catching up to do in the area of design-build and many firms are finding that they are already behind the contractors who are leading the majority of design-build projects. An AIA task force evaluating use of design-build among architects in 1979 reported that of some 3,682 firms surveyed, only 10 percent indicated some experience with design-build, with the remaining 90 percent reporting no experience at all with design-build.[14] Times have changed and according to AIA statistics in 1997, half of all large AIA firms (over 50 employees) were offering design-build services.[15] In just six years, the AIA found that the number of architectural firms doing design-build nearly doubled from 1991 to 1997, jumping from 13 percent of all AIA firms to 23 percent. In the next three years, from 1996 to 1999, the number of firms leaped to 36 percent according to the AIA's survey. What is taking them this direction? For some it is a chance to take back a leadership role that many architects feel escaped the profession years ago; for others it is the challenge of a new method of delivery; still others report that their clients have dragged them into this market, insisting that the architect also become their builder. Brad Buchanan, AIA, owner of an architect-led design-build firm in Denver says, "Design-build is exciting to be a part of. I don't need to set my alarm clock to get up. This keeps me charged up about architecture." Buchanan's firm ranked 12th on the list of Colorado's largest design-build contracts in 1999, and 25th on the list of Colorado's largest general contractors.

In addition to the excitement shared by many architects about the control they have through design-build, others see the process as providing a better solution to the client's needs. As Hill Burgess, AIA stated,

> First, we recognized that our clients were often frustrated during the bidding and construction phases of our projects because the traditional design-bid-

---

[12] *Housing and Public Works Drive Construction,* Engineering News Record, Feb. 28, 2000, p. 27.

[13] *Id.*

[14] AIA Design-Build Contracting Monitoring Task Force Report, May 1981, as reported by H. Block in *As the Walls Came Tumbling Down: Architects' Expanded Liability Under Design-Build/Construction Contracting*, 17 John Marshall L. Rev. 1, 9 n.36 (1984).

[15] *Marked Increase in Design-Build Services*, AIArchitect, Jan. 2000, p.19.

> build project implementation method was difficult for them to understand. Simply put, the tug-of-war between the designer and the builder wore them out. Second, we believed there were certain types of owners that would be attracted to designer-led design-build projects—owners who desired to have the traditional close working relationship with their architect and who would not normally seek a contractor to lead their design-build team.[16]

The AIA's ethical prohibition until 1978 held back many architectural firms from doing design-build, whether as a prime or as a subcontractor. In addition, during a period when malpractice claims rose to a record high in the mid-1980s, risk-conscious design firms pulled back on many services and tried to limit their liability by avoiding high-risk services, especially any extensive role during construction. As one architect stated, "Design-build takes a high risk tolerance. Architects have been backpeddling for 30 years from risk. Developers, contractors and construction managers have all come in and taken on those risks." Now that others have filled the voids that were left in the process, most architects feel that they have lost their leadership role in construction. Design-build, especially if contractor-led, presents a further reduction in the architect's role by placing the designer in a subcontract position, with no direct contract with the client. Designer-led design-build presents an opportunity to regain the leadership role for firms who are also willing to take on more risk.

§ 1.07 OVER HALF OF THE CONSTRUCTION MARKET IS PREDICTED TO GO TO DESIGN-BUILD

There have been numerous predictions about just how much of the market design-build will take up and when. Here are a few of the published predictions:

- Design-build is expected to increase to 50 percent of the market by 2005.[17]
- Design-build "is expected to grow to 60 percent by 2005."[18]
- Design-build is projected to account for more than 50 percent of all projects by the year 2006.[19]
- "The projection is that by Year 2005 there will be as many projects using design-build as the traditional design-bid-build method of delivery."[20]

In addition to these published predictions, in 1996 an AIA panel speculated that design-build would be used for 60 percent to 75 percent of projects in 10

[16] *Award Winners—Great Design on Small Design-Build Projects*, Hill Burgess, AIA, DBIA Conference, Dallas, Texas (Oct. 1999).

[17] DBIA projection, as reported in NSPE's Engineering Times magazine, Nov. 1999.

[18] Seminar Brochure published jointly by AIA and the Canadian Design-Build Institute, Jan. 2000.

[19] Brochure for *Performance Specifying*, Developing RFP Requirements, sponsored by DBIA and the Construction Specification Institute (CSI).

[20] *Design-Build Advisor*, Zweig White & Assoc., Aug. 31, 1998.

years.[21] Eugene Kohn, FAIA stated that he thought design-build would peak at about 70 percent in five years and then decline to about 40 percent. According to the AIA's report of the panel, Kohn's impression is that design-build in the commercial sector is the result of corporate downsizing, and when this cycle is over many of the reasons to use it will no longer exist. Michael Graves, FAIA, declined to give a prediction, responding that growth depends on how well design-build is done.[22]

§ 1.08 TREND PROJECTED TO CONTINUE

A survey by Zweig White & Associates, Inc., published in 1999, reported that the design firms surveyed generated an average of 8 percent of total revenues from design-build projects, up from 6 percent in 1996.[23] The firms surveyed predicted that those numbers would increase to 15 percent by 2004. For firms that provide both design and construction, design-build work increased from 19 percent in 1996 to 22 percent in 1998. One 1999 survey of A-E firms found that 50 percent of design firms said they expected to seek more design-build alliances. Only 3 percent wanted less.[24] According to the survey, design firms expected a medium increase in revenue from design-build work of 10 percent over the next three years, 15 percent over five years, and 25 percent over the next 10 years. Those firms already into design-build, providing both design and construction services, were even more optimistic, projecting growth rates of 20 percent over three years, 30 percent in five years, and a staggering 50 percent over the next 10 years.

"We foresee continued growth in the design-build market, given almost universal acceptance of the project delivery method," said a senior vice president of Opus Group, Minneapolis, one of the largest design-build firms in the country. Echoing this comment was Kevin O'Toole, executive vice president of Hunzinger Construction Company of Brookfield, Wisconsin, who says, "Without a doubt, design-build remains the fastest growing delivery system in the industry. We continue to adjust our strategic initiatives to take advantage of this contracting method."[25]

And what if it does not continue? Brad Buchanan, AIA says, "If the market changes, we could apply what we have learned to any sector that is strong."[26] Having been a contractor makes one a better architect. The knowledge an architect learns about estimating, scheduling, and project management will improve his/her abilities as an architect on traditional project delivery as well. In addition,

---

[21] The AIA panel was composed of James Busam; Michael Graves, FAIA; James Gray; Eugene Kohn, FAIA; and Jack Wolfe.

[22] *Stanford/DBIA Professional Design-Build Conference,* AIA Project Delivery Report, Winter 1996.

[23] *1999 Design-Build Survey*, Zweig White & Assoc., Inc. (1999).

[24] 1999 DFMAR Survey, Design Firm Management & Administration Report, Dec. 1999.

[25] *Masters Raise Bridge to 21st Century,* Design-Build, Dec. 1999.

[26] *How to Increase Your Fees in a Tough Market,* Architectural Record, Nov. 1999, p. 62. Brad Buchanan, AIA, is a partner in the Denver, Colorado design-build law firm of Buchanan Yonushewski Group, L.L.C.

the relationships an architect makes with local contractors, subcontractors, and suppliers should improve the success of projects in other markets.

§ 1.09 DESIGN-BUILD CAN BE FASTER AND CHEAPER

Speed of the entire design and construction process is compressed through design-build. Fast-track construction is the norm, not the exception, with some projects using an even expedited "flash track" or "hyper track" process that is delivering completed facilities to owners months or even years ahead of traditional procurement. The traditional design phases of schematic, design development and construction documents are now blurred into one or two phases of preliminary and final design, with the true "final" design not being complete until construction is finished.

Both public and private owners are using design-build with greater frequency. The Navy reported a 15 percent savings in construction costs through design-build and a 12 percent reduction in facility delivery time. As a result, the Navy reported that its use of design-build would triple from around 10 percent in fiscal year 1998 to 30 percent in fiscal year 1999.[27] A study by the Boston-based Pioneer Institute for Public Policy Research found that significant savings could be realized through design-build.[28] The Florida Department of Transportation found that under design-build, change orders were less than 2 percent of award cost, compared to 8.78 percent, under traditional design-bid-build. The same report showed that design-build construction time was 21.1 percent shorter than normal.

A 1997 study of 351 projects by the Construction Industry Institute compared cost, schedule and quality of design-build projects with those delivered under traditional design-bid-build or construction management (at risk).[29] Projects were submitted by a number of entities, both public and private, from 37 states, spanning a period of seven years. The study found that design-build was overall 6 percent cheaper and 33 percent faster than traditional design-bid-build. Projects were compared based on cost escalation (by change orders), and the results were as follows:

Contracting Method	*Cost Increase*
Design-bid-build	4.83%
Construction management (at risk)	3.37%
Design-build	2.17%

[27] Design-Build Intelligence, RS Means, Nov. 15, 1998, p. 4.

[28] D. Gransberg, *Change Procurement Laws*, Engineering News Record, Dec. 6, 1999, p. 71. The full study is available on-line at http://www.pioneerinstitute.org/.

[29] *Project Delivery Systems: CM at Risk, Design-Build, Design-Bid-Build*, Construction Industry Institute, Research Summary 133-1 (Dec. 1997).

The study found that nearly half of all design-bid-build projects experienced cost growth greater than 5 percent, but only a third of design-build projects had that type of cost increase. The study also found that:

> Having the ability to have construction involvement early in the process was the most common reason for achieving contracted cost. This was particularly evident for design-build, where often the engineer and contractor are the same firm, they have worked together in the past or have previous design-build experience . . . This synergy has proven very valuable in the planning and design phase for value engineering and constructability improvements.

The design-build projects had almost no delays on average, compared with schedule extensions of 4.4 percent on average for traditional projects. In terms of speed, design-build produced a median activity of 9,000 square feet per month, compared with 4,600 square feet for traditional design-bid-build. All three methods had about the same results in terms of quality, performance and cost of operation of the completed facility.

A similar study of 325 projects in the United Kingdom in 1995 found that design-build construction speed was 12 percent faster and total project delivery speed 30 percent faster than that achieved using traditional approaches.[30] The study found that 70 percent of design-build projects were completed on time versus 56 percent

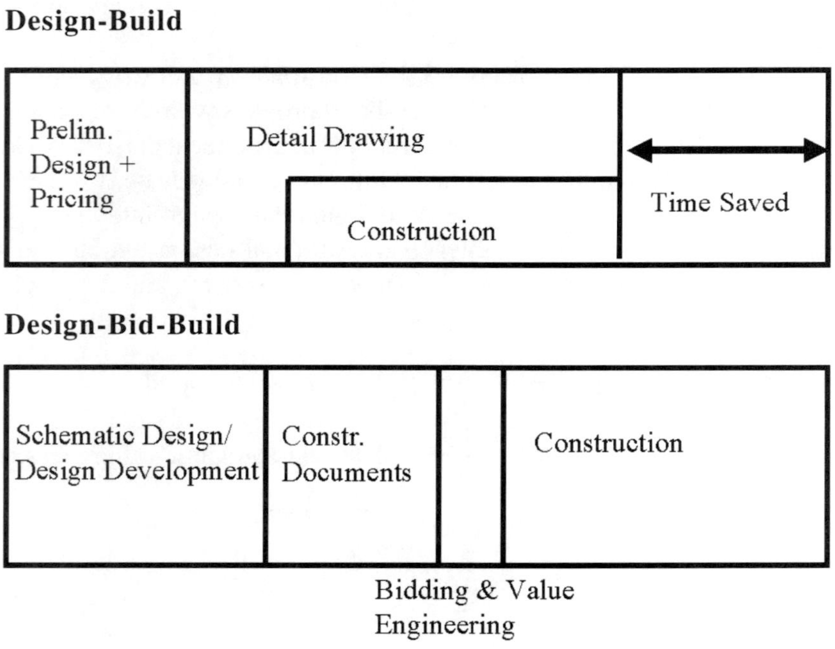

FIGURE 1-3. Comparison of Design-Build to Design-Bid-Build Sequencing.

[30] *Designing and Building a World-Class Industry*, Project Delivery Report, AIA Design-Build PIA, September 1998.

of traditionally procured projects. The reason that design-build can be so much faster is that design and construction phases are blended into one continuous phase, where construction starts before design is completed and design continues on portions of the project as others are being built. (*See* Figure 1-3).

§ 1.10 BUT IS IT BETTER?

This question is often asked by owners who understand the cost and time savings of design-build, but question the quality of construction. Others wonder about the aesthetics of buildings designed by contractors. Bob Degenhardt, president and C.E.O. of Ellerbe Becket says, "You can drive around and identify buildings that are design-build because they're pretty ordinary." While that may have been true in the past, there are an increasing number of outstanding, award-winning designs coming from design-build teams—even when contractor-led.

The quality of the building is assured to be what the owner wanted according to the Design-Build Institute of America. DBIA observes that in traditional construction, the *owner* warrants to the contractor that the design is suitable[31] and the contractor warrants that its work will comply with the design. However, if the design is flawed, then the contractor is not responsible for the end result. With design-build, however, DBIA says that, "The Design-Builder warrants to the Owner that it will produce design documents that are complete and free from error," thereby producing a guaranteed quality project. Some studies that have been done support the concept that design-build can produce projects of equal quality to those procured by the traditional process.

The study of 325 projects in the United Kingdom done by the University of Reading, asked owners and project teams about 16 aspects of quality. The data revealed that the quality of design-build projects was lower than that achieved using traditional project delivery, but not by much. As reported by the study, "Overall, quality targets were met on 60 percent of traditionally procured projects and 50 percent of design-build projects."[32] On aesthetics, surprisingly, design-build was found to provide "superior consistency in aesthetic quality and, overall, marginally better aesthetic quality than traditional approaches."

The 1997 study by the Construction Industry Institute looked at quality of the completed project in terms of turnover-start up difficulty, call backs, and operations and maintenance costs, as well as quality of system performance.[33] Using these criteria, the study found that "Despite the subtle differences, design-build overall scored higher than its counterparts."

Some of the concerns about the quality of design-build come from the architectural profession itself. As two Chicago architects stated, "Architects asso-

[31] The owner's implied warranty of the design is discussed in more detail in Chapter 21, §21.09

[32] *Designing and Building a World-Class Industry*, AIA Project Delivery Report, Sept. 1998, p. 11.

[33] *Project Delivery Systems: CM at Risk, Design-Build, Design-Bid-Build*, Construction Industry Institute, Research Summary 133-1 (Dec. 1997).

ciate design-build with cheap, badly designed buildings. The most arrogant among us believe that only the architect working independently of the builder, and sometimes the client, can protect a project's design integrity."[34] But these preconceptions about design-build are being dispelled by the satisfaction of owners and the repeated use of this method by those same owners. In addition, design awards are being given to design-build projects, showing that good design can come from design-build. At a conference co-sponsored by DBIA and Stanford University in October 1996, Michael Graves, FAIA, gave a presentation on design-build projects his firm had done, including the Portland Building and numerous retail installations for Lenox China. Graves showed that award winning design can be achieved with design-build but he also stressed the importance of working closely with the contractor from the outset to insure that the design can be built within the budget.[35]

§ 1.11 CONTINUED GROWTH LIKELY

A booming U.S. economy sparked incredible growth in the construction industry in the 1990s with an average increase of 8 percent per year, a trend that is projected to continue into the new millennium.[36] The U.S. Department of Commerce business census found that from 1992 to 1997, the number of architectural firms in the United States increased by 15 percent, the number of employees in the field increased by 21 percent, and gross billings jumped more than 50 percent to $17 billion in 1997.[37] During this same period construction owners began to rediscover the benefits of design-build as a method of project delivery.

One reported reason for the increased use of design-build is corporate downsizing. Owners have reduced their number of facilities managers and no longer have sufficient staff to handle "unbundled" services. They are looking for one-stop shopping.[38] Design-build provides a cleaner way to purchase their construction services. At a 1996 design-build conference, a spokesperson for the U.S. Postal Service told attendees that that agency had downsized its facilities staff from 1,100 to 550, and found that they needed to reengineer how they operated to cope with this reduction in staffing. Design-build was one solution that offered a number of benefits, including: (1) reduction in project time; (2) less cost; (3) accountability in one contract; and (4) lower administrative costs.

There is reported growth in parking garages, prisons, process plants, transportation projects, and even sports arenas. Utah's new $75 million state court-

[34] *Architects' Attitudes Toward Design-Build Are Ill-Conceived and Self-Defeating*, Architectural Record, Feb. 1999, p. 22.

[35] *Stanford/DBIA Professional Design-Build Conference*, AIA Project Delivery Report, Winter 1996.

[36] *The Longest Expansion in Construction History: Can It Last?*, Engineering News Record, Oct. 18, 1999.

[37] *Government Documents Recent Evolution of Architectural Profession*, AIArchitect, Feb. 2000, p. 4.

[38] *Design-Build Continues to Grow*, Engineering News Record, June 16, 1997, p. 48.

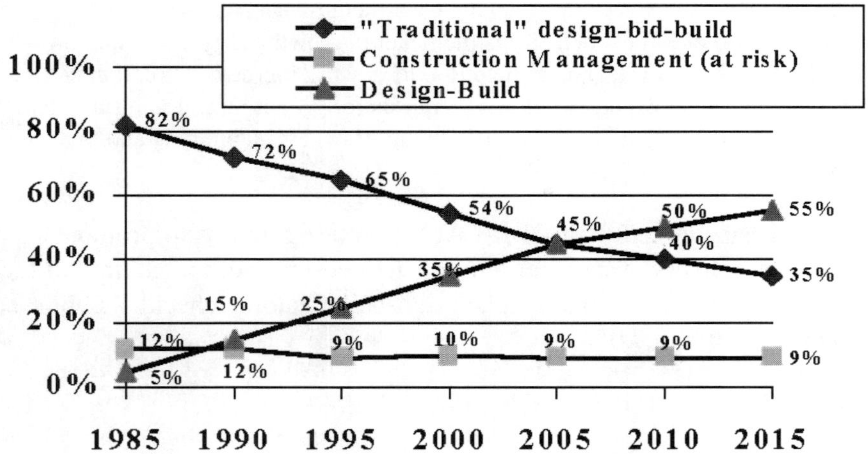

FIGURE 1-4. Projected Growth of Design-Build, Construction Management, and Traditional Method.
Source: The Design-Build Institute of America

house was done using design-build. Prior to the project, Utah law barred design-build. The state legislature changed the law and cut years off the project schedule through design-build. The biggest design-build project in the country in the 1990s was the I-15 project, a $1.325 billion highway reconstruction project in Salt Lake City, awarded in 1997 to Kiewit Construction Group in preparation for the Winter Olympics. The request for proposal for the project was reported to be some 40,000 pages.[39] In addition to I-15, there are 25 federal transportation pilot design-build programs.[40]

As more legislation is passed and as insurance and bonds become available for design-build, we will likely see a continued growth of design-build nationally in both public and private sector work. Reaching 50% of the market by 2006 does not seem unrealistic.

§ 1.12 WHAT HAPPENED TO CONSTRUCTION MANAGEMENT?

While construction management grew in the 1980s, design-build has overtaken it in the 1990s. The Construction Management Association of America (CMAA), like other organizations, is looking at how it can fit in with an industry that appears heading like a freight train toward design-build. According to DBIA projections, design-build will continue to increase in usage, leveling off at about 55 percent of the market, while construction management will hold steady at about 9 percent of the market. (*See* Figure 1-4). Construction managers are finding a new role assisting owners on design-build projects. In a 1999 article in *Engineering News Record*, it was reported that:

[39] Sen. Craig Patterson, Utah State Senator, sponsor of Utah's design-build law, speaking at the DBIA/AIA Annual Meeting, Oct. 1998, Chicago, IL.

[40] *Design-Build Continues to Grow*, *supra* note 38.

> CMAA members believe that teaming construction managers with the design-build method of project delivery is the ticket to growth. They claim that owners lacking in-house technical help due to downsizing of construction staffs need independent experts to prepare requests for proposals and to ensure that plans and expectations match by reviewing drawings, submittals and quality control. [41]

CMAA has worked with DBIA, AIA, and other groups to promote support for all project delivery systems at a time when design-build is getting all the attention. The groups collaberated on a design-build teaming checklist published in 1999. *See* Chapter 7, § 7.05.

§ 1.13 WHERE WILL DESIGN FIRMS BE?

The AIA's 1979 survey of member firms found that though 9 out of 10 firms had no experience with design-build, a large percent of firms thought design-build was a good idea because owners prefer to work with an architect rather than a general contractor, more and more owners favor design-build, and an increasing number of owners favored architect control of design-build projects.[42] Despite the AIA's early optimism for designer-led design-build the workload is decidedly slanted toward contractor-led teams today. Approximately 65 percent of all design-build projects are contractor-led.[43] Most others are led either by an integrated firm, with both in-house design and construction staffs, or by joint ventures of design firms and contractors. At the start of the new millennium, however, a very small percentage of design-build projects are designer-led. This raises the question of where most architects will be if the target prediction of a 50 percent market share of all construction is design-build by 2005 or so. Most design firms will participate in design-build out of necessity, but so far only a select few have ventured out into the lead role of prime design-build contractor. There will no doubt be plenty of business for design firms who want to be subcontractors, but is that really where architects and engineers want to be? Design-build is predicted to be the preferred method of delivery clients will want to use. Architects and engineers will need to decide what level player their firm will be. As Will Rogers said, "You can be on the right track and still get run over if you're standing still." Is a subcontractor the right track for an architectural firm?

Architects Clifford W. Bedar and Michael W. Behm observed that while many architects lament their "dwindling influence on the design and construction industry," few have taken action to change this situation. However, they note that there is hope, stating, "Right now, design-led design-build is whispering in our ears the secret to regaining influence over the built environment. Will we hear the

---

[41] *CMAA Seeks to Broaden Appeal of Construction Management*, Engineering News Record, Sept. 27, 1999, p. 16.

[42] AIA Design-Build Contracting Monitoring Task Force Report, *supra* note 14.

[43] *Designer-Led Design-Build: Why it Works for Contractors*, Design Build Intelligence, Oct. 15, 1999.

message or continue to turn our backs on the opportunities this presents."[44] They observe that design-build offers a chance to recapture professional and financial success and urge that "to sustain the profession and the public interest, architects must take the lead in design-build."

§ 1.14 DESIGN-BUILD IS NOT FOR EVERYONE

Design-build is not for the faint-of-heart. It involves new risks and requires new skills. It is not for everyone. Some design firms have dabbled in design-build, gotten burned, and have quickly retreated. As explained more fully in Chapter 8, § 8.01, there may be certain personality traits inherent in design professionals that prevent them from ever being comfortable in the role of contractor. As one architect put it, "A well-tuned team of architect and contractor will play against each other's strengths, not rub across each other's differences, and take into account the builder's materials and craftsmen. A designer whose self-image would be rattled by this work method should not get involved in design-build."[45]

Frank Gehry, FAIA, winner of AIA's 1999 Gold Medal says that, "Design-build is dangerous. Design-construct is not the way to go."[46] Like Gehry, not all design professionals are embracing design-build with open arms. Past AIA Design-Build PIA chairman Ron Gupta, AIA said that a number of AIA members have not yet accepted the trend toward design-build. AIA statistics in 1997 showed that three-fourths of AIA member firms were still not even offering design-build services to their clients.[47] Many architects see design-build as another specialty that is part of the continuing chipping away at the practice of architecture, and diminishes the value of design, by packaging it in with construction. With contractors taking the lead, and architects reduced to a subcontractor—with no direct relation with the project owner—many design firms are concerned that design-build will cause damage to the relationship of trust and counsel that has existed between the owner and designer for decades. AIA's Gupta says, "Design-build is for entrepreneurial types. If the leaders of your firm are not energized by the challenge of market risk, design-build is not for you."[48]

In contrast to Frank Gehry's comments, renowned British architect, Sir Norman Foster, Hon. FAIA, of Foster and Partners, was quoted as saying:

> I have always believed in the team approach to realizing buildings. Ideally, those concerned with construction, particularly the specialists should be involved from the earliest stages of the project. I believe that is it impossible to design a building without knowing how it goes together and how much it will

---

[44] *Architects' Attitudes Toward Design-Build Are Ill-Conceived and Self-Defeating*, Architectural Record, p. 22, Feb. 1999.

[45] *Design-Build: The New Boys in Town*, The Construction Specifier, June 1990, p. 151.

[46] *Frank Gehry: Plain Talk with a Master*, Architectural Record, May 1999, p. 189.

[47] *Marked Increase in Design-build Services*, AIArchitect, Jan. 2000, p. 19.

[48] *Design-Build*, Ron Gupta, AIA and Paul Doherty, AIA (draft of chapter for AIA Handbook of Professional Practice (1999).

cost. The disciplines are inseparable. But it is essential . . . that this process be led by design.[49]

Commenting on the study done by the University of Reading on design-build, Foster said, "This report sets the scene for best-practice in the field of design-build and shows how it can aspire to world class standards."

In an editorial comment, *Architectural Record's* Chief Editor Robert Ivy, FAIA said that "Design-build is not for everyone, as team member or leader." He continued:

> It is a subject we try to avoid, one we shy away from discussing. Many architects neither understand nor want to understand design-build. Preconceptions and lack of knowledge about this project delivery method abound . . . Architects often lament ceding their traditional, central role in planning and design to someone else, a nebulous "design-build entity" that employs the architect and horns in on what had been the design professional's turn. Some architects fear losing money; others fear the unknown.[50]

With the strong economy as we have had in the 1990s, some design firms see no benefit in jumping on the design-build bandwagon. "It's not an anti-design-build movement. They are asking themselves: 'In a market this good, why should I take risks in design-build?'"[51] If the forecasts are correct, the risk of not getting involved in design-build is that the competition will gain experience with this process and obtain a marketing advantage while others take a more conservative approach, waiting to see if the storm passes before venturing outside of their traditional practice.

For those who move their firms into design-build, they will find a mixed response from contractors to the idea of designers leading the design-build team. As one commentator said: "The reactions of most contractors to designer-led design-build has been more emotional than rational. They denigrate architects' (and to a lesser extent engineers') ability to manage complex construction projects and hasten to warn owners of the risks inherent in the A/E's limited financial capacity."[52] But many design firms are successfully leading design-build teams and finding contractor partners willing to let them lead. NBBJ is ranked as the fifth largest design firm in the country for general building projects. The company's chairman, Freidl Bohm, says that his firm is often in the lead role in design-build and that, "[w]e have no fear of the contractor and are even willing to assume some of the risk, where appropriate." He says that contractors are now more willing to work in partnership with architects, "because [architects] generally have a

---

[49] *Designing and Building a World-Class Industry*, Project Delivery Report, AIA Design-Build PIA, Sept. 1998.

[50] Robert Ivy, FAIA, *Not To Be Ignored*, Architectural Record, Oct. 1998, p. 15.

[51] Fanning, *Design Firms See Design-Build Growth but They Remain Wary*, Engineering News Record, Sept. 27, 1999, p. 15.

[52] *Designer-Led Design-Build: Why It Works for Contractors*, Design Build Intelligence, Oct. 15, 1999.

better sense of the project than the contractor."[53] Another executive of a leading national design firm says that architects and engineers fall into two camps: those who learn to thrive on design-build and those who will avoid it. He adds that, "Those in the latter group will end up falling into a commodity business."[54]

§ 1.15 THE INTERNATIONAL DESIGN-BUILD MARKET

Worldwide construction spending in 1999 was estimated at in excess of $3.5 trillion, an 11.8 percent increase over 1998's estimate.[55] A-E firms based in the United States now earn, on average, 7 percent of their fees from overseas commissions.[56] According to DBIA publications, "Design-build is the project delivery system of choice on more than 50 percent of the non-residential construction projects in the European Community."[57] Firms gaining experience in design-build will find an open market overseas where design-build is the norm, not the new trend.

[A] Asian Markets

DBIA reports that design-build is used on more than 70 percent of the non-residential projects in Japan.[58] Design-build started to develop in Hong Kong in the late 1980s, and today Hong Kong is one of the biggest design-build markets in the Asia-Pacific region.[59] Bridge and highway projects are usually awarded on a design-build basis, as have been railroad projects, residential blocks, and school projects. Government officials seem to have openly embraced the concept of "best value" in awarding public contracts. Most notably, controversy was stirred when the Architectural Services Department (ASD) of the local government awarded a $297 million design-build contract to the *highest bidder*, a joint venture between a local contractor and a British company, whose bid was $39 million above the low bid. The government defended the award, stating that it offered a better design with maintenance savings.

Singapore is said to be Asia's second most active design-build market, with a single source $5.6 billion contract awarded for Singapore's new deep-tunnel sewer network.

[B] The United Kingdom

"It is widely accepted that buildings in the U.K. cost more and take longer to design than those in other major economies. This handicaps the U.K. manufac-

---

[53] *Hot Markets Shift Into Higher Gear*, Engineering News Record, Apr. 10, 2000, p. 56.

[54] *Id.*

[55] *Reports Measure World Market*, Engineering News Record, Dec. 6, 1999, p. 12.

[56] *Global Market*, Principal's Report, IOMA, Inc., Jan. 2000.

[57] DBIA's Introduction to Design-Build (1994).

[58] DBIA's Introduction to Design-Build (1994).

[59] *Hong Kong Still on Fast Track*, Design-Build, Oct. 1999, p. 20.

turing industry when competing abroad and reduces its ability to satisfy domestic needs," says John Carpenter, chair, the University of Reading Design and Build Forum.[60] The University of Reading Forum was set up in 1995, prompted by a need to improve construction performance in the United Kingdom. The Forum published its findings after 18 months of research. The report found that design-build emerged within the United Kingdom building industry during the 1970s and early 1980s to provide a greater degree of certainty for clients who wanted an integrated form of project delivery to overcome cost and time overruns and reduce costly claims and possible litigation—the same concerns that have driven design-build in the U.S. The report found that use of design-build had stalled out at about 23 percent of all new construction and that design-build would decline if the process did not develop to meet the demands of customers. In order to meet those demands, the report surveyed over 500 customers and looked at over 325 projects. The report found that projects using design-build had total project delivery speed much faster than that achieved using traditional approaches. See § 1.09, above.

[C] Other Markets

Design-build is reported to be the most common approach to power plant construction in Latin America. Two large plants in eastern Venezuela, valued at $750 million and $900 million respectively, are being done using design-procure-construct.[61] Roads and highways are also moving into design-build in South America, just as in North America. Two large design-build road projects are being built in Chile and a $380 million bridge project is under way in Argentina over Rio Tarana.[62] Design-build firms in the United States see fast growth ahead for South America in every market sector.

Houston-based energy conglomerate Enron International used design-build to develop a power plant in Dabhol, India. The 826-megawatt first phase was completed in December 1998, followed by a 1,624-megawatt second phase, planned for completion in March 2001. Dabhol Power Co. is half-owned by Enron, but design-build contractor Bechtel reportedly holds a 10 percent share in the company.[63] The total Dabhol project has a $2.65 billion price tag. Using its seven worldwide engineering centers, Bechtel is able to benefit from 24-hour design operations, a system some engineers call "following the sun." As the work day ends in one time zone, CAD files are sent across the globe where other engineers are just starting their day. By repeating this cycle, design work continues around the clock.

---

[60] *Designing and Building a World-Class Industry*, Project Delivery Report, AIA Design-Build PIA, Sept. 1998.

[61] *An Uncertain Embrace*, Design-Build, Apr. 2000, p. 40.

[62] *Id.*

[63] *Id.*

CHAPTER 2

THE HISTORY OF DESIGN-BUILD

§ 2.01 ANCIENT ROOTS

Historians trace the roots of design-build back to at least 2630 B.C. and the great Egyptian master builder Imhotep, designer of the Step Pyramid. A commoner at birth, Imhotep was an architect, builder, and reportedly a genius in many other fields, including medicine, mathematics, astronomy, and poetry.[1] He held the titles of high priest of Ptah, chancellor, administrator of the palace, builder, and sculptor. Imhotep was later deified and worshipped as the god of medicine, not only in Egypt, but also in Greece.

Though not designed by Imhotep, one of the greatest engineering feats in all of civilization is the Great Pyramid at Giza. The pyramid is one of more than 100 monuments commissioned by pharaohs. The Great Pyramid is estimated to have taken 10 years to build, with two to three years of site preparation, five years of actual construction, and two years of delivery ramp removal. It was completed in 2566 B.C.[2] Each side of the Great Pyramid is 756 feet long and aligns perfectly to the compass points. For 43 centuries, its height of 481 feet made it the tallest man-made structure on the planet. The Design Build Institute of America (DBIA) has adopted the pyramid as its logo in recognition of the time-honored role of the master builder. Nobody is certain when the art of architecture was separated from the craft of construction. There have been many master buiders throughout history, known to be both the project designer and builder. Some historians have traced the separation to the 18th century in Europe.

§ 2.02 SEPARATION OF ARCHITECT FROM CRAFTSMAN IN EUROPE

During the Italian Renaissance, Filippo Brunelleschi, sculptor and architect was a master builder of the great dome of the cathedral in Florence, Italy. Born in Florence in 1377, he began his career as a sculptor and goldsmith. In 1401, Brunelleschi tied with Lohrenzo Ghiberti in the famous design competition for the bronze doors of the Florence Baptistry. Though he continued as a sculptor, he made several trips to Rome to study design and construction of early Roman buildings. From 1402 on he devoted his career to the design and construction of the dome over the unfinished cathedral of Florence. In 1420, he was put in complete charge of the project and remained chief architect and director of the work until his death in 1446. Brunelleschi's dome in Florence stands as one of the great engineering accomplishments of the Renaissance period.

As late as the late 1700s and early 1800s in France, it was very common for the architect to hold all the trade contracts and to serve as the general contractor in order to keep peace among the trade guilds. Professional egos and status issues, however, drove some architects away from this contractor role. It is reported that "there were certain architects, particularly those employed by the

[1] *Master Builders of History*, Dateline, DBIA, Mar. 2000, p. 8.

[2] *The Enduring Wonder of the Built World*, Design-Build, Feb. 2000, p. 55.

Government, who held aloof from contracting, and prided themselves on their strictly advisory, or, as we should say, professional character; and this attitude of architects toward their clients came by degrees to be the recognized and usual one."[3] The "usual" method, which we now ironically refer to as the "traditional" method, of separating the design contracts from the construction contracts appears only to have become a tradition in the late 1700s. According to the American Society of Civil Engineers' (ASCE) 1992 Policy Statement on federal use of design-build, "The so-called traditional method was first used during the 18th century and became institutionalized in the 19th and 20th centuries." Two hundred years later, we are returning to our roots.

§ 2.03 HISTORY OF DESIGN-BUILD IN AMERICA

The history of how designers and contractors have struggled with each other over the offering of combined services has roots that predate the forming of the United States. See § 2.04, below. History shows that beginning in the 1800s, professional, trade and technical organizations began to form in both Europe and America, largely as a means to increase the social stature of the various occupations. The architects and engineers wanted not only to elevate their place in society, but to distinguish themselves from mere craftsmen and contractors. As building technology developed, builders organized themselves according to trades and the management of trade contractors became known as general contracting. Specialization has since fragmented the industry into geotechnical, structural, mechanical, electrical and civil engineers, landscape architects, planners, interior designers, and surveyors, leaving some architects to describe what is left of the profession as "exterior decorators."

During the 20th century, design professionals backed away from responsibility for construction, limiting their role to "construction administration," and reacting to increased liability, by limiting services and disclaiming liability for site safety, shop drawings, construction errors, certifications and, in some cases, even overall liability for design errors or omissions. In the period of a liability and insurance crisis in the 1980s, architects and engineers backed further away from the construction phase and construction managers and contractors stepped in to fill the void. This has resulted in the situation at the close of the 20th century where owners are seeking someone to stand up and be accountable, and to pull the whole process back together under one contract—*i.e.*, design-build. Today, the majority of design-build is contractor-led, with architects serving as subcontractors to a general contractor and wondering how they lost their leadership role in construction. The American Institute of Architects (AIA) is even sponsoring leadership seminars to train architects on how to lead. Interestingly, many of

---

[3] T.M. Clark, FAIA, *Architect, Owner and Builder Before the Law*, Macmillan & Co., p. 58 (1894).

these issues are not new, but have lingered for over 200 years, resurfacing every decade or so and, as design-build takes hold time and again.

§ 2.04 COLONIAL ROOTS

If you visit Philadelphia and tour historical Independence Hall, you will notice a small, but architecturally beautiful, brick building just a block away, known as Carpenters' Hall. This building played a significant role in the founding of our nation, but could just as easily be a symbol of the birth of design-build in America. In the Fall of 1774, delegates from 12 of the original Colonies met in Carpenters' Hall to deliberate and determine what measures they might take to redress their grievances against Great Britain and to restore the "union and harmony . . . most ardently desired by all good men." This First Continental Congress was held in the recently constructed Carpenters' Hall because, as a privately owned building, it afforded more neutrality and secrecy to the deliberations. This meeting laid the groundwork for the American Revolution, which would be formalized by the signing of the Declaration of Independence just a block away in Independence Hall in 1776. So what is the design-build connection?

The 18th century master builder, proficient in both construction and design, often began his career as a house carpenter. Though much of colonial architecture was built of brick, the house carpenter, not the bricklayer, usually supervised construction, hiring the other craftsmen as subcontractors—*i.e.*, painters, plasterers, glaziers, and bricklayers.[4] In 1724, a group of these master builders banded together to form the Carpenters' Company of the City and County of Philadelphia, "for the purpose of obtaining instruction in the science of architecture and assisting such of their members as should by accident be in need of support, or the widows or minor children of members." Its members, sometimes referred to by historians as "carpenter-architects" were not literary men, but learned their trade on the job, by apprenticing, not out of books.[5] Members served as city councilmen and even mayor of Philadelphia, as well as designers of Independence Hall, Ben Franklin's mansion, the Carpenters' Hall itself, and other notable buildings throughout the colonies. In 1786, the Carpenters' Company published its price book, entitled *Articles of the Carpenters' Company of Philadephia and their Rules for Measuring and Valuing House-Carpenters Work.*[6] A century later, as the art and sciences of architecture and engineering began to separate themselves

---

[4] History of the Carpenters' Company, published by the Carpenters' Company.

[5] The first school of architecture in the United States was not formed until 1865, when the Massachusetts Institute of Technology started such a program, followed by the University of Illinois in 1867 and Cornell in 1871. Tony P. Wrenn, Hon. AIA, *Our Institute Founders: The AIA's First Steps Toward Leadership*, AIArchitect, July 1999, p. 22.

[6] This book was discovered in the 1970s in the attic of the Carpenters' Hall and was republished in 1993 by the Astragal Press. Copies are available from the Carpenters' Company at (215) 925-0167 or by writing to Carpenters' Company, 320 Chestnut Street, Philadelphia, PA 19106.

from the building trades, design professionals would form their own associations to promote pure design, separate and distinct from construction trades, and in protest against so-called "package builders." The resulting rifts have continued ever since.

§ 2.05 THE ENGINEERING SOCIETIES

The oldest engineering organization in the United States, the American Society of Civil Engineers and Architects (now ASCE), was formed in 1852, though there was only one architect, Edward Gardiner, among its founders. He was elected vice president that year.[7] After Gardiner and others formed the AIA in 1857, the engineering association dropped the "and Architects" from their name in 1869.[8] Societies for mining and metallurgical engineers formed in 1871, followed by mechanical engineers in 1880 and electrical engineers in 1884. In 1905, a group of eight consulting engineers in New York formed the Association of Architectural Engineers to establish standards of professionalism and identify themselves as a source of expertise that was in increasing demand at the turn of the century.[9] This organization later evolved into the American Consulting Engineers Council (ACEC). The Association of Architectural Engineers was open to engineers of all disciplines so long as they were in private practice. The purposes of the organization included promoting the science and efficiency of the profession, "and to foster professional relationships between the architect and the engineer." Design-build was frowned upon by the organization and it was agreed that engineers affiliated with construction firms should not be eligible for membership.

As was common at the time, but now prohibited by anti-trust laws, the consulting engineers association published a schedule of fees for services. The fees could vary depending on the "character, magnitude, and importance of the work or subject involved, and according to the experience and reputation of the individual engineer, from $100 per day to a higher figure."[10] The first code of ethics was adopted by what was by 1909 called the American Institute of Consulting Engineers (AICE).

By 1916, the consulting engineers formed committees to address problems that remain issues even today—*i.e.*, competition from contractors and strained relations with architects. Interestingly, the complaint against architects was their use of manufacturers and contractors to perform "free engineering" services to the exclusion of consulting engineers.[11] A letter to the (AIA) suggested that its members consider:

---

[7] Tony P. Wrenn, Hon. AIA, *Our Institute Founders: The AIA's First Steps Toward Leadership*, AIArchitect, July 1999, p. 22.

[8] *Id.*

[9] Stanley Cohen, Engineering the Future: A History of the American Consulting Engineers Council, ACEC Publication #304/9-93/5M (1993).

[10] *Id.* at 2-3.

[11] *Id.* at 4.

> the proposition that, inasmuch as modern building construction requires engineering skill of high ability, and inasmuch as architectural firms rarely possess this engineering ability, the architectural profession should recognize the necessity of employing the services of independent consulting engineers.

Herbert Hoover was president of a competing organization, the American Federation of Engineering Societies in the 1920s, prior to becoming U.S. President in 1929. During this time several regional engineering organizations sprang up in Chicago, New York, St. Louis and elsewhere. The organizations came together to address an array of professional problems, including the prevailing problem of free engineering services being provided by manufacturers and contractors. AICE adopted an ethical provision that made it unethical for an engineer to submit a bid for construction on work for which the engineer had provided professional services. Today, this prohibition still remains embodied in the law of several states and places a restraint on designer-led design-build.[12]

Turf wars between engineers and architects have been present for decades. In the early 1960s a landmark New Jersey court case drew the various engineering associations together when an engineer was accused of infringing on architectural practice by accepting the assignment as prime professional for the design of a building. AICE, ASCE, and National Society of Professional Engineers (NSPE) all came to the defense of the engineer, arguing that a client should be free to retain the design professional of his choice and that architectural licensing laws were designed to protect the public, not to preserve a market for a specific group. The case, which became known as "the McCamy case," lingered in the courts for five years but was finally determined in favor of the engineer.

It is an interesting historical note that the fight between architects and engineers over who should be prime professional on building projects has raged on among the professions for years, yet neither profession has taken the position that a contractor is unqualified to hold that role. In fact, while NSPE, National Council of Architectural Registration Boards (NCARB), and others do battle and write position papers on whether engineers should or should not be prime on building projects, both the engineering societies (EJCDC) and the architects (AIA) have published form contracts for design-build that put a contractor in the prime position and the architect-engineer firms as subcontractors to the Design-Builder.

In the 1970s the engineers again turned their attention to competition from turnkey design and construction firms, particularly in the field of industrial plant design.[13] These design-build firms, not bound by the engineers' ethical prohibition against advertising, promoted their services through paid advertisements and sponsorship of exhibits at trade shows and conventions. As a result, the Consulting Engineers Council (CEC) produced a set of guidelines that would allow similar efforts at business development, while still maintaining professional standards.

---

[12] For example, Okla. Stat. Title 59 § 46.27. *See also* 48 C.F.R. §§ 36.209 and 36.606(c).

[13] Cohen, Engineering the Future, p. 43.

In the 1980s the growth of design-build posed a threat to consulting engineers, resulting in the ACEC's formation of a Design-Build Response Committee which advocated that member firms form their own design-build teams, led by designers, and compete for the projects head-to-head with the contractors. Some member firms had already formed design-build subsidiaries and seminars were scheduled to educate members on how to enter the design-build arena.

In an article published in *Engineering News Record* in 1974, it was stated that:

> A recent survey by the American Consulting Engineers Council (ACEC) of future prospects for the consultant turned up engineering firms confronted with huge losses due to competition from design-construct firms. "We recognize," one says, for example, "the design-build concept as a means of producing a lower total cost parking structure for the owner. There are several reputable firms specializing in the field that can deliver a high quality product and therefore we can't find fault. Based on research we've done, *we see over 50% of building construction being done by the design-build concept within less than five years*."[14]

While ACEC may have been off by about 30 years in their prediction, the threat they faced then by design-build looks like a bona fide reality at the start of the new millineum as design-build holds a steady 33 percent of the market share and appears poised to reach the 50 percent mark early in the first new decade.

§ 2.06 THE AMERICAN INSTITUTE OF ARCHITECTS (AIA)

[A] Protest Against "Package Dealers"

British architects formed the Royal Institute of British Architects in 1835. Perhaps inspired by that action, a dozen American architects met in New York in 1836 and formed the American Institution of Architects. However, the small organization failed. The AIA was formed in 1857 with 49 founders signing the first AIA Constitution, noting in Article 2, "The object of this Institute is to promote the scientific and practical perfection of its members and elevate the standing of the Profession."[15] Among the AIA's founders was Edward Gardiner, the only architect who was among the founders of the American Society of Civil Engineers and Architects in 1852. *See* § 2.05, *supra*.

The AIA was founded partly in response to what were then called "package dealers," a term that at the time was synonymous with design-builder.[16] In fact, it has been reported that one of the reasons that the architects split away from

---

[14] *Construction's Changing Professionals*, Engineering News Record, special issue published April 30, 1974, p. 275.

[15] Tony P. Wrenn, Hon. AIA, *Our Institute Founders: The AIA's First Steps Toward Leadership*, AIArchitect, July 1999, p. 22.

[16] D. Gordon, *What Is Design-Build and Why Is Everyone Talking About It?*, AIA Memo (February 1994); M. McCallum, *Design-Build: Issues to Consider*, AIA Supplement, 1993.

ASCE only five years after its founding related to claims that some of the engineers were "package dealers," since some of the engineers owned foundries that produced cast iron storefront buildings and iron bridges. These engineers designed the projects, manufactured the materials, and constructed the facilities.[17] To distinguish the two groups, when the AIA adopted its first Code of Ethics in 1909, the Code barred AIA members from engaging, directly or indirectly, in construction, making the client's interest paramount. The AIA's thinking was that a design-build entity that controlled the design decisions and had a financial stake in a project's profitability had an inherent conflict of interest.[18] The result was that the AIA effectively prohibited its members from participating in design-build.

Architects have long suffered from an inferiority complex that has resulted in multi-million dollar advertising campaigns by the AIA to promote the benefits of hiring an architect and seminars on leadership. Architects have lamented in recent years how construction managers, interior designers, landscape architects, contractors, engineers, and so-called "program managers" have whittled away at their profession, each one taking a small piece and leaving architects a reduced role in the construction process. This has been a threat for over 100 years. On the AIA's 50th anniversary in 1907, Glenn Brown, CEO of the Institute, wrote that in 1857:

> educated men in the larger cities knew little of each other's aspirations or desires. The fear of unjust and improper treatment by other practitioners seems to have been widespread. No standard of fees or ethics was recognized. The profession was not appreciated by the public, the fine arts were thought of little importance, and constructors and carpenters, as practical men, were thought far superior to the architect.[19]

[B] Ethics Code Held Back Architects

As mentioned above, when the AIA adopted its first Code of Ethics in 1909, it prohibited its members from engaging in construction due to a perceived conflict of interest where the architect had a financial interest in the project. That ethical constraint remained part of the AIA's Code for nearly 70 years. The AIA's 1977 Code of Ethics stated:

> R. 404 Members may not engage in building contracting where compensation, direct or indirect, is derived from profit on labor and materials furnished in the building process except as participating owners. Members may engage in construction management as professionals for professional compensation only.

---

[17] J. Beard, *Within the Design-Build Procurement Spectrum: Using Negotiated and Competitive Selection,* Cost, Time & Risk, AIA Publication, Nov. 1994.

[18] M. McCallum, *Design-Build: Issues to Consider*, AIA Supplement, 1993.

[19] *Id.*

The process to repeal the rule began in early 1973, when the AIA appointed its first task force to study the design-build concept. The task force reviews found architects were leery of the design-build process and concerned about the potential effect on the integrity of the architectural profession.[20] Some of the concerns were:

1. whether uncompensated competitions would lead to decreased design quality by standardization or use of inexperienced architects with minimum supervision to cut costs;
2. whether the architect losing his or her role as agent of the owner would violate professional ethical standards;
3. whether architects could maintain professional liability insurance as part of the design-build process;
4. whether participation in design-build would constitute illegal fee splitting or aiding others in the unauthorized practice of architecture;
5. whether participation in design-build would open up exposure to product liability laws, warranties and strict liability; and
6. whether participation in design-build weakens the status of the profession because architects are placed in a subordinate position.

These concerns caused a split within the profession as to how to respond to the growing demand among owners for single-point responsibility.

Many architects disagreed with the AIA's position on design-build. One Chicago architect stated: "The AIA says that if the architect could be affected financially by his interest in the construction, he may lose his objectivity in protecting the owner's interests above his own. But it not only is proper but beneficial to the owner for the architect to participate in construction. The closer control that can be exercised gives owners speed, quality and good project cost control."[21] About the same time architects began to rethink the AIA's position barring their participation in design-build, the AIA was sued for antitrust law violations by a member whose membership was suspended over ethical violations under the 1977 Code.[22] The AIA settled that case after a federal court ruled that the AIA had violated antitrust laws.[23] Though the ethical issue in this case did not relate to design-build, the AIA repealed the rule in 1978 for a three-year trial period to al-

---

[20] Leslie A. Nicholson, Jr., *Design-Build—Yes, It's Legal*, CNA 25th Annual Meeting of Invited Attorneys, May 22-23, 1986. Mr. Nicholson was counsel to the AIA Documents Committee at the time.

[21] Sidney Epstein, as quoted in the Chicago Sun-Times, July 27, 1978. H. Block, *As the Walls Came Tumbling Down: Architects' Expanded Liability Under Design-Build/Construction Contracting*, 17 John Marshall L. Rev. 1, 8 n. 31 (1984).

[22] Mardirosian v. American Inst. of Architects, 474 F. Supp. 628 (D.C.D.C. 1979).

[23] Engineering News Record, Aug. 2, 1984, p.19. The *Mardirosian* case was not the end of the AIA's antitrust problems. After a five-year investigation, in October 1990 the AIA signed a "Consent Decree" with the Department of Justice to settle another suit which alleged further AIA violations of federal antitrust laws. United States v. American Inst. of Architects, Case No. 90-1567 (D.C.D.C. 1990).

low its members to participate in design-build ventures. As a result of the antitrust lawsuit, the AIA Board of Directors voted to repeal its entire Code in 1979, replacing it with a "voluntary" code in 1980. The three-year trial suspension of the design-build restraint became permanent when members voted in 1984 to reinstate a mandatory code for AIA membership. In 1986 the AIA adopted its new Code of Ethics but the new Code did not prohibit design-build. With the ethical restraint removed, design-build began to flourish, growing from $18 billion in 1986 to over $73 billion predicted for 2000.

The AIA's Board of Directors had begun to examine its position on design-build several years before the antitrust lawsuit. The AIA's 1975 Policy Statement on Design-Build-Bid noted that design-build constituted a "fundamental departure from traditional methods of practicing as a professional" and advised that "A professional architect is one who gives advice to clients . . . ; that professional architect will have no interest in any project which might influence his judgment and decisions to the detriment of his client." Obviously trying to dissuade architects from engaging in design-build, the AIA's policy statement concluded that,

> a. When an architect commits himself contractually to produce an end product, he invites the courts and other governmental agencies to treat him as a commercially motivated businessman and not as a professional; he may be subjected to strict liability and other disadvantages; b. Such disadvantages may be dealt with as involving warranties and consequently may not be covered by professional liability insurance; c. Such activities may require the architect to furnish surety bonds; d. Such activities may require compliance with controls affecting businesses, from which professionals are relatively free.

[C] Repeal of Ethics Code Opened the Door

In 1983, the AIA's Board of Directors approved a Policy Statement affirming support for qualifications based selection (QBS) on public projects as a method of procuring A-E services on government jobs. The policy stated that competitive bidding for design services is not in the public's best interest, and that if a public owner chooses to contract with one entity for design and construction services, objective criteria should be developed to evaluate the qualifications and competence of the design-build teams.

In 1984 the AIA released drafts of its proposed new design-build contracts. The contracts drew objections from AIA chapters in New York and Wisconsin due to the creation of a new entity—the "Design-Builder"—where that entity is not required to be a design professional. The New York State Association of Architects (NYSAA) urged that either the architect or a joint association between an architect and contractor should be in the position of prime design-build contractor, fearing that allowing non-professionals to broker design and construction services may erode support for architectural licensing among legislators. Wisconsin's objections were that use of design-build in the public sector would provide a way for the state to avoid qualifications based selection.

After the Board debated, the AIA published its first series of design-build contract forms in 1985, keeping the role of prime contractor as a "Design-Builder," with the architect as a subcontractor to that entity. The documents came with advisory instructions that due to insurance, management and licensing laws, "it is expected that in most situations an architect will wish to establish a separate design-build firm. This firm may then subcontract the design work to the parent architectural firm through the use of AIA Document B901 [the subcontract between Design-Builder and Architect]."[24] However, few architects followed the AIA's instructions, and instead of forming new corporations to hold the prime design-build contract, we found that contractors took that role, with architects signing the B901 agreement purely as a subcontractor.

At the request of NCARB's Board of Directors, its Committee on Procedures and Documents held hearings in Chicago in 1985 on the subject of the impact of the design-build process on the public health, safety, and welfare. The committee identified two concerns. First, when the design-build contractor fails to clarify with the owner that the architect is employed by the contractor, not by the owner. The committee felt that the architect's role must be appropriately described. Second, the architect is frequently not asked to perform any construction phase services to see that the intent of the plans and specifications is carried out in the field. The committee recommended changes to the NCARB Model Registration Statute that, "In particular, a person (other than a registered architect) offering design-build services may offer to render architectural services as part of the design-build services, provided he has engaged an architect to furnish those architectural services." To address the lack of construction administration, the committee recommended that the Model Statute require that a registered architect be hired by the contractor "for the purpose of providing periodic on-site observation of the construction" and that this architect certify completion of the project in accordance with the plans and specifications.

The AIA's 1983 Policy Statement was sunset in 1988. In 1988, the policy was remanded to the AIA's Government Affairs Advisory Committee. In 1989, the AIA formed a national task force in response to the increased use of design-build in the public sector. That task force produced an interim report in August 1990. On May 15, 1991, the AIA Board adopted a new policy on public sector design-build, which again emphasized the use of QBS criteria in selecting the design-build entity. That policy recognized, for perhaps the first time that, "The design-build method of project delivery is being used by the public sector, and many AIA members are involved in this work." The AIA's 1991 Policy Statement recommended preselection of a "short list" of firms, based on qualifications, then competition based on qualifications and other criteria, with payment of a "stipend" to compensate the finalists.

In December 1997, the AIA Board adopted a new Position Statement on design-build. The Statement, reprinted in full in Appendix A of this book, recom-

---

[24] Instruction Sheet for AIA Doc. No. A191, *Standard Form of Agreements Between Owner and Design-Builder* (1985 edition).

mends that when dealing with public projects, the public agency should use a preselected short list of design-build entities that is based on competence, plus comprehensive scope of work documents "prepared by licensed architects and other qualified professionals who are retained for the duration of the project," including soil information, outline specifications, budget parameters, and project schedule. The AIA also recommends that the proposals be evaluated by "a jury of qualified professionals (including those licensed professionals who prepared the scope of work documents)," and payment of a stipend is commensurate with the level of information required by the RFP.

From its founding in 1857 in protest to design-build and 70 years of ethical restraints, the AIA has come a long way. The AIA now jointly publishes "how to" manuals with AGC on public design-build, teaming checklists and design-build contracts, plus co-sponsors conventions with DBIA to educate its members on design-build's best practices.

§ 2.07 THE DESIGN-BUILD INSTITUTE OF AMERICA (DBIA)

In response to the emergence of design-build project delivery, the Design-Build Institute of America (DBIA) was founded in 1993 and, like AIA and AGC, it is based in Washington, D.C. DBIA represents design firms, contractors and owners, taking its membership from all phases of the process. DBIA's membership growth has exceeded initial projections. In a 1993 article in NSPE's *Engineering Times*, DBIA's first executive director, Jeffrey Beard, was quoted as saying that the organization hoped to have 50 members after its first year, and "after a couple of years, there could be more than 100." The organization began with just 11 members in 1993. It had 459 members at the start of 1999 and by year-end had 667 members, a 45 percent increase in 12 months. As design-build picks up speed nationwide, membership continues to grow, to 783 as of August 2000. The specific membership categories for firms providing design-build services in the non-residential design and construction markets are the following:

1. **Regular Membership**—Firms that contract "at risk" directly with owner/client as "prime" design builders. This category contains three dues tiers according to overall design and construction volume, starting at $2,500, with annual dues of up to $7,500 per year.
2. **Associate Membership**—Consists of design consultants (design professionals who participate in design-build without "at risk" contract responsibility), subcontractors, vendors and allied professionals involved in the design-build process. Dues in this category range from $500 to $2,000 per year.
3. **Special Individual Membership**—Consists of individuals employed by owners, government agencies and academic and research institutions, with dues of $100 to $500. Individual students are also welcome to join at this level for annual dues of $50 per year.
4. **Student Chapters**—DBIA encourages the formation of formal student chapters. As of January 2000, DBIA had established student chapters at

four universities—Purdue University, University of Florida, Texas A&M University and Iowa State University.

5. **Additional Contacts**—DBIA membership rules permit each member firm to have a specific number of firm members under their corporate membership. However, if a member firm has all of its allocated slots filled, it can name additional contacts at a cost of $100 per year per new representative.

DBIA began issuing charters to regional chapters in 1999 and by March 2000 had five chapters with more in the works. DBIA has also founded a Design-Build Education and Research Foundation which is charged with providing research grants to universities and tuition assistance to graduate and undergraduate students focusing on design-build, and influencing young people at the university level to select design-build as a career choice. Their goal is to see the establishment of an undergraduate degree in design-build at a major university.[25]

[25] Chairman's Message, Goals for the Year 2000, DBIA Dateline newsletter, Jan. 2000.

CHAPTER 3

STARTING OUT IN DESIGN-BUILD

§ 3.01 WHERE TO BEGIN?

Design-build contractors get their start in many ways. Some start by plan, some by client request, and some by a natural evolution of their business. Several contractors have found their way into design-build by being builders of pre-fabricated or "pre-engineered" metal buildings, a form of design-build that has been accepted for years without the special attention that design-build is getting today. Bo Harris of Harris Construction Co. says, "We were a Butler-builder initially. Clients tend to come to a contractor first when they want a pre-engineering building, so we got our start that way." Jim Neenan, president of Colorado's The Neenan Company, says that his company got started as a Butler-builder as well. Neenan now has a staff of 40 in-house architects and has grown to be the largest design-build contractor in the state.

Other firms have gone into design-build as a strategic marketing move, aquiring the needed staff by buying existing businesses. Bob Degenhardt, president and CEO of Ellerbe Becket, says that we will see more consolidation among design and construction companies: "The construction companies will acquire design firms or design firms will acquire construction companies." Omaha's Peter Kiewit Sons, Inc., one of the largest contractors in the United States, purchased engineering firm Bibb & Associates in 1999. Other large contractors have also purchased architectural or engineering firms as the most efficient way to bring design expertise in house. For those design firms without the assets to go out and buy a construction company, there are alternative paths to grow into designer-led design-build.

§ 3.02 START OUT SMALL

Insurance and legal advisors to design firms are consistent on one piece of advice—start small. One bad project can kill a design firm. By starting out on a small project or two, the tools a firm needs to be successful can be slowly developed and the overhead costs spread to a number of projects over time.

Unlike most architects, Dave Shelton, AIA, had experience in construction prior to his career in architecture. He worked for a contractor and for a truss manufacturer prior to starting his own construction company. His construction company started small doing residential additions. That experience made him realize he could be an architect, and in 1990 he enrolled in architecture school at North Carolina State. After working for three different architectural firms, he says, "I never lost the bug to stay in construction." Seeing a market driven by a need among owners who were unhappy with the traditional process, and intrigued by design-build, Shelton interviewed about 20 owners on what they liked best and least on their projects. He found that owners wanted predictability and reliability. "The budget for their project is part of their business plan," Shelton says, "so they need someone who can guarantee the cost. What they kept telling me is that: The price you are given is never the price you get." Shelton then started his own design-build firm to meet the needs that owners felt were not being met.

Architect Larry Maxfield worked for a contractor after graduating from architecture school. He says that his training in construction taught him lessons he could not have learned in school. "Every architect starts his or her career struggling over how to detail things to be built. With design-build, from the moment I designed something I started working with the superintendent in the field on how to build it." Maxfield eventually started his own design-build firm by teaming up with a contractor and a developer in 1981. "At that time, the marketplace wasn't offering this packaged set of services. We designed, built and, in some cases, acted as developer for a fee." Starting out, he says the three firm owners managed their own jobs, even acting as field superintendents. "It was a lean operation. We did the marketing, bookkeeping, design, superintending and project management." Then, about eight months after starting the design-build company, Maxfield says they got a break with a 100,000-square foot shopping center which eventually led to more work for a grocery chain, which led them to even bigger work.

Currently, Maxfield is President of Construction Technologies, L.L.C., a design-build firm specializing in tenant finish work. "We started doing tenant finishes on small jobs of 1,000 to 3,000 square feet. This was low risk, interior finish work. But through that, we eventually landed a $2.5 million shopping center remodel job." His advice to architects going into design-build is to keep it simple and grow slowly as you gain experience. "When we started the new company it was just me and a bookkeeper. We subcontracted 100 percent of the work and didn't buy anything. I did the estimating, design and project management. When we had enough work to justify the salary, I hired a carpenter/superintendent, as a working superintendent." Maxfield recommends that on small jobs, the guy in the field must be a working superintendent, since the overhead is too costly to have a person on salary who is only superintending and not constructing. He warns about the danger of taking on too much overhead quickly, saying that "contractors get in trouble by over-staffing."

Another piece of advice Maxfield gives is "Take no exceptions to whatever your structural engineer says is needed on a design-build job. Don't mess with their recommendations. That is where the biggest risks are, in the structural components."

§ 3.03 START WITH A HOUSE

One way an architect can test his or her skills in design-build and construction is with a small, simple project. Angel Martinez, AIA, started with a house. He bought the lot, obtained the construction loan, designed the house, hired and managed the subcontractors, and eventually sold the house upon completion. Since he was the client as well as the design-builder, this was a very safe way to start out. The risk here, of course, is that one may not be able to sell the house due to market conditions, or one fails to consider real estate commissions that get paid out of the sale price. As far as getting experience in construction, however, this is probably the safest way to begin. The sale of the house could net a profit that could help launch a design-build firm.

We all know architects who designed their own houses and tried to act as their own general contractors. Many report this as the worst experience of their lives. Why? Because they were trying to run their architectural practice at the same time as they were running their construction practice. Lawyers often say that: "A lawyer who represents himself, has a fool for a client." By the same token, acting as your own contractor may not make sense if you cannot devote yourself full-time to the effort. There is no time to call the subcontractors, monitor progress, be on site regularly, and handle paperwork. The subcontractors often place jobs with their steady builder clients ahead of the architect's house, which they see as a one-time job, with no repeat business likely. In addition, there is usually no superintendent on site checking progress and holding the subcontractor accountable. This is not what is intended for a trial run at design-build. If you are committed to being a contractor, you need to focus on that project.

§ 3.04 WHOM TO HIRE?

Angel Martinez, AIA, says that in the beginning he did all the design work and acted as project manager and superintendent. Then, over a three-year period, as work began to pick up, he hired four construction superintendents. "I was fortunate to obtain several contracts early on and I added superintendents as the work grew. I hired a bookkeeper as well, but I did all the estimating, marketing and project management myself. It was just me, the bookkeeper and four superintendents." Today, his design-build firm has a bookkeeper/office manager, two staff architects, one project manager, and four superintendents, in addition to Martinez who oversees all of those positions as company president. His daughter, Cynthia, a recent architect graduate, is getting her training not behind a drafting table like most of her peers, but in the field as a project superintendent.

As with hiring architects, anytime you "staff up" for a particular job, there is a risk of having to lay off workers if work slows down. This results in lack of staff loyalty, lack of familiarity and lack of continuity within the office. Martinez says he has never laid off a superintendent. He tells them that he will carry their salary between jobs if necessary. "My banker and my bonding company think I am crazy, but that is the commitment I make to my employees," he says. While admitting that it can be very expensive, Martinez says that it has its benefits too, including a strong incentive to keep new work coming in. "It builds a strong sense of loyalty with our employees as well."

Design-build jobs usually have short time frames with fast start-up dates. If you staff up and then lay off workers, it will be difficult to respond to owners' needs, especially in a tight labor market like the late 1990s and start of the new decade. There are simply not enough skilled construction people to meet the demands. As Martinez says, "I can't afford to hire someone new and put them on one of my jobs. I only get one chance to succeed with my client and I need people on the job who I trust and who share my vision and philosophy about design and construction. I can't do that if I am always turning over new staff."

Larry Maxfield, AIA, says his design-build firm currently has one bookkeeper and four project managers, who are also superintendents. Maxfield says that he wants the project managers to be in the field, not in the office.

§ 3.05 SELECTING PROJECTS

How does one go about selecting projects for design-build? Projects have varying degrees of risk depending on the project type. For example, at the lower end of the spectrum might be a pre-engineered building for a warehouse. Moving up would be a complex building with curtain walls and significant engineering. At the farthest end would be a project with a guaranteed output requirement, such as a power plant.

In design-build, it is best to start out with a low risk project type, especially one in which there is some prior experience as a pure design firm. By choosing a project type or client with which there is familiarity, the project can be better managed and more focus can be placed on the new tasks of design-build delivery with fewer complications. It is important to use staff with prior experience with that project type. The more familiar ground chosen, the less risk.

One medium-sized engineering firm's patience in selecting the right project shows the benefit of project selection. The engineering firm set up a wholly owned construction subsidiary that shared insurance with the engineering firm for professional liability, general liability, and workers compensation. They then hired three project managers with engineering degrees from a local construction company. While the firm waited for that right project to come along, the three project managers worked doing contract administration for the engineering firm as well as doing some design work, "Which they hated," said the company's vice president. Then they waited a full year for the right project to come along. "Being patient was the hardest part," said the company's vice president, "but it was worth the wait. We were very selective. We wanted our first project to be one that we were familiar with." The waiting paid off handsomely for the firm with a $600,000 profit.

§ 3.06 TEAMING WITH A CONTRACTOR

Naturally, the first step in teaming is for the the design firm to form a strategic alliance with a financially strong, bonded contractor. The terms of the agreement should be that where the architect-engineer firm is lead, the contractor will provide extensive pre-construction services, including estimating, scheduling, constructability, value engineering. In return, the architect-engineer agrees to subcontract 100 percent of the construction work to the contractor. Dave Shelton, AIA, says, "I am not overly concerned about who is in the lead. As long as our roles are well defined, and I am working with people I trust, that is the most important. I want to be comfortable that we will each look out for each other, while at the same time looking out for the owner."

§ 3.07 DEVELOP A DESIGN-BUILD SUBSIDIARY OR "SISTER FIRM"

Many other design firms have decided that subcontracting of the construction work does not give them sufficient control. These firms have formed subsidiary companies to act as the construction division. Firms like HNTB, HOK, Jacobs/Sverdrup, Black & Veatch, and others have all formed design-build or construction divisions. Large firms like these can afford to make the capital investment of obtaining in-house construction staffs and equipment, plus their own bonding line. Smaller firms may want to start a subsidiary, but still enter into a subcontract with a general contractor for the construction work. This may also require the contractor to provide bonding for the project, using a "dual obligee" bond that runs to the benefit of both the owner and the architect-engineer firm.

§ 3.08 DESIGN/CONSTRUCTION MANAGEMENT

One way an architect can test his or her skills at managing a construction project, without diving head-first into design-build is to start by offering construction management (CM) services as an add-on to the standard Owner/Architect Agreement. The AIA's 1997 B141 Owner-Architect Agreement is designed with a generic "front end" that can be used with a variety of "scopes of services," including construction management. By using the AIA's CM-advisor contract, where the construction manager is not responsible for the work of the subcontractors, one can ease into the role of contractor without all the risk. If there are construction defects in the work, the owner's remedy is directly against the subcontractors, who warrant their work to the owner and to the CM. The down side to this delivery mode is the lack of control. As one architect-builder said, "The subcontractors have no contract with you, and you are not the one who pays them. In short, you are not in control of the process. If you are going to do design-build, then get into design-build, not construction management." He warns that with construction management, the architect has no control, but if there is a problem, all the fingers still point to the architect. His recommendation is to be the contractor or do not do design-build, since most other forms involve significant losses of control.

Brad Buchanan, AIA, of Denver's Buchanan Yonushewski Group. L.L.C. says that his firm started with design-construction management. Then the firm broke into full design-build slowly by packaging the work into a few large subcontracts. With experience, his firm began breaking the job down into more trade subcontracts as they gained skill at coordinating and managing subcontractors. He recommends that if one wants to take the next step and venture into full designer-led design-build, "Pick a trusted repeat client on a familiar project type and offer to take on the whole process. Use this opportunity to team up with your favorite contractor and see how much more you have to offer by staying involved from the schematics to the punch list."[1]

Use of design/CM returns some measure of control back to the architect, plus allows for additional fees. Robert C. Mutchler, FAIA says that owners want their architects to provide protection of the owner by regular site visits. "Owners wanted to hear that we would be there full time protecting their interests. Of course that was impossible without increased compensation, and few owners are willing to pay more for AE services. CM solves this problem nicely. Through recovery of the cost for a general superintendent and the mark-up on subcontractor bids, the cost for the CM to represent the owner's interests on at least a daily

[1] *DB: Emerging Trends*, AIArchitect, June 1999, p.14.

basis, is easily covered."[2] He also finds that profits of 30 percent to 40 percent of gross income can be realized through providing construction management services. Mutchler points out that contractors are not at all happy about architects getting into construction management, although contractors do not hesitate to offer design services as part of a design-build package. He notes that by offering CM services, architects are in direct competition with contractors, but "with our education, training and experience, there is no reason why we cannot provide CM services as well, or even better than, a general contractor."

Mutchler also has found that since the architect as CM has more control over the project, the risk is reduced. With the design firm already insured for design liability, he recommends setting up a "sister firm" for construction management activities, which can provide insurance to cover CM activities with a standard construction contractor liability insurance policy.

Be aware, however, that offering construction management services can result in increased exposure to liability for site safety issues and OSHA fines. There have been recent court cases in which OSHA has assessed fines against design firms offering construction management services. The outcome of these cases depends on the contract language. Based on the landmark court ruling in *CH2M Hill, Inc. v. Herman*,[3] however, there is a clear direction away from imposing site safety liability on design firms acting as CM if the contracts clearly allocate site safety liability on the trade contractors rather than on the construction manager. An architect should be sure that this topic is clearly addressed along the guidelines given by that court case and that you have proper legal counsel assist in the drafting of your design/CM contracts.

Aside from concern about OSHA liability, however, daily observation of the jobsite can reduce accidents and improve construction speed and quality by monitoring contractor safety programs and jobsite cleanup. Design/CM offers a great way to obtain experience managing construction projects without taking on all the risk of becoming a general contractor until a firm is ready to take that step.

§ 3.09 BUY A CONTRACTOR

What is the best advice for design firms who want to jump into design-build with both feet? "Buy a contractor," says Barry Dewberry, CEO of architect-engineer firm Dewberry and Davis.[4] This may not be such an outlandish statement considering the difficulty and time it takes to acquire in-house construction expertise, needed equipment, a physical plant, bonding capacity and a portfolio of successful construction projects. In other industries, when a major corporation wants to obtain a market share of a certain business, they merely buy a company

[2] *Construction Management by the Architect*, Project Delivery Report, AIA Construction Management PIA, Sept. 1998.

[3] 192 F.3d 711 (7th Cir. 1999).

[4] *Design Firms See Design-Build Growth But They Remain Wary,* Engineering News Record, Sept. 27, 1999, p.19.

that does that already. Many contractors have bought up design firms rather than start an in-house design section from scratch. Why not buy a contractor?

§ 3.10 "TWO-STEP" CONTRACTING

One variation on design-build contracting that has been tried is a two-step process under which the owner signs packaged, two-part documents. The first part covers only design and is directly with the architect or engineer. The second part is with a construction contractor. This method has not been popular, however, since it lacks the foremost benefit to owners, *i.e.,* single-point responsibility. While the "design-build team" markets the project together and works together to plan, budget and build the project, the owner is still caught in the middle between designer and builder if there is a dispute.

§ 3.11 TEAMING WITH A CONTRACTOR

This is discussed in more detail in Chapter 7, but is probably the most common way that architects and engineers start out in design-build. An experienced contractor can teach an architect much about construction, while at the same time learning from the architect about design. Teaming can be a one-time effort, or a longer-term relationship. It is important to learn about construction from a contractor's view. Even if the architect's role is as a subcontractor, it is important to take the opportunity to get involved as much as possible in what the contractor does. The architect should set up a desk at the contractor's office so he or she can work side-by-side with the contractor's staff. This will help facilitate better communications on the project, and give the architect an insider's view of how a construction company operates.

§ 3.12 SEEING THE PROJECT FROM THE OWNER'S POINT OF VIEW

Many successful design-builders say that their firm philosophy is to think like their clients. Jim Neenan, President of The Neenan Company in Denver, Colorado, runs a design-build firm that started out as a "Butler builder" and has grown to be Colorado's largest design-build contractor, with over 300 employees including 40 in-house architects. Neenan says, "We try to think of the project from the client's point of view. Our goal is to deliver value to the client." He says that by thinking like the client his company is better able to keep value as the ultimate goal of every decision, a company philosophy that has led to the success and growth of his company.

Brad Buchanan, AIA, of Buchanan Yonushewski Group, L.L.C., a 35-man Denver design-build firm whose owners are both architects, agrees with Neenan. Buchanan says, "Our mantra is—Think like an Owner." He says that owners want architects to put projects together on time and within the budget, but "Owners don't care what you call that, as long as they get their building done in time and on budget." Beyond design and construction services, Buchanan says his firm will offer whatever the owner wants, from design, finance, management, maintenance, and construction. "If we don't have it in-house, we'll go out and get it." He says that some clients come to him with a clean piece of property and ask him to help find the right project type to make best use and highest return on the investment. "We'll do a market study to find the highest and best use. Then design, build, whatever the owner needs."

CHAPTER 4

VARIATIONS IN DESIGN-BUILD

§ 4.01 TYPES OF DESIGN-BUILD

Design-build comes in many shapes and sizes these days. There is some confusion within the industry as to what actually is "design-build." Some architects contend that they do design-build, but when asked whether they were prime or subcontractor, they reveal that the owner had two separate contracts, one with the contractor and one with the architect. Though there may be some "design/assist" role by the contractor, this is not true design-build because it lacks the fundamental point that sets design-build apart—and that is, one contract, with one entity, who serves as a sole source of responsibility to the owner. But even within that narrow definition, there are several variations.

[A] Turnkey

One common form of design-build is referred to as a "turnkey" project. A turnkey project is one where the contractor, often a developer, assumes responsibility for design and construction, plus designated real estate services which may include project financing and site selection. The contractor provides a completed facility for one lump sum price to the point of readiness for occupancy.[1] In some turnkey contracts, the contractor also provides the "F, F, and E" (fixtures, furnishings, and equipment). The contractor then turns over a complete facility at the time of occupancy, so that all the buyer has to do is simply "turn the key" in the lock to open the door.[2]

[B] Super-Turnkey

There is an even more comprehensive form of turnkey called "super-turnkey." In this form, the contractor must design the project, provide initial construction financing, perform the construction, and then operate the facility for a period of time.[3] This method is sometimes called *design-build-own-operate*, or "DBOO." An expansion called *design-build-own-operate-manage* is referred to as "DBOOM." The proposed Model Procurement Code under drafting by the American Bar Association in conjunction with others identifies two other variations: *design-build-finance-operate-maintain* or "DBFOM" under which no funding is paid to the contractor until the project is completed, but the contractor is obligated to maintain the completed facility over a contractually defined period; and *design-build-operate-maintain* or "DBOM" where the project owner provides the funding during the contract period.

---

[1] Mobile Housing Environments v. Barton & Barton, 432 F. Supp. 1343, 1346 (D. Colo. 1977); Gantt v. Van der Hoek, 162 S.E.2d 267, 270 (S.C. 1968); Principal Mut. Life Ins. Co. v. General Services Admin., GSBCA No. 13,088, 97-2 BCA 29,175.

[2] Glassman Const. Co. v. Maryland City Plaza, Inc., 371 F. Supp. 1154, 1157 (D. Md. 1974).

[3] A super-turnkey RFP drew five proposals for a $350 million, 20 mile light-rail project in New Jersey. Engineering News Record, Aug. 21, 1995, p.17.

[C] Design-Build-Lease

A growing method of construction financing is for the developer of a project to be required to acquire the site, finance, design, and build the project, and then lease it back to the customer for a fixed number of years, often with an option to purchase the property at the end of the lease period. Such projects are sometimes bid by competition based on the annual lease price as the primary selection criteria.

[D] Bridging

Discussed in more detail in Chapter 6, a hybrid of pure design-build is "bridging," where the owner selects an architect or engineer to program and partially design the project, usually through design development or 30 percent completion. The design development documents then form the basis for competitive selection by design-build firms or teams who bid to complete the project design and also build it. This allows for all bidders to work from the same base starting point, rather than each team preparing different and costly designs as part of their proposal. The "bridge" architect/engineer then is retained to act as the owner's independent professional to deal with pay applications, project inspections, etc. This removes some of the ethical concerns of design-build where there is no independent design professional advising the owner.

§ 4.02 DESIGN-BUILD VARIATIONS

Discussed below are the four basic models to deliver design-build services: 1) contractor-led, 2) architect/engineer-led, 3) joint venture or new corporation, and 4) the integrated firm. The differences depend on which entity holds the prime contract, whether it is a construction contractor or a design firm. The Design-Build Institute of America (DBIA) states in its Manual of Policy Statements that it expresses no preference for one method over another provided that the following conditions are met:

- The design-builder is a singular legal entity holding the contract with the owner.
- This entity is fully at risk for performance of both design and construction, and possesses the financial management resources to perform the design-build contract.
- The professional responsibilities and obligations of the members of the design-build team adhere to the DBIA Code of Professional Conduct.
- The structure of the design-build team provides for appropriate communication of design, construction, and business issues and decisions between members of the design-build team and the owner.[4]

[4] DBIA Policy No. 21, Organization of the Design-Build Entity, Oct. 1998.

§ 4.03 CONTRACTOR AS PRIME (CONTRACTOR-LED)

This is the most common form, where the architect/engineer is a subcontractor to a contractor who holds the prime contract with the owner. A 1999 study found that contractors lead over half of all design-build work.[5] Some advantages of a contractor-led team are:

1. It is usually easier to get bonding due to the established surety relationship, plus assets and experience that sureties like to see; and
2. the ability to make a claim against the professional liability insurance of architect/engineer subcontractors for design errors. This is discussed in more detail in § 4.05, below.

§ 4.04 ARCHITECT/ENGINEER AS PRIME (DESIGNER-LED)

Less common, but growing, is where the prime contract is held by a design firm. In 1999, this accounted for only one-eighth of all design-build projects.[6] All construction is performed by subcontractors to the architect/engineer. This method is used more for engineering projects than architectural projects. Several engineering firms and architectural firms are on the top 100 list of design-build contractors. For example, the engineering firm Black & Veatch is ranked as the seventh largest design-build contractor in the country. Bechtel and Fluor Daniel, two companies that have diversified from engineering into construction, have been number one and two respectively for several years.[7]

A prime example of a designer-led firm is Lockwood Greene Engineers, Inc., ranked in 1999 by Design-Build magazine as the 85th largest design-build contractor in the United States. The constructed value of its projects annually exceeds $5 billion. Based in Spartanburg, South Carolina, the firm merged in 1999 with J.A. Jones, Inc., a subsidiary of the German construction conglomerate, Philipp Holzmann AG. The Lockwood Greene firm was founded in 1832 and has 44 offices in the United States and across the globe. The firm has a staff of more than 3,200 consultants, engineers, architects, and construction personnel. Lockwood Greene offers design-build, stand-alone consulting, engineering, procurement, construction, construction management, or any combination of these on an integrated basis.

One of the disadvantages of this method, however, is the inability of the design firm to make a claim against its own insurance policy for design errors made by the design firm. Though there is coverage for claims made by the owner or third parties, the added construction costs caused by project delays, design errors or omissions by the staff are not covered by the firm's policy and, therefore, must be self-absorbed by the prime contractor.

---

[5] *1999 Design-Build Survey,* Zweig White & Associates, Inc., p. 25.

[6] On average, 12% of design-build work was designer-led, and the majority of the time the designer is a subcontractor. *1999 Design-Build Survey,* Zweig White & Associates, Inc., p. 25.

[7] *The Top 100 Design-Build Firms*, Engineering News Record, June 15, 1998, p. 56.

§ 4.05 COMMON MODELS FOR DESIGNER-LED DESIGN-BUILD FIRMS

There are four common models used for designer-led design-build. Figure 4-1 shows a model under which the architect acts as its own general contractor, providing the design and construction services in-house.

Some design firms, especially starting out, will hire a project manager from a construction company rather than develop that expertise with current staff. This is the purest form of design-build, with one entity offering the owner design and construction services in-house. The design-build firm will subcontract consulting services to engineers for those specialties that the firm does not have in-house, and will subcontract directly to trade subcontractors and suppliers for labor and materials. Some firms bond trade subcontractors to assure performance, or at least those whose contracts are over a certain dollar threshold. The design-build firm is responsible for all scheduling, design, estimating and project management. Where there are construction errors or estimating errors by the trades subcontractors, the design-build firm will look to those subcontractors to cover the loss, or to their sureties if bonded. Keep in mind that some bonds state by their terms that they do not provide protection beyond one year after project completion, even though the design-build firm has liability to the owner longer than that. On the design side, the design-build firm may be able to make a claim against the professional liability insurance of its engineering subconsultants to the extent that their errors cause a loss. However, there will likely be no coverage under a standard policy for losses that the design-build firm suffers due to its own design errors or omissions. A claim made by the owner would be covered, subject to policy exclusions, but if the design-build firm absorbs the cost of a design error and does not default on the contract with the owner, it will not be able to make a

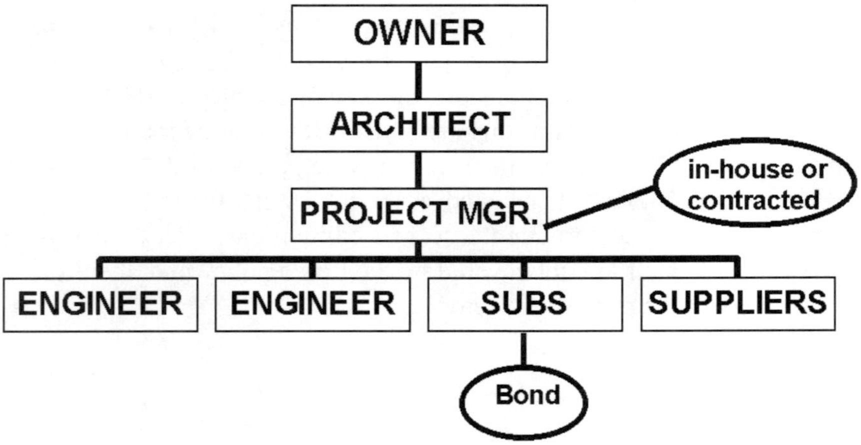

FIGURE 4-1. Architect-Led Design-Build Where Architect Acts as General Contractor.

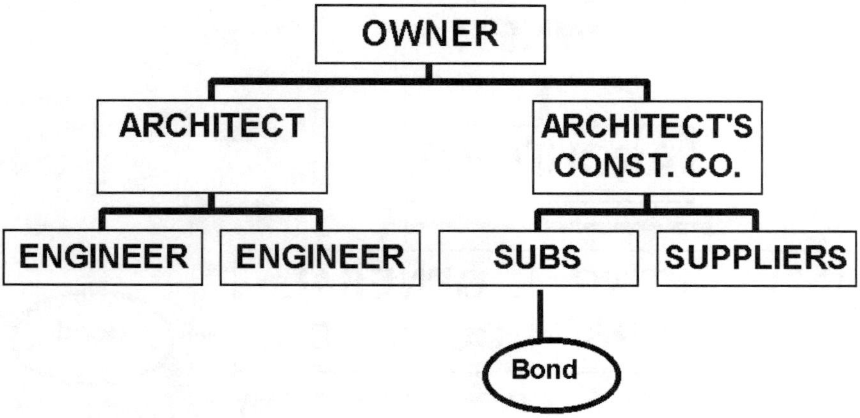

FIGURE 4-2. Architect-Led Design-Build Where Architect Has Construction Subsidiary.

claim against its own professional liability policy to reimburse itself for the costs incurred in excess of the contract price.

Figure 4-2 depicts an arrangement under which the design firm forms a construction subsidiary to hold the construction contract and take on the construction risks. The design firm enters into a pure design services contract with the owner, for example an AIA B141 Owner/Architect Agreement. Then, the construction subsidiary enters into a construction contract with the owner, for example an AIA A101 or A111 Owner/Contractor Agreement. The design firm is able to offer construction services "in-house" through its subsidiary, and to coordinate both services, with input from its construction staff during the design phase. The subsidiary may bid on traditional work designed by other firms, and the design firm can handle either design-build or traditional design-bid-build jobs. However, where the owner wants both services from one entity, this model offers a close equivalent. The benefit of this model is that the subsidiary absorbs the risk of a construction loss, *e.g.*, a cost overrun or an uninsured loss, without putting the design firm at risk for those losses. The construction subsidiary may even be able to make a claim against the professional liability insurance of the design firm by special endorsement, depending on the percentages of ownership and the insurer's exclusions, thereby covering some losses in a way not possible by the design-build firm shown in Figure 4-1. However, the two-contract model does not fit the common definition of design-build and some owners may not accept this model if they are truly looking for design-build under a single contract.

Figure 4-3 shows a design firm holding the prime contract and subcontracting all of the construction work to a general contractor, who then subcontracts with trade contractors. This fills in the construction experience needed by the architect through a "teaming relationship" with an experienced general contrac-

[8] Mark C. Friedlander, *Designer-Led Design-Build Teams*, Construction Briefings, No. 97-9 (Aug. 1997).

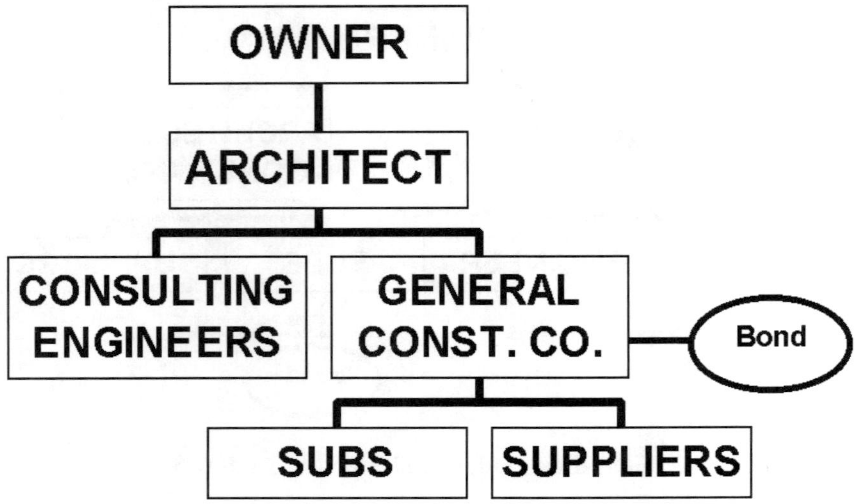

FIGURE 4-3. **Architect-Led Design-Build Where Architect Subcontracts 100 Percent of Construction to a General Contractor.**

tor. By making the owner a third-party beneficiary of the prime subcontract (with the general contractor), bonding the general contractor, and providing the owner with assignment rights to all subcontracts, the owner gets the protection it needs, but the design firm avoids some of the risk.[8] The owner can usually be named as a "dual obligee" with the architect on the contractor's bond, thereby providing protection to both the owner and the architect. However, it is hard to imagine as much profit being made when a general contractor has to be hired to manage all the subcontractors. The design firm seems to be only a "paper leader" under this model. By contrast, if the architect can eliminate the use of a general contractor and act as its own "general," as shown in Figure 4-1, the profits increase. The architect, as prime, could not make a claim against its own professional liability policy, but the contractor-as-subcontractor may be able to make claim under the policy carried by the architect for costs it incurs due to design errors or omissions, thereby holding to the guaranteed contract price and covering the loss with insurance.

§ 4.06 JOINT VENTURE/NEW CORPORATION

The fourth of the common models is the formation of a joint venture, a new corporation or a limited liability company (LLC), owned by the architect and the contractor either as partners or as shareholders in the new entity. Figure 4-4 shows the new joint venture (or corporation) holding the prime design-build contract with the owner and then subcontracting back to the two owners with separate contracts for design services and construction work. This is the model that the AIA had envisioned when it first published design-build contracts with the architect as a subcontractor. *See* § 2.06[C]. As with some of the other models, the design-build entity will not be able to make a claim against the professional lia-

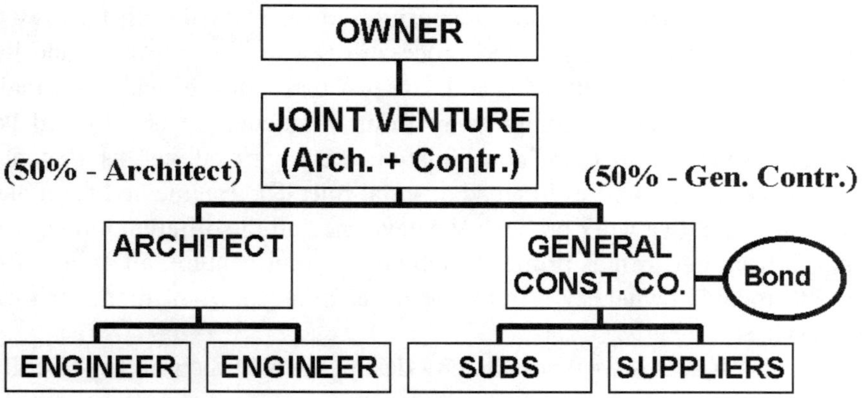

FIGURE 4-4. **Joint Venture (or LLC) Formed Between an Architect and a Contractor. Shown with 50/50 Ownership.**

bility insurance carried by the architect-partner depending on the percentage of ownership of the architect in the joint venture.

The joint venture or new corporation is a popular method for single project undertakings, but accounts for only 8 percent of all design-build projects according to a 1999 survey.[9] Joint ventures are often used when neither firm has the qualifications to perform the entire contract, but does not want to be a subcontractor to the other. This is a workable approach, but has many licensing, insurance, and liability issues that must be thoroughly studied, understood, and managed to make it work. The design firm and the contractor effectively become partners for that one project.

§ 4.07 SINGLE ENTITY WITH IN-HOUSE DESIGN AND CONSTRUCTION (INTEGRATED FIRM)

This form, also called "integrated design-build," is practiced by some large contractors who have in-house architects or engineers on staff. It is also used by some large architectural and engineering firms. The construction trades are subcontractors. There is a move among contractors in this direction, avoiding the need for outside design partners and providing a "seamless" operation. It accounted for about 26 percent of design-build work in 1999.[10]

Ellerbe Becket is one of the largest architectural firms in the United States and one of the largest design-build contractors as well. *Design-Build* magazine placed the firm 102nd in its list of the top 150 Master Builders based on 1997 revenues. Bob Degenhardt, P.E., president and CEO of Ellerbe Becket says,"We call it integrated services," referring to the way his firm does design-build by having both design and construction staff in-house. Ellerbe Becket had been doing design-build since the mid-1970s through a separate subsidiary. After 25 years of that model, the firm decided there had to be a better way. Degenhardt talked to

---

[9] *1999 Design-Build Survey,* Zweig White & Associates, Inc., p. 25.
[10] *Id.*

owners and found them becoming increasingly disenchanted with the way construction is done, with a fragmented process driven by specialization and litigation. He looked at other industries and saw just the opposite, with auto makers, health care, and banking moving away from fragmentation and toward "integrated" companies, with more consolidation. "Disaggregation," as Degenhardt calls it, "places all the parties in an adversarial role. The fragmented team has resulted in an ineffective work process. We have tried all the different models of design-build, from our firm as prime, as sub and as joint venture, all in an effort to respond to today's owners. We think we have found the right model with integrated services."

Degenhardt sees a weakness among design firms in their ability to collaborate with contractors. In collaboration, "everybody gives up something in order to get something." It is a compromise. "If you are all worried about suing each other, or just doing your part, that is not collaboration," he said. With integrated service, the focus is not on protecting yourself, but on the success of the team. Owners get higher value and change orders are virtually eliminated (other than for scope changes).

Degenhardt observed that consolidation in construction has been mostly within disciplines, with roll-ups of trade contractors and mergers between architecture firms; but there have not been many cross-discipline mergers—*e.g.*, a contractor and a design firm. At Ellerbe Becket, there are design and construction members on staff, but each does its own part. The architects do design, the engineers do engineering, and the estimators do estimating. "Problems arise when you put people into a different role," he said. "We manage risk by having the right people in-house to do the job they are trained for." When you break down the barriers between design and construction, he said that it is amazing what can happen. "Removing those barriers opens the door to a high level of trust, everyone working toward the same goal."

Degenhardt says that the claims are far less, since "you do not sue yourself." If they cannot do integrated services, he feels there is an increased risk of litigation. In three years of offering integrated services, he says that Ellerbe Becket has not had one claim by an owner, nor has the cost exceeded the GMP.

Ellerbe Becket has in-house scheduling, estimating, project managers, site superintendents, plus project engineers. Each job has a trailer on site, plus clerical staff. Ellerbe Becket has not yet included trade subcontractors in its integrated practice because owners still want participation from local vendors and subcontractors to be sure the local community benefits from its construction projects.

§ 4.08 THE MULTIDISCIPLINARY PRACTICE

There is a growing trend for companies of all types to attempt to be all things to all people. McDonald's hamburger chain is a prime example of a company known for one product—*i.e.*, hamburgers—but it has expanded to offer everything from pizza, chicken, salad, soup, burritos, and hot dogs in different markets. Likewise, several consulting firms have expanded their practices beyond their traditional roles offering everything "from soup to nuts" for owners building a construction project. National accounting firms like Arthur Anderson and PricewaterhouseCoopers now offer owners financial planning and facility operations services. For businesses that want to outsource their construction and real estate services, the accounting firms have expanded their operations into design-build by offering total "turnkey" services for clients seeking a true one-stop shop for all their construction needs. These firms can plan, finance, design, build, own, operate, lease back, maintain, or whatever the customer wants. It has been reported that in 1999, the company that hired the largest number of new architects was not HOK, SOM, or Ellerbe Becket, but the accounting firm of Arthur Anderson.

In promotional materials, PricewaterhouseCoopers states that for one U.S. telecommunications company, "[w]e provide design, engineering, construction, procurement, and facilities management services to support more than 5,000 buildings with 50 million square feet of space in five states."[11]

There is also a prediction that we will see mergers between law and accounting firms in the first part of the millennium to handle global consulting services for their clients.[12] These multidiscipline practice firms (MDP's) are currently restrained by laws that prevent offering such joint services, with the District of Columbia perhaps the only jurisdiction that permits it. There are MDPs in France and talks of such mega mergers in the United Kingdom where, it is rumored, the Big Five accounting firms have made merger overtures to many of the United Kingdom's leading law firms. Imagine legal, accounting, and design-build services all offered by one firm—a McDonald's for project owners.

§ 4.09 HOW ARE DESIGN-BUILD FIRMS SELECTED?

Design-build contracts are obtained in much the same way as traditional construction projects with one exception: Owners award most design-build contracts by direct selection or negotiated selection.

[A] Direct Selection

Direct seleciton is where the owner hires the design-builder on traditional factors such as reputation, experience, technical and managerial qualifications, past performance, and prior association.

[B] Negotiated Selection

In this method, a number of design-build contractors or teams are prequalified or interviewed, with selection based upon the same factors used in

[11] PricewaterhouseCoopers' Web site at <http://www.pwcglobal.com>.

[12] *The Future of the Multidisciplinary Practice*, London Calling, Supplement to The American Lawyer (Nov. 1999), p. 15.

direct selection. Architects are used to this type of selection and are skilled at presentations to owners when going head-to-head with other firms. Design-build brings the contractor partner into the picture at this early stage, with selection made based on not only reputation and experience, but on cost and possibly design solutions. DBIA recommends that in a negotiated selection, the owner use a qualitative method of evaluating each team using a set of weighted criteria. Figure 4-5 shows an example of how such criteria might be weighted.

Team	Design Solution (weight 10)	Financial Capability (weight 6)	Project Approach (weight 8)	Management Plan (weight 10)	Total Points
A	4 × 10 = 40	5 × 6 = 30	4 × 8 = 32	3 × 10 = 30	132
B	5 × 10 = 50	5 × 6 = 30	2 × 8 = 16	5 × 10 = 50	146
C	4 × 10 = 40	4 × 6 = 24	2 × 8 = 16	3 × 10 = 30	110

Grading Scale: 5 = Excellent; 4 = Good; 3 = Fair; 2 = Poor

FIGURE 4-5. Sample Weighted Negotiated Selection Criteria.

Source: DBIA's Design-Build Process Using Negotiated Selection (1994).

§ 4.10 COST/DESIGN COMPETITIONS

Architects are also familiar with selection based on design competitions, though they usually are reserved more for big budget, high profile projects which justify the financial investment to prepare designs at risk. With design-build, however, cost is added as a factor to the mix, with the selection often made using some weighted points scale based on the quality of design and the low bid.[13] Growing in popularity is the "two-envelope" system, in which qualifications and designs are submitted separately and evaluated prior to opening the bids. The award does not necessarily go to the low bidder, but to the team with the highest overall score which is based on qualifications, design and price. *See* Chapter 14 for more discussion.

§ 4.11 PURE COST COMPETITION

This method is rarely used, but resembles more traditional public procurement based on low bid. With design-build, a request for proposal is prepared with design criteria and often some preliminary design prepared by a consultant to the owner. *See* Chapter 6, Bridging, for more discussion. Prequalified teams compete strictly on price with the award going to the low bidder. On public projects, in states that have not adopted legislation to amend competitive bidding laws, this is the only way that design-build contracts can be awarded.

[13] *See* Chapter 15 for more detailed analysis of this approach and weighted selection criteria.

§ 4.12 PARADIGM SHIFT

Design-build puts many traditional concepts and attitudes on their ear. It reverses the roles and risks of architects and contractors. The adversarial relationship between the design professional and the builder that too often accompanies the traditional method is replaced by teamwork. Teamwork is critical to the success of design-build. The traditional checks and balance is replaced with trust and fair contracting among the design-build team. The old view was separate duties, separate responsibilities, and separate goals. The new view is mutual duties, shared responsibilities and goals.

Also changed is the level and type of risk accepted by the design professional. Where AIA contracts traditionally state that the architect is *not* liable for the means, methods and techniques of construction, or for contractor errors or safety programs, this has turned completely around with architect-led design-build.[14] Architects who become design-build contractors need to begin to think like contractors. With risk comes "reward." Accepting more risk is part of the new paradigm, but so are bigger profits. It is reported by some that design firms who lead a design-build project can make up to four times the profit normally realized on design-only contracts. This margin can be increased even more by including a "shared savings" clause in cost-plus (GMP) contracts, which gives the design-build firm an incentive to bring the job in under the owner's budget price. Some owners will even pay a "bonus" for early completion.

Profits can also be increased by not over-detailing the project. Since the architect will have a direct hand in construction, design details can be limited and worked out in on-site meetings. Sketches and on-site discussion can be done with the trade contractor actually doing the work. The level of detailing is a judgment call, but experienced design-builders report that they get by without full-blown sets of drawings and specifications, often loaded with disclaimers and "CYA" provisions.

This bigger profit is not without a price. In the traditional mode, the architect's biggest concern is keeping the client happy. With architect-led design-build, there are lenders to satisfy, sureties, dozens of subcontractors and suppliers to oversee, schedule and pay, public inspections, etc. Someone compared it to "being nibbled to death by ducks" or "death by 1,000 cuts." Architect-led design-build requires tough skin and an adventurous spirit.

---

[14] *See*, e.g., B141 Owner/Architect Agreement, Paragraph 2.6.2 (1997 edition).

CHAPTER 5

DESIGNER AS SUBCONTRACTOR TO A CONTRACTOR

§ 5.01 WHAT NEW RISKS ARE THERE AS A SUBCONTRACTOR?

Most design-build projects are done with a contractor as prime and the design firm as subcontractor. Statistics show that over half of all design-build work is done by contractors leading as prime contractor.[1] Design firms surveyed in 1999 reported that 95 percent of their design-build work was done as a subcontractor to a contractor.[2] It is surprising that in a profession that has complained about losing its leadership role in the industry that so many architectural firms are willing to be subcontractors to general contractors. But putting egos aside as to who is leading the team, contractors have the bonding capacity and experience in pricing and managing multiple trade subcontractors, so it is no surprise that contractors are in the early lead and may always enjoy that position in design-build. Here are some of the issues a design firm needs to consider if it is acting as a subcontractor.

§ 5.02 LICENSING ISSUES

As discussed in greater detail in Chapter 12, it is important for contractors who are offering design-build services to check the licensing laws of the state in which the project is located. In some states, it is illegal for a contractor to provide both design and construction under one contract unless the contractor is authorized to practice architecture or engineering in that state. So why is this a concern to a licensed design firm acting as subcontractor? Two reasons come to mind. First, it is a violation of state licensing laws to enable or assist an unlicensed party in the practice of architecture or engineering. If the architect or the firm acts as a subcontractor to an unlicensed contractor, it is possible that the state licensing board could impose sanctions against the architect personally or the firm for assisting someone in the unauthorized practice of a licensed profession. Second, if the prime contractor is not licensed in a state where such a license is required, the contract is usually illegal, void, and unenforceable by that contractor. Even though the design firm may be properly licensed in that state and have a valid subcontract, what are the odds of it getting paid if the prime contractor cannot collect from the owner? Unless the contractor is bonded (and the bond is found to be legal), the design firm may have difficulty enforcing and collecting from a contractor who is barred from collecting from the owner due to an illegal prime contract.

§ 5.03 PAYMENT BONDS

Architects and engineers who serve as subcontractors on design-build projects may be surprised to find that they are not protected by the design-build contractor's payment bond. In a 1994 Arizona case, an architectural firm was denied recovery of over $14,000 in fees under the contractor's bond since the statutory

[1] *1999 Design-Build Survey*, Zweig White & Assoc., Inc., p. 94.
[2] *Id.* at 51.

payment bond covered only "labor, materials, or construction equipment." The court held that professional architectural services are not "labor or material" as defined in the statute.[3] But in a Connecticut case, just the opposite happened. The contractor and surety argued that the architect's services were not covered by the statutory bond, which protected only "persons supplying labor or materials." The court held that contracting parties who execute a payment bond can expand coverage beyond the statute and that the bond in this case expressly covered "services."[4] These cases point out the importance of the bond language.

The two Associated General Contractors of America (AGC) Design-Build Payment Bond forms are different only in the inclusion of the word "services" in one (where the surety is liable for the design costs of the work), and the exclusion of that word in the other (where the surety is *not* liable for design costs). When a firm is acting as a subcontractor providing design services to a bonded contractor, it is important to ask for and review a copy of the payment bond to make sure that it covers "services" or the firm may be at risk of non-payment, with no ability to recover against the surety bond.

§ 5.04 PAYMENT CLAUSES—"PAY WHEN PAID"

A hot topic in construction contracts today is the "pay when paid" clause used in subcontracts. These standard clauses state that a general contractor will pay its subcontractor within a specified time after the owner pays the general contractor. A variation is the "pay if paid" clause, which puts the entire risk on the subcontractor if the owner never pays the contractor. If the owner does not pay the contractor on time, a question is raised as to how long the subcontractor must wait to recover from the contractor. The same issue is present in consulting agreements between the prime contractor and a design professional on a design-build job.

It has been held that courts disfavor "pay when paid" clauses and will not construe them as conditions precedent unless the language is plain and unambiguous.[5] Many courts have refused to strictly enforce a "pay when paid" clause, holding that a subcontractor need not bear the burden of the owner's failure to pay unless the contract is expressly clear that payment is "directly contingent upon" the owner's payment as a "condition precedent" to the subcontractor's right to recover.[6] Some cases have held that clauses that require the contractor to pay a subcontractor after the owner makes payment only delay payment for a "reasonable time." In a 1987 case, the federal district court held that despite a "pay when paid" clause, a reasonable time for payment to be made is three months (or one quarter).[7]

---

[3] Fields Hartwick Architects v. Capitol Indem. Corp., 884 P.2d 198 (Ariz. Ct. App. 1994).

[4] Herbert S. Newman & Partners v. CFC Constr. Ltd. Partnership, 674 A.2d 1313 (Conn. 1996).

[5] Byler v. Great Am. Ins. Co., 395 F.2d 273 (10th Cir. 1968) (decided under Oklahoma law).

[6] American Drilling Serv. Co. v. City of Springfield, 614 S.W.2d 266 (Mo. Ct. App. S.D. 1981); Main Elec., Ltd. v. Printz Services Corp., 980 P.2d 522 (Colo. 1999).

[7] Havens Steel Co. v. Randolph Eng'g Co., 613 F. Supp. 514 (D.C. Mo. 1985), *aff'd*, 813 F.2d 186 (8th Cir. 1987).

[A] The AGC Subcontract Forms

The design-build trade subcontracts published by the AGC come in four separate variations, two with "pay when paid" clauses, and two without. Those are:

1. AGC Doc. No. 450 Agreement Between Design-Builder and Subcontractor;
2. AGC Doc. No. 455 Agreement Between Design-Builder and Subcontractor (with contingent payment clause);
3. AGC Doc. No. 460 Agreement Between Design-Builder and Subcontractor (where subcontractor provides a guaranteed maximum price); and
4. AGC Doc. No. 465 Agreement Between Design-Builder and Subcontractor (where subcontractor provides a guaranteed maximum price plus contingent payment clause).

The AGC subcontract between design-build contractor and architect/engineer contains a "pay when paid" clause. It states that, "[t]he Design-Builder shall submit approved applications for payment together with its own to the Owner. Payment shall be made within seven days of the Design-Builder's receipt of payment from the Owner for the amount approved on the Architect/Engineer's application."[8] The clause goes on to require the architect/engineer to pay its lower-tier subconsultants "promptly" after it receives payment from the prime contractor. Better language is used in AGC's trade subcontract form no. 450, par. 8.2.5, which states that payment is to be made within seven days (like the architect/engineer contract), but then goes on to add that, "[i]f payment from the Owner for such Subcontract Work is not received by the Design-Builder, through no fault of the Subcontractor, the Design-Builder will make payment to the Subcontractor within a reasonable time for the Subcontract Work satisfactorily performed." Design firms should consider adding this language to subcontracts to avoid having to litigate the meaning of a subcontract's payment clause. Even better is to define the phrase "reasonable time" in the subcontract itself to establish an outside date for payment to be made.

AGC's trade subcontract no. 455 has an express contingent payment clause, which adds this sentence: "The Subcontractor hereby acknowledges that it relies on the credit of the Owner, not the Design-Builder for payment of Subcontract Work." Be sure to watch for clauses like this, or clauses that say that payment to the architect/engineer is "contingent upon" payment by the owner, which is an "express condition precedent" to the design-build contractor's obligation to make any payment. This is more likely to be upheld by a court as enforceable.

[B] The DBIA Subcontract Form

DBIA's Agreement Between Design-Builder and Designer, Doc. No. 540 (1999 edition) has several payment clauses of concern to architects and engineers. First, ¶ 6.2.1 requires the architect to submit its "certified" application for pay-

[8] Par. 6.3.1, *Agreement Between Contractor and Architect/Engineer for Design-Build Projects*, AGC Doc. No. 420 (1999 edition).

ment monthly. There is no form of pay application published by DBIA, so the certification is not specified. The inference is that the designer certifies its own invoice, an odd requirement. Next, ¶ 6.2.3 and 6.5.1 are "pay when paid" clauses, which condition any payment to the designer on the owner having first paid the design-builder. While DBIA takes the view these are not "pay if paid" clauses, and are not meant to be taken literally (as the courts often do not do), ¶ 6.6.1 makes it clear that "*all payments to Designer hereunder . . . shall not be due until after Design-Builder actually receives payment*" from the owner. How the courts in each state will interpret this clause is unknown, but deleting this provision is the only safe course.

What if a dispute between owner and contractor, such as termination, defective construction, etc.—holds up payment? Paragraph 7.1.1 states clearly that if the prime contract is terminated, the designer only gets paid "*those amounts Design-Builder actually receives from Owner*," which reads like a "pay if paid" clause. These clauses, while invalid in some states, should be amended to state a drop-dead date by which the design-builder must make payment regardless of whether the owner has paid the contractor. Finally, ¶ 6.3.1 permits the design-builder to hold retainage until the owner releases that money to the contractor. Again, disputes between the owner and the contractor over liquidated damages, defects, etc. could delay or prevent release of retainage, putting the designer at risk of ever seeing that money.

§ 5.05 DBIA SUBCONTRACT BETWEEN DESIGNER AND DESIGN-BUILDER

In addition to the payment clauses, the 1999 subcontract between the design-builder and designer (DBIA Doc. No. 540) has a number of other sections that may be of concern to design firms. With the majority of design-build work being contractor led, and the forecast that 50 percent of all non-residential construction will be design-build in the next five years, it is likely that architects and engineers will be asked to sign this contract sometime soon. DBIA's documents committee attempted to draft fair contracts that do not favor the owner, contractor, or design professional. Being the "new kid" in the contract form market, DBIA's authors agree that their contracts, while fair, are not perfect. Of course "perfection" is in the eye of the beholder. Design firms will have to judge for themselves and make modifications where needed.

Consistent with the drafters' goal, the design subcontract is fairly balanced in many areas. However, from an architect/engineer's perspective, there are many risks that are not covered by insurance, plus payment contingencies that should be amended to establish a more fair balance of risk. In addition to the payment clauses discussed in § 5.04, the following are some of the key problems with the new DBIA form.

[A] Conflicts Between Subcontract and Supplement

Paragraphs 1.3.1 and 1.4.1, when read together provide that when in conflict, the DBIA form 540 takes precedence over written supplementary conditions executed by the parties. This is contrary to the very purpose of such conditions, which are intended to alter the form contract. This should be revised to make clear that supplementary conditions will always govern over the DBIA form. *See* Appendix B for proposed language.

[B] No Direct Communication With Owner

Paragraph 2.1.4 bars direct contact between the designer and owner, unless the design-builder agrees otherwise. This is not an uncommon construction subcontract provision but will run contrary to the way in which architects and engineers normally deal with end users of their buildings. The architect/engineer must rely totally on the prime design-builder to pass on the concerns and comments of the owner and designer. This separation between design professional and owner was one of the AIA's early concerns over design-build. *See* § 2.06 [B], above.

[C] Standard of Care and Performance Guarantees

DBIA has included an insurable standard of care in ¶ 2.2.1, consistent with state law in most jurisdictions holding the designer to that level of care and skill "ordinarily used by members of the design profession practicing under similar conditions at the same time and locality of the Project." However, the clause goes on to add that the designer shall comply with any performance standards required by the prime contract (*i.e.*, regardless of the standard of care). This may create a "warranty" or guarantee not covered by professional liability insurance. Be sure to watch for any such performance standards that may be incorporated into the design contract. Consider adding language that modifies this clause to clarify that in any event, this contract does not create any warranty, express or implied.

[D] Estimating Errors Corrected at Designer's Cost

Although the design-builder is responsible for all cost estimating, including "mistakes or miscalculations of market conditions" under ¶ 3.3.3, the architect/engineer is required under ¶ 2.6.6. "at its own cost" to revise the design as needed to achieve the budget. It would seem more equitable that if the architect/engineer is relying on the design-build contractor to price out the design as it progresses, then the contractor (who is responsible for its mistakes) would bear part or all of the cost of design revisions needed due to estimating errors.

[E] Site Observations and Notice of Defects in Construction

Paragraph 2.7.6 requires site visits by the designer at appropriate intervals "to determine if the construction is proceeding in accordance with the Construction Documents." Absent full-time resident services, lots of work will go unobserved by the designer. As a result, the architect/engineer should agree only to observe the progress of work for "general" conformance to the Documents. Watch for any amendments to this clause that require the architect/engineer to guard the contractor against defects in its own work. In one court case, *C.L. Maddox, Inc. v.*

The Benham Group, Inc.,[9] the trial court held that such a clause meant that the engineer was responsible for the contractor's own errors during construction. That decision was reversed on appeal but goes to show the possible risk that such a clause could pass on to a design firm. The DBIA form requires the designer to "promptly notify Design-Builder of any defects, deficiencies, deviations, omissions, or violations observed by Designer in the construction of the Project, and make recommendations to Design-Builder on how to proceed." This seems to be a reasonable requirement for a design-build team whose mutual goal is to resolve problems quickly and efficiently.

[F] Certifications by Designer

A vague, yet broad sentence in ¶ 2.7.8 requires the designer to provide "certifications as may be necessary relative to Substantial Completion." This may include certifications to lenders, sureties and government officials dealing with codes, ADA, and environmental risks. With each third party certification comes increased liability for the architect/engineer. At a minimum, the designer should have the right to: (1) approve the form of certificate in advance; and (2) issue only those certificates it can reasonably make based on its level of involvement during construction. Watch for certifications that might create an uninsurable warranty or guarantee.

[G] Limited Reliance on Contractor-Furnished Information

The prime design-builder will usually furnish the architect/engineer with surveys, geotechnical data, easements, zoning, as-built drawings, environmental and inspection reports, among other data. While the designer is permitted under ¶ 3.3.1 to "rely upon such items," that reliance is limited to the level that the design-builder can rely on such information under the prime contract. Since the architect/engineer (as a subcontractor) most often cannot negotiate the terms of the prime contract, it is left with whatever deal, good or bad, that the contractor makes with the owner. If this type of data is provided by the owner or design-builder, the architect should be entitled to rely on the completeness and accuracy of such data without limitation. The designer must insist on seeing and reviewing the prime contract and all exhibits, conditions, etc. prior to signing any subcontract that incorporates or refers to the terms of that contract.

[H] Ownership and Reuse of Documents

DBIA's theme for this topic is that the designer retains ownership of its documents, but grants a license to the design-builder and owner to use them for this project. While that sounds fine, ¶ 4.2.1 requires the architect/engineer to agree to whatever licenses are given to the owner in the prime contract. This may include broad rights to reuse or alter the documents, or transfer/assign such license to others without the designer's review or involvement. The contract should be modified to prohibit: (1) future reuse on other projects; and (2) alteration by de-

[9] 88 F.3d 592 (8th Cir. 1996).

sign-builder, owner or others on "this or any other project," without an agreement to defend and indemnify the designer and its consultants. The indemnity provided by the DBIA form in ¶ 4.3.1 is limited to use on "other projects," which does not cover changes the design-builder or owner may make on "this project," including completion of the documents by others in the event of termination of the design subcontract for convenience or for cause.

[I] The Seven-Day Meeting

Under ¶ 2.1.5, the parties are to meet within seven days after execution of the contract to discuss the project, schedule, payment, etc. This "2.1.5" meeting is referenced throughout the contract and the meeting agenda should include every topic tied to that paragraph in the subcontract. Be sure to review the subcontract for each reference to that meeting.

[J] Delays, Claims by the Designer, and Dispute Resolution

A standard subcontract provision is to require subs to give the prime contractor notice of delays or claims in sufficient time to permit the contractor to make such claim to the owner. Paragraphs 5.4.1 and 11.1.1 pass this requirement onto the designer. It is therefore important to read the prime contract to understand the time limitations on claims. Of key importance is ¶ 11.3, which states that if a claim involves the owner, the architect/engineer agrees to resolve the dispute per the terms of the prime contract. What if that agreement says that the owner gets to decide disputes, and that the owner's decision is final? Or if disputes are resolved in some far off jurisdiction? Be careful here not to fall into a trap unaware. Under ¶ 11.3, the designer agrees to be bound by the outcome of the dispute resolution as "full and final resolution" of such dispute. What if the owner has claims against the contractor that offset the designer's claim, resulting in no recovery to the designer? Or if the design-builder does not follow strict notice requirements of the prime contract, resulting in a waiver of the claim? Modifications should be considered here. Suggested wording is included in Appendix C to this book.

[K] Insurance

One of the most overlooked traps for design professionals is buried in ¶ 6.5.1, which requires as a condition of final payment that the designer provide "*certificates of insurance* confirming that required coverages will remain in effect consistent with the requirements of the Contract Documents."[10] If the owner requires that insurance be maintained for several years after substantial completion, the designer will be unable to furnish this certificate of insurance in most cases, thus delaying final payment. Since professional liability policies are written most often for one-year terms, with no guarantee of renewal, the most an insurer will certify is that the policy will remain in effect for the remainder of the current one-year term. In absence of a project policy, designers will not be able to meet

[10] *See also* ¶ 9.1.5.

this condition, which should be deleted. The specific insurance required under the subcontract is left open in ¶ 9.1, to be agreed in an exhibit to the contract. Paragraph 9.1.6 requires the designer to name the design-builder and others as additional insureds except as stated in the insurance exhibit. Be sure to exclude professional liability and workers' compensation policies, which normally do not permit additional insureds. Finally, the waiver of subrogation in ¶ 9.2.1 only applies to "property insurance," which is not normally carried by designers or contractors, but by the owner. Discuss these clauses with a knowledgeable insurance agent before signing an agreement not compatible with available insurance.

[L] Release of Claims at Final Payment

Paragraph 6.5.1 also requires a general release of all claims as a condition of final payment. This should exclude claims that first arise after the submission of the designer's final invoice, such as a claim for indemnity dealing with reuse of documents, site safety, etc. The intent is surely to prevent "sandbagging" by holding back on claims until after payment is released, but the broad language could have the unintended effect of barring future, unknown claims as well.

[M] Indemnity Clauses

There are several indemnity clauses sprinkled throughout the contract. Paragraph 6.8.1 requires the designer to indemnify the owner and design-builder for claims of payment or liens by subconsultants, without any requirement that the design-builder has first made payment to the designer. Paragraph 10.1.2 requires the designer to indemnify the owner and design-builder for copyright and patent claims with some exceptions. Missing from the exceptions is the situation where the owner or design-builder provide the design (like a site adapt) in which case it is *the designer* who should be indemnified by those parties, not the reverse. Paragraph 10.2 is a mutual indemnity section under which both parties indemnify each other. Due to differences between insurance coverage provided to designers and contractors, ¶ 10.2.1 should be modified to delete the words "defend," "claims," and "liabilities." Also, without using the term "indemnity," the architect is required in ¶ 11.3.4 to pay damages due to a "breach" without any reference to "negligence" as needed for insurance coverage for this clause. This clause should be modified to tie it to negligent acts, errors, or omissions. Again, consultation with insurance counsel is recommended here.

[N] Dispute Resolution Procedures

Paragraph 11.4 is a multi-tiered disputes clause, with claims between the parties resolved by negotiation, followed by mediation and, last, by arbitration if not settled at the lower levels. One should be sure that subcontracts with consultants adopt this same process and permit joinder of all needed parties into any mediation or arbitration. In any design-build scenario, it is important that all those who might be responsible to indemnify the prime design-build firm, whether they are a subcontractor or supplier, are required to resolve their disputes in the same forum the prime firm is required to do. Otherwise, full relief cannot be granted in one hearing, whether that is in court or arbitration.

[O] Liquidated Damages

Paragraph 11.7.2 passes on to the designer any liquidated damages the owner assesses against the design-builder due to the designer. Remember that claims involving the owner like this are resolved under the prime contract, which result is final and binding. Check the firm's insurance policy and consult a knowledgeable agent to see if liquidated damages are insurable, especially without any reference to negligence in this clause. While special, costly insurance is available to cover liquidated damages, that coverage is not likely available under standard professional liability policies.

[P] Miscellaneous

In ¶ 7.2.1, delete the word "ensure" which may create an uninsurable warranty. Also, ¶ 12.3.1 makes the law of the place of the project applicable to the contract, rather than the law of the designer's home state. Be sure the designer's attorney understands that state's law and that the subcontracts contain the same provision.

The DBIA contracts are intended to be fair. Despite the numerous concerns above, the DBIA documents are very good, although not perfect, but a good balance. It does not appear that there was much input from the design community in their drafting. Balancing a contract is a difficult task, since it means shifting risks equally to all parties. Architects or engineers using this form need to determine whether the risks are balanced by the rewards offered on each specific project. Obtain proper legal and insurance advice before signing this or any other contract.

§ 5.06 INSURANCE ISSUES

Contractors will usually want their subcontractors, trade or professional, to be bound to the same requirements that the contractor has with the owner. The owner may insist on this in many cases. Through what is known as a "conduit clause" or a "flow-through" clause, the subcontract between the prime contractor and the design subconsultant will incorporate all of the terms of the prime contract and make them applicable to the subcontract as well. One must be very careful here. Construction contracts will contain broad indemnity clauses, warranties and performance guarantees that are excluded from coverage under a designer's professional liability policy. As a result, while the contractor may have made that agreement with the owner, it does not mean that the design firm has to accept it, especially if the firm's fee is premised on having traditional insurance protection to cover contract risks.

§ 5.07 HIGHER OR DIFFERENT STANDARD OF CARE

Architects and engineers are insured for professional negligence, which is defined as a deviation from the standard of care that is ordinarily exercised by other professionals under the same or similar circumstances. This is not a standard of perfection and there is usually an exclusion in the policy for express warranties or guarantees. The contractor, on the other hand, will be required to warrant that the roof will not leak, or guarantee some other performance criteria. That is a different, and higher, standard of care under which the contractor promises that there will be no problems and that the contractor will fix those that arise. Watch for contracts requiring the "highest" or "best" standards of the industry, since these subtle words create risks not covered by insurance. Other words to watch out for include: ensure, insure, assure, warrant, guarantee, and certify, all of which can be interpreted to add uninsured risks.

§ 5.08 CLAIMS BY OWNER DIRECTLY AGAINST DESIGN SUBCONTRACTORS

Because the prime contractor will not usually have a corporate certificate of authority to practice architecture or engineering, and cannot get one because of its ownership structure, owners may be agreeable to putting a clause in the prime contract that recognizes this fact. In some contracts, the owner will want the right to pursue claims directly against the design firms in order to get to their insurance. Though this defeats the entire concept of "single-point responsibility," one such clause, used by AT&T, reads as follows:

> *Limits to Contractor Liability for Professional Services.* AT&T acknowledges that Contractor is not a licensed architect, engineer or lawyer, and therefore, is not authorized by law to perform design, engineering or legal or other licensed professional services. Contractor shall not perform design, engineering, legal or other licensed professional services in connection with the Project, but will furnish such services by appropriate agreement with a licensed professional Subconsultant or with Subconsultants who, in turn will retain licensed professionals, or shall manage professionals retained directly by AT&T. AT&T agrees that the liability of Contractor for any claims, damages, errors and/or omissions arising out of or in connection with professional or other services to be performed by any such Subconsultants (excluding construction contractors) under this Agreement, whether through indemnity or otherwise, shall rest solely with the appropriate Subconsultant, and Contractor assigns to AT&T all rights, title and interest that Contractor has to pursue such claims against the Subconsultants or against the Subconsultants' insurance. In consideration of this assignment, AT&T agrees to pursue any such claim directly against the Subconsultant and not against Contractor. Contractor will include the terms and conditions of this paragraph in its contract with the Subconsultants (excluding the construction contractors).[11]

It is surprising to see this kind of clause in a design-build contract, especially an owner-drafted form, but this shows that some owners are concerned enough about licensing issues, for which the contractor may not have insurance, that they are willing to bypass the prime contractor and go straight after the design firm for relief.

---

[11] Construction Management Contract, AT&T Supplier Management Division, Contract No. LEI-338-E (Aug. 1999).

CHAPTER 6
BRIDGING

§ 6.01 WHAT IS BRIDGING?

On design-build projects, the architect or engineer most often works for the contractor. The owner does not have what is considered an "independent" design professional watching out for its interests, acting as owner's agent. In addition, the cost of preparing designs sufficient to submit cost proposals in a design-build competition often limits the field of teams. To solve both issues, some owners (primarily public owners) retain an independent architect or engineer to prepare preliminary designs, assist with the request for proposal (RFP) and selection process, and stay on board during the construction phase to review pay applications, review the work, and certify the completion date. "Bridging" is the name given to this process. A graphic depiction of bridging is shown in Figure 6-1.

As shown in Figure 6-1, the owner hires a design consultant to prepare design documents, usually to a level about 30 percent complete. Those schematic design documents are then used to obtain competitive proposals from design-build teams. Those teams include architects and engineers who will complete the technical drawings with the remaining 70 percent of information needed for construction. The design consultant remains as the owner's representative to provide limited construction administration services, including final inspections and closeout.

[A] Industry Definition of "Bridging"

The American Institute of Architects (AIA) and the Associated General Contractors of America (AGC) define bridging like this:

> Bridging is a hybrid of the traditional design-bid-build process and design-build. An owner selects an A-E to develop a project design through design development (approximately 30 percent–50 percent of the design work), and prepares scope of work documents which form the basis for competitive se-

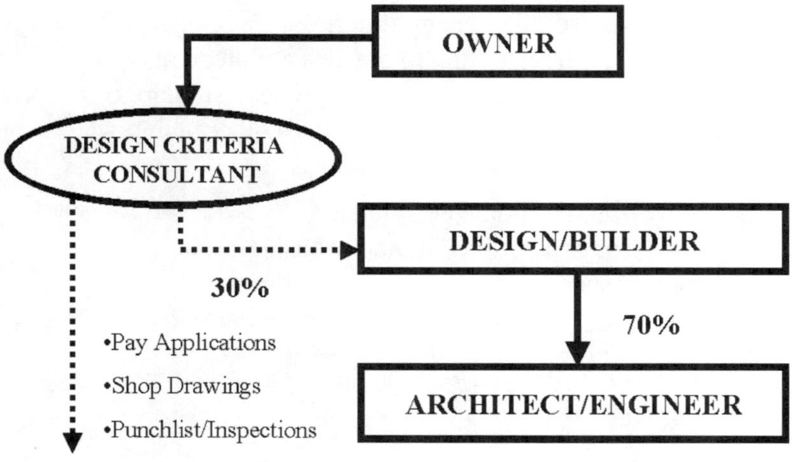

FIGURE 6-1. **Bridging Diagram.**

> lection of the project delivery team. The AE specifies the project's functional and aesthetic requirements but leaves the details of construction technology up to the contractor. Construction technology is specified with performance specifications. The project delivery team then has single-point responsibility for final design and constructing the project.[1]

Bridge firms do limited preliminary design work and then hand that off to other firms to do the technical drawings for construction. But for a bridge firm to serve the client during construction, it needs to know the construction documents well. Though another firm may prepare them, the bridge firm must have enough technical knowledge of the final documents to give the owner an opinion on the quality of the materials, details, and equipment specified by the design-build team, and to detect problems during construction.

[B] Design Criteria Consultants

States with fully developed design-build statutes or regulations often write the role of the bridge firm into the law. Florida was the first to do so. Florida's Consultants' Competitive Negotiation Act is set out at Fla. Stat. § 287.055. *See* Appendix E for the full text of that statute. Under the Florida Act, a "design criteria professional" is selected to prepare a "design criteria package" used to furnish sufficient information to permit design-build firms to prepare a bid or a response to a RFP. The design criteria package is to include:

> performance-based criteria for the public construction project, including the legal description of the site, survey information concerning the site, interior space requirements, material quality standards, schematic layouts and conceptual design criteria of the project, cost or budget estimates, design and construction schedules, site development requirements, provisions for utilities, stormwater retention and disposal, and parking requirements applicable to the project.

Under the Florida Act, the design criteria professional must be a firm that has a current certificate of registration to practice architecture or engineering. The firm is selected using a qualifications-based process similar to the procedures used to hire design professionals for state work. In determining whether a firm is qualified, the agency is to consider such factors as:

1. the ability of professional personnel;
2. whether a firm is a certified minority business enterprise;
3. past performance;
4. willingness to meet time and budget requirements;
5. location;

[1] *AIA/AGC Recommended Guidelines for Procurement of Design-Build Projects in the Public Sector*, p. 5 (Jan. 1995).

6. recent, current, and projected workloads of the firms; and
7. the volume of work previously awarded to each firm by the agency, with the object of effecting an equitable distribution of contracts among qualified firms.

[C] Mini-Bridging

When a trade subcontractor is hired to design-build their portion of the work, it is often not known who will do the actual design. Will it be done in-house, or subcontracted out? What are the qualifications and experience of that designer? Are they even licensed or insured? These questions need to be asked up front.

A variation used by some design firms that lead design-build projects has been called "mini-bridging." In this model, the design-build firm hires engineering consultants to prepare design criteria and performance specifications. Those preliminary designs are then given to trade subcontractors who act as design-build subcontractors for their respective portions of the project—*e.g.*, mechanical, electrical, or even structural components. The appeal of this model, as shown in Figure 6-2, is that the architect can select trusted engineering consultants to prepare design criteria that the architect knows will meet the project needs. This gives a comfort level when working with design-build trade subcontractors with whom the architect has not worked before, or those in whom the prime design-builder does not have complete confidence due to the nature of the project. The trade subcontractors must prepare designs that meet the performance criteria. Architect-led design-build teams often require performance bonds from these trade subcontractors to give more comfort that the subcontractors will perform. Requiring professional liability insurance from these subcontractors is recommended since the surety bond may only guarantee that the work will be completed if the subcontractor defaults, but it will not pay damages due to design errors or omissions by the subcontractor.

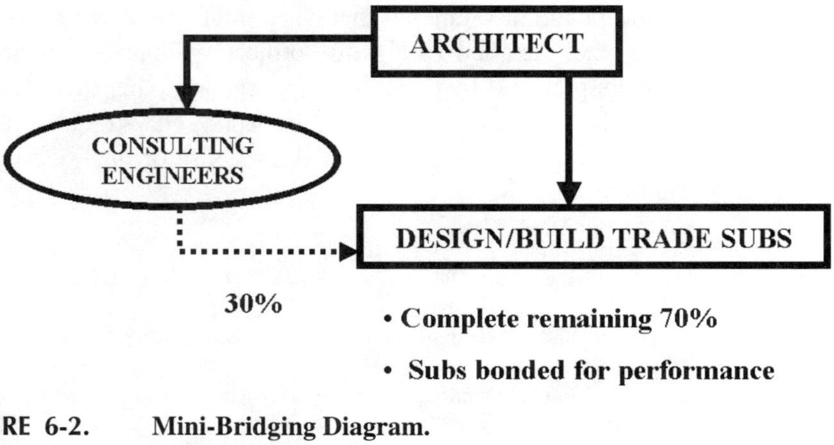

FIGURE 6-2. **Mini-Bridging Diagram.**

§ 6.02 DBIA'S CRITICISMS OF BRIDGING

The Design-Build Institute of America (DBIA) refers to this method of delivery as "Design/Design-Build" or two other equally awkward names "Design-Build/Bid" and "Design-Draw/Build."[2] DBIA discourages this method of procurement, stating that it "eliminates many of the advantages inherent in true design-build." DBIA's primary argument is that the real advantages of design-build are lost when the design is handed to the team already 30 percent complete. Some of DBIA's criticisms of bridging (as it is more commonly known) are:

1. This method precludes the design-build team from any significant creativity and innovation, since basic solutions and concepts are determined before the design-build team begins.
2. Work performed in the first phase determines many of the design and functional aspects of the project for which the phase two team has to take responsibility (and liability), without any real input, leading for greater potential for disputes.
3. True design-build permits fast-track construction with the ability to procure long lead time items and begin field operations well before a project is at the 35 percent point of design.

Dr. James Broaddus, former President of DBIA, said that when 35 percent or more of the design is provided in the RFP, the selection tends to be price oriented. "When less design is provided, there is more emphasis on qualifications-based selection."[3] DBIA urges selection of design-build teams based on a combination of qualifications and price, but feels that under "bridging" or "design/design-build" the selection is based solely on the basis of low price. Case studies of actual public sector projects would show otherwise, since bridging is regularly used in public procurement, yet qualifications of the design team remain a large portion of the selection criteria.

§ 6.03 AIA'S POSITION ON THE ROLE OF THE BRIDGE FIRM

The AIA's official Position Statement on design-build wholeheartedly endorses the bridging concept and recommends that when public agencies use design-build, they should put out a detailed RFP with "project specific comprehensive scope of work documents prepared by licensed architects and other qualified professionals *who are retained for the duration of the project*."[4] The AIA also recommends that the design-build proposals be evaluated by a jury of qualified profes-

---

[2] DBIA's Policy Statement Regarding Design/Design-Build, *Date Line*, DBIA Newsletter (Nov. 1999).

[3] Comments of Dr. James Broaddus at Design-Build conference, Kansas City, Missouri, Jan. 11, 2000.

[4] The full text of the AIA Position Statement is reproduced as Appendix A to this book.

sionals, "including those licensed professionals who prepared the scope of work documents." The AIA recommends that the design-build entity be required to retain and use the design architect of record "throughout the duration of the project to maintain design integrity, functional and technical responsiveness, and to conduct on-site construction observation." This is, of course, intended to make sure that the contractor does not hire an architect only to prepare construction documents, with no role during construction.

§ 6.04 HOW MUCH IS TOO MUCH PRELIMINARY DESIGN?

Certainly DBIA's point is well taken and there is a break point when there is no ability for creativity if the bridge firm performs too much of the design. For example, the City of San Francisco used a "bridge" architect for a city jail project; however, the bridge firm prepared documents that were 90 percent complete, including a model of the jail. The architect contended that the documents helped to define the scope, quality and performance, but left sufficient flexibility to allow bidders discretion to be creative and innovative.[5] The 90 percent bridge documents included performance specifications for quality standards and systems required, and even some prescriptive specifications for items the owner wanted without variation. This project, while successful from the bridge architect and owner's viewpoints, eliminated much of the benefits of design-build and would likely support DBIA's opposition to "Design/Design-Build." Most agree that 30 percent or so is that break point at which enough information has been given to the competing teams so that they can compete on a level field, yet have enough flexibility to be creative themselves.

§ 6.05 WHY OWNERS LIKE BRIDGING

Corporate owners have been downsizing in-house construction staffs and outsourcing more of those functions. With design-build, where the project's program and design-build selection are important to success, owners need help because they have more responsibility for defining and procuring than in traditional construction where an architect/engineer firm traditionally helps with programming and contractor selection. This presents opportunities for a new type of owner-consultant, such as program managers, and even new roles for traditional design or construction management firms.

Owners are used to having an architect or engineer to discuss the project with, to advise them and to "keep an eye on the contractor." Design-build without a bridge consultant requires a great deal of trust. Some owners are just not ready for that level of trust and a bridge consultant helps make the transition to design-build.

---

[5] Presentation by architect Beverly Prior and Lt. Michael LaVigne at AIA "Doing Justice to Design-Build," Washington, D.C., Mar. 18–20, 1999.

§ 6.06 THE BRIDGE ROLE AS A MARKET OPPORTUNITY

Bridging is a target market for some design firms. In an interview in 1999, the CEO of Fairfax, Virginia-based architect/engineer firm Dewberry and Davis said that his firm's main focus is to be an "A/E-1," handling conceptual design and then acting as the owner's agent supervising the design-build team.[6] With firms targeting this role for their business niche, expect more competition. Firms will begin to compete on who is the most qualified "bridge" firm. This raises the question of who is best to serve as a bridge firm, a pure traditional firm, or one that has done design-build work in the past? According to Jeffrey L. Beard, Executive Director of DBIA, Washington, D.C., "Design firms that have worked in design-build are in the best position to advise clients as owner agents."[7]

While fees for "bridge" architects are generally less than those for traditional full-scope services, there is an attraction toward this service for at least three reasons. First, most architects will agree that the "fun" part of architecture is the design phase. Construction documents and specifications—the technical side of design, and preparing "working drawings"—are the less glamorous aspect. Bridge architects get to do concept designs, the way architects did in studio in college, leaving the remaining technical drawings and detailed specifications to the design-build team to develop. Second, the bridge architect retains a close relationship with the owner, preserving the role of agent, advisor and trust that is enjoyed on traditional project delivery. Last, there is less risk for the bridge firm. Insurance statistics show that the majority of claims against architect/engineer firms are for design errors or omissions. With the bridge firm doing only a 30 percent or so set of schematic or design development drawings, it is expected that the design-build team will prepare the technical details used for construction and seal those documents. If there is a design error or omission in the drawings or specifications, liability will most likely fall on the design-build team's architect or engineer who prepared those final documents, not on the bridge firm.

Of course, there is a trade-off. By passing off a design at 30 percent complete, the bridge firm may lose design control in terms of what the final product looks like.

§ 6.07 CAN THE BRIDGE ARCHITECT/ENGINEER PARTICIPATE ON THE DESIGN-BUILD TEAM?

Several states have laws that prohibit architects and engineers from bidding on plans they prepare.[8] South Carolina law bars architects and engineers who perform design services from performing any work on the same project as a contractor, either directly or through a business in which the architect or his firm has

---

[6] *Design Firms See Design-Build Growing But They Remain Wary,* Engineering News Record, Sept. 27, 1999, p. 19.

[7] *Id.*

[8] Okla. Stat. titl. 59, § 46.27.

"greater than a five percent interest."[9] The intent of such laws is to prevent an unfair competitive advantage to the firm most familiar with the design. It would certainly defeat the purpose of having an "independent" architect/engineer if the bridge firm could prepare the design criteria package and then link up with one of the design-build teams competing for the project. In addition to giving that team an unfair competitive advantage, there are a host of ethical issues, which are discussed in more detail in Chapter 24. The better approach is to bar the bridge firm from participating on any of the design-build teams.

Florida's Consultants' Competitive Negotiation Act,[10] makes it clear that, "[a] design criteria professional who has been selected to prepare the design criteria package is not eligible to render services under a design-build contract executed pursuant to the design criteria package."

At least one federal case has ruled that in this situation, the bridge firm cannot be part of a contractor's design-build team. In *Matter of SSR Engineers, Inc.*, the Navy awarded a contract to SSR Engineers to prepare a long-range master plan for certain electrical work at Keesler Air Force Base in Mississippi.[11] The government issued an RFP for a design-build contract to perform a portion of that work, taking the statement of work directly from the SSR master plan. SSR was sought out by several contractors to be part of their design-build teams, but the Navy said the firm was ineligible due to a conflict of interest. SSR protested that decision, but the Comptroller General denied SSR's protest, holding that Federal Aquisition Regulations (FAR) barred such participation and that with SSR's participation, that team would have an unfair competitive advantage and an "organizational conflict of interest." The regulation at issue, FAR, part 9.505-2, states that, "[i]f a contractor prepares, or assists in preparing, a work statement to be used in competitively acquiring a system or services—or provides material leading directly, predictably, and without delay to such a work statement—that contractor may not supply the system, major components of the system or the services [with certain exceptions]." The engineer argued that FAR 36.302 specifically addresses design-build projects and use of consultants to develop a scope of work, and that the regulation does not expressly preclude the firm developing the scope of work from being part of a design-build team. The engineer also argued that it had not obtained any competitive advantage based on its prior work on the project. The Comptroller General rejected these arguments, holding instead that the organizational conflict precluded the engineer from participating on any of the teams due to an unfair advantage.

FAR 9.502 states that an "organizational conflict of interest" may result when there are factors that create "an actual or potential conflict of interest on an

[9] S.C. Code § 11-35-3245.

[10] Fla. Stat. Ann. § 287.055.

[11] Comp. Gen. No. B-282244, 1999 U.S. Comp. Gen. Lexis 139; 99-2 Comp. Gen. Proc. Dec. P27 (June 18, 1999).

instant contract, or when the nature of the work to be performed on the instant contract creates an actual or potential conflict of interest on a future acquisition."[12] Participating on a competing team may give that team an advantage in terms of qualifications since no other team could match the level of familiarity that the developer of the scope of work would have with the project. It would also effectively prohibit the owner from using that same design consultant for construction administration services during construction, to the owner's disadvantage.

§ 6.08 WHO IS THE ARCHITECT/ENGINEER OF RECORD?

It will be interesting to see how design professionals handle the credit for projects in which one firm does preliminary design and another serves as the design-build team's architect. Many of the AIA ethical complaints relate to one firm not giving proper credit to another firm when both play a role in the project. This comes up in awards competitions, in marketing brochures, and in subsequent head-to-head competition and interviews. Two firms showing the same project in their qualification statement and brochures will confuse owners and hurt credibility. There is also the conflict as to who is the "architect of record" or "engineer of record" for the project. These labels, which have no legal meaning or definition in standard contracts, often cause confusion when multiple firms are involved. Who is the true project "architect" on a project in which one firm does the bridge package and another firm does the remaining 70 percent or so of the design work?

The AIA's ethical Rule 4.201 requires members to "accurately state the scope and nature of their responsibilities in connection with work for which they are claiming credit." The official commentary to this rule says that the intent is to prevent members from claiming credit for work they did not do and "denying other participants in a project their proper share of credit." Seven of the AIA's ethics decisions between 1987 and 1992 dealt with the giving or taking of credit for work. With bridging injecting two design firms into a project, we can expect more ethics complaints unless firms come up with a way to credit a competing firm with its role as either bridge architect or project architect or some other terminology.

---

[12] 48 C.F.R. § 9.502.

CHAPTER 7

TEAMING FOR DESIGN-BUILD PROJECTS

§ 7.01 TEAMING RELATIONSHIPS

These are exciting times to be involved in construction. Relations are improving between design professionals and contractors for the first time in decades. Though formal "partnering" was a valid attempt to get everyone working toward the same goal, it has taken design-build to truly get everyone on the same team. Those who are more proactive in marketing have located and signed up the strategic partners with whom they want to work. Some have gone so far as to merge companies. Industry observers say that early on, design-build teams were thrown together quickly to pursue individual projects. Now, firms are developing long-term strategic alliances. Many of the "best" players are being picked, and teams are being formed. Companies need to be pro-active here, because a firm that is waiting until a design-build request for proposal (RFP) hits the street will be too late.

§ 7.02 A NEW WAY OF THINKING ABOUT EACH OTHER

Teaming with contractors may run counter to the instincts of many architects due to years of confrontational, almost adversarial relationships between the two businesses. John Gaunt, FAIA, says that some walls in the industry will need to be broken down to make design-build successful. He says:

> Architects must be willing to build partnerships and alliances with former adversaries in the construction industry. The architect/contractor partnership is fundamental in this, and has a long history of confrontation that must be replaced by cooperation. The entire industry stands to profit enormously from the thoughtful partnering of architects and contractors that build mutual trust and open communication, and who are willing to share the risks of equity partnerships. There is a large attitudinal shift in this for both sides of the partnership—but a positive view of its possibilities suggests it will be well worth the effort involved.[1]

[A] More Satisfaction Through Teaming

Teaming produces a more satisfactory business relationship according to at least one study. The 1997 study by the Construction Industry Institute looked at 351 construction projects done by the traditional design-bid-build, construction management (at risk), and design-build methods.[2] The study concluded that: "The data received in the lessons-learned recognized the quality benefits in an experienced team, particularly a team that has worked together before. Companies experienced in design-build showed better performance, even with teams that have no prior working relations." Therefore, when a firm is just starting out in

[1] Gaunt is the former CEO of Ellerbe Becket and now the dean of the University of Kansas School of Architecture.

[2] *Project Delivery Systems: CM at Risk, Design-Build, Design-Bid-Build*, Construction Industry Institute, Research Summary 133-1 (Dec. 1997).

design-build, it would be prudent to find a teammate who has done a design-build job or two before.

[B] Teaming Is Not for Everyone

Some of the large, sole-source design-build firms are critical of the teaming trend. "We think that design-build works best when the design is performed in house," says William Melsop, CEO of The Austin Company, one of the largest design-build contractors in the United States. Agreeing with this view was Fred Kummer, CEO of HBE Corp., another integrated design-build leader, who said, "These teams are building a Tower of Babel. You can't just take an architect, an engineer and a contractor, put them in a bag and shake them up and produce a design-build team. Too often, it is the participant with the loudest voice that is the one heard, and that isn't the way design-build is supposed to work."[3]

Of course not everyone can be an Austin Company or an HBE Corp., two of the largest design-build companies in the nation. And so, for many small to medium-sized design firms, there is teaming. Fred Kummer is right, of course, that one has to be careful in selecting a teammate and that the wrong team can result in failure just as the right team can mean success. Among the comments by respondents to the 1999 Zweig White & Associates' Design-Build Survey were these from design firms that noted disadvantages of design-build:

- "Our firm being sub to contractor/builder firm."
- "Magnifies contractor verses engineer problems or disputes."
- "Loss of control by A-Es."
- "Engineers become hired help."
- "Designer loses control."
- "Design may be sacrificed."
- "Best design is often not used."

The 1998 Zweig White Survey found similar comments from design firms, such as:

- "Aggravation and disagreements."
- "Contractors have 'short-term, good-enough' solutions."
- "Making sure that the entire team is contributing."
- "If team doesn't work well together, it becomes difficult to complete [the] project."
- "You have to work very hard to make it work well."

[3] *Design-Build Explores New Markets as Public Sector Shows Interest*, Engineering News Record, June 15, 1998, p. 55.

These candid comments from firms engaged at least 10 percent of the time in design-build show that not all marriages are made in heaven. The personalities, the philosophies, the values of the teammates are all critical to success and can be the reason behind failure.

§ 7.03 STRATEGIES FOR TEAMING

The new paradigm means a departure from the adversarial relationships of the past between architect/engineers and contractors (and subcontractors). One design-build firm holds an annual golf outing, not for its clients, but for its subcontractors and vendors, realizing that when the firm is in a jam, it is this support team they need to rely on. Firms can build these relationships by stressing mutual reward and trust. Then firms need to practice what they preach.

Barry Dewberry, CEO of the architect-engineer firm Dewberry and Davis, says that his firm will participate as part of design-build teams, but "we only work with contractors we've had long-term relationships with."[4] Trust is an important factor in building a strong team. The Construction Industry Institute did a study of 262 projects which indicated that projects with a high degree of trust have the lowest cost. Conversely, projects with low trust have the highest cost. The study also correlated owner satisfaction as being highest on projects with the highest trust.

If two firms have not worked together before, they will have to work harder at developing that level of trust. References should be exchanged so that both parties can check up on one another. Some firms engage in more personal contacts, such as golfing, fishing, or sailing together just to see if the personalities and company cultures will fit. Partners should visit one another's offices several times and get to know the staff who will be working on the project. Mid-level staff should be encouraged to do the same, and to check out who their counterpart will be at the other partner's office. It is important to learn about each other's work ethic.

The prime contractor in a designer-led design-build project should be prepared for some hard questions from teammates about payment. The president of one of the largest design-build contractors in the United States said that his biggest concern in working for an architect would be his or her ability to pay him if the owner does not. Design firms do not usually have the kind of financial statements that make contractors comfortable. Confidentiality is a concern for many team partners, especially in financial matters. Design-build partners should require a confidentiality agreement with teammates prior to exchanging financial statements. If the firm's plan is to have the contractor-partner provide the bonding, as a subcontractor to the design firm, the lower-tier trade contractors have protection for their payment, but the contractor does not, since he cannot make a claim against his own bond.

James Bradburn, FAIA, of Fentress Bradburn Architects in Denver says that, "good team relationships, including designer-contractor relationships for-

[4] *Design Firms See Design-Build Growing but They Remain Wary,* Engineering News Record, Sept. 27, 1999, p. 19.

malized in design-build agreements, are like any other relationship—they must be founded on clear expectations between the members with defined roles and responsibilities monitored continually and consistently by the team members themselves."[5] He indicated that most team relationships break down because one party did not understand its role and what was expected. He recommends including partnering sessions with the owner and the other major players with each party's expectations clearly articulated and memorialized in a partnering charter. Then the parties can develop teaming agreements that outline responsibility for budget control, design error and omissions, insurance, and other major issues.

Designers need to find builders who respect the integrity and value of design. Too many contractors see the design professional as just another subcontractor, performing a task that can be done by many others in a subordinate role. Design firms should steer clear of this mentality. A contractor partner should care as much about quality as he or she does about price. Quality might mean different things to different people, however. There should be a commitment that no design changes will be made, no matter who is the lead contractor, without agreement of both parties.

Partners need to know how one another resolves conflicts. Design firms could talk to local subcontractors, since they can give some of the most candid input on just how particular contractors treat their teammates and handle disputes. The contractor they describe is not always the same person the architect or engineer has come to know through office visits and teaming discussions.

§ 7.04 BENEFITS TO CONTRACTORS WHO TEAM WITH DESIGN PROFESSIONALS

Teaming relationships with design firms have numerous benefits to contractors. By establishing teaming relationships with contractors, architects and engineers can bring work to contractors they might not otherwise get if the project was competitively bid. Architects are still the first ones an owner contacts about new projects. Owners rely on their advice about project delivery methods and contractor recommendations. Through teaming (whether as a joint venture or subcontractor) the contractor is guaranteed the job up front through early contact with the owner. This should appeal to contractors who will want to pair up with the best and most aggressive design firms pursuing design-build contracts. As pointed out in one publication, contractors should pair up with design professionals for these reasons:[6]

- *No Marketing Overhead.* Marketing and sales overhead are a significant expense for contractors. With designer-led design-build, the design firm takes on the marketing role necessary to bring in the project. The contractor-teammate is awarded the project often without having to expend its own resources to

---

[5] *Stanford/DBIA Professional Design-Build Conference*, AIA Project Delivery Report, Winter 1996.

[6] *Designer-Led Design-Build: Why It Works for Contractors*, Design Build Intelligence, Oct. 15, 1999.

obtain it. As a result, the portion of overhead that would normally be allocated to sales and marketing instead becomes profit.

- *No Uncompensated Estimating*. The "hit rate" for many contractors is one out of every three bids. This means that the contractor's estimating staff spends two-thirds of its time on projects that go to some other firm. However, with design-build, the team structure is often set up so that there is a "stipend" paid to the unsuccessful team to help defray the cost of preliminary design and estimating services. Also, with the design firm "bringing in" the project, the contractor only estimates jobs that the design partner has already locked up with the owner, meaning that all estimating time is productively used.
- *More Accurate Estimating*. On traditional projects, with bids due three to four weeks after receiving the bid documents, and much of the pricing actually done in the last few days before the bid date, there is lots of room for errors, misinterpretation of plans, and omissions. With design-build, however, the contractor-partner participates in pricing at the earliest stages, a role that continues as the design is being developed, with less rush, more accurate information, and better pricing.
- *Friendly Administration*. A teaming relationship between contractor and designer on design-build should result in more cooperation and less friction, since both parties are on the same team, working for the same goal, and trying to save costs and work efficiently. Substitutions may be more liberally allowed, rather than viewed with suspicion, requests for information (RFIs) handled informally, and the shop drawing process streamlined. Design or construction problems in the field become "our" problem, not "your" problem, with the parties working together to solve it efficiently.
- *Shared Losses*. With a team approach, any savings and any loss are shared per some agreed ratio. While profits are hoped for, a team relationship may mean that losses are not absorbed 100 percent by the contractor, but shared by the team. This also creates a strong incentive for the design partner to work to avoid losses, be more creative in solving problems, and more cooperative in ways to expedite construction.
- *Obtaining Experience*. Most contractors lack design staff and cannot afford to start up an in-house design practice. Teaming on design-build enables contractors without much design experience to build up a resume of design-build projects. The contractor not only gains the experience of working directly through the design process, but compiles a portfolio of design-build work to be able to compete in this growing market.

§ 7.05 THE AIA/AGC TEAMING CHECKLIST

In June 1999, the American Institute of Architects (AIA) and the Associated General Contractors of America (AGC) published a document entitled Design-Build Teaming Checklist, a 10-page booklet designed to "promote valuable and timely discussion of key points of departure between the design and construction partners within a single-source team, regardless of leadership structure." The checklist is broken down into eight subparts entitled:

1. Team selection and corporate cultural values. This includes the compatibility of corporate philosophies, intuitive feelings, conflicts of interest, and problem-solving techniques.
2. Legal considerations. This includes the corporate structure of the design-build entity or team, contracts, licensing and other applicable laws.
3. Getting the job. This section is meant to prompt candid discussion of who will absorb the cost of the marketing process, and whether the teammates are exclusive or can withdraw at any certain time.
4. Risk management. This covers insurance, bonding, how errors in design or construction will be handled, cost overruns, and other risks.
5. Value engineering. This section discusses how the parties will handle cost-savings ideas in the design and construction of the project.
6. Financial considerations. A topic that is awkward, but needs to be discussed early is the cash flow of the design-build entity, team financing, shared savings, retainage and other issues.
7. Definition of roles. This section covers a variety of areas from start to finish of the project, so that each party knows who will fill what role during each phase.
8. Owner's considerations. Topics like owner's insurance, financing, and communication are addressed here.

The joint AIA/AGC checklist is endorsed by nine other industry organizations, including Design-Build Institute of America (DBIA) and Construction Management Association of America (CMAA) as well as the engineers' contracts committee, EJCDC. It serves as a good communication device to see if the contractor and design firm are really compatible. As one of its authors said, "It is too late if you find out during the honeymoon." The better that teammates know each other up front, the more likelihood of project success.

§ 7.06 TEAMING AGREEMENTS

A firm should take the time to develop a written "teaming agreement" with a contractor and, possibly, prime subcontractors before the proposal is even made to the owner. Such agreements permit the parties to establish basic rules for their working relationship during the proposal stage and, if successful, after contract award. This agreement should also address payment of design fees and other costs incurred in the proposal phase. There should be agreement on sharing these costs as well as reimbursement if the contract is awarded to your team. One should be candid and honest in these discussions. Whatever each party wants out of the project should be put into writing. These "bullet points" can be part of the formal teaming agreement.

A teaming agreement can take several forms. At its essence, it is a letter of intent in which the parties set out the groundwork for their relationship, agreeing that they will work together on future projects if their teaming results in a contract from an owner. DBIA and 10 other industry associations are drafting a model teaming agreement form. A sample agreement is set out as Appendix B in this book. No matter what form is used, a teaming agreement should address the following topics.

[A] Exclusivity—Non-Compete Clause

In order to develop the trust that is so needed, the parties need an assurance that each will remain loyal during this project, or this project type, so while working to market the team to the owner or target group there is no fear that one party might change horses when it finds another party that is better qualified, better connected, or with cheaper fees. Teammates should pledge to keep the team together until they decide otherwise by mutual agreement.

[B] Duration of the Team

The teaming agreement should cover the period of time that it remains in effect. If the team is not successful in marketing together after some agreed lapse of time, it may be time to change to a new team, with a more marketable mixture. Events that trigger a termination before the time expires should also be stated along with the right to withdraw upon breach by the other side.

[C] Project Types for the Team

The agreement should set out the types of projects that the team is best suited to pursue. It may be that a design firm has several teaming agreements with different contractors, depending on the project type. This permits both architect-engineer firms and contractors to pair up with partners who have the best credentials for particular project types. No two firms can be competitive for every project. Architects and engineers are usually skilled at putting teams together to pursue specific projects and design-build presents an opportunity to use that skill to add construction companies to the team, matching the right contractor with each design-build project.

[D] Prime/Sub Relationship

The teaming agreement should address which of the teammates will be the prime contractor and which one will be the subcontractor. This may depend on the project and the client being pursued and which firm has the stronger ties to that client. It may be that the teammates do not have a prime/sub relationship and, if not, the agreement should discuss joint ventures or new entities to be formed on a project-specific basis.

[E] Disclaimer of Partnership

The teaming agreement itself is not a partnership agreement. In order to avoid any argument to the contrary, a simple disclaimer in the agreement can dispel the notion that the parties have formed a partnership merely agreeing to be teammates.

A partnership exposes each of the partners to "joint and several liability" for the acts of the other partner.[7] You do not want to fight that fight if not necessary and a simple disclaimer can save thousands of dollars in legal fees down the road.

[F] Compensation

One of the key provisions is how the parties will be compensated. The agreement should state how costs will be shared during the preliminary marketing, proposal, and eventual contract stages. This should include reimbursement for marketing efforts, preliminary design work, printed materials, and travel expenses. In the event that the team is successful in its efforts, the agreement must cover how project profits and losses will be shared. Many agreements state that each party will bear its own costs during the marketing phase, but if successful, those costs will be reimbursed from the design-build contract funds. If not, then each party bears its own costs. Other agreements state that the contractor, for example, will reimburse the design partner for a portion of its costs since it is generally recognized that preparing designs is more costly than the estimating done by the contractor partner.

Some teams get into such detail as to what cost multipliers will be used on direct labor and what costs will be charged to overhead rather than to the job. This will depend on the team and, maybe, the size of the job and type of competition.

[G] Voting and Decision Making

Many teams enter into "50-50 deals" with the idea that each partner will have an equal say in the team. However, each party needs to think how disagreements will be handled—how will a tie vote be broken to avoid a deadlock? The parties need to consider whether there will be a management committee and a project manager; how these people are selected and what they will do.

[H] Legal Form: What Kind of Business Entity?

The design-build team itself is not a legal entity, normally. But if the team is successful in being awarded a contract, the owner will contract with the team. At that point, the team will need to be formalized into either a joint venture, LLC, new corporation, or a prime/subcontractor relationship. The teaming agreement should either discuss this or give some direction about when and how that decision will be made. If there is a need for licensing in the project state, who will obtain that? Will that restrict how ownership is structured? Other legal issues deal with taxes and accounting and these should be considered.

[I] Insurance, Bonding, and Claims

The agreement should cover which insurance policies each party will carry, which ones will need endorsements, which agent will be used if the "team" needs

[7] *See* Chapter 11.

its own insurance, and how premiums will be paid. If bonding is an anticipated need, the agreement should discuss which of the parties will provide the bonding capacity; usually it is the contractor. For claims that are insured, especially design errors, there is going to be a deductible payment made by someone. The teaming agreement should consider how risks like this will be shared for losses not covered by insurance. Also, how will disputes among the team be resolved? Mediation or arbitration clauses should be considered for disputes solely between the team's partners, not involving third parties.

[J] Client Communication

Even though the team shares work and profits equally, there should be only one spokesperson to the owner. As long as both teammates are kept apprised of communications with the owner, and consulted about important decisions involving the owner, it can still work to only have one spokesperson. The owner needs the confidence to know that there is just one phone call to make (remember that single-source responsibility is still the biggest selling point). In deciding who that person will be one should consider who is the best communicator and who has the best relationship with the owner.

[K] Confidentiality

Since many architect/engineering firms and contractors pair up with different partners on different projects, there is a real concern that trade secrets and proprietary information will be revealed to the competition by a design-build teammate. To protect against this, teaming agreements usually include a confidentiality clause that prohibits revealing confidential information to others.

[L] Subcontracting and Assignment

Since teaming agreements are very personal, and are based on the qualifications of the team members, it is reasonable to expect that the players on the team will not change without consent of the other members. The agreement might prohibit the assignment of the agreement and of any design-build contract won by the team to others, as well as subcontracting most or all of either partner's obligations without written consent of the other teammate. Obviously, a design professional may rely more on the contractor's judgment in awarding subcontracts, but the designer should have some say in whom the work goes to.

[M] Ownership of Documents and Designs

The team should discuss who will own the documents, the design (including the copyrights in the design and any derivative works) and what rights, if any, either party has to reuse it on another project without consent of the other. This topic is covered in all of the standard form contracts for design-build and is an important topic to get agreement on up front. *See* Chapter 16 for a more detailed discussion of this topic.

One risk in design-build is that the owner will take the design-build team's design and use it as a basis to get bids from other contractors. For example, in an Oregon court case,[8] one party took the design prepared by its design-build contractor to several other contractors for pricing, eventually hiring one of them to build it. The design-build contractor sued the owner for breach of contract and for fraud, recovering $51,000 in punitive damages for fraud. By including a provision in the teaming agreement, an architect can prevent its partner from dropping the architect from the team and using the design with someone else.

§ 7.07 JOINT VENTURES

Many teaming relationships are formalized into a joint venture. Sometimes this is requested or even required by the client, who wants to see the joint risk or joint strength that comes with a true joint venture. Even integrated firms with in-house design and construct abilities will team or joint venture on particular projects when they need some specialized expertise or qualifications, or a regional partner. The benefit is joining expertise to win the contract award, plus added financial strength and local knowledge. The downside, of course, is loss of sole control, which will be shared with the joint venture partner.

An architect too small or inexperienced to be the prime contractor on a design-build project might be tempted to act as subcontractor. But as a subcontractor, the architect loses that direct contact with the client. Joint venturing has the benefits of higher visibility, direct contact with the owner, shared control, and a voice in contract negotiations. It also allows the architect to spread the risk out over two companies and not have it all rest with his or her firm. As an architect starts out in design-build, there will be a lot to learn. The contractor partner will be the architect's best teacher or mentor in the construction business. One should not be surprised, however, if the contractor wants to learn as much about the design business as the architect wants to learn about construction. Joint ventures can be mutually beneficial.

The joint venture should have a project manager who oversees both design and construction. Part of the job functions of either that person or some other designated financial manager must be monthly financial reports to the joint venture, showing how the team's costs compare to those budgeted and forecast. Many of the disputes that lead to litigation between partners relate to one party's failure to warn about estimating errors, design omissions, and cost overruns until it is too late. If the partners are committed to a successful project, they may be able to turn a losing project around through joint efforts but only if the bad news is shared among the team in sufficient time to do something about it.

---

[8] Building Structures, Inc. v. Young, 883 P.2d 1308 (Or. Ct. App. 1994).

CHAPTER 8

WHAT CONTRACTORS SHOULD KNOW ABOUT TEAMING WITH ARCHITECTS AND ENGINEERS IN DESIGN-BUILD

Design-build puts contractors and design professionals into some different relationships. As discussed in more detail in Chapter 11, the variations on design-build can result in either joint ventures between contractors and design firms, or prime/subcontractor arrangements, where either contractors are in the prime role or as a subcontractor to a design firm. The most common business arrangement is for the contractor to hold the prime design-build contract and the architect/engineer to be a subcontractor. For contractors who have not worked this closely with design professionals before, there are many legal, business, and psychological issues to take into consideration before formalizing this relationship.

§ 8.01 PSYCHOLOGICAL DIFFERENCES BETWEEN CONTRACTORS AND ARCHITECTS

How does a contractor know if the architect it selects as its design-build partner is cut out for design-build? We continually hear people, both architects and contractors, say that architects are different from contractors. "They are not risk takers," say some. "They are artists, designers. Contractors are not afraid to take a risk. They are focused on the bottom line—time and money," say others. Contractors seem to accept conflict as part of their business with the challenges of dealing with a dozen or more trade subcontractors raising daily conflicts on site and in the office. Contractors tend to be more aggressive in pursuing claims for more time or money during the project, and claims against owners by contractors by far outnumber claims made by architects and engineers combined. It seems that even historically, design professionals have been less confrontational than contractors. In an 1894 law book on construction, the author (an architect) observed that out of perhaps a million building contracts in the world every year (even at that time), there were very few court cases involving architects, and that "architects are not only less extensively employed than builders, but it appears to be even more peacefully inclined, and cases to which architects appear are rare."[1]

Contractors teaming with architects probably have their own preconceptions about design professionals, especially when it comes to deciding who should lead the project. It is important to find a partner who is willing to take the risks involved in construction. Robert Ivy, FAIA chief editor of *Architectural Record*, wrote that "[m]any architects, as comfortable on muddy job sites as in corporate boardrooms, already share a strong interest in the craft of building. Those who would like a more hands-on relationship with construction are likely candidates to lead design-build ventures."[2] Just how many architects actually feel comfortable on muddy job sites is in question, as is the number who prefer a "hands-on" relationship with construction. Those who enjoy this direct contact with construction may be naturally inclined that way, while other design professionals may have an inherent resistance to that level of involvement.

[1] T.M. Clark, FAIA, *Architect, Owner and Builder Before the Law*, Macmillan & Co., 1894.

[2] Robert Ivy, FAIA, *Not to Be Ignored*, Architectural Record, Oct. 1998, p. 15.

[A] The Meyers-Briggs Type Indicator Test

One contractor said, "[a]s architects, you were either born that way, to be comfortable with design-build, or you weren't." Are architects and engineers really "born that way"? Are they more "peacefully inclined"? Psychologists use a standard personality test, called the Meyers-Briggs Type Indicator (MBTI) to assess natural preferences and analyze how those preferences relate to one's personal and professional life. The MBTI, which was developed in the 1940s by Katherine Cooks Briggs and Isabel Briggs Meyers, breaks personalities into categories based on individual preferences on four scales:

- Extroversion (E) vs. Introversion (I)
- Sensing (S) vs. Intuition (N)
- Thinking (T) vs. Feeling (F)
- Judging vs. Perceiving (P)

Thus, for example, a person with an INJF personality type would rank the highest in the following traits: introversion (I), intuition (N), feeling (F), and judging (J).

[B] Architects: Thinkers and Planners

In his book, "Looking at Type and Careers," Charles Martin, Ph.D., analyzes which personality types are drawn to various professions, as well as those professions least likely to attract certain personality types. The results are very enlightening for architects considering design-build. Dr. Martin's book indicates that extroverts are rarely drawn to careers in architecture. MBTI types starting with "I" (for introvert) ranked architecture very high as a career choice. INFJ personalities are drawn to careers in architecture and fine arts. They are attracted to professions that require deep concentration, creativity, and organization. They are full of idealism and lofty goals, but are intensely individualistic and private persons. They need significant time alone in their work to focus on their ideas. According to Dr. Martin, "INFJs are found much less often in careers that are characterized by a great deal of technical work, attention to detail, work that requires realistic precision or production . . . They are also found much less often in careers that require more practical hands-on or mechanical work, or careers that may involve a significant amount of inter-personal conflict."[3]

Also drawn to architecture and engineering are INTJs, who find satisfaction in careers that use depth of concentration, logic and analysis, task orientation, and ability to organize. These qualities often lead to executive and management positions. INTJs rank architecture as their top career choice, with careers in law a

[3] Charles Martin, Ph.D., Looking at Type and Careers, Center for Applications of Psychological Type (1995), p. 27.

close second choice. Like INFJs, the INTJ needs quiet, even philosophical reflection prior to engaging in external tasks. According to Dr. Martin, "[t]hey pursue competence and mastering . . . whatever career they choose must have opportunities for learning." INTJs are less likely to choose careers in construction or work "that requires practical, routinized production or delivery of services." They are found less often in hands-on work, or work that depends on "adherence to structures imposed by others."[4]

[C] Contractors: Doers and Problem Solvers

Both introverts and extroverts are drawn to careers in construction. The key traits for construction people are the "S" and "P" factors. Dr. Martin's study shows that ISTPs are found in careers related to building and production. They are curious about how and why things work and about their application. They desire "a wealth of hands-on and sensory experiences" often leading to a high level of skill with tools. They are found in jobs that involve a measure of adventure. "They tend to resist too much structure."[5] ISTPs are skilled at making the most effective use of what is actually available, and they make very good trouble-shooters. This sounds a lot like value engineering done by contractors. According to Dr. Martin, people with an ISTP personality are least likely to select a career in the fine arts or architecture. Like ISTPs, Dr. Martin found that ISFPs and ESFPs are naturally drawn to construction careers, but least likely to become architects or engineers. (ISFPs prefer hands-on detail work that may require solitude—*e.g.*, machine operator, carpenter, or construction worker.) They feel quite comfortable working outdoors, and have a tendency to express through action rather than words. They prefer a fair degree of freedom from restricting structures and rules. "ISFPs report being attracted to careers where they can deal with facts rather than theory, where they feel their work contributes to something that they care about . . . where they can make use of practical action skills."[6] Though they enjoy working in groups, they are not inclined to want to manage or lead groups of people. Like ISFPs, the ESFPs share the desire for hands-on, group work, but their extroverted nature draws them to supervisor roles. They are not likely to select a career in the arts or architecture.

[D] Distinguishing Traits

So what personalities are drawn to architecture? INFJ and INTJ. Disregarding the "I" versus "E" factor (introvert vs. extrovert), we find that architects differ from contractors in these two traits: "N"—intuitive perception, which means one is imaginative, insightful, focuses on future, values knowledge and education, and long range planning. "J"—judging lifestyle, which means one prefers order,

[4] *Id.* at 30.
[5] *Id.*
[6] *Id.* at 33.

structure, completion of tasks, takes deadlines seriously, and enjoys routines. These people are drawn to management jobs and those where they can assume responsibility. Personalities drawn to construction on the other hand are ISTP and ISFP, who have exact opposite traits: "S"—sensing perception (opposite of "N"), which means one is realistic, practical, focuses on present, prefer hands-on work, solves immediate problems, and does not want to rush closure. "P"—perceiving lifestyle (opposite of "J"), which means one who likes spontaneity, the unexpected, leaves options open, prefers troubleshooting, flexibility, and innovation.

[E] Design-Builders: Thoughtful Problem Solvers

Who is most likely to do well at design/build? It would likely be someone with a combination of these traits, such as "N" and "P." Not surprisingly, according to Dr. Martin, someone who has an INTP personality is likely to be drawn to be an architect, engineer, surveyor or fine artist as well as site supervisor, electrician, or machine operator. These "split personalities" enjoy deep concentration, use of logic and analysis, and creative solutions and remedies to complex problems; perfect qualities for a design-build contractor. They are known as problem solvers. A potential drawback is that INTPs are less inclined to supervise and organize others, but if they do work with others, "they prefer working with persons they see as skilled and competent."[7] An INTP would make a good design-build partner, up to the challenge if you are. An ENTP (extrovert) adds the outgoing personality needed to lead the design-build team successfully.

§ 8.02 FINANCIAL CONSIDERATIONS

For years, design professionals have been told to keep their insurance limits high and their retained earnings low. This advice was an effort to protect the assets of the design firm from creditors. Low insurance deductibles and broad insurance coverage limit the financial risk of the design firm and its owners. Keeping earnings low has not been difficult for many firms whose low fees have resulted in low profits for the firm as well. Contractors, on the other hand, are required by their bonding companies to keep a certain level of assets in the bank to give the surety comfort that the contractor is solvent and able to pay a bond claim without a loss to the surety. When exchanging financial information with one's design and construction partners, this difference in financial philosophy will become apparent.

While the general public lumps design professionals in with doctors, lawyers, engineers and accountants, most are shocked to learn that architects' salaries do not keep pace with these other professions. Starting salaries for architects are generally the same as starting salaries for teachers in public schools, and we know how teachers complain about being underpaid. Engineers with a four-year college degree can typically start anywhere from \$10,000 to \$20,000 higher

[7] *Id.* at 40.

than graduate architects with a five-year degree. Doctors and lawyers fresh out of graduate school draw salaries three to four times higher than architects. So be prepared for a financial statement that reflects a lean business operation. But what architects lack in finances they make up in passion and dedication to the built environment. Architects do what they do because they love to design and see their communities improved by their efforts. Architects will make committed design-build partners, with profits secondary as a concern to their passion for quality, aesthetics, and a satisfactory solution to the owner's needs.

§ 8.03 PRIME-SUBCONTRACT RELATIONSHIPS

[A] Design Professionals as Subcontractors

Contractors are used to leading projects as a prime contractor. They usually have experienced project managers who have worked with dozens of trade subcontractors on a single project and know how to manage and motivate people to get work done on schedule. But design professionals do not see themselves as merely another "trade subcontractor." While an electrician who falls behind schedule or does poor or unsafe work can normally be replaced on short notice with another electrical subcontractor, it is not so easy to replace one architect with another. First, contractors may find that there is a close "fraternity" among design professionals and it may not be as easy to replace architects or engineers as it is a trade subcontractor. Design professionals in most cities belong to professional organizations such as local chapters of the American Institute of Architects (AIA) or the National Society of Professional Engineers (NSPE). They serve together on committees for professional, design, and legislative matters and often socialize together. It is very common in most major cities to find that the architects and engineers working there are mostly graduates of the same local universities and have personal relationships going back to college design studios, where close personal relationships were formed. It is also quite common to find that architects in any major city have worked for four or five different firms in that city over their careers, and they will have relationships with all of those former firms. The idea of "taking" work from another firm does not sit well with many design professionals—even though they are competitors—due to these personal and professional bonds.

For years, it was considered unethical for an architect to solicit work for which another architect was under contract, and many architects would not even consider taking over a project unless the first architect had been paid in full. Though the AIA no longer has any such ethical restraints along those lines, the "old school" thought process is still very prevalent in the profession to look out for one another. A contractor should not be surprised to be asked by an architect who is approached to take over the work of a defaulting design firm for proof that the first firm has been paid, and that the first firm's contract has been formally terminated before they will consider taking on the job. This can cause delay, unlike an electrician who can offer to mobilize upon a day's notice to replace a defaulting subcontractor.

Also, as mentioned above, architects are by nature risk-adverse and they have been coached and taught for years about avoiding and limiting professional liability. Insurance and legal advisors will tell their architect clients that before taking over the work of another design firm, they should have certain agreements in place, including:

1. Proof that the contract of the original design firm has been formally terminated, such as a written termination letter or consent agreement signed by the original firm. This avoids the risk of a lawsuit by the original design firm for the tort of intentional interference with contract, a claim that may not be covered by professional liability insurance and one that can result in punitive damages against the successor design firm.[8]
2. Proof that the first design firm has been paid in full. Though no longer an ethical issue, a design firm that knows that its predecessor firm was not fully paid will wonder about its own ability to collect. A contractor who fires one architect without paying for services may just as easily repeat that action with a second firm. This may also be required to preserve relationships that architects have with other firms in their city. While two electrical subcontractors may not team together often, if ever, architects often pair up to go after bigger projects where some specialization or added staffing is needed. These relationships run deep and architects and engineers will be concerned about burning bridges with other firms simply to help a contractor with a troubled project.
3. Proof that the contractor seeking a replacement firm has the legal copyrights in the design and drawings, and can furnish a diskette of the original firm's work without risk of being drawn into some action for copyright infringement. There have been several cases in which an original design firm has sued and won against a successor firm that copies or completes a design done by the first firm, with the courts finding that such action constitutes intentional copyright infringement.[9] A contractor should be prepared to be asked not only for proof of ownership or license to use the design and drawings, but to sign a written indemnity agreement under which the contractor will indemnify, defend, and hold harmless the replacement architect for any claims by the original design firm for interference with contract, trademark or copyright infringement, conversion, or other related claims. The last thing a new design firm wants is to walk into a lawsuit by taking over a troubled project.
4. A contractor may also find that the firm he or she contacts about taking over the project will contact the original firm to find out themselves what the "real story" is behind the termination and what kind of partner the contractor is to work with. If the contractor has not been up front

[8] *See* Chapter 13, for more discussion of this issue.

[9] *See* Chapter 16.

and open with the replacement firm as to the reasons, the replacement firm is more likely to believe a "brother" architect's or engineer's version of the story. This will destroy the type of trust and confidence that design-build partners need to have to work successfully in this market.

5. A replacement firm will also likely ask the contractor for a release of any liability for design errors or omissions by the original firm, since the replacement firm is not insured for design errors of other firms. By furnishing a diskette or set of incomplete documents for a replacement firm to finish, the contractor is asking the new design firm to either: (a) do a complete and comprehensive verification of every detail, calculation, code review, and specification; or (b) accept the design and documents "as is" and to work from that point forward. This may mean accepting some risk for work done by others, a risk that most design firms will not accept without some limitation or indemnity agreement to protect themselves.

[B] Contractors as Subcontractors

Architects typically hold the prime design contract with the owner and hire a team of consultants to provide structural, mechanical, electrical, and plumbing engineering services. While architects refer to these engineers as "consultants," they are in law simply "subcontractors." In effect, architects routinely act as prime contractors in their relationship with the owner. But there is a world of difference between working as part of a design team with consultants the architect knows and trusts, and holding the prime construction contract and managing a team of low-bidding subcontractors and material suppliers, never before assembled. As a result, many design firms going into design-build as prime choose to hire a general contractor as a subcontractor to let someone with experience and relationships with the subcontracting community manage that part of the project. So what can a contractor expect from a design firm that acts as a general contractor?

A contractor can expect design firms to want lots of information about every decision made on the project. Architects and engineers value quality of the completed project highly, and they will want to make sure that their contractor-partner does not cut corners or vary from the plans and specifications without involving the design professional in those decisions. Design firms tend to want to study alternatives prior to making a decision in the field, where time constraints may call for quick field action to avoid added costs or schedule delays. Most architects have never had a course in project scheduling, nor ever prepared a critical-path schedule. The design schedule for an architectural project is a series of bar charts, with lags at several intervals for owner approvals, followed by the next intensive phase of design work by the architect and its consulting engineers. The idea of "float" and "early start/late start" is more an aspect of construction schedules than design phases, where every day of design is critical and delays the release of bid documents and the start of construction. Design professionals may be looking to learn from a contractor-as-subcontractor so that their own staff can become trained in the role of managing as trade subcontractors, eventually leading to an integrated design-build firm that does not need a general contractor to perform that role. Contractors may want to include in their contracts with design firms-as-prime a non-solicitation clause that prevents the design firm from hiring away the contractors' project managers during the project or for some reasonable period after completion of the project. Rather than develop these management skills from within, architects and engineers may be looking at the contractor's staff as a source of talent to bring in-house to help develop a sole-source design-build firm that does not need to pay a percentage of the fee to a contractor for management services.

For many design firms, the need to hire a general contractor-as-subcontractor is driven by the owner's requirement for surety bonds. As stated in Chapter 13, many design firms simply cannot get bonded due to their limited financial resources and inexperience serving in a general contractor role. In those situations, a contractor can expect the design firm to request that both the owner and the design firm be named as "dual obligees" on the performance bond, thereby permitting either party to make a claim against the bonding company if the contractor defaults on some obligation. While design-build's biggest selling point is "single-source responsibility," many design firms will name the owner as an intended third-party beneficiary of the architect/contractor subcontract so that the owner can pursue a construction claim directly against the contractor, despite there being no direct contract between owner and contractor. Just as an architect or engineer would be leery of a contractor who only hired a design firm because they needed a "seal on the plans" or because they needed professional liability insurance, contractors should be concerned about design firms that do not see the contractor as a true partner, but merely an unfortunate necessity to get bonding. Contractors can bring value to the design process and should be dealt with as a teammate, not just another "consultant."

[C] Termination of the Contractor

When a design firm hires a general contractor as a subcontractor, usually the trade subcontractors will work for the contractor, not the design firm. This causes a real dilemma when the general contractor does not perform suitably and the architect contemplates firing the contractor. Surety bonds do not solve this situation and anyone who has gone through the process of terminating a bonded contractor can confirm that business usually comes to a screeching halt if the termination is contested or the surety has to find a replacement contract or negotiate a "takeover" agreement to finish the job. Design firms will want the contractor to agree that upon termination, all of the trade subcontracts and purchase orders are immediately assignable to the design firm, upon the design firm's election, so that the designer can keep all of the subcontractors working together and for the same price as they agreed to with the general contractor.

Probably the biggest concern that a general contractor will have when serving as a subcontractor to an architectural firm is any assurance that the architect can pay the bills if the owner does not. Though design firms oppose "pay-when-"paid or "pay-if-paid" clauses in their own contracts when acting as subcontractors,[10] a contractor should not be surprised to see those very clauses used in subcontracts between a design firm and a general contractor-as-subcontractor. Design firms simply do not have the assets to finance a project in the event that the owner does not pay. And many design firms cannot get bonded, so the general contractor has no protection of a payment bond to assure payment. This may mean that the only security that a contractor-as-subcontractor has is a mechanic lien against the owner's property. As a result, contractors should neither agree up front to waive lien rights as a condition of the subcontract, nor to sign lien waivers (other than "conditional lien waivers") prior to receipt and clearance of the monthly construction draw. A letter of credit drawn on the architect's bank may suffice to give some level of comfort that any given month's draw is secured. While the letter of credit may not be equal to the total construction contract, it may be sufficient to cover the amount of work performed in any given month during which the general contractor is at risk.

[10] *See* § 5.04.

CHAPTER 9

BUSINESS ISSUES FOR DESIGNERS IN DESIGN-BUILD

§ 9.01 WHERE TO GAIN BUSINESS KNOWLEDGE

Most architects will agree that traditional architectural education does not prepare architects to run a business. A student may have one semester of "professional practice," but no classes in marketing, accounting, finance, real estate, management, insurance and other skills that are needed to own and run a company. Whether one is starting out in a new architectural practice or a new design-build company, it is important to get some education in business first.

J. Angel Martinez, AIA, President of the Kansas City firm of Martinez & Associates Architecture and Construction jokes that, "Contractors did not just fall out of the sky with bags of money. They had to start from scratch at some point." Angel, a licensed architect whose design-build firm builds 100 percent of what it designs, says he did not learn his construction skills in architecture school, but from experience. Brad Buchanan, AIA, partner in the Denver, Colorado design-build firm of Buchanan Yonushewski Group, L.L.C. says he also did not get a business degree, but, "I learned at the school of hard knocks."

Asked for his advice to architects starting out in construction, another design-builder said plainly, "Get business education." Design professionals do not need an MBA, but it is possible to take classes in business management, finance, accounting, and marketing at local community colleges and through the Small Business Administration. Being a good architect or a good builder is not enough if you are a poor businessperson. University architectural programs are so packed with studio courses, structures, materials, history, theory, and other required courses that there is little room for business classes unless taken as electives. Yet most architects will admit that they wish they had a better business education to prepare them for the "real world." Some are picking up these courses after graduation to fill in what their undergraduate courses left out.

One architect was quoted in the *Los Angeles Times* as saying "You are either a good businessman or a good designer—thanks, but I would rather be a good designer."[1] That type of attitude may win AIA design awards, but it will not mix well with design-build where both design and business skills are needed to succeed. John Gaunt, FAIA, former CEO of Ellerbe Becket says, "The architect needs to regain the role of synthesizer in the total construction process. The development of design-build ability can enable this. The new architect must have greater depth of knowledge across disciplines, multi-specialty expertise, communication and leadership skills, and fundamental understanding of project delivery systems." A solid understanding of business, real estate financing, and business administration will help to ensure success in design-build.

§ 9.02 FINANCIAL REQUIREMENTS

Most design firms have limited physical assets. Many rent their office space, lease CAD equipment, and have employees drive their own cars to job sites. There

[1] *What Is Driving the Shift in Project Delivery: Private Owner/Architect*, by John E. Sicard, AIA, published in Cost, Time & Risk, AIA, Nov. 1994.

are not large bank accounts holding retained earnings. By contrast, contractors often own their buildings, own and provide pick-up trucks for superintendents, own backhoes and other equipment, and maintain a financial statement sufficient to keep their bonding company writing bonds and their banker advancing credit. So how does a start-up design-build firm develop capital? Considering that it may take a few months of marketing to land the first project, architects or engineers starting a design-build firm from scratch are usually advised to capitalize sufficiently to cover six months' expenses, plus start-up costs. Under normal conditions, with average collection rates, financial advisors recommend at least three months' cash is required if the company has contracts under way. Some sources of start-up capital include:

- personal savings
- personal credit cards
- equity in personal real estate
- Small Business Administration (SBA) loans
- loans from friends and relatives
- commercial loans from a bank
- investment from partners
- line of credit from bank

A start-up design-build firm will want to talk to a banker about establishing a line of credit. Banks do not want to be investors in a company. Their business is based on the certainty of a return on capital in the form of interest on loaned funds, plus a reasonable assurance that the loan will be repaid. A banker will want to see a business plan, including financial projections. Banks will want to see that the firm has invested some of its own capital, often as much as 50 percent, so that the bank is not the major stakeholder in the new business. A surety will want to know how much trust the banker is willing to put in the firm. A firm will need this line of credit to purchase materials, vehicles, and equipment that will be needed to start up in construction. It is important to establish a relationship with a good construction accountant who can help plan the new business venture. The accountant can help prepare the business plan before a new firm approaches a bank about extending credit.

§ 9.03 DEVELOPING A BUSINESS PLAN

Design-build partners should spend some time putting together a business plan before launching into this new market. There are many books available on starting a new business, written by business experts and financial planners. The Small Business Administration also offers advice through staff, seminars, publications, and even over its Web site at <http://www.sba.gov> on topics such as business plans, financing, and marketing. No matter where one goes for advice, here are some topics to include in the design-build business plan:

1. Objective (Mission Statement). What is the target market? What message is to be conveyed to clients?
2. Corporate Culture
   i. *Integrity and ethical beliefs*. Set out the principals to be adhered to.
   ii. *Conflicts of interest*. Are there any at present? Any foreseeable? How will those that come up be handled?
   iii. *Non-compete terms*. This should address exclusivity of design-build team members, as well as restrictions on departing members, owners, and employees that are reasonable in scope and time, yet protect the company's investment in training and marketing.
   iv. *Growth, new owners, new employees*. How big will the company be? How will the company take on new partners and shareholders? Who does the hiring?
3. Legal
   i. *Name*. This may seem simple, but is often the most difficult decision for two companies or a group of design professionals to come up with. Is someone else already using that name? Are there any licensing board restrictions on the name? Does the name have to be registered with the state?
   ii. *Structure*. Today, there are a host of corporate options, from simple corporations, to limited liability companies, partnerships of all types, S-corporations, etc. (*See* Chapter 11 for more discussion.)
   iii. *Licensing*. Are there any contractor or architect/engineer licenses that the business will need from the state or city? One needs to find out if the state licensing laws restrict who can be an owner, officer or director, and what percentage of ownership must be licensed professionals.
   iv. *Officers and titles*. Who are the initial officers or managers and what offices will they hold? What are their duties?
   v. *Contracts, subcontracts*. There are many different forms on the market. Decide which ones will be the company standards, and develop a standard template of supplementary conditions or modifications that suit your needs, insurance and lawyer's advice. (See Chapter 18 for more discussion.)
   vi. *Profits and Losses*. How will these be shared?
4. Start-Up Capital
   i. *Who contributes and how much*. Remember that bankers want to see a personal financial commitment by firm owners.
   ii. *Salaries*. What will be budgeted?
   iii. *Employment agreements*. These are becoming more common for any business, dealing with the terms of employment, benefits,

compensation, company policy, termination and non-solicitation of firm clients or firm employees.

   iv. *Equipment needed.* What needs to be purchased or leased to start up? What does the company hope to acquire over time?
   v. *Line of credit.* The bank will want to see the business plan before opening a line of credit. How much credit is needed?
   vi. *Personal guarantees.* Who will sign at the bank, on the lease, etc. whenever personal guarantees are required?

5. Disputes
   i. *How to resolve them.* There are many options, including mediation, arbitration, and litigation. What city or county will the hearing be held in? The trend is toward mediation and other alternatives to court.
   ii. *Termination.* What grounds terminate the agreements and the business?
   iii. *Buy Out.* What options are there and how is the interest of each owner valued? If one person wants out, does that terminate the whole business? There should be a plan for continuation or buy/sell agreement that allow continuing members to buy out a departing member. An important topic to include in any agreement is how to value the stock or interest of a departing firm owner.
6. Insurance and Bonds
   i. *What needed. See* Chapter 13 for a more detailed discussion.
   ii. *Who provides what?* Some insurance can be better obtained by the contractor partner, and other by the design professional.
   iii. *What agency.* The contractor and architect may use different agents. Will all business be with one agency, and can that agency handle the business? Do they write all the companies that insure the two firms, contractor, and design firm?
   iv. *Personal indemnity on bonds (who will sign).* This may be even more of an issue than the line of credit for two reasons. (1) The company controls whether it draws on the line of credit, but others determine who makes claims upon a bond; and (2) most sureties will require not only the business owners, but also their spouses, to sign the bond application unless there are sufficient assets in the company to cover foreseeable bond losses.
7. Marketing
   i. *Who does it?* Does one firm have more experience or more qualified staff?
   ii. *Who selects projects?* This is a key decision.
   iii. *Who pays for it?* The design team normally has more expenses than the construction/estimating team. How will these costs be handled? Does each party bear their own cost or is a portion reimbursed?
   iv. *Who creates marketing materials?* Again, who is best able to handle this in-house, or should outside consultants be used?

   v. *Who selects team members?* When composing a team of subcontractors and design consultants, who picks the team, and who gets to approve or object to a proposed team member?
   vi. *Announcement.* How will the company market itself? Who will handle the direct mail, print ads or the open house to promote the new business?
8. Risk Management
   i. *Design and construction errors (who pays).* This is another key decision. Who pays for deductibles, as well as uninsured risks?
   ii. *Liquidated damages (whose risk).* These are normally not insured, so how will the company pay for those if they cannot be passed on to subcontractors?
   iii. *Cost overruns (who pays).* Are there any margins of error here? Will these be borne by the party responsible, or split in some percentage fashion?
   iv. *Contingencies.* How much should be built into the bid to cover the unknown? This is a big subject of dispute, especially when design details are incomplete at the time proposals are submitted.
   v. *Bonding of subs.* Will this be required and, if so, for which ones, or for what dollar threshold?
9. Procedures
   i. *How to approach projects*? How will the jobs be managed?
   ii. *How will design and estimating work together?* In the same location, by internet, or video link-up; frequency of meetings. Will the staff move from one office to another during portions of the project? Communication at this stage is essential.
   iii. *Whose decision governs design issues?* Is it a joint determination, or does the design partner or construction partner get final say? Does it matter if it is a life safety issue, or merely a cost issue?
   iv. *What level of documentation needed?* Will it be sufficient to use limited plans, elevations, typical wall sections, then issue details on an as-need basis? What will the city require for permitting? What documents will the subcontractors provide?
   v. *Any design-build subcontractors*? HVAC, sprinkler, landscaping, electrical, or others. Are they insured for design errors? Who does their design work?
   vi. *Self-perform construction.* Is either party able to provide labor, even temporary labor, for rough carpentry, flat work, clean up, other work more economically than by using subcontractors?
   vii. *Who has the client contact?* Clients may want just one contact, which avoids any mixed messages or inconsistencies. Who should it be? How will that person keep others informed?
   viii. *Who resolves disputes with clients and subs?* Will this depend on whether this is an owner, design, or trade subcontractor? Con-

sider a dollar threshold of authority for settlement of small claims without needing a board meeting.

ix. *Verification of client finances.* Will this be a company policy, or just for new clients?

x. *Use of other architects and engineers.* If other consultants are needed, who will pick them?

10. Other

i. *What minimum level of staffing needed?* Full-time and part-time staff.

ii. *Where to office?* Will the design-build firm office out of an existing office and share space, or will it have a separate office?

iii. *What rent charge?* If the company leases space, or shares an office, will the new firm pay rent and, if so, how will it be financed?

iv. *Whose payroll and who handles timekeeping, billing, payroll?* Is one partner better set up to handle this? One should seek tax advice here from a lawyer or accountant.

v. *Health and disability insurance for employees.*

vi. *Car or truck.* Will employees or superintendents get a company truck, car, or an allowance? How much?

vii. *What bank will be used?*

viii. *What accountant will be used?*

ix. *Lawyer.* Which lawyer or firm will be used to set up the business, handle contracts and licensing, or other issues?

x. *Professional organizations (AIA, NSPE, DBIA, AGC).* Which ones does the company belong to, which ones make sense to join? What level of involvement?

xi. *Expenses and entertainment.* What allowances, what budgeted?

xii. *Any projects as owner/developer?* Does it make sense to look into taking an equity position in projects as owner?

xiii. *Union/non-union.* Is this an issue in the market area the company targeted? Are project labor agreements used in your area, or the project city?

§ 9.04 DEVELOPING A BUDGET AND A PROFIT PLAN

Dave Shelton, AIA, left a medium-sized architectural firm in 1999 to start his own design-build firm. He contacted the Kauffman Center for Entrepreneurial Leadership and reviewed the AIA Handbook of Professional Practice for guidance on setting up his own business, and read several books. He worked out of his home for two months, then rented 900 square feet of office space, enough for six people, which is what his initial growth plan covered. Shelton's plan included a staff consisting of one principal, one clerical/reception, and four staff (one project manager, one project architect with three or more years of experience, and two production staff)—with the office fully staffed within one year and annual revenue of $430,000. He then talked to his banker and got a line of credit sufficient to fund a staff of four (75 percent of his projected staff) for a period of four months with no income. He then talked to his insurance agent about professional liability insurance and contractor's insurance policies. He did not plan to undertake projects requiring bonding initially, recognizing that it may be difficult to qualify for bonding. Shelton then incorporated his company and got business cards printed. "The next thing I started doing was to develop a network of contractors. I don't have any formal teaming agreements yet, but have lots of breakfasts and lunches with contractors." This type of networking can lead to business opportunities without formal teaming agreements, and leaving options open to teaming with many rather than just one partner.

The AIA Handbook of Professional Practice recommends establishing a "profit plan," which is an estimated operating budget with a profit goal built into the plan. Figure 9-1 shows the example used in the AIA Handbook.

FIGURE 9-1. Sample Profit Plan for a Design Firm. (Excludes Consultants and Reimbursable Expenses).

Expenses			
Salaries			
Principal	(1@$75,000)	$75,000	
Registered architect	(1@$45,000)	$45,000	
Technical employees	(2@$30,000)	$60,000	
Secretary	(1@$20,000)	$20,000	
Total Salaries			$200,000
Payroll taxes and benefits (@25%)			50,000
Office expenses			
Rent		$12,300	
Utilities		5,500	
Telephone		2,500	
Equipment purchase, lease, maintenance		8,000	
Postage, shipping		1,500	
Publications		1,000	
Insurance (auto, office, liability)		16,000	
Office supplies		2,500	
Travel		5,000	
Printing		3,500	
Marketing tools		3,200	
Professional development: dues, training, etc.		5,000	
Legal and accounting fees		3,000	
Other expenses		1,000	
Total office expenses			70,000
Total expenses			$320,000
Profit goal @ 20% of net revenues			80,000
Net revenue goal			$400,000

Using the AIA's example, knowing that to cover overhead and profits the company has to take in $400,000 per year, a design-build contractor would have to estimate what level of construction volume would be needed to net $400,000 in revenues other than labor and material. For example, $2 million in construction direct expenses might be needed to support this five-person office and net the profit goal. See the sample projections in Figure 9-1. For a design-build firm, net revenues would include not only that related to professional and clerical salaries, but construction labor, project managers, estimators, superintendents, construction insurance, and equipment. Material purchases and payments to subcontractors and engineering consultants come out of the design-build contract funds at 100 percent of the amount agreed to be paid to those entities. Here is an example:

Projected project revenue	$2,000,000	design-build contracts
Minus material purchases	(225,000)	
Minus subcontractors	(1,250,000)	
Minus consultants	(125,000)	
Minus billable staff	(200,000)	
Projected fees	$200,000	cost plus 10% fee basis
Less: Projected expenses		
Direct salaries	(50,000)	non-billable staff
Overhead	(70,000)	
Profit target	$80,000	

Note that while this model profit plan represents a 10 percent fee, it is only a 4 percent profit based on gross revenues from design-build contracts.

§ 9.05 FINANCIAL MANAGEMENT

[A] Sweep Accounts

The most popular cash management technique among contractors in the United States is reported to be "zero balance arrangements" or "sweep accounts." A sweep account earns a better return than interest on checking accounts by "sweeping" excess money into higher-yielding vehicles, such as money market funds or other funds. The bank usually gives the customer a choice of accounts into which the money can be moved. Sweep accounts were used by 82 percent of those contractors responding to the CFMA's 1999 annual financial survey.[2] While a money market fund may offer one or two percentage points over the interest paid on ordinary checking and savings accounts, a contractor can earn a higher rate of return by sweeping funds into a mutual fund.

[2] *Construction Industry Annual Financial Survey*, Construction Financial Management Association (CFMA), reprinted in Engineering News Record, Nov. 8, 1999.

Some banks will sweep the customer's entire balance overnight and on weekends and return it for the next business day. Another option is just to sweep amounts over a certain threshold either set by the bank or the customer. One source reports that with a sweep threshold of $25,000, about two-thirds of businesses with sales between $1 million and $10 million have enough money to qualify for a sweep account.[3] Monthly fees to maintain a sweep account can vary from $20 to $40 on the low end to as high as $150. Some banks use computers to analyze customer use of deposits and sweep money from the checking account, reclassifying the transaction deposits as savings deposits at the close of business on Friday, then move the funds back to the transactional checking account on Monday. If your business is not currently using a "sweep account," talk to your banker.

[B] Fiscal Years and Accounting

Ninety-four percent of the contractors surveyed use the same year-ending date for both tax reporting purposes and for financial reporting. Most contractors report their earnings on a calendar-year basis.

The majority (87 percent) of respondents use the percentage-of-completion method for recognition of contract revenue and gross profit for tax reporting purposes. Nearly all respondents (96 percent) recognized revenue for financial statement purposes using the percentage-of-completion method, with 71 percent recognizing income as soon as costs are charged to the job. Very few (6 percent) defer profit recognition until the ultimate profit can be reasonably estimated.[4]

About a third of contractors surveyed recognized estimated revenue and cost associated with unpriced (but approved) change orders in the current period, rather than deferring recognition until the change order was finalized. Another third recognized the estimated cost in the current period, but not the estimated revenue. Half of the contractors preferred to recognize contract claim revenue when settlement was probable and the amounts could be reasonably estimated, rather than when the claim was settled or paid.

[C] Financial Statements

Three-fourths of general contractors surveyed in 1999 reported having audited financial statements.[5] This is largely due to the requirements of lenders for a contractor's line of credit and the requirements of bonding companies. Most contractors use an outside accountant to prepare their financial statements. An independent, audited financial statement has more credibility with bankers and surety companies. There are many factors that come into play in hiring professional

[3] *Sweeps Aren't Just for the Big Guys Anymore*, Business Week, May 28, 1997.

[4] *Id.*

[5] *Id.*

consultants, whether they are lawyers or accountants. Contractors responding to a 1999 survey said that when selecting an auditor for their company, they relied upon: (1) the personal relationship with the auditor [40 percent]; (2) the auditors' knowledge of the construction industry [20 percent]; and (3) competitive bidding [19 percent].[6]

[D] Taxation

Internal Revenue Service (IRS) examinations of contractors are very common. A 1999 survey found that only about three out of ten contractors said they had never had an IRS examination, and only seven out of ten have been audited in the last six years.[7] The good news is that the majority of companies reported that their last IRS examination resulted in either an immaterial increase or no change at all to their tax liability for the period examined. Nearly half (49 percent) of those surveyed filed IRS Form 8697 (Look-back Interest Computation) with their last tax return.

[E] Accounting and Bookkeeping Personnel

Architectural firms often operate as very small businesses and most are sole proprietors. The receptionist or secretary often also functions as bookkeeper in the one to five person firm. Contractors deal with many more bills each month, with larger payrolls and dozens of supplier and subcontractor bills each day. To keep up with the increased accounting, most contractors report having a staff of at least eight persons dedicated mainly to accounting functions within the company. Very few actually have CPAs on staff however. Contractors reported that about one in five accounting personnel hold four-year accounting degrees without a CPA certificate.[8] A business plan will need to include sufficient bookkeeping staff to handle projected workload. These non-billable salaries must be figured into a firm's overhead and paid from fees earned on cost-plus-fee contracts.

§ 9.06 GETTING A LINE OF CREDIT FROM A BANK

A credit line is a formal loan agreement that permits borrowing of funds at will, up to an agreed upon limit. Like a credit card, there is a cap on how much one can borrow and a set rate of interest. The line of credit is usually secured by personal guarantees by the principal owners of the company, who put their personal assets at risk as security for the line. In addition, the lender may take a security interest in the accounts receivable of the company. Depending on the lender, one may be required to pay down the line of credit to zero at least once yearly, after which time the line of credit becomes available at the full limit again.

---

[6] *Id.*
[7] *Id.*
[8] *Id.*

Lenders generally look at two things when granting a line of credit to assess the borrower's ability to repay the loan. One factor is the current ratio, which is determined from this formula:

$$\text{Current ratio} = \frac{\text{Current assets}}{\text{Current liabilities}}$$

Lenders want to see sufficient current assets to cover the firm's current liabilities, plus a significant cushion. Assets are made up of cash, work-in-progress, and accounts receivable. Current liabilities would be amounts owed to consultants, subcontractors, material suppliers, and other short-term creditors. Lenders usually want a current ratio greater than 1:1, and prefer a ratio closer to 2:1, indicating assets are double liabilities. The second criteria used by lenders is the debt-to-equity-ratio, determined by this formula:

$$\text{Debt-to-equity ratio} = \frac{\text{Long-term debt}}{\text{Owner's equity}}$$

Bankers usually want to see a debt-to-equity ratio to be at least 1:1, which means that the owners of the company have an appropriate level of personal investment in the company relative to outside lending.

§ 9.07 SALARIES, OVERHEAD, AND PROFIT MARGINS

Actual profits in construction, other than from national studies, are often tough to pin down since contractors are notorious for their reluctance to share this information or allow it to be publicly quoted. However, it is no stretch to say that profit margins among contractors are greater as a rule than they are among design firms. Architects are less shy about lamenting how underpaid they are as a profession. Government caps on architect fees of 6 percent have held architects back from realizing more profits on design contracts.[9] But design firms that have gone into design-build are reporting that the profits in that market are significantly better than traditional practice. Brad Buchanan, AIA, partner in the Denver, Colorado design-build firm of Buchanan Yonushewski Group, LLC says that his firm's profits are five times what they would be if they did just the design. "The profit margin for construction work is much, much larger than it is for design work. This allows us to do more up-front work and spend more time on the design without worrying about losing money."[10]

Design firms doing design-build claim that profits are higher because:

- They can more easily build in adequate fees.
- They have more control over subs, time cost—the entire project.

---

[9] *See* Chapter 15.

[10] *How to Increase Your Fees in a Tough Market*, Architectural Record, Nov. 1999, p. 63.

- There are fewer changes, less documentation, and fewer problems.
- There is less time wasted in meetings.
- The system is more efficient.[11]

One engineering firm's first design-build project as a contractor was so successful that the profits exceeded an entire year's worth of profits for the design side. A company officer said, "I am kicking myself for not getting into this earlier." The firm, who did not want to be identified for this book, said that they made the decision to get into design-build despite the perceived risks and determined to take the lead as prime, rather than act as someone else's subcontractor. "We did all the design work in-house and acted as our own general contractor. It was a $4 million project and we made $600,000 in profits. We're looking for more like that." The profits on this first project helped the company to fund the start-up costs and provide the working capital that their bonding company wanted to see. "Like most engineering firms, we had little assets and could not get bonding. We decided to keep all the profits from the first project within the company. Now we can bond projects up to $2.5 million each." With a significant retained earnings from that first project and bonding capacity, the firm is looking forward to the next project when they can actually pay some bonuses and reap the rewards.

[A] Profits and Work Are Up

The 1990s were a period of strong construction growth in the United States. The pattern is predicted to continue into the new millennium. A survey by the Construction Financial Management Association (CFMA) predicted an average increase of 22 percent in contract volume in the year 2000.[12] More than half of those contractors surveyed reported an average backlog of nearly eight months of work. Architects report almost seven months of backlog, with more than 10 percent of the firms surveyed reporting backlogs of a year or more.[13] Profit margins are largely misunderstood in construction. A 10 percent fee does not mean 10 percent profit. A 1999 survey of over 700 contractors found that net income after taxes was only 2.0 percent of revenues.[14] Return on assets was 5.9 percent and return on equity was an average of 21.1 percent. The PSMJ Resources annual *Financial Performance Survey* of the nation's A-E firms found that profits reached a 1999 median (before discretionary bonus/incentive and income taxes) of 12.9 percent of net revenues, up from 11.4 percent in 1998 (a 13.2 percent change).[15] However, PSMJ's research director, William Fanning, says that the survey shows that increased profits, "appear primarily related to doing more rather than doing it

---

[11] *1999 Design-Build Survey,* Zweig White & Assoc., Inc., p. 72.

[12] *Id.*

[13] *Growing Billings, New Inquiries Keep Firms Busy,* AIArchitect, July 1999, p. 5.

[14] *Construction Industry Annual Financial Survey*, Construction Industry Annual Financial Survey, (1999).

[15] 1999 PSMJ *Financial Performance in Design Firms Report.*

better. This leaves the profession vulnerable, as always, to the volume of work available." The *1998 Financial Performance Survey* published by PSMJ, found that design firms were more profitable than they had been in over 15 years, reporting profits over 11 percent.[16]

While these numbers sound good for architects, one should take a practical look at $10 million worth of construction work. If a contractor's net profit is 2.0 percent after taxes, that is $200,000. Figure an architect's fee at 8 percent on $10 million, or $800,000 in fees. If the average firm makes a 13 percent profit, that nets $104,000 on $800,000 in fees. The contractor's profit is only 2 percent, but that nets him nearly double the architect's profit at year end on the same $10 million in project dollars. With design-build, architects can look to realize a profit from the design fee as well as the construction work, potentially netting the combined over $300,000 in profit on $10 million of work.

As this example shows, by realizing profits on the construction side of the project, including participation in construction savings, architects and engineers can make substantially more money than they make offering pure design services. It is not uncommon to hear architects talk of making gross profit margins 50 percent higher on design-build jobs than on their traditional work, and the example shows why. Some firms have even used their profitable design-build divisions to offset the losing design divisions.

[B] Labor Shortage

All this construction work has created a labor shortage, with construction industry unemployment in 1999 reaching its lowest point in 30 years according to the Department of Labor.[17] The pinch is felt by architects, engineers, and contractors who have all struggled to keep enough staff to handle the workload. Many firms have engaged in nationwide searches, or global searches, to find qualified staff. The 1999 survey by CFMA of over 700 contractors showed that the shortage of trained field help and trained project managers ranked first and second respectively, as the top challenges that will be faced in the next five years.[18] The result is that salaries and hourly wages have become more competitive. Like professional athletes, some design and construction firms are paying "signing bonuses" as an inducement for new hires. According to PSMJ's 1999 study of management compensation, "Up to 30 percent of top firms are paying signing bonuses."[19] These bonuses are often disguised as "inflation relocation al-

[16] *Firms Report Higher Profits and Bigger Bonuses*, AIArchitect (Oct. 1998), p.21. A quarter of the firms surveyed reported profits of 19% and higher. Smaller firms with 10 or fewer employees had the highest margins of 22%, where large firms of over 100 employees reported profit of under 10%. Medium-sized firms reported profits of 14% to 16%.

[17] Engineering News Record, Dec. 20, 1999, p. 31.

[18] CFMA *Construction Industry Annual Financial Survey* (1999).

[19] *Staff Shortages and Hot Markets Create a Recruiting Feeding Frenzy*, Engineering News Record, Sept. 13, 1999, p. 57.

lowances," with a housing or auto allowance given during the initial term of employment in order to keep base salaries down.

Dr. Roger W. Liska of Clemson University's construction science and management program says that graduates are "getting two-to-three job offers, many of which have signing bonuses, moving and living expenses, and extremely good benefits. And we're hearing stories about firms that are offering to pay the remaining years of tuition for interns, in exchange for two-to-three year commitments after graduation."[20] These are good times to be a talented employee but tough times for employers to hire and to retain those talented people.

[C] Contractors

One survey showed that among contractors that gave wage hikes in 1998, average salary increases were 4.91 percent for staff-level employees, marking the sixth straight year of escalating raises for contractors.[21] The same survey reported that contractors planned to increase salaries again by 4.34 percent in 1999. Executives at construction companies saw their salaries increase by an average of 5.9 percent in 1999, with plans to raise them again in 2000 by 5.1 percent.[22] Contractors, like design firms, report that the biggest challenge they face in a strong economy is that there are not enough people to do the work. Estimators are reported to be in the highest demand. "Not necessarily the conceptual estimators, but ones who can do takeoffs from a full set of plans and determine where value engineering can be applied," says one executive search consultant.

The high demand for qualified personnel and competitive salaries has prompted many to switch jobs. A survey by the management consulting firm PAS, Inc. of over 23,000 construction employees found that engineers and construction superintendents had the highest rate of turnover, followed by estimators, showing that good project managers are hard to find and even harder to keep. Salaries among construction professionals vary by discipline and by level of experience. But almost uniformly, contractors report that they are paying more than ever to keep qualified staff, including executives, as well as to recruit new employees. "Seven out of ten firms recognize the difficulty. The rest think they can still get by paying third-world wages. They can't," says Michael L. Ketner, a Pittsburgh-based executive search consultant.[23]

To get a feel for the types of compensation being paid by the competition, the tables in Figures 9-2 and 9-3 reported average incomes for various construction positions in 1999.

---

[20] *Commercial Outlook,* Design-Build, Dec. 1999, p. 45.

[21] *Staff Shortages and Hot Markets Create a Recruiting Feeding Frenzy*, Engineering News Record, Sept. 13, 1999, p. 57.

[22] *Contractor Executive Salaries Soar As Demand Reaches Frenzy*, Engineering News Record, Jan. 24, 2000, p. 19.

[23] *Id.*

FIGURE 9-2. **Average Contractor Executive Salaries.**

Position	Median Base Salary	Median Total Compensation
President	$150,000	$202,250
Chairman	$180,000	$275,100
Executive Vice President	$120,000	$169,220
Senior Vice President	$107,150	$145,000
Vice President (Operations)	$96,495	$123,003
Vice President (Estimating)	$96,236	$114,937
Vice President (Bus. Development)	$97,450	$120,000
Vice President (Administration)	$91,500	$102,000

Source: PAS, Inc. Survey (1999)

[D] Engineering Firms

The annual salary survey done by the National Society of Professional Engineers (NSPE) showed that average compensation for engineers increased by 4.7 percent in 1998. The survey by PAS, Inc. of engineering-design staff showed that for staff-level engineers in construction, the average salary increase in 1998 for firms that gave raises was 4.93 percent, and that those same employers planned average salary increases of 4.64 percent.[24] Figure 9-3 shows a comparison of median engineering salaries based on discipline and level of experience.

[E] Architectural Firms

The 1999 PSMJ Resources' annual survey of financial performance among architect-engineer firms found that median direct labor costs per hour rose 4.4 percent over 1998. According to the study, firm managers reported paying staff a median of $50,926 per person, including incentives and bonus payments. William Fanning, PSMJ's research director reports that the average design firm employee saw close to a 5 percent increase in 1999.

The AIA Handbook of Professional Practice contains the results of a salary survey among 350 AIA members who in 1992 worked for contractors. In 1997, the AIA published another report on Compensation at U.S. Architectural Firms based on 1996 salaries. Though the two surveys were done four years apart, the

[24] *Staff Shortages and Hot Markets Create a Recruiting Feeding Frenzy*, Engineering News Record, Sept. 13, 1999, p. 57.

FIGURE 9-3. Average Contractor Personnel Compensation.

Position	Average total compensation
Superintendent	$54,831
Project Superintendent	$65,196
Project Manager/Estimator	$63,634
Project Manager	$81,296
Field Engineer (experienced)	$45,137
Field Engineer (inexperienced)	$36,549
Estimator (chief)	$89,905
Estimator (senior)	$66,924
Estimator (experienced)	$51,044
Estimator (inexperienced)	$39,109
Scheduling Engineer (manager)	$74,038
Scheduling Engineer (senior)	$63,957
Scheduling Engineer (experienced)	$49,942
Contract Administrator (senior)	$57,521
Contract Administrator (experienced)	$41,154
Contract Administrator (inexperienced)	$30,994
Safety Director	$61,536
Assistant Safety Director	$45,700
Business Development (manager)	$70,405
Business Development (director)	$93,774
Accounting Manager	$60,297

Source: PAS, Inc. Survey (1999)

figures show that the salaries of architects working for contractors in 1992 were generally higher than their counterparts in traditional architectural practices four years later. Figure 9-4 shows some examples.

Looking at the numbers in Figure 9-4 would lead to the conclusion that salaries are higher at construction companies than at architectural firms for similar positions. Salaries among construction executives in 1999 ranged from a median of $91,000 to $97,000 for a vice president and to $150,000 for a company president. (*See* Figure 9-2). These are salaries that are simply not available at the typical architectural firm. Higher profit margins no doubt give contractors the ability to pay better salaries than architectural firms. This makes design-build look all the more attractive.

FIGURE 9-4. Comparison of Architect Salaries of Those in Traditional Practice vs. Those in Construction

Title	Mean Salary (1992) (Arch. at Constr. Co.)	Mean Salary (1996) (Arch. in Traditional Firm)	
President	$107,000	$75,000 to 100,000	(Principal)
Vice President	77,538	not available	
Construction Coordinator	76,356	not available	
Director	70,200	58,900	(Associate)
Manager	61,064	53,100	(Manager)
Senior Architect	59,167	45,400	(Architect III)
Design Architect	43,384	38,900	(Architect II)

[F] Salaries and Bonuses

In an article published by *Engineering News Record* in September 1999, Bill Fanning, the director of research for PSMJ, said that, "Salary structures should be adjusted about 5 percent higher this year, well above inflation."[25] Bonuses are becoming common for middle management as well, with surveys showing the average project manager receiving a bonus of 7 percent to 8 percent of salary, with the top quarter of those surveyed reporting bonuses averaging over 15 percent of salary. With increased pressure for performance and incentives from owners for cost and time savings, some construction employers likewise tie compensation bonuses to performance by staff.

§ 9.08 SHARING IN THE SAVINGS

Whether leading the design-build team as prime contractor, or by joint venturing, or even acting as a subcontractor, there are opportunities for sharing in the savings that are not available to design professionals in traditional construction delivery. Much of design-build is contracted for under a "cost-plus" type of contract with a guaranteed maximum price (GMP). If the actual cost of the work comes in under the GMP, then the contract can provide for a share in the savings in some split between the owner and the contractor.

In traditional construction, where the architect is only the designer, with a lump sum or percentage-based fee, there is no extra payment to the architect when the job comes in under the GMP. Only the contractor reaps the benefits of value engineering ideas that save the owner money. But with design-build the architect/engineer can participate in the savings. As the prime under a fixed-price construction contract, the architect/engineer firm can retain all the "savings" and realize greater profits. As prime under a GMP construction contract, the firm will usually split the savings with the owner, but still make money.

---

[25] *Id.*

Many firms overlook the opportunity to share in savings even when the design firm is acting as a subcontractor in a design-build setting. Why shouldn't the design firm get to reap some of the benefits from assisting in value-engineering, constructability review, and making revisions and substitutions that save on construction costs? Some firms think it is unethical to participate in savings as an architect. While not barred under the AIA's Code of Ethics, some professionals think that one loses the trust of the owner if there is the inference that one might make design decisions that are driven more by a desire to make a profit, rather than what is in the owner's or the project's best interest. This type of thinking will only hold architects and engineers back from realizing one of the greatest benefits of leading a design-build project—*i.e.*, increased profits.

§ 9.09 INCENTIVE BONUSES IN DESIGN-BUILD CONTRACTS

Bonuses for early completion can be another source of increased profit. Some design-build contracts provide for liquidated damages for late completion, and a bonus for early completion. The bonuses can be huge in some contracts. For example, the Illinois Department of Transportation agreed to pay early completion bonuses ranging from $10,000 to $18,500 per day for work on the $567 million Stevenson Expressway reconstruction project, a 15-mile stretch of I-55. Liquidated damages for late completion were for the same amounts. If all teams finished early, individual contractor bonuses would double for the remaining days.[26]

On another highway project, in New Jersey, the state Department of Transportation worked with the contractor to find creative ways to save time. The contract completion date is April 30, 2001, but the state will pay a bonus up to $3.5 million, with $9,000 per day for each day finished early after November 23, 2000, and $35,000 per day for each day prior to November 23, with a $1.4 million lump-sum payment due to the economic benefit of having the highway interchange completed before the holiday shopping season. Liquidated damages for late completion are $22,000 per day.[27] The low bidding contractor, a joint venture including Joseph M. Sanzari, Inc., bid $49.7 million with the highest bid at $72.6 million for the same work. "We bid the job at cost and we're going for the bonus," said Mr. Sanzari, confident of his company's ability to make the bonus. In addition to the bonus, the contractor suggested a design change from steel piles to sheet piling, resulting in a net savings of $288,000 and 10 weeks of construction. The state Department of Transportation split the savings equally with Sanzari, saying this was a "win-win" situation for both parties.

Similarly, when the Maine Department of Transportation needed the $46.6 million Sagadahoc Bridge project completed by July 1, 2000 due to tourist traffic over the Independence Day holiday, they included a $1,500 per day liquidated damages clause for late completion, but gave an incentive bonus of $3,000 per day for each day early, with a cap of $1 million.

---

[26] *Chicago Contracts Cruise On*, Engineering News Record, Mar. 6, 2000, p. 24.

[27] *Incentives Drive Fast Track*, Engineering News Record, Oct. 25, 1999, p. 34.

In yet another example, a design-build team received a 25 percent bonus from a manufacturer of disk drives for the fast completion of an $87 million plant in Fremont, California. The owner was able to start partial production just seven months after design began. Traditional methods would have involved at least six months of design, followed by 12 months of construction.[28] If a firm is passing up on incentive bonuses due to ethical concerns, that firm is passing up on an opportunity to make more profits than most architects ever thought possible.

§ 9.10 RETAINERS

Though the AIA standard forms of Owner-Architect provide for an "initial payment" at signing of the contract, many architects fill that in with a zero and finance the design work themselves until the owner pays. Considering that the first bill does not get sent out until after a month's worth of work is done, and has payment terms of 30 days, the architect has performed 60 days worth of work before any payment comes in from the owner. If the owner is a slow paying client, it may be more like 90 days that the architect must essentially finance the project for the owner. Other professions, like lawyers, are not so shy about asking for a retainer from their clients at the initial meeting to help finance the initial work until the first month's billing is sent out and paid. Residential contractors routinely ask for a "deposit" at contract signing to help buy materials.

It never hurts to ask for a retainer. According to one architect with a large firm, his company was hired for a new multimillion dollar stadium project. At the initial meeting with the client, the firm's senior partner told the owner, "If you want us as your architect, then I want a check for $1 million today, before we start any work!" The somewhat startled client reached into his suitcoat, took out the checkbook and wrote the check. Without asking, one will never know what the client may be willing to do.

According to design-builder Angel Martinez, AIA, "I always ask for an up-front retainer to get us started on a construction project. I am not a banker. The client should finance the job, not me." Contractors may have to pay cash for materials, as well as for permits, copies of plans, and other start-up costs. Especially when starting a new company, with no credit history with subcontractors and suppliers, payments may be expected up front or on short terms. As with traditional architectural services, a firm may be 60 to 90 days into a project before the owner makes its first payment. All during that time the firm is incurring costs, and paying bills, eating into either retained earnings, or into its line of credit, resulting in interest charges to the firm. In short, the firm is financing the owner's project. Consider asking for a retainer, but allowing the owner to hold retainage on each draw to make sure that the owner does not overpay for the work.

---

[28] *Design-Build Survey*, Zweig White & Assoc., Inc., 1998, p. 20.

§ 9.11 CASH FLOW AND COLLECTING RECEIVABLES

Design firms are well aware of the problem of collecting payment for their services. As a contractor, however, there are more dollars at risk, plus numerous lower-tier subcontractors and suppliers who look to the contractor for payment. As a result, cash flow is critical and collecting receivables in a timely manner can make or break a contractor. The 1999 CFMA Survey found that generally, a ratio of 45 days or less in accounts payable was considered adequate. However, if the ratio is lower than the days in accounts receivable, cash flow will be negatively affected.

§ 9.12 PURCHASING MATERIALS

Purchasing materials for a construction project is not quite as simple as it may seem. Prices can be affected by when items are purchased, how much is ordered and when payment is made. One must also think about where materials will be stored once they are delivered, and whether one should take delivery of all the material at once, or parcel out an order over time. Does the firm have a priority relationship with its suppliers? When a firm is starting out new in the construction business, it will not have the business volume or the relationship with suppliers to give it what it needs ahead of other experienced contractors. This points out one of the advantages of teaming with a contractor to do design-build. Another advantage of teaming is having an experienced procurement agent who buys the material and negotiates the prices. In one survey, three-fourths of contractors reported that their project managers perform the construction related purchasing function in their companies, while less than a third (29 percent) have dedicated purchasing agents.[29]

[A] Scheduling and Storing Materials

Quantity take-offs are an important part of estimating and one of the major sources of disputes between design-build partners. Once a firm feels that it has an accurate handle on the quantity of materials required, it needs to determine if it has storage facilities on the job or at another location if all the materials are shipped at the same time. Many contractors have secured storage yards and warehouse space where they can hold material until it is needed. A firm may need to rent a storage trailer on-site to keep material dry until needed. Scheduling of materials, like drywall board, is important. Often contractors hold delivery of drywall until the building shell is completed and weather-tight. Some windows may need to be taken out or their installation delayed so that the drywall can be loaded through a large opening, even on upper floors.

Prices for certain materials can rise unexpectedly during the course of a job due to market conditions and demand. A firm should consider whether it should negotiate a price early on for materials to be delivered at a later time when needed, to avoid cost escalation.

---

[29]*Id.*

[B] Discounts from Suppliers

Quantity discounts can affect cash flow. If a firm purchases the entire allotment of material at the start of the contract, it may be able to take advantage of supplier discounts, avoiding price escalation and multiple delivery and shipping costs. It is important to make sure that the contract permits the firm to bill for materials stored on or off site, though not yet incorporated into the project.

A firm needs to know when it makes direct purchases whether the supplier will require payment up front, upon delivery, or will wait until the firm gets reimbursed by the owner following the next application for payment. Price discounts for early payment can be a big factor and another profit source. Many suppliers will give a percentage discount for payments made within a few days or weeks of delivery. Some owners insist that they get the benefit of such discounts (in cost-plus work), but others do not. If a firm chooses to advance payment to the supplier on its own credit, and finance that purchase before getting reimbursed by the owner, many contractors retain the discount to offset the finance charge and risk, charging the owner the full material price. Of course if the owner wants the discount passed on to it, the owner should provide the funds to make the early payment as well.

The AIA A491 Design-Builder-Contractor Agreement (1996 edition) states in Article 8 that all cash discounts obtained on payments made by the trade contractor shall accrue to the design-builder if the trade contractor has applied for and received such payment before making payment, or if the design-builder has deposited funds with the trade contractor with which to make such payments. The AGC contracts handle this differently. AGC's 1999 version of Doc. No. 410, the Cost-Plus a Fee Design-Build contract, states in paragraph 8.3 that all discounts for prompt payment accrue to the owner, "to the extent such payments are made directly by the Owner." However, if payments are made with funds of the design-build contractor, then all cash discounts accrue to the contractor.

[C] Material Price Fluctuations

World events can have a dramatic effect on material prices. Wars, hurricanes, and other hazards can shut down or delay production of materials across the ocean which causes an impact on the availability and price of materials in any city. Flood damage repair on the east coast of the United States has been known to cause a shortage of drywall and plywood throughout the country, affecting prices. Costs can fluctuate greatly even over the period of one project. For example, 2 × 6 dimensional lumber (standard or better) cost about $400 per thousand board feet in Kansas City in 1993, but a year later the price was over $550, a jump of 37.5 percent. National sales of gypsum board rose 21 percent from January 1999 to January 2000. Locking in prices early on avoids such cost escalation.

[D] Reaction to Material Shortages or Price Increases

One of the great advantages of design-build is that one can make changes in materials without going through a complicated change order process, and without the delays that accompany that process on traditional projects. The designer knows best whether one material will work just as well as another. If a designer has a choice between two or more materials, and one is easier to assemble, has less waste, or is more readily available, the designer can simply make the change as long as it does not reduce the quality of the project he or she agreed to build. A designer may find that after it has locked in a lump sum contract, or even a cost-plus with shared savings clause, the material it planned to use costs more than an equally good material and can make a switch that will result in increased profits and no detriment to the project or the owner.

[E] Shipping Charges

Shipping charges must also be factored into material prices. Will the firm pick up the materials, will they be delivered by the supplier, or shipped direct from an out-of-state source by truck or rail? One may find that small shipments cost more than large ones, and that breaking an order up into multiple shipments will be more costly than if delivery is taken all at once. Again, this gets into whether one can store and bill for materials in advance of their use.

[F] Leftover Materials

Overages of materials, those amounts left over when the job is completed, can be a source of profit. Due to anticipated waste, contractors order more than what they need for each job. On lump sum contracts, the cost of such materials is paid from the contract price. The contractor can either return it for a credit, after paying restocking fees, or retain it and store it. Materials that are left over are usually stored by the contractor and can be used on other jobs. Floor tile, ceiling tile, metal studs, drywall, and many other materials may be suitable for use on other jobs, but at no cost since they were paid for on a prior project. Of course, if the owner requires that all such surplus, especially floor and ceiling tile, be returned to the owner, one needs to be sure to order sufficient amounts as required by the owner so that color runs match what is installed.

§ 9.13 CONSTRUCTION EQUIPMENT

What equipment does a design-build contractor need? Should a firm rent or buy that equipment? Most jobs require the basic pick-up truck to haul small tools and materials, a backhoe, trenching equipment, boom truck, and other pieces. When subcontracting most of the work, a firm can rely on the subs to provide whatever equipment they need. However, there are more contractors self-performing such things as concrete flat work, rough carpentry, even masonry work. This may require skid-steer loaders, backhoes, forklifts, scaffolding, shoring materials, and form work, among others. Portable toilets, job site trailers, and ice water are commonly supplied to the site by the prime contractor. When a firm is starting out, renting will be the only option to provide this equipment. However, as the company grows, there will be advantages to owning equipment and renting it to the jobs versus renting from outside sources.

By having both owned and rented equipment, a contractor has the option to bid and self-perform more portions of the project, or to subcontract out work.

[A] Renting Equipment

Whether to rent or to buy equipment is a question faced by most contractors, especially those starting out in the business. Industry experts say that not too many years ago, the decision to lease or buy was fairly straightforward. According to Russ Agosta, with Grant Thornton, "It was always more expensive to lease than to buy. But the big benefit of leasing was that you didn't have to make a down payment."[30] Today, however, leasing companies have developed new options and reduced interest rates, converting capital leases into operating leases and leases with options to buy. Renting is easy to do and easy to estimate. Renting has many advantages such as: (1) knowing the cost when bidding; (2) there is no need to carry insurance; and (3) one can rely on the rental company for repair or replacement if equipment breaks down on the job. The tough part is determining productivity—*i.e.*, the length of time a piece of equipment will be needed on the job. Once on the job, will the equipment be used fulltime? Are there breaks when it makes sense to return it to the rental company and check it out again later, or does it make more sense to pay for the equipment even though idle, just to have it available? These factors all play into the decision of whether to purchase or rent equipment.

Another factor in deciding whether or not to rent is the location of the project. For example, on a $30 million office facility in a Des Moines, Iowa suburb, the design-build contractor from Minneapolis decided to rent much of its equipment since it was not feasible to bring it to the job site, even though the company owned that equipment back in Minnesota. The type of equipment being rented included lifting and hoisting equipment, concrete forms (for self-performed concrete work), backhoes, dozers, mini-excavators, and skid-steer loaders.[31]

Rental rates are readily available merely by checking a local rental company, or by using R.S. Means rental rate cost data, or the AGC "Blue Book" to check national averages.[32] Means' editors contact equipment dealers throughout the country to determine current rental rates on thousands of pieces of equipment. Rental rates are listed by day, week, and month with hourly operating costs for each piece of equipment listed. Another reference book is the AED "Green Book."

Many companies obtain equipment on short-term rental, or rent with a purchase option (RPO). The percentage of equipment owned or rented varies from contractor to contractor. One large $200 million per year heavy construction con-

[30] *Contractors Weigh Benefits of Leasing or Buying Equipment*, Rent Right, Jan. 2000.

[31] *Fast Tracking for John Deere Credit*, Rent Right, February 2000, p. RR6.

[32] *See* AGC's "Rental Rate Bluebook for Construction Equipment," commonly called the "Blue Book."

tractor in Wisconsin reported, for example, that during the peak workload season, about 5 percent of its monthly equipment bill goes for short-term rentals, with 15 percent to RPO equipment.[33] Most contractors advise that a firm is better off to do a minimal investment in equipment when starting out. This is important because of the need for cash, and not to have needed money tied up in equipment. Renting equipment permits the firm to control risk, keep equipment debt off of the company's balance sheet, and improve the company's return on investment figure. It is an "off-balance sheet" item. If business drops off, a contractor who rents equipment can downsize quickly. The contractor won't have its money tied up in non-liquid assets that are not making money for the company.

Short-term rentals are used by most contractors to meet peak demands for business, whether that is seasonal or just a major project, or for specialized pieces of equipment. According to the equipment manager for a large California highway contractor, "If you're not renting, then you've probably got too much equipment. We want to keep a fleet we can support, but it's an inventory kind of game. You don't want to supply all your needs from your own fleet. We probably tend to rent more of the smaller equipment. Normally what drives our rentals is the lack of availability from our own fleet."[34]

Whether to rent or to buy is a tough decision and there are no hard formulas for this. A firm should look to see what types of equipment it is using most often, and whether business has grown in that area to the point that owning a particular piece makes sense. Many contractors feel that in the long run it costs less money to own than to rent. Some measure the breakpoint between rental and ownership at about a year to 14 months. If a firm thinks that demand for the piece will last more than a year, or it has several long-term jobs, it might want to buy. But if the firm has a 14-month job and won't have a project for that piece after that, it may be better off renting.

Renting frees up capital. It requires no cash outlay other than monthly rent payments. As a result, it makes the contractor's financial statement look better. Owning equipment can actually have an adverse impact on the contractor's banking relationship, according to the accountants. If a firm borrows money to finance a purchase, the debt shows up on the balance sheet, affecting the debt-equity ratio, which is looked at by banks in determining loan terms. Bonding companies also look to the debt-equity ratio, as well as some owners such as state highway departments. These entities do not like to see a contractor overburdened with debt to finance equipment.

[B] Renting to Buy

Leasing companies can offer very favorable payment terms, far better than a bank generally offers for financing an equipment purchase. Since the leasing

---

[33] *Equipment Rentals (Special Advertising Section),* Engineering News Record (1999).

[34] *Id.*

company owns the equipment, it can depreciate it over time and pass the savings on to the contractor in terms of a lower payment rate, and over longer periods than banks like to give. Daily or monthly rentals often turn into sales after the contractor gets to the job site and discovers that they need the equipment more than originally planned, or that the job is lasting longer than anticipated.

By renting to buy, a contractor can obtain lower initial payments over time, thereby allowing use and acquisition of equipment that would be too expensive to finance and buy outright. Some rental companies will allow a contractor to convert from a lease to a rent-to-own arrangement, applying the rent payments toward the purchase price. One vendor, for example, allows contractors to apply 100 percent of their rental payments toward the purchase price for the first few months of the lease, then crediting 80 percent for the next few months before the purchase credit drops to 50 percent thereafter.

[C] Buying Equipment

One of the biggest advantages of design-build is speed of construction. Owners want their jobs done right away. By owning equipment, a contractor can respond quickly, especially on emergency work, mobilizing equipment without delay.

When the decision is made to purchase equipment, the next question is whether to borrow or pay cash for equipment. There are several publications to help with this decision. The best known are the Rental Rate Blue Book and Contractors' Equipment Cost Guide.[35] The rates in these publications are determined from ownership costs including purchase price, resale value, depreciation, maintenance, and anticipated replacement costs.

Current cash flows are the primary source of financing for equipment purchases.[36] A spokesman for a very large general contractor said, "We pay cash for everything. I try to stay away from lease-purchase agreements. If I know the need is there for a piece of equipment, I'm better off to buy it outright. I can get a better-defined purchase price upfront."[37] In addition to the financial decision, it is nice (and impressive) to have a piece of equipment on the job with the firm name and logo on it. It shows that the firm has made a commitment and investment to being a contractor. One of the biggest impressions a firm can make is buying a job trailer with the company name and logo prominently displayed on the job site for the world to see. Many contractors park their trailers in visible locations on the site to serve as an advertisement as well as a functioning field office.

There are two risks with owned equipment that can be covered by insurance—theft and destruction. Theft of construction equipment is a very high risk,

---

[35] Rental Rate Blue Book, published by the AGC. Contractors' Equipment Cost Guide, published by Dataquest.

[36] Construction Industry Annual Financial Survey, 1999.

[37] Frank Kazenske, equipment manager for J.S. Alberici Constr. Co., Inc., Equipment Rentals, Engineering News Record (Special Advertising Section) (1999).

especially equipment that is easily transported. It is very common to see two-wheeled electric generators suspended by a job site crane after hours and on weekends to prevent theft. Be sure to coordinate any equipment purchases with the firm's insurance agent.

[D] Storing, Managing, and Maintaining Equipment

Most contractors have an equipment "yard" where they can store materials and equipment. This means having an office property large enough to store owned equipment onsite, or owning or leasing secured property on which to store such equipment. This property should be located near a highway so that the equipment can be easily moved as needed. It should be secured—*e.g.*, chain link fence, barbed wire top, padlocked—to prevent theft or vandalism. Some contractors go so far as to have remote television cameras providing 24-hour surveillance of the materials and equipment yard.

Once purchased, it is important to keep the equipment working so that the firm can bill it out to projects. Keeping equipment working is a key factor for owned equipment. Many project delays are caused by the simplest things such as a blow out on the backhoe tire, replacement of teeth on trench digging equipment, engine malfunctions, etc. The larger the fleet of owned equipment, the more maintenance responsibility falls in-house. This is another overhead expense to be considered.

Very large general contractors have full-time equipment managers on staff and own most of the needed equipment. One of the largest contractors in the United States, J.S. Alberici Construction Co. of St. Louis, Missouri, owns an equipment fleet with a replacement value of about $120 million, including 81 cranes.[38] Internal rental rates for equipment are billed by the equipment department to the various projects, with the project managers charged with the task of keeping the company's fleet busy. According to Frank Kazenske, Alberici's equipment manager, "If a project manager calls me up and tells me he's got a machine available for a week, 9 times out of 10, I'll go pick up the machine and use it on another project. If you own a machine, you can move it around easier without spending as much on transportation." Kazenske says, "I've got to move it around to keep it working," which permits the firm to bill the equipment out and avoid non-billable down time.

This raises the other issue of how to transport equipment from job to job. If a firm makes the decision to buy small tools, these can be hauled in the back of a pick-up truck. But if the firm buys a backhoe, this means that it will also need a flatbed trailer and truck capable of hauling the loaded trailer to the project site.

---

[38] *Engineering News Record,* Equipment Rentals (Special Advertising Section) (1999).

CHAPTER 10

MARKETING FOR DESIGN-BUILD FIRMS

§ 10.01 CHALLENGES FOR DESIGN PROFESSIONALS

At the start of this new millennium, there are still few contractors and even fewer design firms that have design-build experience. According to one construction and engineering recruiter, only one out of 100 contractors had design-build experience prior to the year 2000. As a result, the market is wide open. However, as the design-build market continues to increase, competition will get tougher, says Mark A. Swatek, CEO of CDM Engineers & Constructors of Denver.[1] With increased use of design-build and competition from not only contractors but other design firms, clients will ask, "Why should an owner hire a designer-led design-build team?" This is the most difficult question a designer will need to be able to answer. If a designer cannot answer this question convincingly, he or she should reconsider whether to get involved in design-build. Contractors can show a resume of years of experience managing projects of varying size, an insurance and bonding program, and experienced staff with construction science degrees and actual field experience. So how does a designer get in the door, let alone get hired, for a design-build project?

Clients may be surprised to learn that some of the largest design-build contractors in the nation started out as traditional engineering or architectural firms. Among the largest design-build contractors are names like the Bechtel Group, Inc. and Fluor Corporation (annually the first and second largest such contractors), plus companies like Jacobs Sverdrup, Raytheon Engineers & Constructors, Black & Veatch, the Austin Company, HNTB Corporation, Lockwood Greene Engineers, Inc., CH2M Hill, and Ellerbe Becket. This is proof positive that designers can lead design-build projects successfully.

§ 10.02 PUBLIC PERCEPTION

A question that each architect or engineer considering the design-build market must ask is "What do we want to be known as?" Do we want clients to hire us because our firm is an AIA-award winning design firm, doing cutting edge design, with talented technical and design staff? Or do we want to be known as a construction company, known for tight estimating skills, project management, safe job sites and quality construction? Can we really be both?

Some firms struggle with a concern over client confusion. Clients may perceive a move to design-build as a shift in firm direction, rather than just another service the firm can offer to its clients. How will peers look at the firm? If the firm now gets referrals from contractors on design opportunities, or if it joint ventures with contractors on projects, will those referrals dry up when the construction community sees the firm as a competitor? Or will the firm get even more referrals from contractors once they realize that the firm is a serious player in the design-build market, someone whom they would like to team up with on their next project?

---

[1] *Masters Raise Bridge to 21st Century*, Design-Build, Dec. 1999.

§ 10.03 WHERE TO START

Who better to start with than past clients with whom the firm already has a relationship of trust. The firm should send out an announcement to clients it has worked successfully with explaining its new "full service" design-build services. Clients have, no doubt, heard about design-build, but may not know whom to contact about this type of service. A firm should make sure clients know that their firm is the place to call for the client's needs, even if they just want to discuss the pros and cons of design-build. Clients who already trust the firm's design judgment and professional advice may be more willing to trust the firm as a contractor as well.

A firm can also approach its current clients. A design firm can explain that although other architects or engineers will not guarantee the cost or the completion date, this firm will if it is hired as a design-build contractor. And the firm will sign a contract that says that. A firm should inform the client that there is no need to solicit bids and worry about the costs and delays of re-bidding and value engineering, or the qualifications of a low bidder. As lead design-build contractor, the firm will offer to lock in a price early on and guarantee a job built within the budget. This is music to a client's ear.

Some design firms have stumbled into design-build not by plan, but due to the request of their client. The Philadelphia architectural firm of Kling-Lindquist ventured into design-build when the owner requested it. CEO Mel Slotnick says, "As it turned out, the owner preferred to deal with the architect and engineer as long as there was adequate protection and coverage for him on the construction side."[2] The general contractor was a subcontractor to the Kling firm on the $10 million project, with Kling-Lindquist holding the prime contract.

Once a firm has impressed its client with its ability to manage a construction project as lead contractor, repeat business from this client will follow. "Approximately two-thirds of our business is done with repeat clients. We do a variety of commercial, industrial, retail and institutional facilities for them," said a senior vice president of design-builder the Opus Group, "New clients are secured through normal marketing efforts including referrals."[3]

§ 10.04 WHAT IS NEEDED?

Marketing for design professional has changed in the last 20 years. Most architectural firms are small, and it was very common for the principals of the firm to handle all the marketing efforts themselves. However, architecture schools do not include marketing courses as part of the basic undergraduate degree program. So architects spent a great deal of non-billable time pursuing business, with no training on how to do this. That has changed, not only for design firms, but for contractors as well. Architects and engineers now find that the market is more competitive because design firms are today competing with contractors, construction managers,

---

[2] *Special Delivery?* Building Design & Construction, May 1998, p. 62.

[3] *Masters Raise Bridge to the 21st Century,* Design-Build, Dec. 1999.

and others for the same services. As a result, more and more firms are hiring full-time marketing professionals in-house.

[A] Marketing Staffing

A 1999 study found that nearly three out of four contractors report that they employ a full-time marketing professional.[4] The study reported that on average, a contractor's marketing department consists of four individuals, including secretarial personnel. Another study of design-build firms done in 1999 found that the average firm has one marketing director, one marketing secretary, two business development representatives and a total marketing staff of four persons.

[B] Average "Hit Rate" High for Design-Build

How big is the competition? Half of the firms surveyed had net revenues of just $3.5 million or less in 1997, increasing to $5 million or less in 1998.[5] In 1998, design-build firms surveyed submitted an average of 27 proposals, being short listed on 14 of those and actually being successful in winning the contract on 13 projects.[6] The same survey taken in 1999 found that design-build firms submitted an average of 17 proposals, being short listed on 12 of those, and winning the contract an average of 8 times.[7] Not a bad "hit rate" for any contractor, with surveys showing a near 50 percent success rate two years back-to-back. This may be reflective of the few number of design-build firms currently in the market, or the high cost of submitting proposals, which keeps the field small.

§ 10.05 SPECIALIZATION/MARKET NICHE

As design-build becomes more competitive, firms will need to develop specialties, just as they do in the traditional design market. Robert Ivy, FAIA, says that, "[a]rchitects with specialties in a particular subject matter make another group of strong candidates to lead design-build teams." He discussed the success of the firm known as IDEA, which programs, designs, and constructs buildings. The firm grew out of the experience of Cambridge Seven Associates, authorities in aquarium design. Ivy says that, "Their success in one building type, which led to a separate design-build-management enterprise, provides an example for design professionals schooled in other specialized building types, including laboratories, hospitals, sports facilities, and theaters."[8] Another good example of success in specialization are the sports design firms like HOK Sports, HNTB and Ellerbe Becket, all of whom offer design-build services to their clients. HNTB is

---

[4]CFMA Construction Industry Annual Financial Survey (1999).

[5]*Design-Build Survey,* Zweig White & Assoc., Inc., 1998 and 1999.

[6]*Design-Build Survey*, Zweig White & Assoc., Inc., 1998, p. 39.

[7]*Design-Build Survey*, Zweig White & Assoc., Inc., 1999, p. 116.

[8]Robert Ivy, FAIA, *Not to Be Ignored*, Architectural Record, Oct. 1998, p. 15.

the prime design-build contractor with joint venture-partner Turner Construction on the new Mile High Stadium, a $366 million project in Denver, Colorado, and the first NFL stadium to be constructed using design-build procurement.

§ 10.06 QUALITY OF DESIGN WHEN DESIGN-LED

One of the criticisms of design-build, especially when contractor-led, is that it produces bland architecture. "Vanilla boxes" are how much of design-build is characterized—metal buildings, acceptable for prisons or parking garages, but not for more sophisticated projects. These myths are being broken down by AIA-award winning designs and attractive projects being done by design-build. The new Denver aquarium, Ocean Journey, a $93 million project and winner of an AIA Colorado design award is an example of outstanding design coming from design-build. Designer-led, architectural and engineering firms bring something to the table that owners still want—attractive and functional design. Ray Braswell, vice president of HOK says: "Good design still sells. Architects who can produce good design, guarantee the cost and the schedule will dominate the market." Brad Buchanan, AIA, principal of a Denver architect-led design-build firm says: "There is no question that our buildings are better because of our designer-led construction method. We can subsidize extra design with the construction dollars. In addition, the people doing the design work closely with those who know how the building will go together. Then, because we are the ones designing and building, we essentially give our clients free on-site construction administration. It has had an unbelievably positive impact on the quality of the final product."[9]

§ 10.07 MARKETING TO CONTRACTORS

Contractors who think that design professionals cannot lead design-build projects are missing a great business opportunity. Owners still talk to architects first about construction projects, before talking to a contractor. This is changing slightly as contractors become more aggressive in marketing full services to owners. But with owners and design firms making the early decisions about project delivery, contractors would be smart to pursue business alliances with design firms. By pairing up with the architect/engineer firm early on, a contractor has a better chance of taking the project "off the street" and locking up either a negotiated teaming relationship with the architect/engineer, or being the prime subcontractor to the design firm if the job goes design-build. This should appeal to contractors, with the prospect of a steady source of non-competitively bid work, with the design professional as an ally and teammate rather than an adversary. These relationships can be formalized into "teaming agreements" by which each firm agrees to hire the other for design-build work that the other is leading. *See* Chapters 7 and 8, and Appendix D.

---

[9] *How to Increase Your Fees in a Tough Market*, Architectural Record, Nov. 1999, p. 63.

Teaming partners are in business to make a profit, so they should explore design and construction alternatives that will reduce cost, time, or complexity. The contractor partner, especially, should review the plans to spot omissions or ambiguities that might delay or interfere with construction or result in subcontractor claims. To give each party an incentive to work creatively, design firms should be prepared to agree on a split of savings or profit sharing in a ratio that makes the contractor feel the venture is worthwhile. When the contractor is prime, the design firm should negotiate a split of savings as well, since design innovation results in savings of time and cost as often as construction ideas.

§ 10.08 ARCHITECT/ENGINEER STRENGTH IN MARKETING ON QUALIFICATIONS

Architects and engineers are used to qualifications-based selection, contractors are not. Ask a contractor what "QBS" stands for and they might say it is one of those home-shopping channels on television. Design professionals have competed for years on the basis of assembling teams that are hired based on qualifications alone, not price. Most design firms are very comfortable in the interview process selling their team's strengths and concepts to a selection panel. For contractors, this is a whole new world. Many are used to hard-bid projects, where the only appearance they have to make before getting the contract is to attend the prebid meeting. Not surprisingly, some of the strongest opposition to design-build legislation comes from contractors who are uncomfortable competing on the basis of something other than price. This is where design firms can sell the value of teaming. Architects and engineers are skilled at the interview process, pulling together teams and preparing strong graphic and oral presentations. Architects and engineers are usually better at communication and maintaining client relationships.

Some contractors are even recruiting people with design backgrounds simply to handle client relations and marketing due to their "people" skills. According to one recruiter, design employees at construction firms must first overcome skepticism from their construction colleagues, but those who are successful at that advance quickly. Architects have experience even in college with presentations, including answering tough questions about the design, whether they are from a professor, design jury, or a client. Contractors recognize these skills and are bringing design professionals in-house to help with marketing as well as design. A design firm can market itself to help contractors prepare the proposals as well as be the design partner.

§ 10.09 MARKETING TO OWNERS

Design firms have two chances to market design-build services to owners. For some projects, there is the opportunity to function as the "bridge" architect or engineer, preparing the preliminary design, the request for proposal (RFP), and assisting the owner through the selection process and then during construction as well. The second chance is as a member or lead contractor for the design-build contract itself.

If not successful in competing for the "bridge" position, the A/E firm can team up with a contractor to pursue the prime contract. Design professionals have numerous advantages to market to owners in leading the design-build project. One of the foremost is the level of trust that owners place in the design firms.

[A] Building Off a Trust Relationship

There is a built-in level of *distrust* between owners and contractors. Whatever the historical reasons, few people can deny that the distrust exists. This results in architects, lenders, codes officials, and others inspecting the work periodically to make sure the job is being built as intended. Design professionals, on the other hand, have served as the owner's advisor, counselor, and agent on construction projects and have developed a relationship of trust with the owner. Rarely does the owner hire someone to check the designer's work. Design-build presents an opportunity for the architect/engineer firm to use that level of trust to sell construction services. "Who better to build this job for you than the one who designed it, who knows the owner's needs and who will make sure the plans are followed?" That would be a tough question to answer other than to acknowledge that a trusted design firm can fill the role of contractor best.

Another approach might be to ask if the owner has been satisfied with the traditional method of project delivery. If not, it is time for a change. "If you always do what you always did, you will always get what you always got!" is one way of saying maybe it is time to try something new. Designer-led design-build is something new that just might change the way owners do business in the future.

In addition, an architect may find that he or she is involved in decisions that the contractor does not normally participate in. Brad Buchanan, AIA, says: "The earlier the architect can get into discussions with the owner, the better the chances of being selected and paid well for a job. [As a design-builder], we have the ability to speak to our client in languages other than the language of design, so we are sometimes invited into the boardroom to help make decisions—about site selection, facilities, financing, and other front-end issues. We become the client's partner and are there when the project actually happens."[10]

[B] Two-Stage "Sequential" Design-Build Services

One of the advantages of designer-led design-build is that the owner does not have to commit to a particular form of project delivery at the outset. The owner can hire the architect or engineer to do preliminary design, programming, feasibility, planning, and budgeting work before it must commit to handing the project as design-build or more traditional design-bid-build or negotiated contracts. The design professional, most familiar with the program and owner's needs, is then in a great position to approach the owner about being the contractor for the same project. A contractor, by contrast, cannot offer this advantage to an

[10] *How to Increase Your Fees in a Tough Market*, Architectural Record, Nov. 1999, pp. 62-63.

owner, since hiring a contractor to help plan, schedule, and budget for the project does not make sense if the owner later decides to put the job out for competitive bids, thereby losing the early project partner in the process.

This delivery method permits the owner to begin the design immediately, while reaching some internal decision on the project delivery method best for the project. The owner's project manager may need time to build consensus within the company or the governmental agency and time to sell his superiors on design-build. Awarding a design contract to a firm that is also capable of being the contractor leaves that option open while the owner's management weigh the benefits of design-build. Of course, some benefits of design-build are lost if the decision (and start of construction) is delayed too long. The further along the design is done, the less benefit in terms of speed of the overall project. In a "sequential design-build" project, the process is usually handled with the design firm signing a series of contracts over time, starting with a design contract, which may include preliminary budgeting and scheduling services—then a construction contract. This may be done with the same firm (true single point responsibility). A popular variation with design firms, however, is to hold the construction contract through a subsidiary design-build construction company.

[C] Assist Owner with Contracts

Owners may need the firm's help in developing contract forms for design-build. If they have never used the process before, they may be tempted to try to use their old contract forms and merely relabel them as "design-build" contracts. This does not work however. The firm should be proactive here by offering to provide forms that it has worked with before, whether they are AIA, AGC, DBIA, or some other. The client will appreciate being steered in the right direction. The owner should be encouraged to have its own lawyer review the contracts. If the design firm is the one pushing for design-build, it might even offer to pay for a limited amount of legal fees for the owner's attorney, say $500, and include that in its costs for the project.

§ 10.10 WHAT THE OWNER IS LOOKING FOR

[A] The Design-Build Team's Reputation

A team that has worked together before will be particularly attractive, since the owner will not have to deal with a learning curve. Especially if the individuals at each firm have worked together, there is the advantage of familiarity with each other's working style, a trust and history of successful project completion. Design-build involves a great deal of trust on all sides. Owners will be more comfortable with a team that trusts each other. When meeting with clients, tell them that if you were building your own architectural office today, this is the contractor you would hire for your own work. If you have that level of trust, the client will be more comfortable hiring your team.

[B] Quality Assurance/Quality Control

Having a designated quality assurance or quality control manager will give the owner comfort that quality will not be sacrificed to pressures of time and cost. Describe how quality will be handled, by whom, and their experience and training. A certification, such as ISO 9000, is evidence of a tested QA/QC program that will impress owners.[11] Many industries are moving toward ISO certification and will appreciate what that commitment to quality means coming from their design-build team. A firm should consider letting the owner participate in the quality control process to the extent necessary to establish firsthand that project quality is important to the team.

[C] Strong Design and Technical Proposal

AIA's Design-Build PIA Chair Paul Dougherty, AIA says that, "Clients are looking for the 'wow factor' of experience—architecture that will impress."[12] With so much emphasis from each team on their ability to design, manage and build, plus similar qualifications, it may be the design itself that sets the teams apart. If a proposal uses innovative and economical, structural, mechanical and electrical systems, money should be spent on architectural features that make a strong visual presentation. Money on these features will be seen by the client and by the public, and will enhance the proposal. As Hok's Ray Boswell says, "Good design still sells." According to one consultant, "[t]he areas that usually draw public attention are the exterior skin or finish, public lobbies, and public restrooms . . . Design-build projects are similar to a beauty contest and many more times than not, a more sophisticated proposal will win the day."[13]

[D] Personal Attention to Their Project

Owners want to know that the firm thinks their project is important, that they will get personal attention. One small architectural firm seizes upon that aspect when competing against larger firms for the same project. Dan Scott, AIA says one advantage he has in interviewing is to ask the selection committee to think about who is present at the interview versus who will manage their project. "Are the people who interviewed for the larger firms the same people who you'll be working with? Because I am!" This kind of personal relationship is important to develop the level of trust that is needed in design-build.

---

[11]ISO 9000 is a standard for quality management systems developed by the International Organization for Standardization, a Geneva, Switzerland-based federation of the national standards organizations of 91 countries. Refer to Section 20.12 of this book.

[12]*Marked Increase in Design-Build Services*, AIArchitect, p. 19, Jan. 2000.

[13]William Carr, P.E., quoted in *Design-Build Proposals Favor Experience, Creative Concepts,* Engineering Times (Dec. 1999).

§ 10.11 RESPONDING TO THE RFP

[A] Study the Owner's Request for Proposals

The team will need to read the Request for Proposals (or for Qualifications) thoroughly and determine whether this is a job worth pursuing. Architects are skilled at responding to RFPs and RFQs and routinely respond to government solicitations found in Dodge Reports, Commerce Business Daily and other sources, preparing impressive proposals with resumes, photos and references. This is not so different with design-build, but will take some creative "team building" with the addition of construction to the owner's needs from the firm. Start the contacts and research to put together the team capable of gaining the most points available in each category of qualifications. Some experienced design-build firms say that, "If you do not think you can get the most points available, in every category, then don't bother submitting. If you do not think you are the best, the client won't either."

[B] Take the Lead in Preparing the Proposal

The design firm should offer to take the lead on the team in assembling the proposal documents. Most likely, the design firm will have graphics capabilities in-house that a contractor will not, and will think more visually about what proposal will make an impact, be eye-catching, and distinguish the proposal from the rest. A Project Director should be assigned to head up the proposal phase. A key contact person from the contractor and any other team members should be designated. A schedule of what needs to be done should be developed in order to meet the proposal deadlines. The schedule should be posted in the "war room" where the proposal is being put together so everyone can stay on track.

The key project selection factors will need to be analyzed. Is this a two-phase selection process? How will the teams be short-listed, on what factors? What are the factors used in evaluating technical proposals, and are there some weighted criteria that determine who is selected? Is the owner's project criteria realistic in terms of time and budget? Who can be added to the team to pursue this; are they interested; are they already on someone else's team; do they see this team as the best group assembled? The design firm may have to do some selling up front to a contractor partner to convince the one identified as the best that the firm is their best chance of getting this job. Then, each must decide to pass or to play.

[C] Organizational Chart

An organizational chart should be developed that shows what specialty players are needed for a particular job—*e.g.*, parking consultant, security or detention specialist, local associates with ties to owner. Are there minority business enterprises (MBE) and women business enterprises (WBE) goals required? Even if not, will that give you an advantage with this particular client? Even if not a public project, many private corporations stress the use of MBE and WBE consultants. This could separate one firm from the pack even if not required by the terms of the RFP. Positions should be filled in, matching capabilities with who is available and interested. If there are several consultants who fit the project needs, the firm may have to pre-qualify, screen, and interview several before deciding who will be on the team. The contractor partner should be involved in this process, and the design firm should tap into their connections and recommendations. Ask candid questions about workload, manpower, etc., so you can determine if this team can actually meet the owner's requirements if you are selected. It is a good idea to have a letter of intent or teaming agreement with each key player so they do not jump teams to a more attractive competitor or, worse yet, show up at the interview on more than one team.

[D] Team Meetings During the Proposal Stage

The RFP should be distributed to the entire team once it is assembled. A kick-off meeting should be held with the team to discuss how to best approach the submission package. Who will do what, and on what schedule? The Project Director should run the meeting and get commitments from everyone on their respective tasks. A key date schedule should be issued to the group. Then, written minutes and reminders of important milestone dates should be distributed to keep everyone on schedule. The group should be asked about their concerns over the RFP and what questions they want answered at the pre-bid meeting. One spokesperson should be designated for the group at the pre-bid meeting, most likely the Project Director. If anyone on the team has a personal or business contact with the client or the selection team, consider using that relationship by having them attend the pre-bid meeting and including them in the interview process, if there is one.

[E] Hold a Design Charette

When the submission involves some form of design competition, many firms hold a design charette, bringing the key team members together for a full day or more to sketch out solutions. The contractor-partners can give important input on constructability, availability and cost of materials, and structural systems that expedite construction. Even specialty subcontractors can contribute to the charette with ideas on cost and timesaving ways to design the project. There should be a review of the budget constraints, time, site, weather, and other factors such as need to demolish an existing facility or work around an ongoing part of the client's operations. It has been often repeated that, "many hands make light work" and that is certainly true at this stage of the proposal process. Each element of the RFP should be brainstormed, with emphasis on the points allocated by the owner to every weighted factor. Again, the proposal should be approached as if the team can get the maximum number of points in every category. Golfers know that the 250 yard drive off the tee box counts the same as that two-foot putt. It is that one point that separates the winners from the losers. If a firm knocks them dead with a great design, but blows some other area of the proposal criteria, it loses critical points. Avoid missing the two-foot putts in the proposal.

Especially when selection criteria are based on a not-to-exceed budget, one can figure that the prices are all going to be close to each other. As a result, the difference between proposals that will win the day will likely be in the technical submission and qualifications proposals. Though it is trendy to say "think outside the box," the team needs to focus on what will impress the owner. As Paul Dougherty, AIA says, the "wow-factor." One should avoid preconceived designs. Many instructors in architectural school tell their students that their first idea is great, "[n]ow throw it away and think of another one!" The immediate solution is probably one that the competition will think of as well. What will take the owner and the competition by surprise? What will blow them away? Cost saved in one area like mechanical or structural could support spending more on architectural elements that will catch the client's eye. Good design still sells.

[F] Continual Contact, Checks and Balances

The contractor partner, or the design firm's in-house estimators, will need to participate in conceptual estimating during the charette, keeping the "idea" people firmly planted in the real world of cost and budget constraints. After the charette, the team should continue to gather cost data as the design is refined. Continue value engineering at each stage and prepare quantity takeoffs. This is one of the most frequent sources of litigation between design-build partners, where the fixed price is based on an erroneous quantity take off prepared during the proposal stage, resulting in cost overruns and a loss on the job. It is important that the team players who are most qualified to do these takeoffs are the ones doing them. The design team may not be the best ones to do this. Then, as the design takes shape, continue to refine the estimate, preparing detailed estimates for each component with sufficient contingencies to account for designs that are not complete or not fully priced out.

§ 10.12 COMMUNICATION WITHIN THE TEAM

Designer and contractor communication is important. Some teams network their computers with cameras mounted on the designer's and the estimator's desks, so that those two can stay in constant visual and voice communication, sharing design information and pricing through a dedicated Web site or other computer network system, even if they are not at the same physical location. Others move the team together during this entire process, with the contractor's staff setting up a home base in the architect's office as the drawings and estimates are being prepared. An architect can learn a lot from watching how the contractor puts together a cost proposal.

[A] Electronic and Computer Communication

Technology is advancing so rapidly that to even comment on what is available today risks being out of date when this book goes to print. However, some sophisticated design-build contractors assemble their teams based on each company's experience with using 3D CAD technology on past construction projects and working together. For example, on a 1997 California pharmaceutical project, the design firm of Flad & Associates and the contractor, Hathaway/Dinwiddie Construction Co., put together a team that included plumbing, HVAC, and electrical subcontractors with the goal that each team member would explore ways to use existing software to integrate CAD technology with cost and scheduling software.[14] The architect developed a 3D architectural model which then served as the base for all estimating and scheduling functions. The CAD estimate link developed by Ketiv and Timberline supported design-cost integration. A Schedule Simulator by Jacobus Technology provided design and schedule integration. Figure 10-1 shows how the system was integrated.

To perform design-cost integration, the team linked estimating items and work packages from Timberline's Precision Estimating database with graphical objects created with AutoCAD R13 and Ketiv's Archt Architectural Drawing software package, so that quantity takeoffs were calculated from within the CAD environment after all work packages were attached to CAD design elements. The Ketiv system provided electronic verification that all objects in the CAD model had been included in the estimate.[15]

The architect members of the team reported spending more time orchestrating the collaborative design process and less time performing detailed design. However, they worked closely with specialty contractors who performed detailed design for their disciplines. The team reported that the process was challenging, but with more experience, the time to set up the system will diminish and the benefits will increase. Hathaway reported that with quantities being automatically calculated and inserted into the estimate, they reduced estimating time by 25 percent.

A similar system was used by The Neenan Company of Denver, Colorado on a $31 million office building in lower downtown Denver known as the 1899 Wynkoop Building. The project architect and the structural engineers collaborated on a 3D design, building a prototype of the project in the computer. The different trade subcontractors transferred their drawings electronically to Neenan, using software by Bently Systems, maker of MicroStation, instead of two-dimensional hard copy drawings. If mistakes are made in coordination of piping and ductwork, the computer catches those instantly rather than waiting for those problems to first surface in the field.[16] Another benefit of this system is that pieces of steel can be fabricated on the computer and a 3D shop drawing can be sent to the fabricator, eliminating the need for shop drawings by combining de-

[14]*Industrial Case Study of Electronic Design, Cost and Schedule Integration*, 1998 DBIA/AIA Convention, Chicago, Illinois (Oct. 1998).

[15]For information on the systems used by this team, *see* these Web sites: Autodesk-AutoCAD R14: <http://www.autodesk.com>; Jacobus Technology-Schedule Simulator: <http://www.jacobus.com>; Ketiv Technologies-Archt: <http://www.ketiv.com>; Microsoft-Project: <http://www.microsoft.com>; and Timberline-Precision Estimating: <http://www.timberline.com>.

[16]*Getting the Big Picture*, Colorado Construction, June 1999, p. 19.

FIGURE 10-1. Schematic Diagram of Software Used to Perform Design, Cost, and Schedule Integration—Flad & Associates (designer) and Hathaway/Dinwiddie Construction Co. (contractor).

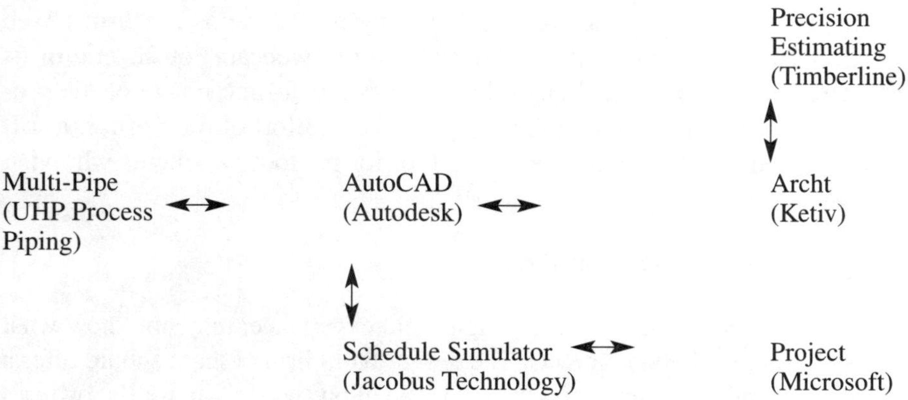

sign with fabrication drawings in one system. The project team used a dedicated Web site for all communications and an on-site "web cam" to monitor progress, with updated pictures every two minutes.

[B] Do Not Miss the Deadline

As the deadline nears for submission, the team should finalize pricing, check and re-check everything and be sure to deliver the proposal exactly where it is required, when required by the terms of the RFP. It is amazing how many bid protests and court cases there are because a delivery service took the submittal package to the wrong room or the wrong building, or left it at the front gate for delivery, or missed the deadline by a few minutes. This should never happen. If necessary, take the proposal there personally, leaving sufficient time for delays in flights, traffic, etc. to make the deadline. Delivery services on occasion lose packages or confuse addresses.

§ 10.13 SPICE UP THE OFFER WITH SOMETHING THE OWNER DID NOT ASK FOR

All of the teams competing will be able to meet the owner's requirements, so how does one team distinguish itself from the others. One could offer some warranty beyond what the RFP required, or a different type of warranty, such as an operations/maintenance warranty. What about technological advances specific to the owner's industry?

[A] A Project Web Site

This should be offered as part of the base bid even if not requested, so that the owner can track progress, communicate quickly, access daily and field reports, even see the progress through photos posted regularly. Owners like to watch their projects go up and a web-cam gives the owner a view from their desk top. If the budget will support it, an on-site camera should be provided that supplies "real-time" photo images of the project. One of the most successful design-build firms in the country is the Neenan Company in Colorado. The firm's Web site at <http://www.neenan.com> has hot links to its "webcam" on several of its projects, offering the clients and any other visitors a real-time image of the project, so that owners can check on progress from the comfort of their office, as often as they like. It is also a good marketing tool for prospective clients who visit the firm's Web site.

[B] Contract Forms and Legal Review

First-time owners with no prior design-build experience, may not know what form to use for design-build contracts. The design-build firm or team should offer a form of contract and consider giving the owner an allowance to pay for the owner's legal fees in reviewing the form. If there is a choice of governing laws or place for arbitration, mediation or litigation, make the form favor the owner with the owner's state's laws, the owner's city, etc. Most contractors and design firms use forms that favor them with their state law and city even if they have no relation to the project. This is a way in which a firm can set itself apart from the competition.

§ 10.14 KNOW THE COMPETITION

Making the short list is great. But who else is going after the job? The team should ask itself: "Can we really beat these guys?" When selection is based on factors other than price, the team needs to know whom it is up against. Is there a past business relationship between members of the competitor's design-build team and the client? Does their team have some specialized experience with the project type, particularly with design-build that the firm cannot match? Does anyone have a geographic advantage? While David sometimes can beat Goliath, you have to evaluate the odds before putting the investment into a proposal, especially one with costly design components and low—or no—stipend. If your competition has something you do not, go get it. Architects are great at assembling teams of design consultants with specialized expertise in a particular project type. As a design-builder you can build that type of team and include contractors and specialty trade subcontractors with unique experience that set your team apart from the rest.

§ 10.15 KNOW THE OWNER EVEN BETTER

John H. Patton, senior vice president of The Haskell Company says that when design-build selection involves more than price, one must be more creative.

[17]Comments at 1998 Joint DBIA/AIA Convention, Chicago, Illinois on Design-Build 101.

[18]J. Sanders, *Winning Presentations*, The Handbook for Marketing Professional Services, 1994, Chapter 28.

"Be thorough, out think your competitors, and know your client."[17] Janet Sanders of Sverdrup Civil, Inc. claims a 70 percent "hit rate" on competitive marketing of projects totaling over $23 billion. She recommends "[a]dvance intelligence gathering, what the military calls G2," as the most important step in creating a winning presentation.[18] Most contractors are used to preparing bids from the comfort of their offices, talking only to their staff, subcontractors, and teammates. On one architectural project, there were three firms being considered for a Catholic school addition. One firm's principal was a parish member, with children at the school; another had done lots of schools. The third firm, realizing it lacked the inside track, visited the school several times prior to the formal presentations, talked to the school principal, spent time studying how the school operated, solicited input from the teachers and staff, plus did research from materials recommended by the staff. When the selection time came, all three firms had strong designs and qualifications—but the selection team felt only one firm understood them, and took the time to listen, then incorporate their ideas into the proposal. The choice was clear and the contract went to the third firm without hesitation.

As Sverdrup's Janet Sanders says, "Clients select firms based on their specific needs, goals, preferences, and constraints. If you don't know those things, you don't know what you must present to convince them to select your firm. Telling about *you* doesn't sell them on selecting you; selling *your firm's solutions* to meet their own needs, goals, preferences, and constraints does." She says that based on the firm's "G2" on the client, one must identify at least two (in case the client rejects one), and no more than five, strategic selling points, which should then be the major topic areas in the firm's presentation.

§ 10.16 SHARING THE SAVINGS

There is a wide difference on how this handled. Most contractors want to participate in any savings under the guaranteed maximum price (GMP) cost. Bob Degenhardt, CEO of Ellerbe Becket, says that his firm prefers not to participate in the savings since it creates a level of distrust with the owner. "Are we making decisions that are in the best interest of the owner, or to increase the savings split? Incentives should not be there to cheapen the project. Fair compensation is all we are after." This is, of course, a business decision, but it can be a marketing one as well—whether to share in the savings. Some contractors tell the owner they do not want the savings, but ask that it be placed into an interest-bearing account for one year after completion. When there are warranty problems with the project, the cost to repair or correct the work comes from the savings. At the end of the year, the owner gets to keep it all. This is a clever way to have the owner pay from budgeted construction dollars to fix defects in the project, without tapping into the contractor's profit margin. Alternatively, the money in that account can be split between owner and contractor after the year is up.

§ 10.17 OPENING THE BOOKS TO THE OWNER

Many design-build contractors find that owners can overcome the distrust that exists about contractors by opening their books during the project so that the owner sees all their costs. One design-build contractor calls this "getting naked," with nothing hidden from the client. Another major designer-led design-build firm operates on an "open book" basis, and clients are welcome to look at the books at any time to see where their money has been spent, and just what the firm is making as profit. This leads to more trust with the owner and may set one team above the other.

§ 10.18 A "MEET THE PROJECT STAFF" VIDEO

One large contractor pursuing a project for Disney prepared a marketing videotape in which all of the project staff who would be assigned to the project gave brief personal interviews. All the teams have key resumes in their proposals, but the owner only got to "put a face with" the staff of one bidder. Each staff member started with, "Hi. I'm John Doe, and if we are hired, I'll be your project manager . . . My experience includes [blank] years of [whatever]. I am looking forward to working with you on this project." The extra cost put into marketing may be just the thing that separates one firm from the stack of resumes from equally qualified firms.

§ 10.19 THE ORAL PRESENTATION OR INTERVIEW

Design-build partners need to carefully plan the presentation as a team. Janet Sanders, Ph.D., vice president of business planning and marketing for Sverdrup Civil, Inc. says that the purpose of the presentation is to sell the decision-makers on why they should select the firm, not to tell them about the firm or that the firm can do the project.[19] She advises that telling how the firm approaches programming, design, project management, or construction usually doesn't sell because everyone in the industry does about the same thing. A firm is not giving the client a reason to select it if it is saying the same thing as everyone else. She says that what does sell is talking about what the firm will do to meet the client's specific needs. She says that the firm should emphasize what it is going to do for the client, rather than what it does in general.

[A] Plan, Rehearse, and Execute with Enthusiasm

One should re-examine the RFP, to make sure that the team hits all the high points of importance to the client and the selection committee. The entire presentation must be planned out, including who will be the lead spokesman, who else will speak, and in what order. Janet Sanders says that many firms have a standard

[19] *Id.*

formula to presentations, which includes: an executive opening by stating how happy the team is to present its credentials, followed by the firm's history and a string of comparable projects; an introduction of the team; a detailed description of the project approach and the various people who will handle each aspect, with their experience; with the executive asking for the award of the contract before closing. She says that if this formula sounds familiar, then the client has heard it all before too. If a firm gives the same presentation as the competition, the firm is not going to get the project. Sanders recommends that each firm tailor its presentation to its team, to the competition, the client, the job, and the selection panel.

Sanders says that in the first 30 seconds of each presenter's talk, "he or she must establish rapport with the listeners and win their attention." She says that standard opening lines like "Good morning, we're pleased to be here . . . " do not do the trick. Opening statements must refer to what the client cares about. One should say something to show the client that the firm understands and values what it does. Her rule of thumb is that 60 percent of the content of the presentation should be about the client, with 40 percent of the content about the firm.

One engineering consultant stresses not relying on notes in the interview. People should rehearse enough that they can do their presentation from memory. "In the interview, if you can answer the owner's inquiries without having to refer to notes, it will indicate your familiarity with the project. This also enhances your chances of success."[20] One should anticipate questions from the owner and make sure someone is there to respond. A format needs to be chosen for the presentation, whether powerpoint presentation, slides, video, boards or a combination. Then one must practice and critique each presenter. Architects are used to making presentations in settings like this, but many contractors are not. They are used to turning in a bid on paper and never having to show their face prior to award. The team needs to be prepared to change speakers depending on how the practice session goes.

It is important for the presenter to be enthusiastic about what is being sold. Finish with "an energy surge," says Janet Sanders. Lines like: "We want this project. Thank you. Do you have any questions?" is the standard closer, and it is a weak one. The close should be brief, with a memorable review of the strategic selling points and an energetic finish that carries emotional impact. One design-build team, for example, closed its presentation by saying: "We guarantee that we will finish this project on time and within the budget. You have our bond to guarantee this, but you also have our word." This connotes a message of personal pride, reputation, and trust that stays with the selection committee.

[B] Hire a Marketing Consultant to Coach and Critique

One should consider hiring a marketing consultant to help with the oral presentation. The investment can really pay off. For example, on the Scott M. Math-

[20]William Carr, P.E., quoted in Design-Build Proposals Favor Experience, Creative Concepts, *Engineering Times* (Dec. 1999).

eson Courts Complex project in Salt Lake City, the team of Big D Construction, MHTN Architects and HOK Architects hired a professional consultant to coach and critique their presentation before the interview. This investment paid off and the team was awarded the $68 million project.

§ 10.20 MARKETING DRAWBACKS

Going in for that first interview, or even the first of many, a design firm with little or no prior construction experience is at a disadvantage going head-to-head with a contractor. While the contractor may show dozens of projects to its credit, just like the one being considered, the design firm likely has built none itself. This suggests two approaches. First, a design firm should team up with a contractor, by joint venture, so that it can show prior construction experience on the "design-build team." Second, the design firm should subcontract the construction to general contractors until it has built up a portfolio of work to compete on its own merits. When the owner asks how many projects the firm *has built* like this one, the firm representative might ask how many projects the competing contractors *have designed.*

§ 10.21 MARKETING ON THE WORLD WIDE WEB

The internet has added a new vehicle for marketing. A 1999 survey of architect/engineer firms found that 71 percent of design firms have a Web site and only 29 percent do not.[21] Twelve percent planned to add a Web site in the year 2000. These statistics will increase each year as the Internet becomes a primary mode of communication. Millions of people surf the Internet looking for what they need. Some may even look for design-build contractors this way. If a firm is not on the Web, it misses those leads. Posting and maintaining a Web site is fairly inexpensive, but allows clients to check out a firm, see its projects, your awards, even real-time pictures of other projects under construction. If 7 out of 10 design firms had a Web site in 1999, how many more have one today?

[21] 1999 DFMAR Survey, Design Firm Management & Administration Report, Dec. 1999.

CHAPTER 11

CHOOSING A BUSINESS FORM

§ 11.01 WHY A SEPARATE CORPORATION?

In order to protect the assets of the design firm, many architects and engineers getting into design-build set up a new corporation to hold their design-build contracts. Under this method, the design firm is operated as a separate legal entity and often subcontracts with the new corporation to provide design services. The new corporation contracts with the owner and then subcontracts with trade contractors, just as a general contractor does. This approach minimizes the risk of an uninsured loss, bid mistake, or other situations that might otherwise put a firm out of business.

Many of the largest design firms have gone this route when setting up their design-build practice. HNTB Corporation, for example, formed HNTB Design-Build, Inc., a subsidiary that handles the company's design-build work on such projects as Denver's new $366 million Mile High Stadium and the $712 million Alameda Corridor railroad project in Los Angeles. Once the decision is made to set up a new legal entity to hold design-build contracts, a number of other decisions must be made, starting with which legal entity to choose.

[A] Which One to Choose?

There are many forms of business today, more than ever before. When planning a new business, many considerations come into play, such as ownership structure, licensing requirements, tax laws, and liability. Historically, the majority of architectural firms has been sole proprietors: individual architects with no partners, just "hanging out their shingle" and practicing architecture. But as tax laws and professional liability have become more complex, many design professionals have sought the protection of other forms of business organization.

[B] Business Failure Rates

In 1997, there were 166,740 new businesses formed, according to Dun & Bradstreet.[1] However, business failures increased by nearly 16 percent in that same year, with over 83,000 businesses failing in 1997.[2] Statistics show that in 1997, just over 1,500 new architectural or engineering firms opened for business nationally, with 374 firms failing—even in such a strong economy, an increase of 15 percent over 1996. These statistics show that new businesses must plan for not only success, but for failure, since half of all new businesses fail within the first four years of operation. Those businesses failing in 1997 left an average unpaid liability of $448,970.[3]

---

[1] Dun & Bradstreet, *Business Starts Record, 1996-1997*. In a June 8, 1998 press release, Dun & Bradstreet reported nearly 799,000 new for-profit corporations were formed in 1997; however, D&B only reports a new business "start" as one added to their files because they are trying to sell products or services, or seeking commercial credit. D&B recognizes that many people start new businesses and never move beyond the initial filing of incorporation papers.

[2] Dun & Bradstreet, *Business Failures—Industry Analysis, 1997 Year-to-Date*.

[3] Dun & Bradstreet Corp., *Business Failure Record*, Final 1996 & Preliminary 1997 (1998).

Architect/engineer firms are not the only ones who have seen increases in failures. In 1997, the construction industry had over 4,200 general contractors and over 6,300 subcontractors fail, increases of 10 percent and 11 percent respectively. If this many design and construction firms failed during a period of record unemployment and increased profits by the majority of firms, there is clearly still a great deal of financial risk that must be anticipated in construction businesses, including design-build. This chapter looks at the most common forms of business organizations used by design professionals and design-build firms and the legal benefits and concerns with each one.

§ 11.02 SOLE PROPRIETORSHIP

This is the most popular among architects because it is the easiest way to do business for a small firm. Under this form of business, there are no complicated filings with the state, no partnership agreements, no bylaws—one can just print up some business cards and start to work. The liability associated with design-build, however, makes this a very risky choice, and one almost never seen outside of residential construction. With a sole proprietorship, the personal assets of the firm owner are at risk for a business failure and for any liability not covered by insurance.

§ 11.03 PARTNERSHIP

"Partner" is a term used very loosely in business. Many design professionals, even lawyers, refer to their co-owners as "partners," though very few professional firms actually practice as legal partnerships. We should all be more careful about throwing the term "partner" around, since clients and the public may be misled into thinking that the firm is something that it is not. The term "principal" is even more loosely used by design firms and has no real legal meaning.

[A] What Is a Partnership?

A partnership, also called a "general partnership," is defined in the Uniform Partnership Act as, "an association of two or more persons to carry on as co-owners a business for profit."[4] There does not have to be any written agreement between partners to make a legal partnership, but one is strongly recommended. There is no formal filing with the licensing board in many states. An architect and a contractor who decide to go after a design-build project jointly and to split the profits 50/50 may have become partners without even realizing what they have done, or the liability they have assumed. Very few design firms practice as partnerships anymore, less than 10 percent. Many of the large architect/engineer firms that practiced for decades as partnerships have abandoned that structure and are now incorporating. Why do so many firms avoid partnership?

---

[4] Uniform Partnership Act § 101(6) (1997 Act). Note that there is also a 1914 and a 1994 version of the Uniform Partnership Act. The 1997 Act had been adopted in nearly half the states as of January 1, 1999. The 1914 version has been adopted by most states.

FIGURE 11-1. Joint and Several Liability Among Partners.

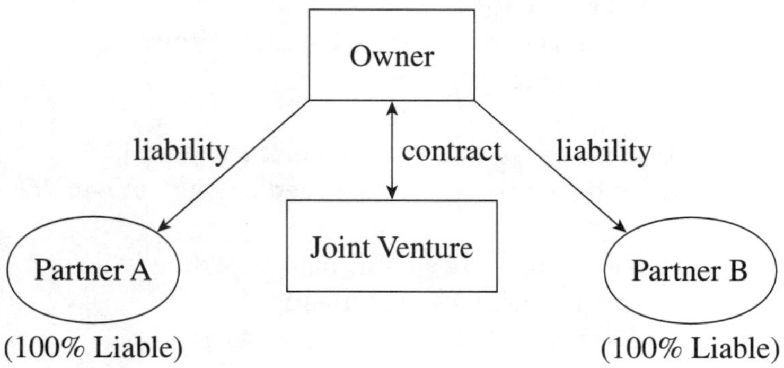

[B] Joint and Several Liability

Partners have what is called "joint and several liability" for actions of the partners. This means that if Partner A commits a negligent act, the plaintiff can sue Partners A and B jointly, or the plaintiff can sue *either* partner, since they are both separately liable for the acts of each other. Each partner is 100 percent liable for the actions of his or her co-partners.[5] (*See* Figure 11-1). As a result of this expanded personal liability, partnership agreements often contain "indemnity" clauses which require any partner to reimburse his or her co-partners for any damages they may be compelled to pay when they were not at fault. Professional liability insurance for a partnership will cover the negligent acts of any partner, but problems come up when there is a claim that is in excess of policy coverage limits, or excluded from coverage, such as construction errors by a contractor partner. In addition to professional liability, partners are jointly liable to their creditors, such as landlords, project owners, subcontractors, suppliers, employees and others who might be owed money for a partnership debt. If a design firm fails to maintain insurance and has limited assets, an owner could collect 100 percent of a design liability from a contractor-partner who is still solvent. The reverse is also true of the design firm's liability for errors of the contractor-partner in a joint venture.

[C] Partnership Agreements

Due to the failure rate of new businesses, it is advisable to have a written partnership agreement with your partners. In addition to naming the partners and naming the business itself, the agreement should address:

[5] Uniform Partnership Act § 306(a) (1997 version).

1. initial capital contributions of each partner;
2. subsequent contributions;
3. percent of profit and loss attributed to each partner;
4. naming a "managing" partner;
5. term or "life" of the partnership;
6. establishing grounds for termination of the partnership;
7. the basis for taking in new partners and buying out withdrawing partners;
8. governing law and methods for dispute resolution;
9. death of partners; and, most importantly,
10. valuing the ownership interest of each partner.

There is generally no requirement that partnership agreements be written. In fact, partnership agreements may be expressed orally or implied from the acts and conduct of the partners.[6] This is one reason that a Teaming Agreement should be clear on this point and should disclaim any intent to form a partnership absent the parties' express intent to do so.[7] If there is no written agreement, the rights of the parties are governed by the applicable state partnership laws.

[D] Dissolution of the Partnership

The biggest disputes arise when one partner wants out, or when the partners decide to close shop and there is no agreement on how to pay the withdrawing partner(s) for their share of the business. Also, someone has to "wind up" the affairs of the business. Even though a partnership is dissolved, it has a continued existence until its assets are collected and distributed to creditors and the "winding-up" of partnership affairs is completed.

Under state statutes, a partnership is dissolved automatically with:

1. a partner giving notice of intent to withdraw;
2. the death of a partner;
3. the express will of at least half the remaining partners;
4. the completion of the undertaking such as completion of a project by a joint venture;
5. a terminating event as stated in the partnership agreement;
6. an event that makes it unlawful to continue the business (such as failure to comply with licensing laws for partnerships); or
7. a court order.[8]

---

[6] Uniform Partnership Act § 101(7) (1997 version); Warren v. Warren, 784 S.W.2d 247, 251 (Mo. Ct. App. W.D. 1989).

[7] Section 202(a) of the Uniform Partnership Act (1997 version) states that "the association of two or more persons to carry on as co-owners a business for profit forms a partnership, whether or not the persons intend to form a partnership," unless their association is formed under some other statute.

[8] Uniform Partnership Act § 801 (1997 version).

[E] Taxes

A partnership does not itself pay any taxes. All income flows through the business to the individual partners, who declare income and pay taxes on their personal returns. The partnership files an informational return and each partner gets a Schedule K-1, which tells him or her what the taxable share of profit or loss was for the year. This is one of the biggest advantages of a partnership—the avoidance of double taxation.

§ 11.04 JOINT VENTURES

Some design-build projects are handled by joint ventures between a design firm and a contractor. A true "joint venture" is a partnership set up for a limited scope and duration. It is treated as a partnership for all legal purposes, and the joint venturers can be sued jointly or severally, just like partners. This is why it can be very risky to enter into joint ventures for projects without full analysis of the legal and insurance implications. Professional liability insurance policies sometimes exclude coverage for work done as a joint venture, requiring special endorsements or a separate policy. A written Joint Venture Agreement is highly recommended, covering many of the same issues of a typical Partnership Agreement. The AIA publishes a Joint Venture Agreement, Doc. No. C801 (1993 edition), which covers some of the important issues. It is not written for design-build joint ventures, however, and does not cover all the key issues. The Federation Internationale des Ingenieurs Conseils (FIDIC) also publishes a Joint Venture (Consortium) Agreement and companion guidebook, but these are intended only for joint ventures between consulting engineers.[9]

When two firms talk about setting up a "joint venture" for a project, they may mean a partnership, or they may mean a new corporation, owned jointly by the two firms. A common practice in design-build is to set up a brand new company to hold the prime contract, and then to subcontract work back to each of the two firms. *See* Figure 4-4 for an example. In addition to insurance coverage, the joint venturers need to agree on management of the venture, how billings and accounting will be handled, distribution of work, etc. When setting up a joint venture for minority set-aside work, the SBA, city or other governmental agency certifying the minority status will usually require the minority member to own at least 51 percent of the business and, often, be in control of the management of the business.[10] When setting up an architect-led design-build firm, the design partner needs over 50 percent interest to be in control. This may even be a requirement under state licensing laws.

Joint ventures will put the architect-partner at risk for construction defects, for which the architect may be uninsured. Indemnity agreements may help the ar-

[9] FIDIC Joint Venture (Consortium) Agreement, Doc. No. W-0068 (1996 edition); Guide to Joint Venture and Sub-consultancy Agreement, Doc. No. W-0069-G (1996 edition).

[10] *See, e.g.*, 49 C.F.R. § 23.45.

chitect-partner develop the comfort level needed to enter into the joint venture with a contractor and vice versa. A "limitation of liability" clause may also reduce the architect-partner's exposure. The idea of both clauses is to protect the design firm from construction liabilities not covered by insurance.

In an Illinois case, the $6.4 million hospital design-build contract stated that the architect was employed by the contractor. The owner sued both the architect and the contractor for defects in the work. The court held that the owner could not sue the architect-subcontractor directly and had to pursue its claims for design errors against the design-build contractor, with whom it had a written contract. Had the architect been a joint venture partner, however, it would not have been dismissed from the suit and could have had liability for design errors and construction errors.[11]

This points out the need to choose partners carefully. Whether forming a joint venture or subcontracting, the design firm is at considerable risk in design-build for construction errors. Architects and engineers should avoid projects where the owner wants to select the partner. "Shotgun" marriages are never a good idea in business.

§ 11.05 LIMITED PARTNERSHIPS

A limited partnership is a hybrid of a partnership and a corporation. Limited partnerships are common for investor groups for the development of real estate, but rare for professional firms. There must be two or more persons, one of which is the "general partner," with the others "limited partners." The general partner is in charge of running the business and is personally liable to creditors, with unlimited liability. The limited partners (investors) contribute capital and share in the profits, but cannot participate in the management or operation of the business. A limited partner is not liable to third parties for the obligations of the business. This is why this form of business is so popular among investors, who want to finance a business and share profits as a "silent partner," but have no liability. The immunity from liability can be waived if a limited partner takes part in control of the business, thereby becoming a "general partner."

This form of business is never used by design-build firms since most partners want to be active partners, not just silent investors.

§ 11.06 LIMITED LIABILITY PARTNERSHIPS

This is a fairly new business form, created by statute, known as "LLPs," as a way for professionals to retain all the benefits of a partnership for tax purposes, without the joint and several liability of general partners. Because it is still a new animal, LLPs are not recognized in all states, and it is unclear how other states would treat such a business, whether as a general partnership or an LLP.

So what is the benefit of forming an LLP? Section 358.150.2 of Missouri's version of the Uniform Partnership Act states:

---

[11] Kishwaukee Community Health Servs. Ctr. v. Hospital Bldg. & Equip. Co., 638 F. Supp. 1492 (N.D. Ill. 1986).

> Subject to subsection 3 of this section, *no partner* in a registered limited liability partnership *shall be liable* or accountable, directly or indirectly, *including by way of indemnification,* contribution, assessment or otherwise, *for any debts, obligations and liabilities of,* or chargeable to, *the partnership* or each other, *whether in tort, contract or otherwise,* which are incurred, created or assumed by such partnership while the partnership is a registered limited liability partnership.

This creates virtual immunity for the individual partners, much like shareholders of a corporation, who are not liable for the debts or obligations of the corporation. The limited exposure of partners includes a co-partner's negligence, wrongful acts, omissions, misconduct or malpractice, or that of any person under the partner's direct supervision and control, or the partner's liability for any taxes. The partners of an LLP do not escape professional liability for their own negligent acts, but they do not have liability for the negligence of their partners, nor are they liable to creditors, owners, subcontractors, suppliers, etc. for general business debts, unlike partners of a general partnership. With this type of express immunity, we may see more design-build firms choosing to operate as LLPs in coming years.

§ 11.07 CORPORATIONS

Corporations continue to be the form of choice for many design professionals nationally, with most larger architectural firms over five employees electing to incorporate their business.[12] Most design-build firms incorporate. According to a 1999 survey, 93 percent of design-build firms incorporate, with 43 percent choosing to be C-corporations, 36 percent become S-corporations, and 14 percent are professional corporations (PC).[13]

Corporations are separate legal entities, with their own liability both for business obligations and taxes. The main advantage of forming a corporation is that, as a general rule, the officers, directors, and shareholders are not personally liable for the acts or debts of the business. By shielding personal assets of the owners from creditors, only the corporate assets are at risk. If all the corporate formalities are not followed, however, or the corporation is an undercapitalized "sham," used to defraud creditors, the courts may "pierce the corporate veil" and allow creditors to sue the shareholders personally. This requires keeping corporate minutes, adopting resolutions, obeying company bylaws, electing officers annually, and filing all state reports and paying corporate fees.

[A] Professional Liability Not Avoided

While most other industries choose to incorporate in order to shield the business owners from personal liability, design professionals cannot avoid personal liability for their professional conduct merely by incorporating. Most

[12] AIA 1997 Firm Survey, p. 14.

[13] *1999 Design-Build Survey*, Zweig White & Assoc., Inc., p. 33.

state licensing statutes provide that individual architects or engineers are not relieved of responsibility and liability for services performed as employees of a corporation.[14] This is where insurance becomes important in order to provide protection for the personal assets of a design professional.

[B] Tax Laws and S-Corporations

One of the perceived drawbacks of practicing as a corporation is the possibility of double taxation of corporate income. When fees are paid by a client to the corporation, to the extent not otherwise deductible, the corporation reports the fees as income, which is taxed at the corporate level at corporate tax rates. If the firm pays a dividend to its shareholders, based on revenues, the shareholder then pays taxes on the dividend as personal income. The same dollar of fees is therefore taxed twice, first at the corporate level, and then at the shareholder level. One way to avoid this is to practice as an "S-corporation." An "S" or "sub S" corporation refers to a subsection of the Internal Revenue Code. Corporations that qualify for treatment under this section are taxed like a partnership, with all income flowing through the business to the shareholders, taxed only at the personal level. Of those contractors and subcontractors responding to a 1999 survey, 51 percent operated as S-corporations.[15] There are some restrictions on "sub S" corporations, but for most firms, these do not restrict the planned practice.

Another way to avoid the double taxation is to merely pay out the remaining corporate revenue at year end as a bonus to the key employee-owners, which is then deductible as salaries. This is a common practice and generally permitted, but tax advice should be sought on the size of bonuses paid out. Also, the design-build firm's bonding company may want to see a certain amount of retained earnings held in reserves, subject to corporate income taxes.

[C] Professional Corporations

In some states, design professionals may incorporate only as "professional corporations," also called "professional associations" in some states. The limitation on professional corporations is that the directors, officers, and shareholders must all be licensed professionals. This severely restricts potential ownership in a design-build firm and, as a result, is hardly ever used for design-build.

In some states, when filing the Articles of Incorporation as a professional corporation, a certificate must be filed from the regulating board of the profession involved stating that each of the incorporators is duly licensed to practice that profession, and that the proposed corporate name has been approved. For all other business purposes and taxation, a professional corporation is treated like a general business corporation.

---

[14] *See, e.g.,* K.S.A. § 74-7036; R.S. Mo. § 327.401; Okla. Titl. 59 § 46.9.

[15] Construction Industry Annual Financial Survey, Construction Financial Management Association (1999).

Another drawback to the professional corporation law is that employees can have personal liability for the negligent appointing, supervising, or participating in the activity in question with another employee. Beyond this "supervisory" liability, however, the statutes go on to clarify that the personal liability of a shareholder of a professional corporation in his or her capacity as such shall be no greater in any respect than that of a shareholder of a corporation organized under general and business corporation law.

§ 11.08 LIMITED LIABILITY COMPANIES[16]

The limited liability company (LLC) combines the protection from personal liability offered by a corporation and the tax benefits of a partnership. This form of business is one of the latest developments in the search to find the optimum form of legal entity through which business should be conducted. Not all state licensing boards have yet permitted design professionals to practice as LLCs and, as a result, only about 1 percent of architectural firms are organized this way.[17] However many design-build firms are formed as LLCs to take advantage of the favorable tax treatment and limited personal liability of the project owner.

An LLC is a separate legal entity, separate and distinct from its owners who are called "members."[18] This protects the business owner from personal liability just as if the owner held stock in a corporation, but allows the owner to receive profits from the corporation in the form of distributions just like a partnership. This means that the business owner has the benefits of the limited liability of a corporation without "double taxation" at the corporate and shareholder levels.

A corporation or a general partnership may be an LLC member. This membership flexibility is one of the elements that distinguishes an LLC from, and is a greater benefit over, an S-corporation. In order to form an LLC, Articles of Organization must be filed with the Secretary of State.[19] Management of the LLC is handled through drafting an "operating agreement" among the members. The operating agreement sets forth the rights, duties, and obligations of the members and managers, if any, and provides for the location of the principal office and for formal meetings of the managers and members. Such an agreement takes the place of bylaws, buy/sell agreements, and other corporate documents typically used between corporations and their shareholders. Usually, the unanimous consent of an LLC's members is required to initially adopt an operating agreement.

Though taxed like a partnership, the members of an LLC do not have personal liability for debts of the business, whether under contract, tort or otherwise, and there is no joint and several liability as there is in a partnership.[20] (*See* Figure 11.2.) Since LLCs are not yet authorized to practice architecture or engineering

---

[16] Thanks to Randal L. Schultz, Esq. for his assistance in writing this subsection.

[17] AIA 1997 Firm Survey, p. 14.

[18] Uniform Limited Liability Company Act § 201 (1995).

[19] *Id.* at § 202.

[20] *Id.* at § 303.

FIGURE 11-2. No Joint and Several Liability with LLC.

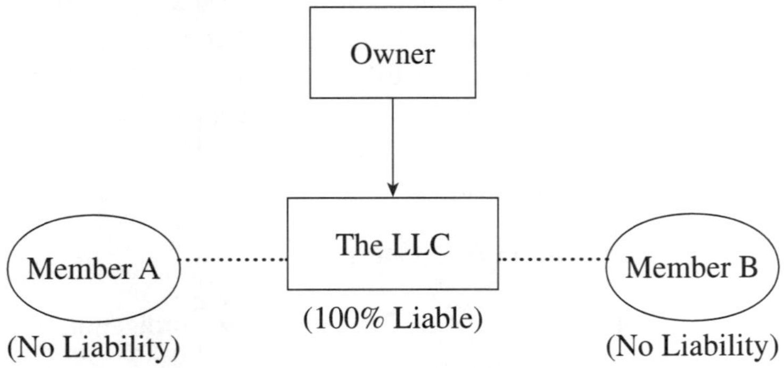

in every state, a design-build team must be concerned about performing work as a limited liability company in states that do not have a limited liability company law, or whose licensing boards do not recognize LLCs.

The big advantage of the LLC is the flow through of income to the members without being taxed at the corporate level. Although the tax characterization is much like that of an S-corporation, an LLC can have corporate shareholders, unlike an S-corporation. As a result, the LLC is a much preferred vehicle for structuring joint ventures where corporate owners are involved. Also, the LLC can have an unlimited number of owners whereas the S-corporation is limited to a set number of shareholders. This element alone allows a larger number of firm owners than is available through an S-corporation.

§ 11.09 OWNERSHIP RESTRICTIONS

Many states require that corporations, partnerships, or limited liability companies that offer architectural or engineering services obtain a certificate of authority from the state licensing board. These certificates are occassionally overlooked by corporate lawyers whose practices do not regularly include representing design firms. Failure to get this certificate, when required, can have drastic consequences.

[A] Certificates of Authority

One example of a certificate of authority is the National Council of Architectural Registration Boards (NCARB's) Model Regulations. The regulations state that:

> Any corporation (which term as used herein shall include any professional corporation), any limited liability company and any partnership (which term as used herein shall include any registered limited liability partnership),

whether organized under the laws of this or any other jurisdiction, may not offer to engage in or engage in the practice of architecture in [name of state] until such corporation, limited liability company or partnership has obtained a certificate of authorization issued by the Board.[21]

[B] State Variations

The NCARB model has not been adopted in every state and, as a result, there are many variations on who can be owners, and who must be owners, in a firm offering design services to the public. State licensing board regulations in some states limit who can have ownership in a firm that offers design services and require proof of ownership before issuing a certificate of authority. The percentages of ownership vary greatly from state to state. In Kansas, for example, the law requires that LLCs practicing any regulated profession can only be owned by members who are licensed in that profession.[22] This means that 100 percent of the owners must be licensed design professionals. This restriction virtually eliminates the use of an LLC by an architect and contractor who want to form a new firm to handle design-build work, since the contractor will often be a 50 percent owner and rarely will be a licensed design professional. It also prevents non-licensed spouses from owning stock in a design firm. Even NCARB's official commentary on its model licensing laws states that requirements that 100 percent of the owners of a firm be licensed in a particular state is not recommended, noting that, "Expense and unnecessary processing to achieve that goal is out of proportion to any useful purpose it serves."[23] NCARB instead takes the official position that two-thirds of the managers of the firm (*i.e.*, the board of directors) should be licensed as architects or engineers, since the board controls the firm's operations, not the shareholders.

At one end of the spectrum are states like Colorado, Louisiana and Texas that do not license corporations to practice architecture. At the other end of the spectrum, Ohio law requires that more than 50 percent of the shareholders and directors be licensed professionals *in Ohio*, and that they own more than 50 percent of the stock of the company.[24] In many other states, it is a requirement that two-thirds of the shareholders be licensed professionals.[25] States like Nevada require that those two-thirds of the owners are not only licensed, but licensed in Nevada.[26] In New York, only architects may own architectural firms.[27] It is therefore important when setting up a design-build firm to determine which states the firm is going to work in, and research what ownership restrictions must be met. It

[21] Rule 100.901(A), *Legislative Guidelines and Model Law*, NCARB, July 1999.

[22] K.S.A. § 17-7668.

[23] *Legislative Guidelines and Model Law*, NCARB, July 1999.

[24] Oh. Rev. Code Ann. § 4733.16.

[25] Ark. Code § 17-14-303; Iowa Code § 544 A.21.

[26] Nev. Rev. Stat. § 623.349. This was a reduction made in 1997 to the state's prior requirement that 100% of the directors, shareholders, officers, and partners be licensed in Nevada.

[27] N.Y. Educ. Law § 7307.

may be that one model of ownership and management will fit the requirements of most of the states you plan to work in, but that a separate entity will be needed to qualify for a certificate of authority in one or two particularly restrictive states.

[C] The National Council of Architectural Registration Boards (NCARB)

The NCARB is an organization that represents the interests of the various state licensing boards. NCARB publishes model licensing laws and regulations to guide state boards in making their own laws. The organization has published a model law that is directly aimed at preventing the practice of architecture by construction companies who have only one architect on staff, by requiring that the controlling body of the corporation consist of licensed design professionals. NCARB's 1999 *Legislative Guidelines and Model Law* recommends that two-thirds of the partners (if a partnership), two-thirds of the managers (if an LLC), or two-thirds of the directors (if a corporation) are registered as architects or engineers, with one-third registered as architects. NCARB also recommends that the person in charge of the practice of architecture be a partner, manager, or director of the firm and licensed in the state in question. In states that adopt the NCARB model, this will prevent design-build companies that practice architecture in-house from getting a corporate certificate of authority unless two-thirds of the directors are licensed professionals. In the commentary to the model rule, NCARB states:

> For example, it has become common for an architect associated with a corporate firm to take a contract in a restrictive state in his/her individual name, thus technically complying with the law although the work is being done by an out-of-state corporate entity. Under this practice, *a construction company employing a single architect is able to evade the restrictions* . . . Strict enforcement to prevent such practices in order to ban the company not in control of design professionals proves impossible as a practical matter, because such enforcement would result in wholesale barring of responsible corporate architectural firms in many states. The [proposed rule] in contrast, could be used successfully to bar entry to corporate or limited liability company firms not controlled by professionals while not forbidding practice by the vast majority of legitimate corporate or limited liability company firms who should be able to meet the requirements.

Under the NCARB model, a design-build firm could still be owned 50 percent by architects and 50 percent by contractors, but the board of directors would have to be controlled by architects and engineers, with one-third of them licensed as architects. Check the requirements in the state where the project is located to see if a corporate certificate is required and, if so, how that will affect the ownership structure of the design-build firm. In some restrictive states, designer-led design-build will be the only way a firm would offer design and construction services to its clients.

CHAPTER 12

CONTRACTOR AND PROFESSIONAL LICENSING LAWS

§ 12.01 LICENSING ISSUES AFFECTING DESIGN-BUILD

One of the most important legal issues related to design-build is licensing. There is no "design-builder" license issued by any state. But mixing design and construction together in one contract requires thinking about what licenses are required for each of those services. There are essentially three licenses that might be needed to engage in design-build: (1) a contractor's license; (2) an occupational license; and (3) a Certificate of Authority to practice architecture or engineering. This chapter discusses all three of these and the legal ramifications of not having a proper license when one is required.

[A] Contractors May Not Have the Necessary Licenses

Whether or not a design-build contract is legal often depends on the licensing law of the particular state in which the project is located. Architects and engineers are required to be licensed in all 50 states, as well as the Virgin Islands, Puerto Rico, Guam, and the District of Columbia.[1] Where the architectural or engineering firm is a corporation, most states require the corporation to obtain formal certificates of authorization to offer or render design services. However, most general contractors are not licensed or authorized by the state licensing boards to render architectural or engineering services. Design-build contractors frequently enter into the business of designing and building structures without considering whether the state licensing statutes apply to them. In offering this dual service to property owners, contractors often overlook the fact that they are contracting with the owner *to design* the building. By the same token, design firms need to be aware that contractor licensing laws may apply to them, depending on their role in design-build. *See* §12.03, below.

In a typical design-build scenario, the design-builder (contractor) contracts with the owner to provide both design (architectural, engineering or both) services, as well as the construction work. *See* Figure 1-2. Although the standard design-build contracts published by the American Institute of Architects (AIA), Design-Build Institute of America (DBIA), and Associated General Contractors of America (AGC) state that the design-builder will contract with a licensed professional to actually *perform* the architectural services, the architect is nonetheless a subcontractor of the design-build contractor. The contractor is still the "prime contractor" as far as the owner is concerned, providing both design and construction services to the owner. If there is a design error or omission, the owner can bring an action directly against the design-build contractor for breach of contract to recover damages caused by the acts of the architect/engineer (subcontractor). The owner looks to the prime design-build contractor as the designer, regardless of whether the design work is done in-house or subcontracted out.

---

[1] *Summary of Architectural Licensing Laws*, AIA Press (1996).

[B] Why Design Professionals Should Care

A design firm that has the needed authorizations or licenses in the project state, might be inclined to ignore the issue of licensing as being only a contractor problem. But what if the prime contract is declared illegal for lack of a needed certificate of authority? Or what if the court orders that the contractor pay back to the owner all moneys paid for design services since the contractor was not properly authorized? The design firm should have two concerns. First, if the prime contractor cannot collect from the owner, what are the chances of the design firm collecting from the contractor? In most cases, if the contractor cannot get payment from the owner, it cannot pay the subcontractors, including the design firms it has hired. Second, the licensing board in the state may investigate who assisted the contractor in the unauthorized practice of architecture or engineering. If the design firm is found to have enabled an unlicensed company to render design services, the licensing board may take action against the design firm. So, design professionals need to understand the licensing risks and inquire about whether the proper licenses are in place and, if not, help to obtain them prior to signing any contracts for a design-build project or even making a proposal.

§ 12.02 PERSONAL ARCHITECT LICENSES

The discussion of licensing and design-build starts at the earliest stages—that is, obtaining a personal license as an architect or engineer. State licensing laws require that architects and engineers serve several years in an internship after graduation before being eligible to sit for the licensing examination. With the growth of design-build, graduates may find that their first job out of school is for a design-build firm or a contractor. Be sure to check with the licensing board in the state where you plan to take the licensing examination, since some states do not give full credit for years spent working for a contractor.

For example, Nevada licensing law requires that interns obtain their experience under the supervision of an architect practicing as a principal. That state's licensing laws disqualify architects who are employed by firms engaged in design and construction of buildings, whether by the same or affiliated organizations. Further, training received in firms engaged in design and construction is not considered to be training in the office of a registered architect who is practicing as a principal.[2] In a 1981 Georgia court case, the state licensing board only gave an architect two years of credit for his three years of experience working for a design-build firm. The architect appealed the ruling but lost. The court explained that the requirement that an architect working for a design-build firm must obtain some full-time experience working for an independent architect was not unreasonable, stating:

> The job of an architect is to ensure that his plans are followed precisely, irrespective of the additional cost to the contractor. In many respects, the architect is

[2] Nev. Stat. § 623.495.4.

> seen as an antagonist to the contractor, as the contractor is seeking the maximum profit, while the architect is seeking the best financial product possible. Individuals working in the setting of a 'design-build' firm experience a constant conflict of interests not normally present in the setting of an independent architect.[3]

What a shock it would be to an architect to find that after spending three years preparing to take the licensing examination he or she is still a year short of fulfilling the requirement, since all the experience was gained at a design-build firm or at a construction company.

§ 12.03 CONTRACTOR LICENSES

Thirty out of fifty states require general contractors to hold a contractor's license in their state in order to contract for work there. This usually requires taking an examination in that state, though in some states a license merely requires payment of a license fee or filing required papers with the state. The process to obtain a license can be quite difficult in some states. In 1973, in Florida, 2,149 general contractors took the state licensing exam, and 2,149 contractors failed. After protest from the contractors who failed, the licensing board reversed its decision and graded on a curve so that 88 percent were allowed to pass and granted licenses.[4] Some states require that the person who takes and passes the examination be a full-time employee, and prevent "borrowing" a licensed employee or a license. Not having the license can prevent filing a lien for work done, void the contract or, worse yet, result in criminal charges including fines and jail time.

[A] State Licensing Statutes—50-State Survey

The following is a list of states that license contractors:

Alabama—Contractors must obtain licenses before undertaking public or private projects costing $50,000 or more. (Ala. Code. § 34-8-1).
Alaska—Construction contractors are required to be registered. (Alaska Stat. § 08.18.011).
Arizona—Contractor's license is required before submitting bids or performing work. (Az. Stat. §§ 32-1101, et seq.).
Arkansas—Contractor's license required in order to bid on and perform public or private projects of $20,000 or more. (Ark. Stat. §§ 17-25-101, et seq.).
California—Contractor's license required of general contractors. (Cal. Bus. & Prof. Code §§ 7025, et seq.).
Colorado—No state license required.
Connecticut—No state license required.

---

[3] Wise v. State Bd. for Qualification & Registration of Architects, 274 S.E.2d 544, 546 (Ga. 1981).

[4] J. Sweet, Legal Aspects of Architecture, Engineering and the Construction Process, 4th Edition, p. 164.

Delaware—Contractor's license required of all contractors doing business in the state. (Del. Stat. 30 § 2502(a)).

District of Columbia—No license required.

Florida—Contractor's license required by examination. (Fla. Stat. §§ 489.101, et seq.).

Georgia—No state license required.

Hawaii—Contractor's license is required to bid on all work, public or private. (Hawaii Stat. §§ 444-1, et seq.).

Idaho—Contractor's license required of contractors on public works over $5,000. (Idaho Code §§ 54-1901, et seq.).

Illinois—No state license required.

Indiana—No state license required.

Iowa—Contractors must obtain a license from the state. (Iowa Code §§ 91C.1, et seq.).

Kansas—No state license required.

Kentucky—No state license required.

Louisiana—Contractor's license required to bid on most public and private work in excess of $50,000. (La. Civ. Code §§ 37:2150, et seq.).

Maine—No state license required.

Maryland—Contractor's license required of all contractors, public and private. (Md. Bus. Reg. §§ 601, et seq.).

Massachusetts—No state license required, but construction supervisors must be licensed. (Mass. Stat. Ch. 143 § 94(I)).

Michigan—Contractor's license required only on residential work. (Mich. Stat. §§ 18:425 (2401), et seq.).

Minnesota—Contractor's license required only on residential work. (Minn. Stat. § 326.83).

Mississippi—Contractor's license required on public projects over $50,000 and private projects over $100,000. (Miss. Code Ann. §§ 31-3-1, et seq.).

Missouri—No state license required.

Montana—Contractor's license is required. (Mt. Stat. §§ 39-9-101, et seq.).

Nebraska—No state license required, but non-resident contractors must register with the Tax Commissioner and must register each contract. (Neb. Stat. § 77-3102).

Nevada—Contractor's license required on public or private work. (Nev. Stat. § 624.230).

New Hampshire—No state license required.

New Jersey—No state license required.

New Mexico—Contractor's license required for public and private work. (N. Mex. Stat. § 60-13-12).

New York—No state license required.

North Carolina—Contractor's license is required for public or private projects of $30,000 or more. (N. Car. Stat. § 87-1).

North Dakota—Contractor's license is required for projects in excess of $2,000. (N. Dak. Stat. § 43-07-02).

Ohio—No state license required.
Oklahoma—No state license required.
Oregon—Contractor's license required for private or public work. (Ore. Stat. §§ 701.005, et seq.)
Pennsylvania—No state license required.
Rhode Island—Contractor's license required on residential work in excess of $500. (R.I. Stat. §§ 5-65-1, et seq.).
South Carolina—Contractor's license required. (S. Car. Stat. § 40-11-5, et seq.).
South Dakota—Contractor's license required only to bid on highway work or street contracts. (S. Dak. Stat. § 10-47B-69).
Tennessee—Contractor's license required on public or private projects over $25,000. (Tenn. Code Ann. § 62-6-102 and § 62-6-103).
Texas—No state license required.
Utah—Contractor's license required for public and private work. Licenses issued in several types, including: (1) General Engineering; (2) General Building; (3) Specialty; and (4) Residential and Small Commercial Contractor. (Utah Stat. §§ 58-55-301).
Vermont—No state license required.
Virginia—Contractor's license required and issued in three classes: (1) Class A, any job over $70,000, or over $500,000 for all jobs within any 12-month period; (2) Class B, jobs less than $70,000 in a single contract; and (3) Class C, jobs over $1,000 but less than $7,500 in a single contract. (Va. Stat. 54.1 § 1100).
Washington—Contractors must register as a contractor with Department of Labor and Industries. (Wash. Stat. §§ 18.27.005, et seq.).
West Virginia—Contractor's license required for all work. (W.Va. Stat. § 21-11-6).
Wisconsin—No state license required.
Wyoming—No state license required.

[B] Local Ordinances May Require License

Even if not listed above, a contractor's license may be required by the municipality in which the project is located. A firm must check local ordinances to see if a contractor's license is needed. For example, Colorado does not require contractors to be licensed, but Denver local ordinance Section 49-596 states:

> It shall be unlawful for any person to engage in the business of general contracting and construction work except as provided in and authorized by this division and without first having obtained a license from the director of excise and licenses so to do.

§ 12.04 CHECK LOCAL OCCUPATIONAL LICENSE REQUIREMENTS

Generic "occupational licensing" is also required by some municipalities. This has nothing to do with being a general contractor, or having a professional license, but is merely a way for cities to raise revenue from businesses. A firm needs to check local ordinances to see if an occupational license is needed from the city in which the firm is located.

§ 12.05 PROFESSIONAL LICENSES AND CERTIFICATES OF AUTHORITY

The next "license" is extremely important, and often overlooked. In nearly every state, corporations that practice architecture or engineering, or "render" those services, must obtain authorization from the state licensing board.[5] Illinois adopted the first architectural licensing law in 1897, followed by Arkansas and California in 1901, and New Jersey in 1902. The pace was slow, however, with Massachusetts not adopting a licensing law until 1941, Kansas in 1949, and Vermont and Wyoming not until 1951.[6] Today, every state licenses architects and engineers. In many of these states corporations, partnerships, or limited liability companies that render design services must obtain a Certificate of Authority to practice architecture or engineering. Failure to obtain such authorization can prohibit the firm from collecting fees for work done in that state. These laws can be very strict, causing forfeitures of hundreds of thousands of dollars, merely because someone forgot to get the proper paperwork from the state board prior to signing the contract.

[A] Advisory Opinions from State Boards

Though most state licensing boards have not yet dealt with design-build by statute or regulations, some have issued advisory notices giving their interpretation of the requirements in their state. For example, in June 1993, the Pennsylvania State Architects Licensure Board wrote to state license holders advising that "design-build is not legal in Pennsylvania." The board advised that a non-architect cannot offer architectural services and it recommended the formation of a joint venture between the contractor and architect as the way to overcome licensing concerns. Pennsylvania's state board for engineers maintained that design-build could only be offered by a contractor with an in-house engineer, or by a joint venture.

In a 1993 notice to license holders, the Executive Director of the New Jersey Board of Architects gave some advice on design-build. The Board stated that preparation of construction documents "clearly falls within the definition of architectural services as defined in the law. These services may only be advertised or offered by an individual licensed by the Board of Architects as part of a sole proprietorship, partnership, professional service corporation or general business corporation holding a Certificate of Authorization." The Board continued stating that:

---

[5] *See generally Summary of Architectural Licensing Laws*, AIA Press (1996) for an overview of the requirements in each state and U.S. territory.

[6] T. Wrenn, *Our Institute Founders: The AIA's First Steps Toward Leadership*, AIArchitect, July 1999, p. 23.

> Much attention has been given to the term "design-build." In the literal sense, because only architects may advertise architectural services, the concept of design-build is illegal in New Jersey. To protect the health, safety and welfare of the public, Board regulations require that an architect exercise independent, professional judgment that may not be clouded by a financial relationship with the contractor. The only permissible relationship where both architect and contractor provide coordinated services to the public, therefore, is one where the architect has a separate, independent contract with the client.

The New Jersey Board of Architects currently approves of design-build if the prime contract is held by the architect or a corporation that has a Certificate of Authorization to practice architecture. A proposed regulation under consideration by the Board states:

> Nothing herein shall prohibit an architect from entering into a contract with the owner, and the architect may act as the architect, the contractor, the owner or the developer, or any combination of those roles in the same design-build contract.

Projects in New Jersey may have to be designer-led as the only safe way to comply with state law. In a June 27, 1983 memorandum, the New York Department of Education rendered an opinion that the design-build contracts published by AGC might involve contractors in the illegal practice of architecture and engineering. The opinion stated that such a contract "would involve the contractor in what could possibly be viewed as the illegal practice of architecture or engineering since he is undertaking to offer professional services even though he does so through a licensee." The opinion went on to say that since the contractor receives total payment from the owner and then pays professionals he employs, "it would involve a licensee in illegal fee splitting since under [New York law], he is not allowed to share any fees for his services with unlicensed individuals, including the contractor." New York has been a battleground over licensing issues for design-build. *See* §12.15 [B], below.

[B] Board Rule or Regulation on Design-Build

Some other state boards have dealt with design-build either by statute or regulation to clarify what is required for design-build. In August 1991, the National Council of Architectural Registration Boards (NCARB) proposed a model law to be adopted by the states clarifying when a contractor can offer design-build services without violating state licensing laws. NCARB recommended that nothing in the state licensing law be construed to prevent:

> A partnership or corporation from offering a combination of (1) services involved in the practice of architecture and (2) construction services provided that:
>
> (a) a registered architect or person otherwise permitted [by law] to offer architectural services participates substantially in all material aspects of the offering,

> (b) there is written disclosure at the time of the offering that a registered architect is engaged by and contractually responsible to such partnership or corporation,
> (c) such partnership or corporation agrees that the registered architect will have direct supervision of the work and that such architect's services will not be terminated without the consent of the person engaging the partnership or corporation and,
> (d) the rendering of architectural services by such registered architect will conform to the provisions of the Chapter and the rules adopted hereunder.

However, this model law is not included in the 1999 version of NCARB's Model Law and Model Regulations. This proactive approach should be encouraged so that the construction industry does not have to wait for the state legislature to enact a statute on design-build—a process that can take years and hit many bumps along the way. Nebraska, South Dakota, and Illinois are three states that have taken the early steps of enacting licensing statutes or board rules like those proposed by NCARB to address design-build and to clarify when a license is required.[7] Illinois Board Rule 1150.85 reads like this:

> **a) Design-Build.**
> 1) The design-build project delivery process is a method whereby an entity signs a single contract to provide a combination of architectural and construction services.
> 2) The design-build entity will not be required to register as a professional design firm pursuant to Section 21 of the Act only if the services in the design-build project delivery process are provided by the entity in accordance with the following:
>
> > A) An architect licensed or a professional design firm registered in Illinois independently contracts with the entity and participates substantially in all material aspects of the offering and providing of architectural services relating to any bid process, contract negotiations, design, consultation, development, preparation and coordination of construction documents, and verification of adherence to construction documents and completion.
> > B) At the time of offering services, a written disclosure shall be given to the client by the entity identifying the architect who will be engaged by and is contractually responsible to the entity offering design-build project services.
> > C) The entity agrees that the architect will have direct supervision of the architectural work and the architect's services will not be terminated on the project without the immediate replacement by another architect mutually agreed to by the client and the entity.
> > D) A design-build entity shall not offer to provide or provide architectural services, unless the design-build entity is an Illinois licensed architect or professional design firm. Offering to provide architectural services shall include, but shall not necessarily be limited to, any tender

[7] Code 1150.85; Neb. Board Rule 7.2; Neb. Rev. Stat. § 81-3449; S. D. Stat. § 20:30:22:04; 69 Ill. Adm. Code §1150.85.

> of architectural services either independently or in combination with construction services by any sign, card, advertisement or other device which might indicate to the public that the entity is entitled to provide architectural services.

Without any case law or specific statutory guidance, some contractors have taken the steps necessary to obtain a Certificate of Authority to practice architecture or engineering. While this move is usually taken related to the company's design-build work, there is nothing that would prevent a contractor who obtained such a certificate from going after pure design contracts as well. Is this what the design professionals want? All the contractors becoming licensed architectural and engineering firms? Board rules like the one in Illinois are needed to help give some structure to a simple issue that has caused a disproportionate amount of litigation in design-build.

[C] Apply Early for Any Certificates or Licenses

If a contractor's license is required, the test might only be given once per month, and applications may need to be received by a certain date prior to the exam. On the architect/engineer side, the licensing boards around the country vary in the time it takes to process an application for a corporate Certificate of Authority—ranging from several days to several months. Some applications can be approved by board staff on the spot, while other states require approval by the licensing board, which meets only once each quarter. During the time the application is pending, the corporation should not enter into any contracts for professional services, since the contracts may not be enforceable. Court cases have held that if the party to the contract is not properly licensed or authorized *at the time the contract is entered into*, the contract is illegal on that date and it cannot be made legal by later obtaining the needed license. Therefore, if a firm has a project in a state that requires corporate certificates or licenses, it is important for it to check into that issue months before the contract will be signed in order to allow adequate time for the certificate to be issued.

[D] Ownership and Management Restrictions

As discussed in §11.09, corporate planning for design-build requires checking into the licensing statutes in the states in which the company plans to do business. Requirements to obtain a corporate Certificate of Authority to practice architecture or engineering vary from state-to-state. Some of the variations are:

1. No corporate license required.
2. One employee must be a licensed architect in that state.
3. At least one director, partner, officer, shareholder, manager, member or principal must be registered as an architect in that state.[8]
4. One-third of the owners or directors must be licensed in that state.

[8] Kan. Stat. Ann. § 74-7036; Okla. Stat. Title 59 § 46.9.

5. More than half of the shareholders and directors must be licensed professionals in that state and they must own more than 50 percent of the stock of the company.[9]
6. Two-thirds of the shareholders must be licensed professionals.[10]
7. Two-thirds of the directors must be licensed professionals.[11]

It may be that the design-build firm, if made up of a contractor and an architect, cannot meet these requirements in every state. This may dictate who holds the prime contract, or may require setting up more than one new corporation, with different ownership structures depending on the state in which the project is located.

§ 12.06 BIDDING ON ONE'S OWN DESIGNS

In some states, it is illegal for an architect or engineer to act as both designer and contractor for public works. The law prohibits the firm that designed the project from bidding on the construction work. The intent of such laws is to prevent one bidder from having a competitive advantage over other bidders due to the level of familiarity with the design. For example, Oklahoma, licensing statute Okla. St. Titl. 59, section 46.27 states that:

> It shall be unlawful for an architect, at any time, to bid for a contract for the reparation, alteration or erection of a building or other structure for which he has prepared the plans and specifications.

Federal procurement regulations contain a similar provision that states:

> No contract for the construction of a project shall be awarded to the firm that designed the project or its subsidiaries or affiliates, except with the approval of the head of the agency or authorized representative.[12]

Does this merely apply when the architect prepares plans which are then bid on by other contractors, where the architect might have a competitive advantage? Or does it bar an architect-led design-build team from both preparing a design and then submitting a design-build fixed price to a client? In absence of a court ruling in that state, an architect may need to request an opinion from the state licensing board as to whether this prevents design-build teams being selected by competitive bidding, where the price is based on the team's own designs. More than likely, these types of laws would prevent a bridging architect (*see* Chapter 6) from being part of a team whose bids are based on the designs of that firm. The state licensing board and federal agencies should clarify this by regulation given the apparent conflict with the design-build method of project delivery.

---

[9] Oh. Rev. Code Ann. § 4733.16.

[10] Ark. Code § 17-14-303; Iowa Code § 544A.21.

[11] 225 I.L.C.S. 305/21 (Illinois).

[12] 48 C.F.R. § 36.209; 48 C.F.R. § 36.606(c).

§ 12.07 DESIGNERS BARRED FROM SUPPLYING MATERIALS TO JOBS THEY DESIGN

Another wrinkle in the public sector is laws that prevent an architect or engineer from supplying materials to the job. Louisiana Statute Section 38:2190, for example, prohibits architects and engineers from owning a "substantial financial interest," either directly or indirectly (*e.g.*, through a spouse), in any corporation or other organization that supplies materials for construction of a public project where the architect/engineer has performed design services on that same project. The intent of the law is to prevent an architect or engineer from profiting by specifying a material that only they can provide, where the choice of materials is motivated by financial gain rather than what is best for the public. Owning a few shares of stock in a publicly traded company does not disqualify a firm, since the phrase "substantial financial interest" excludes any publicly traded stocks on a major stock exchange. Obviously, such laws could be interpreted to prohibit design firms from leading a design-build team, where the materials are supplied by either their own firm or by a subsidiary or joint venture.

§ 12.08 EFFECT OF NO LICENSE AT TIME OF BIDDING OR CONTRACTING

Even before any contracts are signed, state laws may require a license just to submit a bid or to solicit work in states that require a license to perform. In a 1999 Arkansas case, it was discovered after bids were opened that the low bidder for an arena seating contract did not have a valid Arkansas contractor's license. The bidder paid a $10,000 fine and renewed its license. The owner waived any deficiencies in the bid, accepted the low bid and awarded the contract. However, the second low bidder sued for an injunction against the award, claiming that without a valid license, the low bidder should have been disqualified from bidding. The trial court granted summary judgment in favor of the bidder. On appeal, the Arkansas Supreme Court held that the licensing statute, Arkansas Code Section 17-25-313, makes a license a prerequisite to be considered for an award and a criminal misdemeanor to submit a bid without a license. However, the court held that the licensing statute requires a license *at the time of contract award,* not at the time of bid opening. Therefore, the award was proper.[13] Be aware that the law may differ in other states and that the time and money spent preparing a detailed design-build proposal could all be lost, including any stipend, if state law requires the license merely to bid on or to solicit work in that state.

If the contractor does not have a required license at the time the contract is entered into, then getting the license during the course of the job will not cure the defect according to some court cases. For example, in one North Carolina case, the contractor was not licensed at the time the contract was entered into, but took and passed the licensing exam on his second attempt after completing about two-

[13] Quality Fixtures, Inc. v. Multi-Purpose Facilities Bd. for Pulaski County, 986 S.W. 865 (Ark. 1999).

thirds of the project. The contractor sued when the owner refused to pay for disputed extras. The court held that the time of entering into the contract "is of great significance" since that is when the owner determines if the contractor is qualified. As a result, the court held that the contract illegally entered into was unenforceable and could not be validated by the contractor's subsequent procurement of a license.[14] Recovery was denied to the contractor.

§ 12.09 INABILITY TO ENFORCE CONTRACT OR RECOVER FOR WORK

In most, if not all, states, it is illegal for a person or corporation to contract to render architectural or engineering services without a proper license. Basic contract law prevents enforcement of such illegal contracts. The courts have held that a valid and enforceable contract may not arise out of a transaction which is prohibited by a statutory law.[15] Acts that are forbidden by a statute passed for protection of the public, especially where the statute provides for a penalty, cannot be the foundation of a valid contract.[16]

Since the unauthorized practice of architecture and engineering is prohibited under state licensing laws, and such laws are passed for the protection of the public and provide for a penalty, it has been argued that a design-build contract is illegal and void if the design-build contractor is not itself a licensed architect or engineer. This means that contractors and design-build teams who pair up to offer design and construction services need to learn the licensing requirements in each state they plan to do work in or else they risk being unable to collect under an illegal contract.

Some states put this concept into the statutes by expressly prohibiting enforcement of contracts by unlicensed contractors. For example, Florida Statute Section 489.128 states:

> As a matter of public policy, contracts entered into on or after October 1, 1990, and performed in full or in part by any contractor who fails to obtain or maintain his license in accordance with this part shall be unenforceable in law or equity. However, in the event the contractor obtains or reinstates his license, the provision of this section shall no longer apply.

A similar statute in Nevada, Nev. Rev. Stat. Section 624.320 provides that no unlicensed person or entity:

> engaged in the capacity of a contractor shall bring or maintain any action in the courts of this state for the collection of compensation for the performance of any act or contract for which a license is required.

Arkansas contractor licensing statute Ark. Stat. Ann. Section 17-25-103(d) states that:

---

[14] Brady v. Fulghum, 308 S.E.2d 327 (N.C. 1983).

[15] Greer v. Zurich Ins. Co., 441 S.W.2d 15, 26 (Mo. 1969).

[16] King v. Moorehead, 495 S.W.2d 65, 78 (Mo. Ct. App. W.D. 1973); *see also* Design Dev. Inc. v. Brignole, 570 A.2d 221 (Conn. App. Ct. 1990); Ransburg v. Haase, 586 N.E.2d 1295 (Ill. App. Ct. 3 Dist. 1992).

> No action may be brought either at law or in equity to enforce any provision of any contract entered into in violation of this chapter. No action may be brought either at law or in equity for quantum meruit by any contractor in violation of this chapter.

In addition to contractor licensing statutes, the laws in other states prohibit recovery by persons or companies who do not have licenses to practice architecture or engineering. Missouri's Statute Section 327.461 provides that:

> Every contract for architectural or engineering or land surveying services entered into by any person who is not an architect or professional engineer or professional land surveyor, as the case may be, and who is not exempt from the provisions of this chapter, shall be unenforceable by the unlicensed or unauthorized person, whether in contract, quantum meruit or other legal theory, *regardless of whether a benefit has been conferred.*

The language "regardless of whether a benefit has been conferred" means even if plans were prepared and, possibly, even if a building has been built. The statute declares unenforceable any agreement under which an unregistered or "unauthorized" party contracts to render professional architectural or engineering services. Where the design-build contractor agrees to "perform" architectural services, section 327.461 bars enforcement of the contract if the design-build contractor's corporation is not properly authorized. There have been several Missouri cases holding that out-of-state architectural or engineering firms that contract "to render" professional design services in Missouri before obtaining a Certificate of Authority cannot enforce their contracts.[17] However, where the design-build contractor agrees to merely "furnish" such services, courts across the nation are divided on whether the licensing laws prohibit recovery.[18]

If the state licensing law does not specifically bar enforcement of the contract, there may be an argument that the legislature did not intend that result. For example, an unlicensed subcontractor was allowed to maintain an arbitration action against the prime contractor in a 1997 New York case, *Migdal Plumbing & Heating Corp. v. Dakar Developers, Inc.*[19] The court held that although the licensing law imposes civil and criminal sanctions on unlicensed plumbers, it does not mention depriving them of contractual payment rights. The court held that "[t]o bar recovery against a commercial customer for work done by an unlicensed plumber—a bar not required by the relevant statute of Code provision—would lead to an inequitable result and would be disproportionate to the lack of a license."

---

[17] *In re* Branson Mall, Inc., 120 B.R. 1006, *aff'd*, 970 F.2d 456 (8th Cir. 1992); Haith & Co., Inc. v. Ellers, Oakley, Chester & Rike, Inc., 778 S.W.2d 417 (Mo. Ct. App. W.D. 1989); Hospital Devel. Corp. v. Park Lane Land Co., 813 S.W.2d 904 (Mo. Ct. App. W.D. 1991). Whether such contracts are void has been held to be purely a question of law to be determined by the court. Carson-Mitchell, Inc. v. Macon Beef Packers, Inc., 544 S.W.2d 275, 276 (Mo. Ct. App. W.D. 1976).

[18] *See* C. Halsey & W. Quatman, *Design-Build Contracts: Valid or Invalid*, 9 Constr. Lawyer No. 3, Aug. 1989.

[19] 662 N.Y.S.2d 106 (App. Div. 1997).

§ 12.10 RECOVERY OF MONEY ALREADY PAID

Some courts have permitted owners to recover from contractors money that was paid under an illegal contract, if the contractor was unlicensed. For example, in a New Mexico case, an unlicensed contractor was ordered to repay the entire contract sum to the owner since the contractor did not have a valid contractor's license. The court stated that to permit the contractor to keep the money would be sanctioning the illegal activity.[20]

In a 1992 federal case involving a design-build contract, the contractor had to forfeit the design fees that had been paid by the owner. In *Kansas City Community Center v. Heritage Industries, Inc.*, a manufacturer of prefabricated modular housing contracted to design and build a drug and alcohol treatment facility in Missouri.[21] The contractor actually performed the design work "in-house" using unlicensed staff to do the design work. The design-build contract said the contractor would "complete" architectural and engineering plans and "do" the grading and utility plans. The trial court granted summary judgment declaring the contract void under Missouri licensing law because the contractor lacked a corporate Certificate of Authority to practice architecture or engineering. The trial court also ordered the contractor to repay $9,546 that the owner had paid for architectural plans. The court of appeals upheld that ruling, including the order to repay the design fees, holding that the entire contract was illegal since the architectural and engineering services were an integral part of the builder's duties and could not be separated from the construction obligations. In that case, nothing had been built prior to the lawsuit being filed so the court did not have to address whether the contractor might have had to repay any construction costs.

In a North Carolina case, the court held that the legislature did not intend to authorize the recovery of amounts already paid to unlicensed contractors after the buildings in question had already been built, even though the contract would be unenforceable and illegal.[22] North Carolina law requires a contractor's license if the contract price exceeds $30,000. In that case, the unlicensed contractor was hired to build two houses costing $74,700 and $57,700, respectively. Nonetheless, the court held that the owner could not obtain a refund of the money paid to the contractor.

§ 12.11 WHAT IF THE CLIENT KNOWS THAT THE CONTRACTOR IS NOT LICENSED?

It is recommended that design-build contractors address the licensing issue in their contracts and advise the owner that the contractor is not a licensed design professional and is not practicing architecture or engineering. If, in fact, the contractor is not practicing a licensed profession, then this probably is not even

[20] Mascarenas v. Jaramillo, 806 P.2d 59 (N.M. 1991).

[21] 773 F. Supp. 181, *aff'd*, 972 F.2d 185 (8th Cir. 1992).

[22] Hawkins v. Holland, 388 S.E.2d 221 (N.C. Ct. App. 1990).

needed, but it is still a good provision to prevent any misunderstanding. However, if a license is actually required, it is unlikely that merely advising the owner of that fact will a contractor avoid the licensing requirements. In a 1969 case, the court held that the general contractor's licensing statute does not grant the unlicensed contractor immunity merely because he advised the customer that he was acting in violation of the statute.[23]

In a 1998 case, a construction manager hired by a bank was denied the ability to recover for its fees. The construction manager was held to be a "contractor" within the meaning of the licensing statutes, which define a contractor as any person or entity who "purports to have the capacity to undertake to, or submits a bid to, or does himself or by or through others, construct . . ."[24] Even though the construction manager did not itself perform any of the work, the court found that it had work performed by others. The court further stated that it was irrelevant whether the bank knew of the lack of a license.[25]

Full disclosure was a key issue in a 1961 California case where a licensed general contractor contracted to prepare drawings, provide site work, and "procure engineering." The court held that the contractor was entitled to recover for its services where its employer knew that the contractor was not licensed as an architect or engineer.[26]

These cases show that there is a split on issue of whether knowledge on the part of one contracting party that the other party is not properly licensed will cure the lack of a required license.[27] However, as stated by one court, "Individuals cannot by their conventions derogate from the force of laws made for the preservation of public order and good morals."[28] Contract clauses that warn of a missing license, or agree to waive that as a defense, might not be upheld by the court for reasons of public policy.

---

[23] Ar-Con Constr. Co. v. Anderson, 168 S.E.2d 18 (N.C. Ct. App. 1969).

[24] Nev. Rev. Stat. § 624.020(2).

[25] Interstate Commercial Bldg. Servs. Inc. v. Bank of Am. Nat'l Trust & Sav. Ass'n, 23 F. Supp. 2d 1166 (D. Nev. 1998).

[26] Barondon Corp. v. Kunika Nakawatase, 16 Cal. Rptr. 472 (Ct. App. 2d 1961). *But see* Medak v. Cox, 90 Cal. Rptr. 452 (Ct. App. 1970).

[27] *See, e.g.*, Design Dev., Inc. v. Brignole, 570 A.2d 221 (Conn. App. Ct. 1990), where designer was denied compensation even though people who hired him knew he wasn't a licensed architect; Cochran v. Ozark Country Club, Inc., 339 So. 2d 1023 (Ala. 1976), where owner knew contractor was not licensed and contracts were prepared to avoid license requirements. Court held owner *still* entitled to rely on defense of non-licensure; Millington v. Rapoport, 469 N.Y.S.2d 787 (App. Div. 2d 1983), holding that a party may not waive a licensing statute enacted for his benefit. Even though owner was aware of absence of license, held no exception to statutory defense to lien; Plaza Builders, Inc. v. Regis, 502 So. 2d 918 (Fla. Dist. Ct. App. 1986), unlicensed contractor could not recover even where owner knew that contractor had no license but continued to deal with contractor anyway. *See also* D.R. Gallo Builders, Inc. v. Travelodge Int'l, Inc., 268 Cal. Rptr. 79 (Ct. App. 1990), where an unlicensed contractor alleged that the project owner promised to "waive" the California licensing requirement. Although the contractor was precluded by statute from enforcing the contract, the court permitted an action for fraudulent misrepresentation against the owner.

[28] West Baton Rouge Parish Sch. Bd. v. T. R. Ray, Inc., 367 So. 2d 332, 334 (La. 1979).

§ 12.12 EFFECT ON LIEN RIGHTS IF NO LICENSE

In Arkansas, contractors must be licensed for all jobs over $20,000. If not, the contractor cannot enforce its contract or file a lien. Some cases have held that companies that fail to obtain an architectural or engineering Certificate of Authority, if required, cannot file a mechanic's lien.[29] In a 1991 Florida case, a design-build contractor was prohibited from enforcing a mechanic's lien under an oral contract. The design-builder entered into a Part 1 AIA design-build agreement with the owner for the furnishing of preliminary design. The Part 2 Agreement for final design and construction was never executed. The design-build contractor sued for breach of contract, quantum meruit and to enforce its mechanic's lien alleging that services were provided beyond the scope of the Part 1 Agreement. The court rejected the lien claim, finding that the contractor was not an architect covered by the Florida lien statute, Section 713.03, Fla. Stat. (1987). Further, the plans were not used to improve the property, but only to obtain financing.[30]

§ 12.13 POTENTIAL CRIMINAL LIABILITY

When a license is required, it is often a criminal act to perform work without such a license. In a 1917 criminal case dealing with a design-build contract, the Illinois Supreme Court held that a construction company could contract to *furnish* architectural work without criminal liability as long as the actual design work was done by licensed architects. The court held that "[a]ny natural person, licensed or unlicensed, may make a contract to furnish architectural work to be performed by a licensed architect."[31] In a 1999 Maryland case, a contractor was prosecuted, sentenced to jail, and fined for engaging in construction without a contractor license. Maryland statutes provide that "a person may not act . . . as a contractor in the State unless the person has a contractor license." The contractor in this case engaged in construction contracting on seven home improvement contracts without a license. The Maryland Home Improvement Commission prosecuted the contractor for violation of the law, and the conviction was upheld on appeal, with each project constituting a separate violation of the law.[32] North Carolina statute section 87-13 states that:

> Any person, firm, or corporation not being duly authorized who shall contract for or bid upon the construction of any of the projects . . . without having first complied with the provisions hereof, or who shall attempt to practice general contracting in the State, . . . and any person, firm, or corporation

---

[29] *In re* Branson Mall, 120 B.R. 1006, *aff'd*, 970 F.2d 456 (8th Cir. 1992); Maran-Cook, Inc. v. Purler Excavating, Inc., 585 S.W.2d 38 (Mo. banc 1979).

[30] Miller Constr. Co. v. First Indus. Tech. Corp., 576 So. 2d 748 (Fla. Dist. Ct. App. 3d Dist. 1991).

[31] People *ex rel.* State Bd. of Architects v. Rodgers Co., 115 N.E. 146 (Ill. 1917); *but see* Kennan v. Tuma, 240 Ill. App. 448 (1926).

[32] Huffman v. State, 1999 WL 1102727 (Md.).

> presenting or attempting to file as his own the licensed certificate of another or who shall give false or forged evidence of any kind to the Board or to any member thereof in maintaining a certificate of license or who falsely shall impersonate another or who shall use an expired or revoked certificate of license, and any architect or engineer who recommends to any project owner the award of a contract to anyone not properly licensed under this Article, *shall be deemed guilty of a Class 2 misdemeanor.*

Similarly, Arkansas contractor licensing statutes state that:

> Any contractor shall be deemed guilty of a misdemeanor and shall be liable to a fine of not less than one hundred dollars ($100) nor more than two hundred dollars ($200) for each offense, with each day to constitute a separate offense, who: (1) For a fixed price, commission, fee, or wage, attempts to or submits a bid or bids to construct or contracts to construct, or undertakes to construct, or assumes charge in a supervisory capacity or otherwise, or manages the construction, erection, alteration, or repair of, or has constructed, erected, altered, or repaired, under his or its direction, any building, apartment, condominium, highway, sewer, utility, grading, or any other improvement or structure, when the cost of the work to be done or done, in the State of Arkansas by the contractor, including, but not limited to, labor and materials, is twenty thousand dollars ($20,000) or more, without first having procured a license with the proper classification to engage in the business of contracting in this state.[33]

In addition to the criminal fines, many states permit civil penalties to be assessed for violation of licensing laws. Arkansas law allows for the assessment of civil penalties for violation of the licensing law up to $400 per day of illegal activity, not to exceed 3 percent of the total project cost. The state licensing board can also obtain an injunction to restrain and prohibit the contractor from performing the work being done or the work that is about to commence.[34] Forfeiting the contract money is bad enough, but criminal charges can ruin a career and result in debarment from further state or federal contract work. (*See, e.g.*, Chapter 24).

§ 12.14 CONSTITUTIONAL ISSUES

Some contractors presume that because they have one architect or engineer on staff that they do not need a corporate license also. This dual licensure requirement for both the individual architect performing the services and the

---

[33] Ark. Stat. Ann. § 17-25-103. Architects and engineers are exempt from the licensing requirement for contractors if their only financial interest in the project is the architectural or engineering fees for preparing plans, specifications, surveys, and supervision customarily furnished by architects and engineers. Ark. Stat. Ann. § 17-25-102. However, when a design firm steps outside of the "customary" practice, and engages in design-build as a joint venture partner whose fee is not limited to professional services, the exemption may not apply and a license may be required.

[34] Ark. Stat. Ann. § 17-25-104; *see also* Ala. Code § 34-8-6; Idaho Code § 54-19-20.

corporation who holds the contract has been successfully challenged in one case as unconstitutional.[35] A New York court stated that the requirement of a corporate license would effectively bar a contractor from conducting nationwide business and would render out-of-state contractors non-competitive, calling dual licensing a "tourniquet on interstate activity" and a "local trade barrier."

§ 12.15 COURT CASES DEALING WITH DESIGN-BUILD CONTRACTS

There have been many court cases in which a design-build contract has been challenged based on the prime contractor not having a license to practice architecture or engineering. The cases have split into three groups with: (1) some cases finding the contract valid (usually because the design work was subcontracted out to properly licensed persons or firms); (2) others finding the contract illegal and void; (3) and at least one case in which the contract was split in two, with one part being valid and the other part declared illegal.

[A] Contract Wording Is Important

Although at first glance it may seem to be an exercise in semantics, courts have placed great significance on the specific language used in design-build contracts, primarily whether the contractor has agreed to actually "perform" or simply to "furnish" design services. As long as the "furnished" plans are prepared by a licensed architect or engineer, even though as a subcontractor, some courts have found that the legislative purpose has been met and the public health, safety, and welfare is adequately protected. However, other courts have interpreted the contracts and statutes more strictly, finding that even the "furnishing" of plans and specifications without a license violates the law. In a 1978 Texas case, the appellate court placed great significance on the terms "furnish" and "perform," finding a general contractor's design-build agreement valid and enforceable. In *Seaview Hospital, Inc. v. Medicenter of America*, the contractor entered into a "turnkey" contract under which it did not perform architectural or engineering services itself, but rather procured such services from licensed professionals. When the contractor sued to enforce the agreement, the owner defended by claiming that the contract was void and unenforceable under Texas law since neither the contractor nor its employees were licensed architects or engineers. All of the architectural and engineering work was actually performed by other firms who were duly licensed to practice in the state of Texas, or by individuals working under the direct supervision of licensed professionals. In upholding the contract the court concluded that:

> A general contractor, under the licensing statutes here involved, is not precluded from entering into a contract with the owner, which provides that the contractor: will engage or hire architects and engineers who are duly licensed in Texas. The stated objective of [the licensing statute] is to protect public

[35] Tetra Technologies, Inc. v. Harter, 823 F. Supp. 1116 (S.D.N.Y. 1993).

> health, safety, and the general welfare by ensuring that architectural and engineering work in this State be performed only by qualified persons who are duly licensed.[36]

The court held that contracting to "furnish" design services did not require the contractor to "perform" such services. Therefore, the contract was not unenforceable under Texas law. For this reason you will find that the better design-build contract forms make it clear that the prime contractor, if not an architect or engineer, is providing or furnishing the design services, not performing them directly. For example, AGC's standard form of design-build agreement states that architectural and engineering services shall be procured from licensed, independent design professionals retained by the design-builder or furnished by licensed employees of the design-builder, or as permitted by the law of the state where the Project is located.[37]

DBIA's standard agreement similarly states:

> Design-Builder shall, consistent with applicable state licensing laws, provide design services, including architectural, engineering and other design professional services, required by this Agreement. Such design services shall be provided through qualified, licensed design professionals who are either (i) employed by Design-Builder, or (ii) procured by Design-Builder from independent sources.[38]

[B] Cases Holding Design-Build Contracts Invalid

Design-build cases are usually started one of two ways. First are those cases in which the state licensing board takes action against a design-build contractor who is offering or performing design services without proper authorization. The board usually investigates due to a complaint by an architect, engineer, or a competitor of the alleged wrongdoer. In those cases, the board typically seeks an injunction to stop the unauthorized contractor from continuing the practice. The licensing board may additionally discipline the design professional involved for aiding another in the "unauthorized practice of architecture or engineering." Second, and more common, are those cases in which the owner seeks to avoid payment to a design-build contractor by raising licensure as a defense or by seeking declaratory judgment as to the validity of the contract.

Some courts have taken the position that design-build contracts violate state licensing laws and are therefore entirely void and unenforceable, despite the resultant "unjust enrichment" that may result to the owner. Two of the earliest such

---

[36] Seaview Hosp., Inc. v. Medicenter of Am., 570 S.W.2d 35, 40 (Tex. Civ. Ct. App. 1978). *See also* Lytle v. Godfirnon, 6 N.W.2d 652, 653 (Wis. 1943).

[37] AGC Document No. 415, Standard Form of Design-Build Agreement and General Conditions (Lump Sum version), ¶ 2.2 (1999 edition).

[38] DBIA Document No. 520, Standard Form of Preliminary Agreement Between Owner and Design-Builder, ¶ 2.1 (1998 edition).

cases were both decided in 1939, one in New York and the other in California. In the New York case, *American Store Equipment & Construction Corp. v. Jack Dempsey's Punch Bowl*, a construction company was contracted both to design and build a building. The court denied the contractor recovery *in full* finding that the contractor was not licensed to practice architecture in New York. The contractor argued that only about 5 percent to 10 percent of its contract work consisted of architectural services. However, the court ruled that the entire contract was illegal, stating: "The contract was entire and indivisible: to plan, construct and furnish a complete unit. . . . They cannot be separated into different classes— legal and illegal."[39] The court apparently could foresee that determined design-build contractors might find a way around the licensing issue when it stated that:

> [t]o sustain the legality of the balance of the agreements would lead to widespread disregard of the licensing statutes. It would be easy for any construction contractor to thwart the purposes for which the licensing of architects was enacted, by merely providing in his contract that architectural services would be given gratis, so long as the contractor were awarded the contract itself.[40]

In the California case, *Baer v. Tippett*, the court had nearly identical reasoning. The contractor maintained that the contract was severable and divisible, and that he should be entitled to recover for the services provided as a contractor. However, the court held:

> The contract had but a single object, construction of certain buildings, and the two kinds of services called for were thoroughly intermingled. The parts are interdependent and there is a common consideration which is clearly based upon a completed whole. In our opinion, the contract was entire, and not severable.[41]

More recent New York and California decisions have upheld design-build contracts as valid and enforceable.[42] Courts in California, Ohio, Michigan and Pennsylvania have likewise struck down design-build contracts as illegal when the contractor did not have a license to practice architecture.[43] As mentioned in § 12.10, above, a Missouri federal court not only found a design-build contract to

---

[39] American Store Equip. & Constr. Corp. v. Jack Dempsey's Punch Bowl, 21 N.Y.S.2d 117, 119, *aff'd*, 258 A.D. 794, 16 N.Y.S.2d 702, *aff'd*, 283 N.Y. 601, 28 N.E.2d 33 (1939). *See also* Gordon v. Adenbaum, 567 N.Y.S. 2d 777 (A.D. 1991); Marshall-Schule Assocs. v. Goldman, 523 N.Y.S.2d 16 (N.Y. City Civ. Ct. 1987).

[40] *American Store Equip. & Constr. Corp.* 21 N.Y.S.2d at 119.

[41] Baer v. Tippett, 92 P.2d 1028, 1032 (Cal. Ct. App. 1939).

[42] *See* § 12.15[D].

[43] Elephant Lumber Co. v. Johnson, 202 N.E. 2d 189 (Ohio Ct. App. 1964); McGill v. Carlos, 81 N.E. 2d 726 (Ohio Ct. App. 1947); Wedgewood v. Jorgens, 157 N.W. 360 (Mich. 1916); F.F. Bolinger Co. v. Widmann Brewing Co., 14 A.2d 81 (Pa. 1940). Ohio has since enacted legislation that authorizes design-build as long as the design work is done by licensed architects or engineers. Ohio Rev. Code §§ 4703.182 and 4733.161.

be illegal, but ordered the unlicensed contractor to pay back to the owner all design fees that had been paid.[44]

There is a 1976 federal court decision in New Jersey dealing with unlicensed architects which seemed to indicate that design services could only be offered by licensed architectural firms. The case dealt with a design contract entered into by a Maryland architectural firm, not licensed in New Jersey. The plans and specifications were prepared and sealed by an employee who was a New Jersey licensed architect. However, the individual was not the contracting party. The court found the contract void and against public policy, stating: "A contract to prepare plans and specifications for a New Jersey building can only be made by an architect licensed by New Jersey, or by a professional corporation formed under its laws. When made by others, the contract is illegal."[45] The court stated that there could be no recovery on an illegal contract, whether in contract or in quantum meruit. The court further warned that:

> This result is what the law intends. New Jersey landowners or developers or builders, as well as unlicensed professionals and foreign corporations, are both [sic] to take the consequences of entering into an illegal contract, and derivative claims can rise no higher than their source.

In a 1947 Ohio case, a contractor-partnership was hired to design and prepare plans for a new house for a 5 percent fee although neither partner was a licensed architect. The court conceded that the plaintiffs held themselves out to be builders, not architects. Therefore, the issue was whether the contract constituted the practice of architecture. In holding the contract unenforceable, the court stated that under Ohio law:

> a builder who is not a registered architect may contract *to furnish* plans and specifications for the construction of a building for an owner provided the plans and specifications are prepared by a registered architect . . . [b]ut the Court is of the opinion that under the laws of the State of Ohio, a builder who is not a registered architect may not *prepare* complete plans and specifications for the construction of a building for another. . . .[46]

Ohio has since enacted legislation that authorizes design-build as long as the design work is done by licensed architects or engineers.[47] *See* Appendix E for full text of that statute.

[C] Cases Holding Design-Build Contracts Partially Invalid

The harsh results of such "all or nothing" decisions have caused at least one court to fashion more equitable remedies. In a 1969 federal case based on Iowa

---

[44] Kansas City Community Ctr. v. Heritage Indus., Inc., 773 F. Supp. 181, *aff'd*, 972 F.2d 185 (8th Cir. 1992).

[45] Dalton, Dalton, Little, Inc. v. Mirandi, 412 F. Supp. 1001, 1004-1005 (D.N.J. 1976).

[46] McGill v. Carlos, 81 N.E. 2d 726, 729 (Ohio Ct. App. 1947).

[47] Ohio Rev. Code §§ 4703.182 and 4733.161.

law, the court ruled that design services may be separated from other services when it comes to enforcement of a design-build contract. In *Food Management, Inc. v. Blue Ribbon Beef Pack, Inc.*, the court held that a contract to provide architectural and engineering services, as well as other services, was unenforceable *only* as to the fees for design services where the contractor was not properly licensed under Iowa law. The design services were subcontracted to a licensed design firm, but none of the prime contractor's officers or employees were licensed as architects or engineers in Iowa. While the illegal portion relating to design services was not enforceable, the portion of the contract that related to non-architectural services could be enforced. Therefore, the court permitted recovery for the balance of the contract.[48]

[D] Cases Finding Design-Build Contracts Valid

At least seven states have found design-build contracts to be valid even where the contractor is unlicensed as an architect or engineer. In one of the earliest such cases, *Binford v. Boyd*,[49] the California Supreme Court held that California licensing laws for architects did not prohibit a non-licensed corporation from contracting to "furnish" plans and specifications as long as full disclosure was made to the owner. The court ruled that such contracts are valid if the plans and specifications are to be prepared by a third party who is a licensed architect and the person furnishing the plans and specifications informs the owner that he is not a licensed architect. The court held that the purpose of the law is to prevent the *preparation* of plans by an unqualified person.[50]

In a 1952 Illinois case, *Gastaldi v. Reutermann*,[51] a builder was allowed to recover $2,000 from an owner for the cost of architectural plans and other services furnished by the builder prior to the owner's abandonment of the project. Although the builder contracted directly with the architect for the plans, the court rejected the owner's defense that the builder was engaged in the unlawful practice of architecture under Illinois law.

In a 1961 Missouri case, the state Supreme Court upheld an oral design-build contract as enforceable even though the contractor was not a licensed architect. In that case, a contractor sued to enforce its mechanic's lien against a privately owned building leased to the government as a post office. The owner defended against the lien, in part, by claiming that the contract was void because the builder agreed to furnish the plans. The old Missouri statute at issue provided

[48] Food Mgt., Inc. v. Blue Ribbon Beef Pack, Inc., 413 F.2d 716 (8th Cir. 1969).

[49] 174 P. 56 (Cal. 1918).

[50] *See also*, Walter M. Ballard Corp. v. Dougherty, 234 P.2d 745 (Cal. Ct. App. 1951); Meyer & Holler v. Bowman, 9 P.2d 936 (Cal. Ct. App. 1932); Barondon Corp. v. Kunika Nakawatase, 16 Cal. Rptr. 472 (Ct. App. 2d 1961).

[51] 104 N.E.2d 115 (Ill. App. Ct. 1952). *See also*, Kansas Quality Constr., Inc. v. Chiasson, 250 N.E.2d 785 (Ill. App. Ct. 1969); Continental Paper Grading Co. v. Howard T. Fisher & Assoc., Inc., 120 N.E.2d 577, 581 (Ill. App. Ct. 1954).

that "[a]ny contract entered into for any work in this State contrary to the provisions of this Chapter shall be null and void." The Missouri Supreme Court held that the statute was no defense to the lien claim, stating:

> In the first place we do not find any evidence in the transcript to the effect that plaintiff agreed to *prepare* any plans or *render* other architectural services. Secondly, the plans used by plaintiff were prepared by a registered architect, paid for by plaintiff, and the cost therefor was charged to defendants under the cost-plus contract. We see nothing in the agreement or in plaintiff's method of procuring the plans which was contrary to the provisions of Chapter 327, RSMo 1949, V.A.M.S. (which prohibits the practice of architecture by persons not registered under the provisions of said chapter), as same existed on the date here involved.[52]

In another Missouri case, *Strain-Japan R-16 School Dist. v. Landmark Systems, Inc.*,[53] the court held that a contractor did not violate Missouri licensing laws by entering into a design-build contract to furnish professional engineering services without being registered as an architect or professional engineer. The plaintiff-school district solicited bids for additions to its building complex, accompanying its invitation with a floor plan, an elevation drawing, and design requirements for a "pre-engineered" building. These bid documents were not signed or sealed by a licensed architect or engineer. The parties entered into a lump-sum contract. In the space labeled "architect" the contract read: "*Plans by General Contractor*" with the work to include "*a pre-engineered metal shell . . . to be erected as shown in the drawings*." The contractor, who did not have a corporate Certificate of Authority to practice engineering, was required by the school district to use a specific building system product and to provide signed and sealed construction documents for the building and foundation.

The plans for the pre-engineered building were supplied to the contractor by another firm, which also performed the construction work as a subcontractor of the defendant contractor. The foundation plan was furnished by an engineering firm hired by the prime contractor. All pages of the building and foundation plans were signed and sealed by professional engineers registered in Missouri who were not employees of the contractor. The school district argued that the contract was not enforceable since it is a contact for "*architectural or engineering services*," and since the contractor expressly agreed to "provide" engineering services by some means, it violated state licensing law, thereby precluding recovery. In upholding the contract, the court of appeals stated that it had "no hesitation in holding that there is no statutory violation, and no illegality, in the arrangement these parties entered into." The court said that if the school district's position were correct, "it would be necessary either that owners contract directly with architects and professional engineers, or that all corporations engaged in the

[52] Honig Constr. Co. v. Szombathy, 345 S.W.2d 111, 116-117 (Mo. 1961).
[53] 965 S.W.2d 278 (Mo. Ct. App. E.D. 1998).

general contracting business qualify for registration as architects or engineers." The court added that "[t]he purpose of [the licensing law] is to make sure that buildings are soundly and safely designed. *We sense no purpose of giving registered architects and engineers a monopoly on the construction business*." The court concluded that the contractor acted within the requirements of the law when it used the owner's drawings to bid, entered into a fixed-sum contract with the owner, subcontracted with registered professional engineers for all services requiring their professional skills, and then presented signed and sealed plans for the owner's approval.

Unjust enrichment is a key factor in some cases dealing with enforceability of contracts with unlicensed individuals, with the courts troubled by the windfall that an owner could reap if it did not have to pay for construction work. The issue of unjust enrichment to the owner played a factor in a highly publicized 1988 New York case, in which the state's highest court split four to three on the issue of enforceability of a design-build contract, finding it to be valid, enforceable, and not a violation of public policy. In *Charlebois v. J.M. Weller Associates*,[54] the trial court entered summary judgment, finding the contract valid. That decision was narrowly affirmed in a three-to-two decision on appeal to the Court of Appeals. The contract in *Charlebois* was a standard AGC form contract which required the contractor to "furnish" services of an architect. After construction started, disputes arose over cost, design, building code compliance, and defects. The owner refused to pay $600,000 in costs allegedly owed the contractor and instead brought an action for declaratory judgment that the contract was invalid as against public policy. In upholding the contract, the high court ruled that the design-build contract did not constitute the unauthorized practice of engineering, since the contract expressly required all design work to be performed by a licensed engineer. A strong dissent argued that the contract clearly violated the public policy of New York law and should be declared void and unenforceable.

The *Charlebois* case started a statewide debate over design-build and, in particular, the growing practice of design firms delegating design work and responsibility to contractors. That led to a 1991 complaint by the New York General Building Contractors Association that design firms were impermissibly delegating design responsibility to contractors. In response, the state Education Department issued a memorandum that such delegation to unauthorized firms was "unprofessional" and violated state licensing laws. That ruling was challenged by an engineer-attorney in *Rubin v. New York State Educ. Dept.*,[55] in which the court held that the Department's memorandum was non-binding and a mere guideline to be weighed with other factors. As a result, the Board of Regents adopted Rule 29.3(b) which provides that when a licensed professional delegates work that is "ancillary" to the main project to another licensee through an unlicensed third

[54] 526 N.Y.S.2d 648 (3d Dept. 1988); *aff'd*, 72 N.Y.2d 587, 531 N.E.2d 1288 (1988).
[55] 620 N.Y.S.2d 143 (App. Div. 1994).

party, such delegation is not deemed the practice of professional design work and is not unprofessional conduct.

In a 1997 New York case, *SKR Design Group, Inc. v. Yonehama, Inc.,*[56] a design-build contract was upheld even though the contractor was not licensed to practice architecture on the basis of a contract provision, which stated that the "Design-Builder shall provide or cause to be provided and shall pay for design services" to be performed by a qualified architect. The contractor paid a licensed architect to prepare and seal the plans. Even though the contract required the contractor to retain a licensed architect the court held that provision "[did] not convert that contract into one *for the performance* of those services."

Other cases have found design-build contracts to be legal and enforceable where the contractor was merely furnishing plans, not preparing them.[57]

§ 12.16 LICENSING BOARD CASES AND INJUNCTIONS

There are several reported cases in which state licensing boards have sued for injunctions against unlicensed contractors providing design services. In cases in Arkansas and Connecticut, a company offering design-build services was found to be acting in violation of state licensing laws and was enjoined from further practice.[58] In a Kentucky case, a contractor who designed and built a 78-bed nursing home was enjoined from engaging in the further practice of architecture since it had no corporate license to practice engineering or architecture, although its principal owners were both licensed professional engineers.[59]

§ 12.17 SOME RECOMMENDATIONS

What these cases show is that the law varies greatly from state to state and, sometimes, even within the state depending on the contractual arrangement. Worse yet, is that many states have no case law or statute that deals with design-build, leaving contractors to guess at what might be the law. Some general rules seem clear in a majority of states:

A. If the design work is done in-house, by the contractor's employees, the firm needs a corporate license (or certificate of authority).
B. If the contract says the firm will "perform" or "do" the design work, the firm needs a corporate license.

---

[56] 660 N.Y.S.2d 119 (App. Div. 1997).

[57] Gastaldi v. Reutermann, 104 N.E.2d 115 (Ill. App. Ct. 1952); Kansas Quality Constr., Inc. v. Chiasson, 250 N.E.2d 785 (Ill. App. Ct. 1969); Seaview Hospital, Inc. v. Medicenter of Am., 570 S.W.2d 35 (Tex. Civ. App. 1978); Lytle v. Godfirnon, 6 N.W.2d 652, 653 (Wis. 1943).

[58] Arkansas State Board v. Bank Bldg. & Equip. Co., 286 S.W.2d 323 (Ark. 1956); Connecticut Society of Architects, Inc. v. Bank Bldg. & Equip. Co., *unreported and reversed on other grounds*, 193 A.2d 493 (Conn. 1963).

[59] Dahlem Const. Co. v. State Board of Examiners and Registration of Architects, 459 S.W.2d 169 (Ky. Ct. App. 1970).

C. If the firm contracts to merely "furnish" or "provide" the design, and then subcontract the work to a licensed architect or engineer, the firm will be fine in most, but not all, states without a corporate license.

If in doubt, however, get the needed license or certificate of authority. The risk of not being able to enforce a contract is too great. Check the law in the project state before proposing and, certainly, before entering into any design-build contract.

§ 12.18 SOURCES OF LICENSING INFORMATION

There are several publications available that summarize state-by-state licensing laws for design firms and contractors. Here are some of the best:

1. *State-by-State Guide to Architect, Engineer, and Contractor Licensing*, Aspen Law & Business (a comprehensive two-volume set, with annual supplements).
2. *Licensing Manual*, PSMJ Resources, Inc.
3. *Engineering Licensing Laws, A State-by-State Summary and Analysis*, National Society of Professional Engineers. This is a three-ring binder that summarizes licensing laws for engineering firms in all 50 states;
4. *General Contractor's Manual,* American Insurance Association. This is a three-ring binder with tabbed dividers by state. It is published and updated every year and can be obtained by calling AIA Law Publications at (202) 828-7100. The cost in May 2000 was $117.00.
5. *Summary of Architectural Licensing Laws*, AIA Press.
6. *The Design-Build Process: A Guide to Licensing and Procurement Requirements in the 50 States and Canada*, ABA Forum Committee on the Construction Industry.

While detailed and updated legal research may be needed to verify the law in any given state, these books provide some general guidance and summaries of state licensing requirements.

CHAPTER 13

INSURANCE AND BONDING FOR DESIGN-BUILD

§13.01 DESIGN-BUILD INSURANCE IN GENERAL

A design firm can participate in a design-build project as the prime contractor, a subcontractor, or a joint venturer. In each case, the design firm's insurance needs are different and policies must be reviewed for necessary endorsements or to obtain additional coverage.[1]

Not too many years ago, an architect or engineer wanting to go into design-build as prime contractor could not get insurance. In addition, contractors could not get professional liability insurance to cover their design risks. The insurance scene has improved considerably and the insurance market is rapidly changing to accommodate the growth of design-build. Over a three-year period, the number of firms able to obtain professional liability insurance for design-build projects increased—from 77 percent in 1997, to 87 percent in 1998, to 100 percent in 1999.[2] In the soft insurance market that exists in the first part of the new millennium, firms shopping for design-build insurance stand to benefit from the stiff price competition that has been brought on by the growth in the number of insurers competing for business. In a 1997 interview, James Bechter of RA&MCO Insurance Services said, "Pricing is half of what it used to be five or six years ago. I do not know how much it can decline from this point [and] it would be difficult to imagine how prices could decline much further than they already have."[3] This is good news for design professionals and contractors who are considering expanding into design-build.

The simplest insurance issues occur when the design firm acts as a subcontractor to a contractor. In this model, the design firm is providing only design, not construction, and will have coverage for claims made by the client (in this case the prime contractor) or other third parties without needing special endorsements—as long as the policy does not exclude design-build work in any form. Claims statistics would indicate that the risk is no greater, and possibly less, when doing design-build as a subcontractor, largely because of the team approach to resolving problems that come up during construction. Nonetheless, most firms carry insurance for third party claims and in case the teaming relationship breaks down.

§13.02 WHAT INSURANCE IS NEEDED?

Every contractor, design-build or otherwise, should have certain basic insurance policies as part of an insurance program. Health insurance, dental plans, and disability insurance or other employee benefits are not discussed in this chapter.[4]

---

[1] For more detailed information about insurance for design-build projects, DBIA publishes a good reference manual called *The Design-Build Insurance and Bonding Guide*, DBIA Doc. No. 221 (1999 edition). The guide is meant to provide project owners and design-build firms with practical information on insurance and bonding.

[2] *1999 Design-Build Survey*, Zweig White & Assoc., Inc., p. 29.

[3] *A/E's Find New Insurance Needs Under Design-Build*, Engineering Times, Jan. 1997.

[4] One study found that 9 out of 10 contractors offer health care insurance to their office employees and their families, and two-thirds offered this same benefit to their non-union field employees. With rising insurance costs, getting competitive bids is very common in getting health insurance coverage. Contractors surveyed in 1999 reported use of PPOs (62%) and HMOs (38%) as ways to contain costs of health care coverage.

Because a particular project may warrant higher limits, a firm should talk to its insurance agent about what program is most appropriate for the company. The types and limits of insurance shown in Figure 13-1 are fairly common and shown for example only.

[A] Standard Endorsements

Four of the endorsements to the Comprehensive General Liability (CGL) policy are worth mention in more detail. First, the broad form "Contractually Assumed Liability" endorsement is important, since it picks up risk that the insured assumes in the construction contract for another person's liability, including broad indemnity clauses. Second, the "Products/Completed Operations" endorsement means that the designer has coverage for property damage or bodily injury that occurs after completion of the project. The standard CGL policy excludes coverage for injuries or property damage arising after completion of the work, known as the "completed operations hazard." Third is the standard "Per Project Aggregate" endorsement which gives the contractor the full policy limits on each project. The standard CGL policy has an annual cap on the amount of coverage available for all claims covered by the policy. Therefore, a contractor with a $1 million policy and

FIGURE 13.1. Typical Design-Build Insurance Requirements.

TYPE:	TYPICAL MINIMUM LIMITS OF INSURANCE:
1. Workers' Compensation	Statutory Amount
2. Employer's Liability	$500,000 by disease $500,000 each accident $500,000 each employee
3. Comprehensive General Liability (CGL)	
(1986 ISO Form or Broader)	
General Aggregate:	$1,000,000
Completed Operations Aggregate:	$1,000,000
Limit Per Occurrence:	$1,000,000

The CGL policy should include the following conditions, in the policy or by special endorsement, often required by owners:

(a) Premises and Operations
(b) Broad Form Contractual/Contractually Assumed Liability
(c) Independent Contractors
(d) Personal Injury
(e) Products/Completed Operations
(f) "XCU" Coverage—explosion, collapse, underground hazards
(g) Endorsement for Per Project Aggregate

4. Automobile Liability: (Hired, Owned and Non-owned Included)	
Combined Single Limit per Accident	$1,000,000
5. Builder's Risk	In the amount of the Contract
6. Umbrella Policy	$1,000,000 to $5,000,000
7. Professional Liability (Errors and Omissions)	$1,000,000 per claim and annual aggregate

10 projects can get an endorsement that in effect provides $1 million of protection on each of the projects, for a net of $10 million of coverage. This is often a no-cost endorsement to the policy. A design-build contractor should get these endorsements whether or not they are required by the owner, since they add greater protection to the company for very little premium increase, if any. Finally, the "XCU" endorsement reverses an exclusion in the standard CGL policy to provide coverage for explosion, collapse, and underground hazards.

The owner should be named as an additional insured on the general and automobile liability policies of insurance. Most often the owner cannot be named as an additional insured on the workers' compensation policy, nor on the professional liability policy, if any, carried by the design professional or contractor, though some owners ask for this anyway. The owner will usually require the contractor to furnish the Certificates of Insurance for insurance coverages. Such certificates shall show the owner as an additional insured where required, and normally must provide that the insurance policy will not be canceled without advance written notice, often 30 days, to the owner.

§13.03 INSURANCE RATES

How much should one expect to pay for insurance as a general contractor? When making the switch from pure designer to contractor, one should be prepared for an increase in cost of general liability insurance. Typical design firm coverage under general liability policies is $3,000 to $4,000 per year depending on firm size and policy limits. For a contractor's general liability policy, however, the premiums are more on the order of $30,000 to $50,000 per year due to the increased risk. Workers' compensation premiums will also increase significantly as employees change from officer workers to field superintendents, foremen and laborers.

A 1999 study reported that contractors pay the following rates, generally, as a percentage of payroll: Workers Compensation, 5 percent; Health Care Insurance, 7 percent; and General Liability (CGL), 3 percent.[5] Rates can fluctuate from year to year. In fact, one study of workers' compensation insurance found that rates have been steadily dropping since 1993. Most of the contractors responding to the study (60 percent) said they get competitive bids to reduce insurance costs and the majority (57 percent) also use aggressive loss control programs within the company to lower their rates.

§13.04 INSURANCE SERVICE OFFICE, INC. (ISO)

The insurance industry formed this organization in 1940 to develop standardized language for general liability insurance policies. Many policies and endorsements, as well as other forms for additional insureds, etc. are referred to by their ISO document number. ISO drafts insurance policies and then obtains approval from the various state divisions of insurance. While insurers are not

[5] CFMA's *Construction Industry Annual Financial Survey* (1999).

required to use ISO language, it makes it easier to get approval state-to-state by using an ISO form.

§13.05 WORKERS' COMPENSATION INSURANCE

As discussed below, the price of workers' compensation insurance is based, in part, by the insured contractor's safety record and claims history. What number and kind of claims can be expected as a general design-build contractor? Very few design firms have much claims experience with workplace injuries to their employees. An architect's or engineer's office is usually a safe place to work. By contrast, a construction site is one of the most hazardous places that you can send an employee. The Department of Labor consistently ranks the construction industry as one of the most dangerous industries in the United States. As a result, when one becomes a contractor and takes on a payroll of employees and subcontractors who work all day on a construction site, one can expect a higher frequency of injuries and accidents involving its employees. Contractors responding to a 1999 survey reported an average of 24 workers' compensation claims, with an average cost per claim of $3,629.[6] This was down from an average of 34 claims in 1995. Therefore, it is important for general contractors to carry workers compensation insurance and to make sure that their subcontractors do too.

Most states require that contractors maintain workers' compensation insurance at certain statutory limits. Companies with sufficient net worth can be authorized to self-insure in some states. However, if a subcontractor fails to carry workers compensation insurance and one of its employees is injured on the job, the subcontractor can make a claim against the general contractor's insurance policy by statute. The general contractor is known as the "statutory employer" in this case. Failure to require subcontractors to carry workers' compensation insurance can result in liability to the general contractor in some states. For example, in North Carolina, N.C. Gen. Stat. Section 97-19 provides that:

> *Any principal contractor*, intermediate contractor, or subcontractor who shall sublet any contract for the performance of any work without requiring from such subcontractor or obtaining from the Industrial Commission a certificate, issued by a workers' compensation insurance carrier, or a certificate of compliance issued by the Department of Insurance to a self-insured subcontractor, stating that such subcontractor has complied with G.S. 97-93 hereof [which requires that employers carry insurance or be certified to self-insure], *shall be liable*, irrespective of whether such subcontractor has regularly in service fewer than three employees in the same business within this State, *to the same extent as such subcontractor would be* if he were subject to the provisions of this Article for the payment of compensation and other benefits under this Article on account of the injury or death of any employee of such subcontractor due to an accident arising out of and in the course of the performance of the work covered by such subcontract. If the principal contractor, intermediate contractor or subcontractor shall obtain such certificate at the

[6] *Id.*

> time of subletting such contract to subcontractor, he shall not thereafter be held liable to any employee of such subcontractor for compensation or other benefits under this Article.

This is a fairly strong incentive for general contractors to make sure that their subcontractors carry workers' compensation insurance. In addition, the general contractor's insurance agent will audit its insurance certificates annually to see if it has obtained certificates of insurance from its subcontractors showing that they carry workers compensation insurance. Failure of a subcontractor to carry workers' compensation makes the general contractor liable, under its own policy, in many states. As a result, the general contractor's premiums may increase to the extent that its insurer feels that it has exposure to claims by subcontractor employees due to its failure to insist on certificates of insurance.

§13.06 COMMERCIAL GENERAL LIABILITY INSURANCE (CGL)

This is the core policy carried by contractors and it covers personal injury, property damage and death which results from the negligence of the contractor, other than damage to the work itself. Up until 1986, the CGL policy was known as the "comprehensive" general liability policy, but in that year, the name was changed to "commercial" general liability. The topic of what CGL insurance covers and what it does not is an extensive topic and is not be discussed in detail in this book.[7] What many design professionals do not realize, however, is that a contractor's CGL policy does not generally cover the cost to repair defective work. That risk is typically one that is self-insured.

§13.07 BUILDER'S RISK INSURANCE

This insurance policy provides property coverage during the construction process, protecting the actual structure under construction, including materials, fixtures, and equipment to be used in the construction. "All risk" coverage is the most broad and most common insurance of this type. The Builders' Risk insurance is usually written for the full contract price to cover a loss from fire, lightning, explosion, windstorm, vandalism, sprinkler leakage and, if broad form, falling objects and weight of snow and ice at the project site. This policy is usually provided by the contractor as part of the contract price since the contractor more often has better insurance contacts to price and provide this coverage. The policy covers all contractors and subcontractors, as well as the project owner, if they are all named as insureds.

§13.08 AUTOMOBILE INSURANCE

Businesses that own vehicles carry Business Auto Coverage. The liability protection under such policy can be structured to apply to any automobile, to

[7] For further information on CGL coverage and exclusions, *see* Scott C. Turner, Insurance Coverage of Construction Disputes, 2d Ed., The West Group (1999).

only owned, hired or non-owned automobiles, or to specifically described automobiles. This coverage is standard and the owner frequently asks to be named as an additional insured on this policy.

§13.09 UMBRELLA INSURANCE

The terms "excess" or "umbrella" insurance are often used interchangeably to describe a policy of insurance to add to the policy limits of a CGL general liability policy. Primary general liability insurance is normally written for no more than $1 million, so additional limits of insurance can be obtained by purchasing an "umbrella" policy which adds not only more coverage, but can also add coverage for perils not covered under the underlying CGL policy.

§13.10 PROFESSIONAL LIABILITY INSURANCE (ERRORS AND OMISSIONS)

Commonly called "E&O" insurance, a design-build contractor should carry some type of contractor's professional liability insurance or require that its architect and engineer consultants or joint venturers are adequately insured for professional malpractice. The policies cover the design professional for damages due to negligent acts, errors or omissions, with certain exclusions. A design consultant working under a subcontract to a general contractor can expect to see requirements that the architect/engineer carry specific limits of professional liability (errors and omissions) insurance. The limits of such insurance may be dictated by the owner in the prime design-build contract and flow through to the design professional by reference or by specific limits in the design subcontract. For example, AGC's Standard Form of Agreement Between Design-Builder and Architect/Engineer for Design-Build Projects,[8] has a very extensive insurance section, ¶. 7.2, with blanks to fill in the limits of insurance for each type of coverage. For professional liability insurance, the AGC form even specifies a maximum deductible that can be included in the policy. DBIA's Standard Form of Agreement Between Design-Builder and Designer[9] also has an insurance section. The drafters of the DBIA form said that they "punted" when it came to specific insurance requirements and the document refers instead to an exhibit that is to be prepared by the parties and attached, instead of dealing with the types and limits in the form itself. AIA's 1996 version of the B901 subcontract between design-builder and architect does not even mention insurance at all, in the body or by an exhibit, other than in ¶. 5.4.6, which allows the architect to recover as a reimbursable expense the additional premium for coverage requested by the design-builder in excess of that normally carried by the architect and its consultants.

One problem with both the AGC and the DBIA forms is that they require the design professional to maintain professional liability insurance for a specific

[8] AGC Doc. No. 420 (1999 edition).

[9] Doc. No. 540 (1999 edition).

number of years after substantial completion of the project.[10] Since most professional liability insurance is written for one-year renewable terms, there is no guarantee that the carrier will renew the policy, and other than for project insurance, it may be impossible to get a certificate of insurance that shows that the policy will remain in effect for more than one year.

§13.11 DIRECTORS AND OFFICERS INSURANCE

This type of insurance covers claims made against the officers and directors of a corporation due to their position with a corporation. When outsiders, such as a lawyer, accountant or another advisor who is not an owner of the company serves on the board of directors, that person may be concerned about being exposed to some claim arising out of the company's acts. Providing insurance to defend and protect the board of directors from such claims will ease those concerns. When the board of directors consists of persons who are also shareholders in the company it is less common to carry this type of insurance. According to a 1999 study, only 27 percent of contractors reported carrying directors and officers liability insurance.[11] Project owners do not require this coverage, so it is a business decision whether to obtain this type of coverage to protect the officers and directors from a claim against them relating to the actions of the company.

§13.12 FIDELITY BONDS

A fidelity bond is a surety bond that protects a company against theft and embezzlement by employees of the company who are involved with the handling of money and company finances, including accounting personnel and anyone who disburses or receives money for the business. A blanket form of bond covering all employees is recommended. Other insurance policies do not protect a company from theft by an employee. Studies show that about half (46 percent) of contractors carry at least a $100,000 fidelity bond on their employees.[12] Again, owners do not require this and it is a business decision whether to obtain this type of bond to prevent a loss caused by internal embezzlement of company funds.

§13.13 CERTIFICATES OF INSURANCE

It is very common for construction contracts and subcontracts to require that certificates of insurance be submitted to the owner, architect, or contractor prior to the start of work. Larger contractors have staff whose job is to monitor

---

[10] ¶. 7.2.4, Standard Form of Agreement Between Design-Builder and Architect/Engineer for Design-Build Projects, AGC Doc. No. 420 (1999 edition); ¶. 6.5.1, Standard Form of Agreement Between Design-Builder and Designer, Doc. No. 540 (1999 edition).

[11] CFMA's *Construction Industry Annual Financial Survey* (1999).

[12] *Id.*

and collect these certificates. One recent study reported that over half of all contractors require certificates of insurance before allowing lower-tier subcontractors on site and that 57 percent require a certificate before making any payments.[13] These certificates will confirm that the proper insurance is in place and has not expired. What is often overlooked, however, is the upcoming expiration date. While a certificate may show the policy is currently in effect, be sure to note the expiration date. If the policy expires prior to the completion of the subcontractor's work, the contractor should require that a supplemental certificate be submitted to show that the policy was renewed.

Certificates should also show any additional insureds that are required under the contract. It is normal for owners to require that they be listed as an additional insured under a contractor's general liability (CGL) and automobile policy. It is also recommended that the prime contractor require the subcontractors to list the contractor as an additional insured on their policies as well.

Larry Maxfield, AIA, an architect-contractor, warns: "Do not let a sub on your job site without an insurance certificate for workers comp and general liability." His firm had to pay a claim for a subcontractor whose employee was injured and who did not have workers' compensation insurance. "Though this was paid by our insurer, it affected our company's EMR rating and increased our premiums. Never again." Expiration dates must be noted and, then the contractor should require subcontractors to provide a new certificate showing the renewal date of the policy.

With multiple trade subcontractors on multiple projects, the process of tracking certificates of insurance can be overwhelming if done manually. However, over recent years, a number of options for automating the certificate tracking process have been developed. These systems offer greater accuracy and efficiency than manual methods. Data can be tracked and sorted at the click of a mouse, with data kept for longer periods of time. Software can be loaded onto the user's computer to catalog coverages, verify compliance with insurance requirements, warn of policy expiration dates, and even generate letters requesting new certificates in the event of coverage non-compliance, expiration or cancellation. Some companies even offer "outsourcing" of certificate tracking services, including data entry, compliance verification, managing all related correspondence, and imaging of the original certificates. As a design-build practice grows, one should consider using these software products to help manage the firm's insurance program.

§13.14 BASIC DIFFERENCES BETWEEN CONTRACTOR'S AND ARCHITECT/ENGINEER'S INSURANCE

There is a fundamental difference between the type of insurance and risks covered under an architect's or engineer's professional liability (errors and omis-

[13] *Id.*

sions) policy, and a contractor's general liability (CGL) policy. A design professional takes out insurance with the understanding that, aside from the amount of the deductible, the insurance company will pay all damages that result from a design error caused by the insured's negligence, or by one of the insured's employees or subconsultants. Coverage is written very broadly, with specific exceptions that do not often come into play under normal claims situations. By contrast, a contractor's CGL policy does not cover correcting the contractor's own defective work—the most common type of "claim" made by owners. Further, a contractor's CGL policy only covers damages arising out of death or personal (bodily) injury or property damage. A design professional's insurance policy is much broader and covers "economic losses" in addition to bodily injury and property damage.

[A] Claims Made Versus Occurrence Coverages

Nearly all design professional liability insurance is written on a "claims made" basis. This simply means that the policy covers claims that are made within the policy period, regardless of when the work was actually done. Since these policies are usually written with one-year renewal periods, the claim must be made against the insured architect within that one-year period in order for there to be coverage under the policy.[14] In addition, each claim, as well as the legal fees and defense costs, eat away at the available policy coverage for that one-year period. So a firm with a $1 million policy that has two claims in the same year, one of which incurs $50,000 in legal fees and the other of which pays out a $200,000 settlement, reduces the available coverage for other claims by $250,000, in effect reducing the policy limits to $750,000—despite a certificate of insurance showing a $1 million policy in place.

This is very different from a contractor's CGL policy. Contractor' CGL coverage is written on an "occurrence basis," which means the policy in place when the negligent act occurred is the one that covers the loss, regardless of when the damage happens or when the claim is made. This means that if a contractor is covered by Aetna when the job is built, but switches to USF&G a year later, the claim is still covered under the Aetna policy if it is made even many years after completion of the project, though Aetna is no longer the current insurer for the contractor and receives no more premium payments.

Another key difference is the ability of a contractor to name additional parties as "additional insureds" to insurance policies. This is routinely done in the industry, giving owners, architects and engineers protection from claims that are made against them due to a contractor's negligence. By contrast, a design professional normally cannot have an owner or contractor named as an insured on the architect's professional liability policy.

---

[14] It is possible to purchase an "extended reporting period" endorsement that allows claims to be made and covered under the policy for a limited time after the policy expires. This is usually not available for extended periods longer than one to two years.

[B] Defense Costs and Decreasing Limits

Another major difference is that legal fees in defending a claim against a design professional are paid out of the policy limits, thereby reducing the actual coverage available to pay a settlement or a jury verdict. Some policies have what is called "first dollar defense," which means that the policy starts paying legal fees from day one of the claim, which reduces the policy limits every hour that the defense lawyer bills his or her time to the file. Most policies require the insured to satisfy a deductible and the insured pays the legal fees for the first $25,000 or so, depending on the deductible amount, before the insurer pays those fees from the policy limits. By contrast, a contractor's defense costs are paid outside of, *and in addition to* the policy's limits, so that the legal fees do not deplete the available insurance coverage. On design-build projects, when architects work with a contractor as a partner or as a subcontractor, they need to be sure to require the per project aggregate limit endorsement mentioned above, so that the policy limits are not diminished by claims on other projects.

[C] The Contractor's Faulty Workmanship Exclusion

If the contractor's work has to be replaced or repaired, the contractor usually pays for that out of its own pocket or backcharge to a subcontractor. The contractor's insurance only covers personal injury, death, or damage to property *other than* the contractor's own work. It does not cover repairing the defective workmanship. By contrast, if an architect's design error causes a wall to have to be torn down and replaced, the professional liability policy generally covers the cost of the reconstruction and any consequential damages or injuries, like lost rent, lost income, etc. The contractor, however, is not covered for the cost of replacing his own work, nor for the economic losses. If the defective wall falls over, the contractor's CGL policy will cover injuries to persons hurt and damage to property it fell on. Architects and engineers getting into the design-build business need to understand that unlike the design business, where most risks can be insured, a prime contractor self-insures for many risks, which are paid out of the "bottom line" profits on the project.

§ 13.15 PROFESSIONAL LIABILITY EXCLUSIONS

The typical professional liability policy carried by architects and engineers contains several exclusions to coverage related to design-build. It may be possible to remove these exclusions by endorsement, but this will vary with the insurer and the firm being insured. The standard professional liability policies for design professionals written by companies like CNA, DPIC, Lexington, Zurich-American, and others all contain similar exclusions, many of which relate to design-build projects, especially as a joint venture of some sort. Here are some of the common exclusions that should be discussed with an insurance agent before undertaking a design-build project:[15]

1. Liquidated damages (CNA); [16]
2. Penalty clauses (DPIC);
3. Fines and penalties (CNA, DPIC, Lexington, Zurich); and
4. The cost to repair or replace faulty workmanship in any construction performed by the insured (CNA, DPIC, Lexington, Zurich). Lexington adds to the exclusion any faulty construction by any "enterprise and/or any subsidiary of any enterprise" that the insured controls, manages, operates or holds ownership in, as well as any such enterprise that controls, manages, operates or holds ownership in the insured's firm.

DPIC has a special endorsement for design-build firms that excludes coverage under the professional liability policy for claims related to:

> assembly, construction, erection, fabrication, installation or supplying of products or materials which was performed in whole or in part or provided by:
> a. the Insured; or
> b. subcontractor(s) of the Insured; or
> c. a person or enterprise that wholly or partly owns, operates, manages or otherwise controls the Insured or is wholly or partly owned, operated, managed or otherwise controlled by the Insured; or
> d. an enterprise which is wholly or partly owned by or operated, managed, or otherwise controlled by any present principal, partner, director or officer of any Insured,
> except those Claims related to professional services of the Insured.[17]

This essentially means that the architect is covered for design errors under the policy, but not for construction errors. Before taking on a design-build project, architects and engineers should talk to their insurance agent about available endorsements for any of these exclusions so that they are aware of the risks that are covered and those that are not.

There have been court cases in which insurers have attempted to deny coverage under a professional liability policy on the basis that design-build work was not traditional architectural services. In a 1987 Illinois court case, two insurers denied that they had a duty to defend an architectural firm in an action for personal injuries brought by a construction worker who fell through a skylight.[18] The two insurers carried the architect's general liability insurance and its professional

[15] Based on CNA's Form G-126985-A, 1998 edition; DPIC's Form No. DP-37000-1, 1999 edition; Lexington's Form LEX-CM-PL7, 1997 edition; Zurich's Form No. U-PL-528-A CW, 1996 edition.

[16] There are some policies available to contractors today that will insure for liquidated damages up to $150 million. Minimum premiums for such coverage start at $250,000. *See* § 13.23, below.

[17] DPIC Design-Build Exclusion Endorsement (Form No. DP-37244-0).

[18] USF&G v. Continental Casualty Co., 505 N.E.2d 1072 (Ill. App. Ct. 1st Dist. 1987).

liability insurance. The professional liability carrier argued that it was not required to defend since the architect's design-build activities were more like those of a general contractor and that the policy only covered "traditional" architectural services. The court rejected this argument, however, noting that there was no distinction in the policy between "traditional" and design-build services. Therefore, the carrier had a duty to defend. This case points out a good reason for architects to obtain professional liability insurance and general liability insurance from the same insurer so that there is no dispute that the insurer has to defend the claim under at least one of the policies, thereby avoiding any gap in coverage and lawsuits like this one.

§13.16 THE INDEMNITY EXCLUSION

Construction owners increasingly seek to address risks by indemnity clauses in the design and construction contracts. Architects and engineers are advised (for insurance reasons) to limit such indemnity obligations only to damages "caused by the architect's negligent acts, errors or omissions." Contractors, on the other hand, are often asked to indemnify the owner and a host of others for all claims arising out of the project, not necessarily limited to those caused by the contractor's negligence. For contractors, it is possible to get a "contractual liability" endorsement on the CGL policy which covers the specific indemnity language in the contract. That same endorsement is not generally available to design professionals on a professional liability policy. If the architect takes on liability under the indemnity clause that the architect would not otherwise have, there may be no coverage. For example, the standard CNA Professional Liability policy excludes coverage for:

> liability of others you have assumed under a contract or agreement unless the liability is caused by your wrongful act.[19]

DPIC's exclusion is for:

> Claims related to liability assumed by you under any "contract." This exclusion does not apply if you would have been liable in the absence of that "contract" due to your own error, omission or negligent act.[20]

The Lexington and Zurich exclusions are similar to these two. A more specific exclusion is found in the standard contractor's CGL policy, excluding coverage for damages:

---

[19] The term "wrongful act" is defined in the policy as "a negligent act, error or omission in the performance of professional services for others by you or any entity, including joint ventures, for whom you are legally liable." CNA Policy Form G-126985-A (1998 edition).

[20] DPIC defines "contract" as any agreement, "written or oral, including but not limited to, hold harmless and indemnity clauses." DPIC Form DP-37000-1 (1999 edition).

for which the insured is obligated to pay damages by reason of the assumption of liability in a contract or agreement.

However, contractors can (and all design-builders should) get an endorsement to remove this exclusion from the policy. Again, design professionals should talk to their insurance agent and have the agent and their lawyer review the insurance and indemnity provisions of every design-build contract and advise them on what is insurable and what is not.

§13.17 THE WARRANTY EXCLUSION

Contractors are required to warrant their work for a set period of time. This is a standard part of every construction contract, usually for a period of one year from substantial completion.[21] Warranties can run longer than one year, however. For example, the I-15 design-build highway project in Salt Lake City has a 75 year warranty on bridges. It is typical to see specific guarantees required for roofing and other components for specific periods of time. For architects and engineers doing design-build, this creates an uninsurable risk when the design-build contract contains a broad warranty clause covering all of the contract work (both design and construction). Guarantees and warranties mean that there will be no errors, no leaks, no defects. That is a higher standard of care than architects are held to under the law, which is a standard that recognizes architecture as an "inexact science," and does not require perfection.

CNA, DPIC, and Lexington all exclude coverage for "express warranties or guarantees" under the standard professional liability policy. As a result, an architect needs to either qualify such guarantee as applying only to the construction work, or be prepared to self-insure the risk of a claim that is brought under the contract warranty clause as it relates to design work.

§13.18 THE JOINT VENTURE/JOINT OWNERSHIP EXCLUSION

Often design-build ventures between a design firm and a contractor are handled by forming a new corporation or limited liability company that is partly owned by the design firm and partly owned by the contractor. *See* Chapter 11. One of the drawbacks that must be considered is that such a joint venture can bar insurance coverage under the professional liability policy absent some endorsement. This depends upon just how much of the new design-build firm is owned by the insured engineer or architect. As noted below, the percent of ownership that triggers the exclusion can vary from 15 percent to 50 percent depending on the insurer. The insurer's concern here is that design firms might be more lax in quality control if they know that they can simply make a claim against their own insurance policy to cover design errors that cause a loss to the design-build firm.

---

[21] *See, e.g.*, AIA's A201 General Conditions, ¶. 12.2.2 (1997 edition).

Naming the new design-build firm as an additional insured under an architect's or engineer's professional liability policy can have the unintended result of barring coverage. Most policies flatly exclude coverage for "any claim made by any Insured against any other Insured." This means that the architect's new design-build firm (if named on the policy) has coverage for a claim by the project owner, but would have no coverage for a claim made by the design-build firm against the architect's design firm. If a design error or poor quantity take-off by the architect's design firm causes its contractor partner to underestimate the cost of construction, leading to a loss on the contract, the design-build firm cannot seek damages from the architect's professional liability policy to cover the loss.

CNA's 1998 edition of its professional liability policy excludes coverage for any claim made against you (the insured architect) by any entity that:

> is operated, managed or controlled by you or in which you have an ownership interest in excess of 15 percent[22]; or wholly or partly owns, operates or manages you.

Zurich's exclusion is for any claim by any individual, business or their subrogees or assignees:

> (1) that wholly or partially owns, operates or manages you; or (2) in which you have an ownership interest in excess of 20 percent; (3) that is controlled, operated or managed by you.

Under Lexington's policy, the exclusion is for claims by any company that:

> wholly or partly owns the Insured or which to any extent controls, operates or manages an Insured, or that is wholly owned or in which an Insured has a greater than 25 percent ownership, or which is controlled, operated or managed by an Insured.

The DPIC joint ownership exclusion bars coverage for claims:

> against you by (a) an entity in which you maintain a cumulative ownership interest of fifty percent (50%) or more, or (b) an entity that maintains any ownership interest in you.

This is an increase by DPIC from its 1996 policy provision that limited the percent of ownership to only 10 percent. These types of exclusions will prevent a separate design-build company (or subsidiary) in which the architect is a 50-50 owner with the contractor from being able to make a claim back against the insured architect that would be covered under the policy. If an architect is forming a

[22] This 15% is reportedly being increased to 49%, so that a firm can have a minority interest in a design-build firm or other company, and still be able to have coverage. Check with your agent to verify whether the policy can be endorsed to allow a higher percentage of ownership without losing coverage.

new company to hold design-build contracts, or forming a joint venture, which then subcontracts with his or her firm for design services, it is essential to get an endorsement to remove this exclusion from the policy. Otherwise there would be no coverage for claims made by the design-build firm against the architect subcontractor.

DPIC automatically covers joint ventures, but only for liability related to professional services performed by the joint venture. If an architect is forming a joint venture with a contractor, a specific endorsement should be obtained.

§13.19 THE CGL DESIGN EXCLUSION

Just as architects and engineers are not insured for construction work performed by them, contractors are not insured for design work done by them. The typical commercial general liability (CGL) policy for a contractor contains an exclusion for "professional services," defined as:

> The preparing, approving, or failing to prepare or approve, maps, shop drawings, opinions, reports, surveys, field orders, change orders or drawings and specifications; and inspection, supervision, architectural or engineering activities.

As a result, it is important to understand that insurance for a design-build firm will be split protection, if properly endorsed, with coverage under the professional liability policy for design errors and limited coverage for construction defects under the CGL policy. It is recommended that the contractor's general liability policy be amended by adding endorsements CG 2279 (total exclusion) or CG 2280 (limited exclusion), which are amended versions of the professional services exclusion that offers limited design-build coverage. The key difference between these two exclusions is that the limited exclusion does not apply to contractor operations in connection with construction work performed by the contractor or on its behalf. This endorsement provides coverage for the contractor's professional liability to the extent that the resulting damage is bodily injury or property damage *and* the design services were provided by the contractor.[23] The CG 2279 endorsement was developed in 1994 by the Insurance Services Organization (ISO). It was developed along with CG 2280 in response to a 1990 court case in which the insurer denied coverage when the shoring design on a construction project failed and resulted in consequential damages to an adjacent property.[24] The CGL policy in that case contained an "Engineers and Architects Exclusion" for any plans, specifications, or professional services by the contractor. The contractor's subcontractor hired engineers to design sheet piling and

---

[23] J. Kent Holland, Jr., Insurance, Risk and Surety in Design-Build, DBIA Seminar (August 1998).

[24] Harbor Ins. Co. v. Omni Constr., Inc., 912 F.2d 1520 (D.C. Cir. 1990); Terry R. Tenant, *Design-Build Insurance Issues for Owners*, DBIA/AIA Convention, Chicago, Illinois (October 1998).

shoring. The contractor argued that the exclusion was not intended to apply to the customary design work associated with "means and methods" of construction, but only to "non-traditional activities" such as design-build.

When the ISO endorsement CG 2279 is included there is no coverage for design services that are provided within the state licensing laws, but the policy does provide coverage for the incidental design services related to construction means, methods, sequences, and techniques. When insurance beyond this level of design is wanted, the contractor can obtain ISO endorsement CG 2280. Even with a CG 2280 endorsement, the CGL policy does not cover the cost of redesign or reconstruction (unlike a professional liability policy), but will only cover design-related consequential damages, such as bodily injury or property damage.

The intent of the insurance marketplace, under a CGL policy, is to provide "incidental" design coverage for the construction means and methods of the contractor—*e.g.*, temporary steel or trench bracing, cofferdams, concrete shoring and other engineering needed during the construction process—not to cover architectural and engineering design of the completed structure itself.

§ 13.20 DESIGN-BUILD INSURANCE POLICIES

Insurance representatives suggest that contractors who are serious about design-build should not rely on endorsements to their CGL policies that add back only limited professional liability coverage, since that only covers damages arising out of bodily injury or property damage at best. This leaves a huge gap in coverage for economic losses that are caused by design errors.[25]

Many insurers also offer a "Contractor's Design Liability Insurance Program" that covers design-build contractors for design errors. The policy covers design by: (1) in-house design staff; (2) design done by subcontractors; and (3) joint ventures with design firms. Limits of up to $15 million are available from some carriers. The policies often carry a minimum $25,000 deductible for general contractors.

§ 13.21 VALUABLE PAPERS INSURANCE

This insurance protects an architect or a design-builder from loss of documents by any of the means described in the policy. Documents in storage as well as work in progress can be covered, along with client documents in the care and custody of the architect. A contractor-partner should insist that the design-partner carry this coverage to protect its investment in the design documents prepared for a design-build team.

§ 13.22 SUBCONTRACTOR INSURANCE

It is important to set up adequate insurance requirements for subcontractors, whether they are design consultants or trade contractors. To the extent that an architect subcontracts with a trade contractor to design-build a portion of the

[25] J. Kent Holland, *supra*, note 23.

project (*e.g.*, mechanical, electrical, sprinkler), he or she needs to be aware that all these same insurance exclusions discussed above apply and must be endorsed or else the architect has limited protection in the event of a claim that is the fault of its subcontractor.

§13.23 FINANCIAL CONTINGENCY AND EFFICACY INSURANCE

There are more types of insurance available to cover more risks today than ever before. One of the products that has come on the market covers an area excluded by most insurance products. Financial contingency and efficacy risk (or FC&E) insurance is available from some carriers to cover such risks as force majeure delays and liquidated damages. These policies have very high deductibles and even higher premiums. But for the right project, it may be worth the investment. Some of the risks covered include project schedule delays and late completion; project performance guarantees, such as in output or other performance factors for electrical power plants; termination costs; cost overruns (due to estimating errors); loss of prime tenant; and unanticipated ground conditions, including archaeological discoveries. Such policies can carry limits of $125 million to $200 million and are written by companies like Lloyds of London, Zurich and others.

For example, on one 65 million wastewater treatment plant project, the design-build contractor was quoted a premium of $300,000 for a contractor's liquidated damages insurance policy where the prime contract had liquidated damages of $10,000 per day for each day late in substantial completion, plus performance liquidated damages based on shortfall in production. The policy had limits of $1.8 million on delay liquidated damages and $4 million for performance-related damages, with an aggregate policy limit of $5.8 million which would start only after the contractor had paid $500,000 in the aggregate in damages under either contract section, as a type of deductible. The design-build team decided to increase the cost of work by $800,000 to cover the premium and the deductible. In the end, the owner paid for the insurance through the increased contract price. If there were no claims, then the $500,000 deductible became another source of profit.

§13.24 OWNER-CONTROLLED INSURANCE PROGRAMS (OCIP)

One method of insuring construction projects, used sometimes on larger projects, is a controlled insurance program, also referred to as "wrap-up" insurance. When the owner administers this type of package it is called an "OCIP," short for an "owner controlled insurance program." When the contractor administers the program it is called a "CCIP." The program covers all participating contractors and subcontractors and can result in substantial savings by reducing the duplicative coverage and administrative costs. Typical CIP programs include builders risk, general liability, and workers' compensation. Centralizing the purchase of insurance has the benefit of economies of scale, but such programs are generally used only for very large projects on the order of $75 to $100 million. There is reduced litigation among the participants to the project, especially subrogation claims, since there is a single general liability and workers' compensation insurer for the entire project team. The program is "managed" by a program administrator.

§ 13.25 LIFE INSURANCE

So what does this have to do with design-build? Remember that teaming relationships are often based upon the trust built between the design and construction partners. The trust is often embodied in one individual at each company. What if that person dies? How will that affect the project and the ability to perform as required under the contract? Many companies obtain what are called "key man" plans, a policy of life insurance protecting the company from loss in the event of the death of a principal of the company. If a joint venture's success hinges on one or two key people, one might explore what the cost would be to include a term life insurance policy in the cost estimate for the project with the joint venture as the beneficiary. If an architect is setting up a longer-term design-build subsidiary, such life insurance policies should be part of the benefits packages and business planning that are put together just as they are with any company.

§ 13.26 CONSTRUCTION SURETY BONDS

In the public sector, nearly all construction contracts are required to be bonded by statute. Since the government does not want its buildings subject to mechanic's liens and sold at a foreclosure on the courthouse steps, government contractors must post payment bonds to protect the rights of subcontractors and suppliers.[26] In the private sector too, many owners want their contractors to be bonded, not so much to protect the rights of subcontractors, but to assure that the project will get built for the agreed price. Overall, a 1999 study reported that in construction generally, 42 percent of total contract revenues were bonded.[27] With design-build, some owners are concerned that one-time partnerships and LLCs being formed for a particular project may not be able to deliver on their promises. As a result, owners may require bonds to help assure that the team will perform. The bond guarantees to the owner that the project will be built and the bills will be paid by the contractor. What is of concern to the bonding industry, however, is whether the surety bond also will be held to cover design errors by the bonded contractor. Overall construction bond losses were on the rise in 1999, but an article by two managing directors for Marsh, Inc., stated that "[n]o data is available which indicates how much of this loss activity is from design-build firms or contracts, but sureties continue to be concerned with the long-term tail of exposure associated with bonds that guarantee design performance."[28]

---

[26] *See, e.g.*, 40 U.S.C. § 270a, known as the Miller Act, which requires contractors to post payment bonds on federal government projects. Most states have similar statutes, commonly known as "little Miller Acts" with similar requirements for state construction work.

[27] CFMA's Construction Industry Annual Financial Survey (1999).

[28] *Design-Build Surety and Insurance—Changing Times?* Dateline, DBIA, Mar. 2000, p. 6.

Surveys show that design-build firms are bonded on 20 percent of their work.[29] This, of course, means that by far the majority of design-build work is not bonded. This is good news for design professionals, who generally have difficulty getting bonded. If 80 percent of the design-build work is not bonded, then there should be plenty to go around. As one architect/contractor said, "Architects who want to do design-build can do as much as they want without bonding if they just pick their clients right."

[A] The Bond Application and Indemnity Agreement

Design professionals who have never had to apply for a bond before need to get ready for a new experience. Unlike insurance companies who insure a firm and pay damages that the insured causes in return for an annual premium, the surety wants more from the insured—and from the insured's family. In order to obtain a surety bond for a firm, the architect is required to personally guarantee any loss that the surety incurs as a result of writing bonds for the firm. Many bonding companies want all officers and owners of the company (and their spouses) to sign a General Agreement of Indemnity as part of the bond application process. These are some of the broadest and toughest indemnity clauses known to man, and they are vigorously enforced by most sureties who have to pay a bond claim.

The surety underwrites the bond losses on the presumption that it will recover any claim it pays from either the bonded construction company or, if it is not solvent, then from the personal assets of the indemnitors. The indemnitors do not even have to have any ownership in the company, just a strong financial statement. This personal guarantee is one of the most important legal documents an architect will ever be asked to sign since it allows the surety to sue the architect personally if his or her company suffers a bond loss. As one architect/contractor said from experience, "Concerns about your architect's license are nothing compared to the fear of losing your house." If the surety requires the signature of the architect and the architect's spouse on the indemnity agreement, a bond loss can open up the family's property and possessions to claims by the surety company.

As part of the bond application process, the bond agent will ask for company and personal financial records. The surety will also look at the experience and qualifications of the contractor to perform the type of work required for the project for which bonds are sought. If a firm has no experience managing a construction project, nor any staff with such training, expect a tough time getting bonds. Undercapitalized and inexperienced design-build firms will have difficulty getting bonded. The surety's key question is whether the bonded contractor can complete the project without defaulting. The ability to make the surety

[29] *1998 Design-Build Survey*, Zweig White & Assoc., Inc., p. 40; *1999 Design-Build Survey*, Zweig White & Assoc., Inc., p. 28.

comfortable with that risk can be a challenge when one is starting out in design-build as prime contractor.

It is reported that a bond for design-build costs three to four times as much as a standard construction-only bond. An architect should point out to the owner that he or she may prefer to spend that money on "bricks and mortar" or furnishings rather paying some bonding company.

[B] What the Surety Wants to See

A surety may require a showing of a specific level of net assets in addition to personal guarantees by the firm owners *and spouses* before they will even consider writing a bond.[30] Sureties use what they call a "5 percent case" or a "10 percent case" to establish bonding capacity. This is a mathematical formula based on the contractor's amount of working capital. For example, a contractor with a good history of successful projects and a financial statement showing a positive working capital of $500,000 might qualify for bonds of up to $5 million. On a "10 percent case" this is merely dividing $500,000 by 10 percent (or .10). The better the track record and financial statement, the lower the percentage used, so that a large contractor with good experience might have a "2 percent case." That same $500,000 in working capital could qualify the larger contractor for bonds totaling $25 million. According to a 1998 survey of design-build firms, the average line of bonding credit for design-build firms is $5 million per project. That same survey found that in 1999, the average bonding line had jumped to $20 million per project.[31] That may have been influenced by larger contractors responding to the survey, however.

Bonding agents are paid a commission ranging from 20 percent to 50 percent of the bond premium. A large portion of that commission, often up to half, is contingent upon the absence of losses under the bond.[32] A bond agent can lose thousands of dollars in contingent commissions if a single large loss occurs on one poor risk. As a result, bonding agents are very selective in whom they will write bonds for, and they will want to get to know the architect, the business, the partners, the staff, and the financial statement very well. The agent will want to meet with the bonded design-build firm at least annually and will be anxious to review the firm's financial statements.

One dilemma is that design firms frequently keep their assets low, so that claims are paid out of insurance coverage only, rather than from corporate assets.

---

[30] In the late 1990s, USF&G put together a design-build bonding program specifically for design firms. Some 40 to 50 A-E firms were bonded under the program, mostly engineering firms. It is reported that USF&G wanted to see $1 million of net worth, plus $1 million in current assets to qualify under the program. Audited financial statements and personal guarantees are also part of the bond application. Underwriters met with many engineering firms to assess the risk before agreeing to provide the bonds.

[31] *1998 Design-Build Survey*, Zweig White & Assoc., Inc., p. 40; *1999 Design-Build Survey*, Zweig White & Assoc., Inc., p. 29.

[32] *Obtaining Payment and Performance Bonds*, The Construction Contractor, June 1, 1998.

Sureties, however, want to see some significant retained earnings, working capital in the bank, and some physical assets. Not surprising, of design firms surveyed in 1998 and 1999, only 28 to 31 percent respectively of the design firms reported that they were able to get bonding for their design-build work.[33]

[C] What If an Architect Cannot Get Bonding?

Though most design firms can easily get professional liability and general liability insurance, most cannot get a bond. If the surety will not extend the architect a bond, even with personal guaranties, the architect will need to become creative. Here are some suggestions:

1. **Subcontract the Construction to a Bonded Contractor**—Even if the architect is the prime contractor, the owner will often accept a "dual obligee" bond from the architect's major subcontractors, whether it is one general contractor-sub, or multiple trade subcontractors. Nearly two-thirds of the design firms surveyed in 1999 (67 percent) reported that when they were unable to procure bonds for design-build work, they made arrangements with a contractor who was bondable.[34] Dual obligee bonds were used 75 percent of the time.
2. **Letters of Credit**—Though owners out of habit ask for 100 percent performance and payment bonds, the risk of a loss to the owner is normally far less than the total contract value. An architect should have a candid discussion with the client about what the risk is at any time during the project, especially if the owner is holding retainage (a type of cash bond). Some owners will accept a letter of credit for a percentage of the contract amount, rather than a 100 percent bond. The letter of credit is drawn on the architect's bank and conditioned upon his or her successful performance of the work and payment of all bills. This should be drafted by the architect's lawyer to prevent the owner from too easily making a claim against the Letter of Credit without establishing an actual default and quantifiable damages.
3. **Small Business Administration Bonding Programs**—If insufficient financial strength is holding an architect back from getting a bond, he or she should check into the Small Business Administration (SBA) surety bond guarantee program, which guarantees bid, performance and payment bonds for contracts up to $1.25 million for eligible small businesses that cannot obtain surety bonds through commercial channels.

[33] *1998 Design-Build Survey*, Zweig White & Assoc., *1999 Design-Build Survey,* Zweig White and Assoc.

[34] *1999 Design-Build Survey*, Zweig White & Assoc., Inc., p. 28.

§ 13.27 THE PERFORMANCE BOND

A performance bond guarantees to the owner that the bonded contractor will perform all contract obligations. When the bond is for a design-build contract, this includes both the design and construction work. The performance bond of a design-build contract may require special drafting. The surety industry takes the position that a contractor's bond is to cover only construction defects, not design errors. The sureties argue that a bond is not the equivalent of an architect's errors and omissions insurance policy. However, some courts have held that a performance bond for a design-build contract covers the contractor's entire obligation, including the design negligence of its architect.[35] This may present bonding problems absent special bond provisions tailored to the job. Texas design-build statutes for public work state that payment and performance bonds are not required for the design services portion of the contract.[36]

The AGC publishes four different forms of design-build surety bonds, with alternate payment and performance bond forms where the surety either does or does not include design services within the bond's coverage.[37] Experience has shown that sureties are increasingly willing to underwrite even the design risks if the contractor requires the design professionals to carry professional liability insurance, and if the surety can be subrogated to the contractor's rights against the insured design firms. The AGC Design-Build Performance Bond No. 470 states:

> LIMITED LIABILITY FOR DESIGN. This Bond shall cover the costs to complete the Work, but shall not cover any damages of the type specified to be covered by the Design-Builder's Liability Insurance or by the Professional Liability Insurance required pursuant to the Contract, whether or not such insurance is provided or in an amount sufficient to cover such damages.

In the traditional method, the architect provides some protection for both the owner and the surety by monitoring the contractor's progress and certifying payments. Court cases have held that an architect has duties to the surety when approving work and certifying payment.[38] When the architect is also the contractor, however, the surety feels it is missing a project neutral to protect it from losses. The architect's independent role is somewhat diluted when he or she has a

---

[35] *See, e.g.*, Nicholson & Loup, Inc. v. Carl E. Woodward, Inc., 596 So. 2d 374 (La. Ct. App. 1992).

[36] V.T.C.A., Educ. Code §§ 44.036(j) and 51.780(k).

[37] AGC Documents No. 470 and 472 (Performance and Payment Bonds Where the Surety is Liable for the Design Costs of the Work), and Documents No. 471 and 473 (Performance and Payment Bonds Where the Surety is NOT Liable for the Design Costs of the Work) (1999 edition).

[38] *See, e.g.*, Aetna Ins. Co. v. Hellmuth Obata & Kassabaum, 392 F.2d 472 (8th Cir. 1968).

financial stake in the project outcome, even more so when the architect is a subcontractor to the bonded contractor. Sureties may want to hire their own independent consultant to monitor job progress and payment in these cases.

§13.28 THE PAYMENT BOND

The payment bond guarantees to the owner that the contractor will pay its bills, thereby keeping subs working and avoiding mechanic's liens on the property. As explained in more detail in Chapter 5, the ability to recover on a contractor's payment bond may depend on the wording of the bond or the statute that requires bonds, as to whether the bond covers "services" or just labor and materials. One should check the wording of the bond before starting work. This will be an unusual request from a subcontractor, but it is one that makes sense on design-build jobs.

CHAPTER 14

PUBLIC SECTOR DESIGN-BUILD

§14.01 SOME HURDLES TO OVERCOME

Design-build presents some challenges in the public sector based on traditional methods of procuring construction.[1] Governmental agencies are used to awarding construction contracts based on low bid, with cost being the only factor. This low-bid mentality is deeply entrenched into both the public agencies' and the contractors' way of thinking and their way of doing business. Design-build is strongly opposed in many states by contractors who are not comfortable with subjective factors like qualifications, experience and design entering into the hiring process. They are set up to win jobs based on providing a bid bond and a low bid, based on someone else's design. Not only do contractors need to adapt to this new procurement method, but public agencies and their staff need to learn how to go about hiring design-build contractors. "Most municipalities don't have anyone on staff who has ever done design-build," said Janice Burnett, Executive Director of the Arizona Society of Professional Engineers, following passage of a sweeping public design-build law in 2000.[2]

The ASCE's 1992 Policy Statement on Design-Build in the Federal Sector recognizes that government agencies "will need to have strong in-house architectural and engineering expertise to package the design-build requests for proposals (RFPs); and to evaluate, manage and assess owner/user requirements throughout the design-build delivery process." With the increasing number of states passing design-build legislation, educating public agencies on how to use design-build for their projects will be both a challenge and an opportunity for design firms, contractors and their professional organizations.

Despite the legal and procurement hurdles of the past, there is significant growth in public-sector design-build, drawing many firms to focus on that market. "We finally think design-build is going to emerge on the public side, so we are consciously shifting our focus [in 2000] toward public work," said Ed McSpedden of HNTB Design-Build, Inc., Los Angeles.[3] The volume of new transportation work planned for the public sector is so compelling, according to McSpedden, "it makes it hard to argue that we should focus on other areas." Appendix E to this book shows that a majority of states have passed laws to permit public design-build in some form. As these types of laws catch on, new opportunities will open up for firms that concentrate on the public sector.

§14.02 PUBLIC LAWS ON DESIGN-BUILD

[A] VARIATIONS IN DESIGN-BUILD PROCUREMENT LAWS

State and local laws regarding design-build tend to fit into one of four categories: (1) those that expressly prohibit design-build (of which there are very few); (2) those that pose obstacles to design-build, while not prohibiting it;

[1] *See generally* K. Roberts and N. Smith, *Design-Build Contracts Under State and Local Procurement Laws*, Public Contract L. J., Vol. 25, No. 4, Summer 1996.

[2] *With a Push from PE's, Major Design-Build Measure is Enacted*, Engineering Times, May 2000.

[3] *Commercial Outlook*, Design-Build, Dec. 1999, p. 41.

(3) those that pose no obstacles, though they do not expressly permit design-build; and (4) those that expressly permit design-build. There is an abundance of laws that fit into the second category, in that they place obstacles to the use of design-build. These would include competitive bidding statutes, qualifications-based selection laws, appropriations statutes, subcontractor listing statutes, laws that prohibit awarding a single contract to a general contractor (by mandating separate trade contracts), and licensing laws. Laws fitting the fourth category, expressly permitting design-build, are increasing each year as more legislation is introduced to permit public agencies to use design-build on a particular project, a group of pilot projects, or wholesale use throughout a state or agency.

[B] Model Procurement Acts

The Design-Build Institute of America (DBIA) published a Model Regulation for Design-Build Procurement in October 1996, DBIA Document No. 402. The model law suggests that the public agency solicit qualifications statements by public advertisement, then short-list three to five teams based on those qualifications. A design criteria package is prepared by the agency using a design professional selected pursuant to the applicable professional selection law or a design professional on staff of the agency. The criteria package is then given to the short-listed teams to submit two proposals: (1) a "qualitative proposal" with preliminary designs, outline specifications, schedule and other data; and (2) a "price proposal," submitted in a separate sealed package. The agency then is to review the qualitative proposal and establish a rating for each team's submission. The price proposals are then to be opened publicly and the proposal with the highest combined score, offering the "best value" of quality and cost, being awarded the contract. This "two envelope" system is the most common system used by public agencies for design-build procurement. DBIA's model law also includes recommended procedures for resolving bid protests.

DBIA also publishes a guide for Enactment and Application of the Model Design-Build Procurement Act, DBIA Document No. 401. Included in that guide are recommendations on getting a statute or ordinance passed to permit public design-build by a state or local government. The guide also reprints another model act, the Model Design-Build Procurement Act, drafted by the Building Futures Council, plus a "short form" model act, which would serve as general enabling legislation to authorize design-build, but leave the particulars of procedure up to later regulation. So far, these model acts have not been adopted as written, but portions and variations have been passed in several states. *See* Appendix E to this book.

§ 14.03 APPROPRIATIONS RESTRICTIONS

One of the first drawbacks to public design-build is the public appropriations process used to fund construction. For many public entities, design fees are appropriated in one year and the construction funds appropriated in another, based on the designer's estimate. For design-build to be used, there needs to be a single appropriation of all the funding in the same year. This means a significant change in how construction funding is handled. Many public bodies are used to hiring a design firm who then prepares an "engineer's estimate" of what the project will cost. That figure is used to establish a cost to be plugged into the annual budget approved by the state legislature. With design-build, the public body will need to either hire a cost consultant to help prepare the estimate in advance of hiring the design-build team, or have state staff do rough estimating, even without a design, in order to come up with a figure to be put into the budget for approval by the legislature.

§14.04 THE LEGISLATIVE PROCESS

A recent survey found that about a third of design-build firms have encountered laws preventing them from participating in public design-build projects.[4] Changing the law to permit design-build in the public sector means getting acceptance from the state legislature, the contractors and their state associations, labor unions, and the design community. Getting all these groups to agree on anything is a challenge. The best approach seems to be building consensus by allowing each group to participate in the bill drafting process.

For example in Washington, prior to writing legislation to authorize use of design-build, the state Department of Transportation (WSDOT) solicited input and support from the engineering associations and the state Associate General Contractors of America (AGC) chapter, all of whom were pleased to be part of the development process.[5] As the legislation progressed, these groups were asked to participate and the legislators relied on their input before making decisions. The law passed in Spring 1998, but the public input did not stop there. In October 1998, a meeting was held with some 60 individuals to obtain comments on the legislation and the process to be used by the state, with a report published on these meetings in February 1999.[6] Before doing the first project, the WSDOT got input on an RFP for the pilot projects, publishing a draft RFP on the state's Web site for comment by interested parties. The final RFP was prepared based on those comments. This type of open process helps each group and the public to "buy in" to the process and feel a sense of authorship.

§14.05 THE FEDERAL BROOKS ACT AND QBS

When contracting for design services, the federal government is required to select architects and engineers based on qualifications and level of competency rather than on fees alone.[7] These statutes, known as The Brooks Act, after former Texas Congressman Jack Brooks, have been the model for many state procure-

---

[4] *Design-Build Survey*, Zweig White & Assoc., Inc., 1998, p. 33.

[5] Keith Molenaar, Ph.D., *Stakeholder Input in Design-Build Highway Construction*, DBIA Conference, Dallas, Texas (Oct. 1999).

[6] The input from the construction community at these meetings is available at <http://www.wsdot.wa.gov>.

[7] 40 U.S.C. §§ 541-544.

ment laws, known as "mini-Brooks" Acts. The federal law states that Congress declares it to be "the policy of the Federal Government to . . . negotiate contracts for architectural and engineering services on the basis of demonstrated competence and qualification" with the government to "negotiate a contract with the highest qualified firm" for such services.[8] The concept behind the law is that since federal projects are built by the lowest bidding contractor, the plans for the project should be developed by the highest qualified design firm. Government buildings need to be of high quality to serve the public over time, so we want the most qualified firms to design them. Since the contractor with the lowest bid is required to build the same building as the high bidder, it makes sense to award the construction contract based on competitive bidding. All things being equal, the public should get the same building from either the lowest or the highest bidder.

The concept of Qualifications-Based Selection (QBS) is widely used in the public sector for hiring of professionals. Even the DBIA Code of Professional Conduct states that "[m]embers shall select design professionals, contractors and the other principal teammates on the basis of qualifications and best value."[9] With architects and engineers to be selected based on qualifications, but construction contractors selected based on price, this creates a conflict when design and construction are combined together in one contract, and in one procurement process. If a governmental body solicits competitive bids from design-build contractors, with the low bidder being awarded the contract, the project architect or engineer is selected *by the contractor* rather than by the government. In such cases, the Brooks Act procedures are not necessarily followed in selecting the design professionals. The low bidder may be more interested in hiring the cheapest design firms, not the most qualified. As a result, design-build may run counter to state and federal selection laws.

[A] Brooks Act Challenges to Design-Build

At least one court has found that design-build based on competitive bids violates the state mini-Brooks Act. In *City of Lynn Haven v. Bay County Council of Registered Architects, Inc.*,[10] the Florida Court of Appeals stopped a city from allowing a design-build contractor to select the architect for a public project. In effect, by requiring the successful low bidder to provide architectural drawings, the court found that the city's procedures violated the Florida selection law by allowing the contractor with the low bid to select and hire the architect. A group of architects obtained an injunction to stop the city's practice. On appeal, the court held that, "the City's procedures contravene legislative intent and undermine the effectiveness of the [Procurement Act]."

The Texas Attorney General issued a similar opinion in 1990 that design-build contracts awarded on the basis of competitive bids violate the Texas Profes-

---

[8] *See also* Federal regulations at 48 C.F.R., Subpart 36.6—Architect/Engineer Services.

[9] *DBIA Code of Professional Conduct,* Policy No. 11, Oct. 1998.

[10] 528 So. 2d 1244 (Fla. Dist. Ct. App. 1988).

sional Services Procurement Act. The Attorney General stated that the Act prohibits the procurement of architectural or engineering services by competitive bids.[11] The Arkansas Attorney General issued an opinion in 1993 that design-build violated the state competitive bidding laws and questioned whether it would also violate the state's architect-engineer selection statute.[12]

In 1992, Illinois amended its "mini-Brooks" law to clarify that the state can contract for design-build projects.[13] The state adopted a two-phase procurement process for transportation projects.[14] Texas also enacted a QBS law for public design-build. Texas adopted a two-phase process for design-build contracts for educational facilities.[15] The Texas laws require the school district to contract directly with an inspection firm for field inspections or testing of materials, separate from the design-build contract. These statutes are all reprinted as part of Appendix E to this book. They serve as models of how some states have integrated QBS into design-build procurement.

§14.06 COMPETITIVE BIDDING LAWS

Design-build contracts in the public sector are not normally awarded based on price alone. As described in Chapter 15, public design-build is more often awarded based on a combined ranking of price and qualifications of the design-build team. The team with the best score gets the contract, not necessarily the low bidder. But aside from design-build, public contracts for construction are almost always let by competitive bidding to the lowest bidder. One survey of the industry showed that 58 percent of U.S. construction work was acquired through competitive bidding.[16] As a result, competitive bidding laws are the main roadblock preventing many public agencies from using design-build. The legislatures have expressed a desire for price-based selection as a way to increase competition and to prevent corruption and undue influence. However, when selecting design firms, those same legislatures have expressed a desire to hire the most qualified design firms based on qualifications, rather than on price.

If design-build projects were awarded on competitive bids, the owner would be limited in its ability to weigh the qualifications of the designers as well as the quality of the proposed design in awarding contracts for combined design and construction. The lowest bid may not be the best design, nor in the best interest of the public in terms of quality of materials and equipment, aesthetics and public image, overall cost of operation, and maintenance. If design-build selection was based solely on cost, there would be incentive to cut corners on quality in order to reduce the price and assure the award of the contract. Problems frequently come

[11] Tex. Atty. Gen. Op. No. JM-1189 (July 24, 1990).

[12] Ark. Stat. Ann. § 19-11-801. Ark. Atty. Gen. Op. No. 93-051 (1993).

[13] 30 I.L.C.S. 535/75.

[14] 70 I.L.C.S. 3615/4.06.

[15] V.T.C.A. Educ. Code §§ 44.036 and 51.780.

[16] *Construction Industry Annual Financial Survey*, Construction Financial Management Association (1999).

up when a public agency attempts to award a design-build contract using a combined price-plus-qualifications procurement, without any statutory authority to depart from competitive bidding laws. These awards are often challenged by a low bidder or a member of the public who insists on strict adherence to competitive bidding laws.

[A] Competitive Bid Challenges to Design-Build

While some feel that in design-build "price" should not be used as the basis to award contracts, others feel price should be the *only* basis for selection. In several court cases, the design-build process has been challenged as violating state competitive bidding laws for public projects. These cases are discussed in the next sections.

[B] Cases Finding No Violation of Competitive Bidding Laws

Even in states where design-build is not expressly authorized, some public bodies have found ways to avoid competitive bidding, though they have not always avoided litigation. For example, the use of design-build for an elementary school was questioned in an Indiana case, *Negley v. Lebanon Community School Corp.*[17] A private not-for-profit building corporation was formed to own and construct a new elementary school to be leased by a school corporation. After the state Attorney General issued an opinion that design-build violated state competitive bid laws, the school superintendent stopped the project. The building corporation and the school corporation filed suit and successfully enjoined the superintendent from halting the design-build process. The court held that such private building corporations were not bound by state competitive bidding laws and, therefore, could use design-build for the project.

In a 1999 Alabama case, the state Supreme Court ruled that it was not a violation of state competitive bidding laws to award an engineering contract that also included the "incidental" procurement of $1 million in equipment.[18] A local board of education negotiated a contract to perform an energy audit on county school facilities for a fee of $35,750. The contract was awarded to The Trane Company, which led to a follow-on contract that called for Trane to design, furnish and install systems needed for energy conservation, guaranteeing a certain level of energy savings. A taxpayer challenged the contract as violating state competitive bidding laws. However, the Court ruled that the energy audit contract with Trane was exempt from competitive bidding under the "engineering services" exception. Two judges dissented, questioning how such a large purchase of equipment could be considered "incidental" to such a modest amount of engineering services.

In a 1997 Kansas case, the state sought proposals for a "build to suit" lease. When the contract was not awarded to the low bidder, suit was filed claiming that

---

[17] 362 N.E.2d 178 (Ind. Ct. App. 1977).

[18] Anderson v. Fayette County Bd. of Educ., 738 So. 2d 854 (Ala. 1999).

Kansas law required awarding of construction contracts based on competitive bids.[19] The court of appeals acknowledged that the "build to suit" lease may require some construction, but held that its primary intent was to lease real estate over 10 years, and that the work required to prepare the building would not be performed over a significant portion of those years. The court held that there is no requirement that lease agreements be competitively bid. This may point the way for public bodies to avoid competitive bids by wrapping the construction work into a lease agreement.

Two Wisconsin cases have approved use of design-build on public projects.[20] The court found that the use of design-build on public projects was "innovative" and represented a substantial savings to the state over the conventional system of contracting. Likewise, the Alaska Supreme Court upheld the use of design-build over challenges that a city charter and municipal ordinances required competitive bidding.[21] Courts in New York and Ohio have also upheld public design-build contracts as not violating competitive bidding laws in those states.[22]

[C] Design-Build Held to Violate Competitive Bidding Laws

However, courts in other states have ruled the opposite way. Illinois and Minnesota courts declared design-build contracting illegal and a violation of competitive bidding laws.[23] In several states, the Attorneys General have written opinions that the state could not do design-build because it violated competitive bidding laws. For example, in Virginia, the Attorney General wrote that absent an emergency, a "fixed-price design-build" procedure could not be used for state public works contracts.[24] The Attorneys General of Idaho, Kentucky, Indiana, New York, South Carolina, New Hampshire and New Mexico have all found that design-build was at odds with public bidding laws.[25]

[D] Wicks Law Cases

Since 1921, under New York's "Wicks Law," separate contracts must be awarded for plumbing, heating and air conditioning, and electrical work on state projects over $50,000.[26] Pennsylvania and Massachusetts both have similar laws

[19] Decker v. Kansas Dept. of Social & Rehab. Serv., 942 P.2d 667 (Kan. App. 1997).

[20] Waste Mgt. Inc. v. Wisconsin Solid Waste Recycling Auth., 267 N.W.2d 659 (Wis. 1978); J.F. Ahern Co. v. Wisconsin State Public Bldg. Comm'n, 336 N.W.2d 679 (Wis. Ct. App. 1983).

[21] Breck v. Ulmer, 745 P.2d 66 (Alaska 1987).

[22] Imburgia v. City of New Rochelle, 645 N.Y.S.2d 111 (App. Div. 1996); Greater Cincinnati Plumbing Contractors' Ass'n v. City of Blue Ash, 666 N.E.2d 654 (Ohio Ct. App. 1995).

[23] Smith v. Inter-Governmental Solid Waste Disposal Assoc., 605 N.E.2d 654 (Ill. App. Ct. 4th Dist. 1992); W.V. Nelson Const. Co. v. City of Lindstrome, 565 N.W.2d 434 (Minn. Ct. App. 1997).

[24] 1977-78 Op. Va. Atty. Gen. 91.

[25] *See* 1978 Op. Idaho Atty. Gen. 71; 1980-81 Op. Ky. Atty. Gen. No. 81-381 (Oct. 8, 1981); 1976 Ind. Atty. Gen. Ann. Rep. & Official Op. No. 76-8; 1982 Op. N.Y. Atty. Gen. 54; 1988-89 S.C. Atty. Gen. Ann. Rep. 227 (Dec. 6, 1988); 1985 Op. N.H. Atty. Gen. No. 85-117; 1983 Op. N.M. Atty. Gen. 5.

[26] N.Y. Gen. Mun. Law § 101(1), (2); N.Y. State Fin. Law § 135.

requiring separate contracts to trade contractors. It has been said that these laws were enacted during the Great Depression to protect subcontractors from abusive bidding practices. At least one study concluded that such bidding practices cost the state more money due to change orders, delays, and an adversarial tone. The study focused on over 900 Massachusetts projects compared to states that permit design-build and other alternate forms of project delivery and found Massachusetts was paying 3.5 times more than Texas and 9 times more than Indiana.[27] Nonetheless, without specific statutory authority, design-build might violate these laws since all the trade contracts are rolled into one team and one prime contract.

In a 1995 Pennsylvania case, bids were solicited for a public bus maintenance facility under a single design-build contract.[28] The mechanical contractors association sued for an injunction arguing that state law required that the public entity directly award separate contracts to the lowest bidders for plumbing, heating, ventilating and electrical work (called the "Separation Act" in Pennsylvania), and that the Transportation Authority could not award a single design-build contract and delegate the selection of these trade contracts to the design-build contractor. The Transportation Authority argued that it required the design-build contractor to be responsible for complying with the Separation Act. The court held that the Act requires the award of trade contracts by the public entity, and that this function could not be delegated to a general contractor. Therefore, the design-build project was enjoined from going forward.

§ 14.07 HOME RULE STATUTES

In many states, local municipalities are not required to take competitive bids and can regulate themselves under their local charter, known as "home rule." In California, for example, unless the legislature has imposed competitive bidding requirements on a local agency, that agency is not required to competitively bid its contracts.[29] The California Constitution exempts some charter cities from competitive bidding and permits those cities to elect in their charters to have the exclusive power to legislate their local affairs. However, with respect to matters of statewide concern, even home rule cities are subject to general state laws that preempt municipal regulation.

One example of a home-rule charter city is Los Angeles, California which awarded a $712 million design-build contract as part of a joint agency formed with the City of Long Beach to build a 10-mile-long, double track rail system to be built some 30 feet below street level, called the "Mid-Corridor" project.[30] The two cities formed the Alameda Corridor Transportation Authority (ACTA), an

---

[27] D. Gransberg, *Change Procurement Laws*, Engineering News Record, Dec. 6, 1999 p. 71. The full study is available on-line at <http://www.pioneerinstitute.org/>.

[28] Mechanical Contractors Ass'n v. Southeastern Pa. Trans. Auth., 654 A.2d 119 (Pa. 1995).

[29] San Diego Service Auth., for Freeway Emergencies v. Superior Court, 198 Cal. App. 3d 1466 (1988).

[30] As reported by B. Papernik and N. Smith in *By Design*, Los Angeles Lawyer, July-Aug. 1999, p. 32.

agency that has powers common to both cities, including the procurement requirements of the city charter. The Los Angeles charter generally requires that contracts be let based on competitive bidding but provides an alternate procedure for competitive negotiations based on "ultimate lowest cost." ACTA took steps to use this alternate contracting procedure for the Mid-Corridor project. ACTA adopted a prequalification process to make sure that each design-build team had appropriate experience and qualifications. It also required bidders to identify their design subcontractors as well as other known trade subcontractors, and to submit a plan describing how competitive bidding would be used to select all other major subcontractors. Once a subcontractor was selected, no substitutions would be allowed except as permitted by state bidding laws. This project is a good example of combining qualifications-based selection with price-based selection.

§14.08 PUBLIC/PRIVATE DEVELOPMENTS

One possible method of avoiding Brooks Act, competitive bid and prevailing wage laws on public design-build projects is the use of "turnkey" development by private entities who build and own the project, then lease or sell the completed project and land to the governmental body. However, without specific legislation authorizing such turnkey construction by private developers on public projects, it is questionable whether a public body can contract for design-build and thereby avoid architect-procurement and competitive bidding laws. Even though privately developed, the project may nonetheless be viewed as a "public works," subject to legislative restrictions.

For example, in a Missouri case, a city solicited bids from private developers to purchase city-owned lots, build a firehouse/police station on the property, and then lease the completed project back to the city with an option to purchase.[31] The city had the option to buy back the improved land for $10 at the end of the lease. Even though built on land owned by a private entity, the Missouri Court of Appeals found the project to be a "public works," constructed for public use or benefit, and therefore subject to legislative restrictions. The court stated that "[a] city may not do indirectly that which it cannot do directly," finding that the developer was obligated to pay prevailing wages just as the city would have been required to do. The court stated that to hold otherwise "would lend judicial approval to an easy method of nullifying the [prevailing wage] Act."

In a similar Pennsylvania case in 1998, the City of Harrisburg, Pennsylvania entered into a contract under which the City agreed to remove asbestos from and then demolish a downtown building, then convey the site to an insurance company who agreed to build its new headquarters there rather than move to a suburban location. The new $30 million headquarters building would be financed under various Pennsylvania financing laws. Upon completion, however, the insurance company would own the building. The Prevailing Wage Division of the state Department of Labor determined that the entire project site was a private project, not subject to pre-

---

[31] State ex inf. Webster v. Camdenton, 779 S.W.2d 312 (Mo. App. S.D. 1989).

vailing wage law. Several unions filed grievances with the Prevailing Wage Appeals Board which determined that the asbestos removal portion was a "public works" within the meaning of state law and, therefore, the subsequent construction of the office building was subject to prevailing wage law as well. The state Supreme Court held that there was no question that the site preparation work constituted "public work," but ruled that the subsequent office building construction did not automatically become subject to the prevailing wage law unless the financing constituted "payment out of the funds of a public body." Since there was no clear evidence on that point, the court sent the case back to the trial court for more hearings.[32]

These cases point out that it is sometimes a fine line between what is public works and what is not. Attempts to avoid public laws merely by transferring property to a private entity does not always take the project out of the public sector, and compliance with prevailing wage laws and competitive bidding laws can significantly affect the viability of the project.

§ 14.09 LEGISLATIVE DEVELOPMENTS

To avoid court challenges to design-build, a growing number of state and federal laws have been passed to permit the use of design-build in both the public and private sectors. As Appendix E to this book shows, most states have authorized, acknowledged, or otherwise addressed the use of design-build as a means of construction contracting. Probably the most aggressive state in legislating approval of design-build is Florida. Following the court's decision in *City of Lynn Haven* in 1988,[33] legislation was passed the next year which revamped the state's professional and contractor licensing laws. The new laws provide that licensed general contractors need not be licensed as architects in order to offer design-build services as long as the design services are rendered by a licensed architect.[34] Similarly, architects and engineers are not restricted from offering or rendering design-build services as long as a licensed general contractor actually does the construction work.[35]

For public projects, Florida's "Consultants' Competitive Negotiation Act" now provides that design-build may be used by selecting a design-build firm based on competitive proposals after a package of design criteria has been prepared by a "design criteria professional."[36] The design criteria professional is selected by the public body in accordance with the mini-Brooks law procedures. That firm is thereafter not eligible to render services under the design-build contract. The design-build contract is then awarded based on competitive bids from at least three design-build firms. Florida's model has been the basis for many other public agencies and has influenced the selection process in many states.

---

[32] Pennsylvania Nat'l Mut. Cas. Ins. Co. v. Department of Labor & Indus., 715 A.2d 1068 (Pa. 1998).

[33] *See* § 14.05[A].

[34] Fla. Stat. Ann. § 481.229(3).

[35] Fla. Stat. Ann. § 489.103(16).

[36] Fla. Stat. Ann. § 287.055(10).

§14.10 PREPARING AND RESPONDING TO RFQs AND RFPs

There are several publications on preparing and responding to RFQs and RFPs for design-build work. DBIA publishes a guidebook for owners called *Design-Build RFQ/RFP Guide for Public Sector Projects.* The AIA and AGC jointly published a document in January 1995 called *AIA/AGC Recommended Guidelines for Procurement of Design-Build Projects in the Public Sector*. The Guidelines are endorsed by DBIA. The Guidelines recommend the following basics:

1. Adopt general criteria to determine what projects will be appropriate for design-build.
2. Formally adopt general procedures for selecting design-build contractors.
3. Review local laws and regulations that might limit design-build.
4. Prepare a solicitation that clearly spells out the procedures to be followed in:
   i. Selection of the design-builder; and
   ii. Management of the project.
5. Set out criteria to be used for selection, and identify the composition of the selection panel.
6. Provide assurance that the project is fully funded.
7. Set out the scope of work, program, equipment needs, etc.
8. Provide site information, survey and borings.
9. Provide all budget requirements, MBE/WBE requirements, schedule and an outline specification.

The Guidelines go on to describe the two-phase selection process recommended by the AIA and AGC, similar to that used by the federal government.

§14.11 TWO-PHASE SELECTION PROCESS

In order to market design-build, design firms have to understand how clients go about hiring design-build contractors. A growing method of procurement is "two-phase selection." AIA, AGC, and DBIA all recommend a two-phase process, where design-build teams are first ranked based on qualifications to prequalify up to five teams; then, in the second phase, teams submit cost proposals. The team with the best overall score is awarded the contract. The AIA recommends the payment of a stipend to the unsuccessful teams to help offset the cost of preparing the design and proposals. See § 14.15, below for discussion of stipends.

Under the AIA/AGC Guidelines, selection criteria should state what "weight" will be given to each criterion. In the first phase, after receiving submittals, a "short list" of prequalified finalists is compiled of three to five design-build teams or entities. The short list may be based on written submittals alone, or may involve personal interviews. The criteria used are limited to:

1. the ability to satisfactorily carry out the project design and construction requirements;

2. past performance of team members;
3. relevant experience of the team or team members; and
4. financial capacity to perform.

Price quotations from competitors are not to be considered at this stage, in order that price does not taint the review of qualifications. In the second phase, the competitors submit on design and construction approaches to the project; ability to meet program requirements; management plan; and, finally, cost of the project. The weight given to price varies, but it is recommended that this decision be made prior to the design-build solicitation. The two-phase process is a way to limit the field to only the most qualified teams, who then compete further to see who is most qualified and who offers the best overall value.

§ 14.12 WEIGHTED SELECTION CRITERIA/ADJUSTED LOW BID

There are many variations on how public agencies rank design-build proposals. No single process is right for every project. Therefore, the owner must determine which procedure best fits the particular project. DBIA's publication, *The Design-Build Process Utilizing Competitive Selection* (1994), discusses two methods commonly used—*i.e.*, weighted criteria and adjusted low bid.

With weighted selection criteria, the owner establishes a point range for qualitative factors and price. Using a two-envelope system, the owner's selection team rates the qualifications and design of the team using a group of weighted factors. Price is submitted in a separate envelope and not opened until the qualitative factors are ranked. Opening the cost proposals beforehand might unduly influence the ranking of the more subjective qualitative factors. Points are assigned to the cost proposals as well. DBIA gives an example where the qualitative and design factors are weighted at 60 points maximum, with price given 40 points. In their example table, here is how the teams might fare:

Proposer	*Qualitative Score (60 pts. max.)*	*Price Proposal*	*Price Score (40 pts. max.)*	*Total Score*
Team A	51	$1,629,000	37	88
Team B	53	$1,546,000	39	92
Team C	44	$1,510,000	40	84

Using this method, there needs to be some clear criteria on how a score will be given to the price proposal. For example, does the low bid get the maximum score (here 40 points), and each higher bid have just one point deducted for each place they are in the bid tab sheet? This would result in selection being made more on qualifications, where multiple points are given or taken away on subjective criteria, but only one point taken for off for price. Another approach might be that price scoring is based on a deduction of one point for every 1 percent that a bidder is higher than the low bid, rounded to the next highest percent. Using DBIA's example, the price proposals would be ranked this way:

Low Bid	$1,510,000	Percentage High (0%)	Score = 40
2nd Low	$1,546,000	Percentage High (2.3%)	Score = 37 (less 3 pts.)
3rd Low	$1,629,000	Percentage High (7.8%)	Score = 32 (less 8 pts.)

Another method is the adjusted low-bid process, which uses the same steps as in weighted criteria, except price is divided by the score awarded to the qualitative factors to yield an "adjusted bid." DBIA's example of how this might work is as follows:

Proposer	*Qualitative Score*	*Price Proposal*	*Adjusted Price (price÷qual. score)*
Team A	0.85	$1,000,000	$1,176,471
Team B	0.95	$1,300,000	$1,368,421
Team C	0.50	$ 800,000	$1,600,000

According to this example, the low bid was also the least qualified team. The top qualified firm was also the most expensive. By adjusting the scores, the middle firm, who is neither the lowest bidder nor the most qualified team, has the best adjusted price and is awarded the contract. The contract price is the figure from the price proposal, not the adjusted price.

The following are some actual weighted criteria formulas used on design-build projects to show how public owners have varied the factors and weighting in selecting design-build teams.

[A] $48 Million State Prison in Cameron, Missouri

The five competing teams were rated on the basis of these evaluation criteria:[37]

Quality of Layout, Design, and Space Planning		(35.0 Pts)
1. Space Planning	(10.0 Pts.)	
2. Quality of Proposed Construction	(10.0 Pts.)	
3. Total Square Feet Proposed	(10.0 Pts.)	
4. Schedule of Equipment and Fixtures	(5.0 Pts.)	
Strength and Evaluation of Design-Build Proposer		(35.0 Pts.)
1. Strength of Design Team	(10.0 Pts.)	
2. Experience in Correctional Facilities	(15.0 Pts.)	
3. Construction Organization	(10.0 Pts.)	
Total Cost of Proposal Submitted		(30.0 Pts.)
Total Points		(100.0 Pts.)

[37] On this particular project, the highest ranked team in qualifications was also the most expensive. The low bidder was ranked lowest in qualifications. The contract was awarded to the team that was neither the lowest bid, nor the highest ranked in qualifications, based on overall score. The low bidder filed a lawsuit challenging the award on the basis that Missouri had no statutory authority to award state construction contracts other than to the low bidder. The lawsuit was dropped when the judge ordered the protesting bidder to post a $2 million injunction bond in order to put the construction on hold while the case was litigated. The project was completed by the selected team following about a two-month delay due to the lawsuit.

[B] $11.6 Million, 1,410 Car Parking Garage, University of Missouri

The teams were evaluated on this basis:	
Category	Points
1. Appearance	200
a. Parking Structure	170
b. Site	30
2. Durability/Maintenance	100
3. Functionality	150
a. Site	40
b. Plan	100
c. Systems	10
4. Safety/Security	50
Total Points for Quality	500
5. Cost	500
Total Overall Points	1,000

The RFP stated that a minimum of 350 points were required for a proposal to be deemed responsive. The university also gave additional weight for additional parking spaces over the required minimum of 1,410, subtracting one (1) point for each space under 1,410 and adding one-half point for each added space above 1,410, up to a maximum of 20 spaces (or 10 additional points), with no added points for over 1,430 spaces. In addition, added points for extra parking spaces were not permitted to increase a team's score to achieve the minimum 350 quality points. The project had a not-to-exceed budget of $11.6 million. The evaluation of all proposals was made based on the highest point total using this formula:

$$\frac{(\text{Lowest Cost Responsive Proposal} \times 500)}{\text{Proposer's Cost Proposal}} + \text{Quality Point Value}$$

$$\pm \text{ points for Parking Spaces} = \text{Total Score}$$

The RFP gave these examples of how it might work if the low bid was $11.4 million:

$$\text{Team 001: } \frac{(\$11{,}400{,}000 \times 500)}{\$11{,}600{,}000} + 400 + 10 = 901$$

[Note: this proposal wins the contract, even though it is not the low bid due to high qualifications score and added parking spaces.]

$$\text{Team 002: } \frac{(\$11{,}400{,}000 \times 500)}{\$11{,}600{,}000} + 380 + 5 = 876$$

$$\text{Team 003: } \frac{(\$11{,}400{,}000 \times 500)}{\$11{,}400{,}000} + 365 + 0 = 865$$

$$\text{Team 004: } \frac{(\$11{,}400{,}000 \times 500)}{\$11{,}100{,}000} + 320 + 10 = 843$$

[Note: non-responsive because quality points less than 350.]

$$\text{Team 005: } \frac{(\$11{,}400{,}000 \times 500)}{\$11{,}800{,}000} + 450 + 10 = 943$$

[Note: non-responsive because exceeds $11.6 million budget.]

[C] 500-Bed Minimum Security Prison, State of New Hampshire

Design-build teams were ranked on nine separate factors totaling 200 points, with cost being no more than 50 points.

Cost	50 pts.
Staffing efficiency	35 pts.
Aesthetics	25 pts.
Overall review of spaces and juxtaposition	20 pts.
Overall building security	15 pts.
Quality of materials	15 pts.
Construction methods	15 pts.
Plant operating & maintenance costs	15 pts.
Energy conservation	10 pts.
Total	200 pts.

[D] $46.6 Million Sagadahoc Bridge, State of Maine and FHWA

The teams were evaluated on this basis by a selection committee made up of 19 scorers:

Durability	20 pts.
Quality of Design	17 pts.
Maintenance of Traffic	15 pts.
Maintainability	12 pts.
Quality of Construction	10 pts.
Understanding the Scope of Work	10 pts.
Quality of Schedule	5 pts.
Community Impacts	5 pts.
Aesthetics	5 pts.
Navigational Vertical Clearance	1 point
Total	100 points

The 19 graders were given score cards for each of the 10 categories and suggestions of how to keep score. A sample score card for that project is set out in Figure 14-1.

§14.13 CRITICISMS OF SUBJECTIVE SELECTION CRITERIA

The weighted selection criteria, with subjective factors playing a large role in the selection of the design-build contractor have critics. Many contractors ar-

FIGURE 14.1. Sample Score Card Used By Maine Department of Transportation to Rank Proposals.

Example Proposer Score Card

MAINE DEPARTMENT OF TRANSPORTATION

BATH-WOOLWICH BRIDGE
D/B Technical Proposal Score Sheet

- **D/B Team:** **Control #**
  INFORMATION ONLY

- **Scorer ID:**

- **Date:**

- **Item #2 QUALITY OF DESIGN**

  - Intent: The Design-Build Team has selected and properly utilized design criteria and methods appropriate for their Proposal.
  - Examples that may be considered in scoring this Item include, but are not limited to: Geotechnical/foundation design > Scour minimization and mitigation > Design criteria for conditions, materials, and structure type > Appropriate degrees of redundancy > Proposed specifications and details > Plans (appropriate level of detail)

- **Supporting Comments by Scorer:**

RAW SCORE DEFINITION

0%	10%	20%	30%	40%	50%	60%	70%	80%	90%	100%
I							I			I
Marginal							Average			Exceptional

RAW SCORE THIS ITEM: _____ %

gue that with subjectivity entering into the process, the door is left wide open to favoritism and politics that could steer work to a particular company and away from others. They argue that competitive bidding, based on price alone, was designed to eliminate the fraud and corruption that was associated with the award of public contracts in some cities and states. "Negotiated bidding is not good public policy," according to Texas AGC President, Bob Lanham, "it will inherently undermine the taxpayers' faith in the system because the potential for scandal exists."

Charges of political favoritism were made in 1998 when the City of Atlanta sought proposals for a 20-year, long-term contract for the operation and maintenance of the city's water system. Five teams were ranked on six categories including EEO/EBO, technical, cost, management team, employee plan and past performance. When the team led by CH2M Hill Cos. Ltd. and its subsidiary OMI, Inc. was ranked the highest with a score of 85.7 points out of 100 possible, critics accused the team of being the mayor's preselected choice to win the contract. In an article titled "Charges, Countercharges Fly in Atlanta Water O&M

Bidding," the team's ranking was alleged by George Heery, ex-chairman of A-E firm Heery International, Inc. to "smack of cronyism."[38] Heery advocated final selection based on price and water quality. The city contended that the original proposals were judged by 12 teams of city employees set up to prevent one or two people from having undue influence. Despite apparent best efforts to prevent improper evaluations, systems that permit subjectivity to affect the contract award will always be open for criticism as they were in Atlanta.

§14.14 SET BUDGETS

In some cases, the owner will have a fixed budget as part of the RFP and will ask the design-build teams to work within a set budget, with weight given to proposals according to how far the bids are below the budget. This levels the playing field among the competition. The state of Missouri awarded the Cameron prison project using this basis, telling all five prequalified teams that the prison could not exceed $48 million. One bidder whose price was $49 million was disqualified as "non-responsive." The other four were ranked on qualifications, design and price, with the job awarded to a firm that was not the lowest price, due to weight given to other criteria.

On the Scott M. Matheson Courts Complex project in Salt Lake City, Utah, the project budget was set at $68 million, but bidders who exceeded the budget would lose points. Cost was factored at 20 points out of a possible 100 total points, with weighting like this:

Cost below $68 million	20 pts.
$68 to $70 million	15 pts.
$70 million and over	10 pts.

Obviously, a team with a price of $67.9 million would get the same point score as a team with a $65 million price, so there was no incentive to drop much lower than the target budget. However, a team with a price of $70 million would lose 10 percent of the total possible points automatically, knowing that its competitors would be going for the most possible points in that category. Public owners will need to retain consultants to establish realistic budgets if this method is used.

§14.15 PAYMENT OF STIPENDS

The cost of preparing designs and proposals on competitive design-build projects can be sizeable, and is rarely reimbursed.[39] Many design-build competitions are uncompensated, with the teams totally at risk for the costs of their pro-

[38] Engineering News Record, Aug. 24, 1998, p. 16.

[39] 92% of the firms surveyed reported that their fee was at risk most of the time on design-build competitions. Half of the firms report being reimbursed for at least 50% of their expenses. Zweig White, p. 34. In Utah, on the $1.59 billion I-15 project, each of the three proposers received stipends of just under $1 million for their effort. In return, the state got ownership of all designs and ideas.

posals. The financial impact can be long lasting. For example, a 1993 article in *Architectural Record*, reported that 11 firms submitted "best and final offers" for a $100 million IRS Computing Center in Detroit, after a grueling 14-month competition. James DeStefano, one of the competitors, estimated that each team spent more than $500,000 on their proposals. He estimated that the ordeal would affect the profitability of his 30-person office for about two years, "because we have to take a design to 30 percent completion, architects bear the brunt of the costs in these competitions."[40] These costly design competitions narrow the field of teams able to take the financial risk. One way to entice teams and to reduce the risk is to pay a stipend to the teams whose proposals are not selected.

[A] AIA/AGC Design-Build Guidelines on Stipends

In the *AIA/AGC Recommended Guidelines for Procurement of Design-Build in the Public Sector* (1995), it is stated that a design-build project is "far more expensive to prepare in the pre-selection process" than a traditional design-bid-build and, therefore, it is appropriate to pay a reasonable stipend in order to induce quality teams to compete for such projects. Unfortunately, when Congress was debating legislation to permit a two-phase process for federal design-build projects, a stipend provision was included in the initial drafts but removed prior to passage. Some federal agencies pay stipends, though not required to by statute, in order to help defray some of the costs incurred by teams in design-build competitions.

Due to a lawsuit filed by the Department of Justice against the AIA in 1990, the AIA and its chapters are prohibited from preventing or restraining their members from providing free services.[41] As a result, the AIA cannot formally come out in opposition to free design competitions, but it can recommend to the government that design-build teams be compensated for their efforts. In fact, not surprisingly, the AIA's official Position Statement on design-build recommends that when public bodies solicit proposals from design-build teams, they should:

> Provide proposal preparation compensation to the design-build entities commensurate with the level of information required when a facility design and cost proposal are required by the RFP.[42]

The AIA/AGC Guidelines similarly recommend that a "stipend" be paid to each of the unsuccessful design-build teams that compete in the second phase of a two-phase process to help offset the cost of preparing design submissions.

[B] DBIA's Position on Stipends

Not surprisingly, DBIA's official Policy No. 35, Stipends and Honoraria, recommends payment of a stipend, stating that "DBIA believes that payment of

---

[40] *The New Client, Design-Build*, Architectural Record, Nov. 1993, p. 37; *Big Tests Ahead for Design-Build*, Engineering News Record, June 15, 1998, p. 47.

[41] Consent Decree, U.S. v. The American Institute of Architects, Case No. 90-1567, U.S. District Court for the District of Columbia.

[42] AIA Position Statement Regarding Design-Build, Dec. 1997, reproduced in Appendix A.

honoraria is a positive action on the part of the owner—particularly when the submittal requirements for the proposal and/or the number of final competitors is relatively high. It is also a strong indication to the proposers that the owner is serious about awarding a contract and receiving a quality project."

[C] FIDIC's Position on Stipends

Even in the international marketplace, it is recognized that payment of stipends will encourage competition. In its guidelines for EPC/Turnkey Contracts, the Federation Internationale des Ingenieurs Conseils (FIDIC) states:

> Understandably, tenderers [bidders] are often reluctant, in the face of intense competition, to incur great expense in the preparation of tender designs. When preparing the Instructions to Tenderers, thought should be given as to the extent of detail which tenderers can realistically be expected to prepare and include in their Tenders. The extent of detail required should be described in the Instructions to Tenderers . . . Consideration may be given to offering some remuneration to tenderers if, in order to provide a responsive Tender, they have to undertake studies or carry out design work or a conceptual nature.[43]

[D] Federal Government's Position on Stipends

In 1996, Congress passed the National Defense Authorization Act of 1996 (Senate Bill 1124), which permits the federal government to procure design-build services using a two-phase selection process.[44] Initial drafts of the law contained a provision for the payment of stipends. However, during consideration of the two-phase selection process Rep. Wayne Gilchrest (R-Md.) offered an amendment to remove the stipend provision and committed to hold a hearing on the subject in order to give members of the architectural and engineering communities a chance to argue their views on inclusion of stipends. Although the AIA later testified in support of the stipend provision, Congress adopted the federal design-build statute without a stipend provision.[45]

[E] Stipend Amounts

While there is no set guideline for stipends, here is a sampling of stipends paid on projects of various sizes:

---

[43] Guidance for the Preparation of Particular Conditions, FIDIC Conditions of Contract for EPC/Turnkey Projects (1999 edition).

[44] The portion that adopted the two-phase selection process has been re-named the Clinger-Cohen Act.

[45] *Evolution of Design-Build in the Federal Sector*, Project Delivery Report, AIA Design-Build PIA, Sept. 1998.

Project Type and Location	*Stipend*	*Percent of Project*
$20 million housing project in Texas	$3,000	0.015%
$25.5 million arena in Michigan	$10,000	0.039%
$48 million state prison project in Missouri	$15,000	0.03125%
$68 million courthouse in Salt Lake City	$25,000	0.036%
$124 million city/county office building in Denver	$100,000	0.08%
$1.59 billion highway project in Salt Lake City	$1,000,000	0.063%

Some have suggested that the stipend should be based on what the public body would normally pay for schematic design services. By federal statute, the government cannot pay more than 6 percent for design services provided by architects and engineers, including "production and delivery of designs, plans, drawings, and specifications."[46] The AIA used to publish a proposed weighting of fee percentages for the various phases of basic architectural services. The old schedule suggested that the Schematic Design Phase account for 15 percent of the total design fee.[47] All "design phases" accounted for 75 percent of the total fees, so schematic design would be 15 percent of 75 percent, which is 20 percent of "total design" phase fees per old AIA schedule. If the work required for a design-build proposal is based on schematic design phase services, then a stipend based on the total schematic design fee might be calculated like this:

$$\text{Total design fee } 6\% \times \text{Schematic Design Portion } 20\% = 1.2\%$$

Dividing that stipend equally among four of five proposers would mean that each team received only one-fourth of a normal schematic design phase fee. This would be $1.2\% \div 4 = 0.3\%$ for each team who submits a design that is not awarded the contract. For a $10 million project, this would provide a stipend of $30,000 each. As seen from the examples above, however, typical stipends range from 0.03 percent to 0.08 percent, which are anywhere from one-tenth to one-fourth of what a reasonable stipend would be. An architect should keep this in mind when considering a design-build competition, since stipends do not come even close to compensating the design team for the effort required to prepare a schematic design.

As DBIA's official Policy No. 35, Stipends and Honoraria, candidly states: "A substantial honorarium rarely covers the cost of proposal preparation. In the case where the RFP requires substantial preliminary design work and submittals, the difference between the honorarium and the cost of proposals may become so high as to render the honorarium meaningless." DBIA views the amount of the stipend as a strong indication to the proposers that the owner is serious about awarding a contract and receiving a quality project. Apparently some owners are

[46] 41 U.S.C. § 254(b); 48 C.F.R. § 15.903.

[47] AIA's Instruction Sheet for 1977 edition of B141 Agreement Between Owner and Architect suggested fees be weighted at 15% for schematic design, 20% for design development, 40% for construction documents, 5% for bidding, and 20% for construction administration.

more serious than others, or they are all simply naïve about the amount of effort that goes into a design-build proposal.

[F] Variations on Stipends

The Alaska District of the Corps of Engineers uses an unusual procedure to provide stipends to the unsuccessful proposers. For a project at the Eielson Air Force Base, the Corps proposal documents provided that the aggregate amount for stipends would be divided regressively, depending on where the unsuccessful proposer was ranked. Instead of dividing the stipend equally among the unsuccessful teams, the Alaska District reduces the stipend amount as the proposer's rank decreases. For example, if there are five proposers and $75,000 available in stipends, the second highest ranked proposer would receive $30,000, the next would get $22,500, the fourth $15,000, and the last ranked proposer would receive only $7,500. This method rewards teams who put more effort into a design proposal.

§14.16 SETTING FEES AND EXPENSES

As mentioned above, under traditional design-bid-build procurement, architect design fees are limited to 6 percent under federal regulations.[48] According to researchers, the 6% fee goes all the way back to a lawsuit by Richard Morris Hunt in 1861, in which Hunt sued a client who refused to pay his fee. According to an article in *Architectural Record*, "During the trial, several architects testified that their customary charge was 5 percent of the project's cost, and the jury agreed, setting 5 percent as the standard accepted minimum compensation."[49] Then, in 1866, the AIA issued its first document titled, a "Schedule of Charges," confirming 5% as the proper charge for architectural services, including design and site visits. The AIA's 1888 Constitution and By Laws also included a 5% fee schedule, similar to the 1866 Schedule, and despite increasing scope of work for architects. In 1908, the AIA raised the fee schedule to 6%, a figure that is now part of federal laws which limit the amount that the government can pay architects nearly 100 years later.

Some states similarly cap the fees that can be paid for architectural services on state projects. For example, Kansas procurement statutes cap architectural fees according to a set schedule, ranging from 5 percent to 7 percent.[50] One benefit of design-build is that design fees are wrapped up into one lump sum price, coupled with construction, with no limitation on the design fees. Whether the architect's fee is 6 percent or 10 percent of the overall design-build proposal is irrelevant to the owner whose main concern is the total project cost. Architects can, therefore, charge more for design services when the public agency is not hiring the architect directly, but as part of a package of design and construction for a lump sum price.

---

[48] 41 U.S.C. § 254(b); 48 C.F.R. § 15.903.

[49] *Why Architects Don't Charge Enough*, Architectural Record, Oct. 1999, p. 110.

[50] Kan. Stat. Ann. § 75-1263.

It may have taken over 100 years to get here, but there is finally some relief from the 6 percent fee cap through design-build.[51]

§ 14.17 SUBCONTRACTOR LISTING STATUTES

In some states, contractors are required to identify their major subcontractors at the time bids are submitted. The intent is to prevent bid shopping and bid peddling, in which a contractor gets a bid from one sub, uses that figure in its bid to the owner, then "shops" that price to other subcontractors in order to get them to perform the work for the same price or lower. One such statute is California's Subletting and Subcontracting Fair Practices Act, passed to protect the public from unfair competition and poor quality work.[52] Contractors that violate the law may be subject to disciplinary action with respect to their contractor's license, a penalty of 10 percent of the amount of the subcontract involved, or cancellation of the contract.

In a design-build setting, these laws may be difficult, if not impossible to meet, since the prime contractor has not finalized the design at the time of the selection of the team. Unless the trade subcontractor is, itself, furnishing a design-build portion of the project there are usually no detailed plans upon which to get subcontractor bids. Legislation that authorizes design-build on public projects in states like California need to account for this situation and permit a supplemental listing of trade subcontractors. One method, used by one California agency, requires the design-build contractor to identify the work to be subcontracted at the time the selection, and then requires the contractor to award subcontracts based on competitive bidding.[53] This has a drawback, of course, in that the design-build

---

[51] Consider how an architect's 6% fee rates with the profits made by the contractor who builds the project and the realtor who sells it. Contractor profits are thought to run an average of around 5% of the cost of construction. Architects making a 6% fee pay half of that fee to consulting engineers, netting a tidy 3% fee for their efforts. According to surveys, architects' average profits are around 13% of billings. Though this is higher than the contractor's margin, one has to stop to realize what this is 13% of. This means, of course, 13% of a 6% fee, or a whopping 0.078% of the cost of construction. Take a $500,000 construction project, where the contractor's 5% profit nets him $25,000, while the architect's 0.078% nets a profit of just $3,900. Worse yet is comparing this to the 6% fee charged by the realtor who, unlike the contractor and architect, spends a fraction of the time to show the project and sell it, and who has very little liability exposure. Architects meet with the client, help develop the program, design and redesign the project, help solicit bids, redesign again if bids exceed budget, administer the construction contract, and often get dragged into litigation either during the project, or by any construction worker or building user who falls and gets hurt, causing the architect to pay his or her $25,000 to $50,000 insurance deductible in defense costs. And all this for the same 6% the realtor gets for selling the project. The architect might have to work on six $500,000 projects just to pay for one deductible.

[52] Cal. Pub. Cont. Code §§ 4100 *et seq.*

[53] Cal. Gov't Code §§ 14661 *et seq.* A reported, but unpublished opinion of the California Attorney General concluded that the impossibility of meeting the state listing requirement on a design-build project constituted a "public necessity" for postponing subcontractor identification until the project designs were completed. Cal. Op. Atty. Gen. on Subletting and Subcontracting Fair Practices Act, May 30, 1991 (unpublished), as reported by B. Papernik and N. Smith, By Design, *Los Angeles Lawyer*, July-Aug. 1999.

contractor cannot handpick its team for a project in which it carries more than normal risks.

§14.18 SELECTION OF THE PANEL OR JURY

Selection committees can vary in size and makeup. For example, on one project, the $46.6 million Sagadahoc Bridge for the State of Maine, the teams were evaluated by a selection committee made up of 19 scorers from the Maine Department of Transportation, local cities, the University of Maine, the Federal Highway Administration, and even one "visiting scorer" from the Texas Department of Transportation. For smaller, local projects, selection committees may be comprised of three to five members made up of agency staff.

In the *AIA/AGC Recommended Guidelines for Procurement of Design-Build in the Public Sector* (1995), it is recommended that the selection panel include design and construction staff from within the governmental agency who are familiar with the project. In addition, it is suggested that the panel include representatives of the agency that will use the facility because, "if the ultimate users are involved in the process, changes requested by the users later during the project can be avoided which can save substantial amounts of time and money." The AIA/AGC Guidelines also recommend having outside advisors on the panel, who can bring fresh perspectives to the discussion. The names of the jury should be made public and included in the initial solicitation so that competing teams can evaluate their chances of being chosen based on "design and other predilections the jury members have" and because a qualified panel of jurors will attract a better level of competition.

Just as jurors during a trial are not supposed to communicate with any of the parties involved in the case, it is recommended that the competitors and jurors agree not to have any communication about the project or selection other than as permitted in the selection procedures.

CHAPTER 15

DESIGN-BUILD AND THE FEDERAL GOVERNMENT

§15.01 USE OF DESIGN-BUILD BY THE FEDERAL GOVERNMENT

One of the largest users of the design-build method of project delivery is the federal government. Federal agencies that use or have used design-build include the General Services Administration (GSA), NASA, the U.S. Postal Service, the Corps of Engineers, the Environmental Protection Agency (EPA), the Federal Bureau of Prisons, Department of Housing and Urban Development (HUD), Department of Energy, Federal Highway Administration and the military, including the Air Force and the Navy.

Some of the earliest federal applications of design-build were in the 1940s, when the Naval Facilities Engineering Command used this method of project delivery on housing projects. NASA began using design-build in 1962, followed by HUD's use of turnkey housing projects in 1968. The Military Construction Authorization Act of 1986 permitted each branch of the military to use design-build on three projects a year.[1] The use of design-build in the public sector doubled from 1986 to 1992. In Fiscal Year (FY) 1992, Congress lifted the restrictions and permitted broader use of design-build. Since then, use of design-build in the federal sector has rapidly spread.

In March 2000, the Design-Build Institute of America (DBIA) published a *Guide to the Federal Design-Build Marketplace*. The guidebook recaps the history of design-build in the federal market and contains some examples of federal design-build projects, plus "helpful tips" in pursuing federal projects and surveys of government agencies who have used design-build.

§15.02 TWO-PHASE SELECTION PROCESS

The biggest boosts to design-build came in 1994 and 1996 with the passage of two laws that permit award of construction contracts by the federal government on the basis of "best value." The Federal Acquisition Streamlining Act of 1994 (FASA) and the National Defense Authorization Act of 1996 permitted the federal government to procure design-build services using a two-phase selection process.[2] Under the 1996 Act, federal agencies are to determine if design-build is appropriate for the public project. If so, the first phase of the process involves evaluation of specialized experience and technical competence of the proposers. Cost-related evaluation factors are not permitted in this phase. Then, proposers submit bids for the project. The contract is awarded to the design-build team with the highest overall ranking, based on qualifications and price.[3]

---

[1] P.L. 99-167, Dec. 3, 1985; P.L. 99-661, Nov. 14, 1986.

[2] Pub. L. No. 103-355, 108 Stat. 3243 and Pub. L. No. 104-106, § 4001, 110 Stat. 186, 642. The portion of the 1996 Act that adopted the two-phase selection process has been re-named the Clinger-Cohen Act. There was precedent for awarding federal contracts using a combination of qualifications and price even before the new law was passed. *See* Matter of Robert G. Rupprecht, 1994 WL 68309 (C.G.) and Matter of D & M General Contracting, Inc., 1995 WL 274914 (C.G.).

[3] 10 U.S.C. § 2305a and 41 U.S.C. § 253m.

Following the passage of the 1996 federal design-build statute, the Federal Acquisition Regulations (FAR) were amended in 1997 to incorporate design-build procedures. 48 CFR §§ 36.102-36.104 and 36.300–36.303 outline the process of soliciting and evaluating design-build proposals.[4] These regulations add more detail to the government's two-phase process. The design-build regulations were first published for comment on August 7, 1996 and became final on January 1, 1997.[5] Under these regulations, cost or price data are specifically excluded from the phase-one analysis. This eliminates competition based solely on price and allows the government to weigh cost and quality, using the same method as the private sector owners. Excluding price from round one is intended to reduce the bidder's proposal preparation costs, most of which are generated in the design phase. Another benefit is that the expense of a phase-two design proposal is limited to those bidders with a strong chance of being awarded the ultimate design-build contract, by essentially prequalifying the prospective teams.[6]

The federal government's experience with design-build and "best value" contracting has been generally favorable. One survey found that use of design-build by government agencies resulted in time savings ranging from 15 percent (Naval Facilities Command) to 28 percent (Department of Veterans Affairs).[7] Cost savings of 14 percent were reported by the Department of Defense in that same survey.

§15.03 THE DEPARTMENT OF DEFENSE

The Military Construction Authorization Act of 1986 authorized "one-step turnkey selection procedures" for the military.[8] The 1986 Act authorized three pilot projects for each military agency through October 1, 1990. In 1991, 0.01 percent of the U.S. Navy's spending for construction was for non-traditional procurement methods. The Navy's use of design-build has grown significantly since then to become the leading federal agency in use of design-build, increasing from a reported 9 percent in FY 1995 to an estimated 40 percent in FY 2000.[9] One design-build project by the Navy is a $7.5 million design-build contract for renovation of the Naval Sea Systems Command at the Washington Naval Yard, a 75,000 square foot "mission critical" facility.

---

[4] For a detailed history of this new law, *see* K. Didier, *Construction Contracting and the New Two-Phase Design-Build Selection Procedures*, 27 Pub. Contract L. J. 589 (1997).

[5] 61 Fed. Reg. 41, 212 (1996); 61 Fed. Reg. 224 (1997).

[6] *See generally Best Value in Federal Construction Contracting*, The Construction Lawyer, American Bar Association, April 1999, p. 25; and *Best Value in State Construction Contracting*, The Construction Lawyer, American Bar Association, Apr. 1999, p. 31.

[7] Tanya C. Matthews, Informal Survey of Federal Agency Experience with Design-Build, DBIA/AIA 1998 Conference (Oct. 14, 1998).

[8] 10 U.S.C. § 2862.

[9] Federal Design-Build Survey, Guide to the Federal Marketplace, Appendix D, DBIA, Mar. 2000.

The U.S. Air Force has also used design-build. One project, for example, is the AETC Contracting Squadron's use of design-build for a two-story commissary and exchange mall at Lackland Air Force Base, Texas. The estimated $10 million project will also include demolition of existing buildings, site clearing, utilities, roads, and associated on-site improvements.

§15.04 THE CORPS OF ENGINEERS

The U.S. Corps of Engineers has been authorized to use design-build since the 1980s through the agency's privatization program, known as Third Party Contracting (TPC). The TPC program allowed the military to enter into long-term contracts with private firms to design, build, own, and operate facilities on government land. In 1989, the Corps awarded a $30.5 million contract to Proctor David Ray Consulting Engineers to design, build, and operate a wastewater treatment plant in Huntsville, Alabama. The 2.7 million GPD facility was the military's first privatized wastewater project.[10]

In 1999, the Corps awarded an $11.1 million contract for phase-one design-build of a family housing project at Offutt Air Force Base in Omaha, Nebraska. The project included demolition of eight existing buildings and the design and construction of 82 new family dwellings, comprised of single and duplex units, new site utilities, and roadway. The Corps is also using design-build for an air freight-passenger terminal at Patrick Air Force Base in Florida. The project is estimated to cost between $10 million and $25 million, and includes a 3,200 square meter air freight passenger terminal, a 6,000 square meter base supply traffic management complex, and an 18,000 square meter concrete aircraft ramp.

§15.05 GENERAL SERVICES ADMINISTRATION (GSA)

GSA began to flirt with design-build in 1987, when the agency developed its first design-build contract. In 1989, the agency held a regional conference on design-build. In FY 1991, GSA spent $1.719 billion on capital construction, with $577 million, or 33 percent, accomplished using design-build. GSA published a comprehensive Design-Build Request for Proposal Guide in the spring of 1991, including model contracts for both single and two-phase source selections. In 1993, GSA had its first two-phase selection and design-build accounted for 25 percent of GSA's design and construction budget that year. The agency's use of design-build has decreased since then, according to one publication, with $238 million spent on design-build in FY 1993, $40 million in FY 1994, $14 million in FY 1995 and a low point of just $5 million in FY 1997.[11] Use spiked up again in FY 1998 to over $200 million. The up and down cycle with this agency's use of design-build makes it difficult to project future use of this method of procurement.

---

[10] *Corps Project Goes Private*, Engineering News Record, Oct. 5, 1989, p. 23.

[11] Federal Design-Build Survey, Guide to the Federal Marketplace, Appendix D, DBIA, March 2000.

§ 15.06 FEDERAL BUREAU OF PRISONS (FBOP)

Jails have been a popular building type for design-build at the state and local levels. At the federal level, the FBOP has fully embraced this method and is leading the nation in design-build correctional construction, trailing only the Navy in use of design-build on an annual basis. "We think it's faster, quicker, and puts liability on the contractor, where it ought to be. That's our intention and that's where we're going with it," says FBOP deputy chief of design and construction, John Seeley.[12] The FBOP's first use of design-build was in 1997, when design-build accounted for less than 7 percent of the agency's annual design and construction budget. That percentage jumped to near 30 percent in FY 1998's budget and is projected to increase.[13] The agency's proposed construction budget for FY 2001 is $836 million, and it will reportedly seek $1.3 billion in 2002 and 2003, which means plenty of opportunity for design-build teams in the correctional institution market.[14] Among FBOP's design-build projects is the $139 million, 864-cell high-security facility, minimum-security prison camp, administrative buildings and plant in Martin County, Kentucky, scheduled for completion in November 2002.

§ 15.07 DEPARTMENT OF ENERGY (DOE)

The DOE's first large scale design-build project is its $100 million super computer center at the Los Alamos National Laboratory in New Mexico, awarded to Hensel Phelps Construction Co. The DOE reviewed qualifications from 10 companies before shortlisting a select group. The 267,000 square foot facility has 43,500 square feet of unobstructed computing space and 300 offices. The DOE planned to follow the computer facility with a $60 million design-build office building six to 12 months after the computer center. The DOE's division director for the program management division, Walt Kirchner, said, "We're looking at ways to streamline the process in these times of more aggressive budgets. We're confident this is the best approach for these projects."[15] The DOE was so impressed with its first design-build contract, that it planned to use design-build on the rehabilitation of the department's existing structures at the Los Alamos site.[16]

§ 15.08 FEDERAL AVIATION ADMINISTRATION (FAA)

In 1999, the DBIA lobbied for passage of legislation that recognized use of design-build for projects that use funds from the Airport Improvement Program

---

[12] *Institutional Outlook*, Design-Build, Dec. 1999, p. 64.

[13] Federal Design-Build Survey, Guide to the Federal Marketplace, Appendix D, DBIA, March 2000.

[14] *Prisons: A Positive Signal for Fiscal 2001 Budget Boost*, Engineering News Record, Mar. 13, 2000, p. 11.

[15] *DOE Tries Design-Build at Lab*, Engineering News Record, Apr. 19, 1999, p. 16.

[16] *DOE Selects Three Contractors for Los Alamos Refurbishment*, Engineering News Record, Feb. 21, 2000, p. 39.

(AIP). The pilot program contained in the legislation permitted an airport sponsor to apply for permission to use design-build and then seek reimbursement of design and construction costs for which the sponsor has become obligated when using design-build. Previously, the FAA had required airport sponsors to submit plans and specifications before an AIP grant was made as part of the approval process, thereby restricting design-build. The pilot project in the legislation was limited to just seven projects.

§15.09 THE FEDERAL HIGHWAY ADMINISTRATION (FHWA)

Much of the growth in design-build is by state highway departments. There is also growth being pushed by the Federal Highway Administration, such as the Transportation Equity Act for the 21st Century (TEA-21, Section 1307).[17] *See* §17.02, in Chapter 17. Congress has mandated that the FHWA develop limited design-build regulations by July 2001. According to a survey of government agencies, the Highway Administration spent $631 million on design-build projects in its 1998 budget, up from $110 million in 1997.[18] With more highway work planned to be done using design build, this appears to be a strong market.

§15.10 FEDERAL TRANSIT ADMINISTRATION (FTA)

The FTA started looking at combining design with construction in the late 1980s. More recently, under a demonstration program, the agency authorized two pilot design-build projects, including the $300 million Union Station Gateway terminal in Los Angeles, and a $160 million extension of Baltimore's light rail system, two of five planned design-build projects. The agency is gathering data and analyzing the lessons learned to offer a variety of examples of how design-build can be done on transit projects.[19]

§15.11 VETERANS ADMINISTRATION

The U. S. Department of Veterans Affairs has used design-build since 1994. From that year to 1998 its use of design-build went from zero to approximately 20 percent of its design and construction budget.[20] One agency project done using design-build is a $39.4 million hospital project at the Sacramento VA Medical Center in California. The contract calls for construction of a five-story building and a new electrical energy center and mechanical energy center. The contract was awarded to the team of ERHDR and John F. Otto, Inc.

---

[17] Pub. L. No. 105-178, § 1307 (1998), codified at 23 U.S.C. § 112(b)(2)(D).

[18] Federal Design-Build Survey, Guide to the Federal Marketplace, Appendix D, DBIA, March 2000.

[19] *FTA Demonstrations Eye Time, Cost and Innovation Benefits of Design-Build*, Design-Build, Feb. 2000, PG. 28.

[20] Federal Design-Build Survey, Guide to the Federal Marketplace, Appendix D, DBIA, March 2000.

§ 15.12 PROHIBITION AGAINST DESIGNER-LED DESIGN-BUILD

In order for design firms to lead design-build teams, there is still need to update Federal Acquisition Regulations that prevent architects and engineers from bidding on the projects they design. FAR 36.209 states:

> No contract for the construction of a project shall be awarded to the firm that designed the project or its subsidiaries or affiliates, except with the approval of the head of the agency or authorized representative.

By contrast, some agencies have made specific allowances in their regulations for designer-led design-build. The U.S. Department of Agriculture's regulations permit the head of the contracting activity "to approve the award of a contract to construct a project, in whole or in part, to the firm (inclusive of its subsidiaries or affiliates) that designed the project."[21] This would allow a design firm to set up an affiliate company to hold the construction contract. More regulations like these are needed.

§ 15.13 ASCE'S STATEMENT ON DESIGN-BUILD IN THE FEDERAL SECTOR

In 1992, the Board of the American Society of Civil Engineers (ASCE) approved a Policy Statement on Design-Build in the Federal Sector which recognized that federal agencies desired to use non-traditional methods of construction, including design-build. ASCE's Policy Statement stated that, "Today's owners are requiring more attention to cost effectiveness, higher quality product and service, and fewer legal concerns. Federal owners are no exception to this trend. Design-build is one of the ways for an owner to obtain a single source of accountability and responsibility for its project." ASCE recommended that whenever the federal government chooses to use design-build, it should consider the following factors:

1. Preselection of design-build offerors based on the qualifications of the team, in the spirit of the Brooks Act.
2. Owner scope of work information that is of sufficient detail to allow competing teams to cost effectively produce their proposals.
3. Selection criteria that is communicated clearly and structured to reinforce the qualifications-based selection process.
4. Predetermined reimbursement for the "short-listed" firms that are making full-fledged design proposals (those firms that were preselected in accordance with number one above), for the purpose of enhancing the quality of the proposals.
5. Assurance of direct line of communication between the owner and design professional through specific provisions in design-build contracts, such as those developed by the Engineers Joint Contract Documents Committee.

[21] 48 C.F.R. § 436.209.

CHAPTER 16

OWNERSHIP OF DESIGN PROPOSALS, TECHNICAL SUBMISSIONS, AND COPYRIGHTS

§16.01 WHO OWNS THE DESIGN?

A common source of questions by design professionals is ownership of drawings and copyrights. With increasing frequency, clients modify the contracts to grant ownership of the documents to the client. The topic today includes rights in the physical documents, reproductions, electronic media versions, and the building itself. The Design-Build Institute of America's (DBIA's) official Policy No. 41, entitled "Who Owns the Design?" raises the question that comes up with some frequency in design-build projects. Architects, trained under the American Institute of Architects (AIA) contracts and philosophy, believe that they should own the designs they prepare, including the copyrights. The client only pays for and owns the original drawings, with permission to use them on the one project they are developed for. This concept is well entrenched in the AIA documents, which grant the owner a license to use the design and terminate that license when the owner fails to pay the architect for any services or terminates the contract.[1] Today, however, there is competition for ownership rights among the designer, the owner and even the contractor.

§16.02 OWNERSHIP OF THE DOCUMENTS

The general rule on ownership that has been developed by the courts is that:

> In the absence of a contrary agreement, *the architect is not the owner of plans* prepared by him which are accepted and paid for by his employer. As a general rule, one who employs an architect to prepare plans and specifications for a building and pays him therefor becomes the owner of the plans. An exception, of course, arises where the contract specifically states that all plans and specifications are considered the property of the architect and are to be returned to him.[2]

Stated otherwise: "In the absence of any agreement to the contrary, where the architect has prepared plans, superintended construction, and been paid for his work, *the owner of the building is entitled to the plans*."[3] In short, the client who pays the architect for its services becomes the owner of the drawings upon payment. That can be changed, of course, by contract and often is.

Although the general rule is that *the client* owns the documents he pays for, there are contrary cases holding that *the design professional* owns its documents. In a 1990 New York case, the architect and client had no express agreement on ownership, each thinking that they owned the drawings. The court held that "[a]n architect owns his drawings, unless expressly agreed otherwise by the parties," citing no legal authority for this statement.[4]

[1] *See, e.g.*, ¶1.3.2.1, Standard Form of Agreement Between Owner and Architect, AIA Doc. No. B141 (1997 edition).

[2] 6 C.J.S. Architects § 19.

[3] 5 Am. Jur. 2d Architects § 11.

[4] Kunycia v. Melville Realty Co., Inc., 755 F. Supp. 566 (S.D.N.Y. 1990). *See also* Aitken, Hazen, et al. v. Empire Constr. Co., 542 F. Supp. 252 (D. Neb. 1982).

§16.03 THE STANDARD AIA DOCUMENTS

Under the standard form AIA documents for traditional projects, the architect owns the documents it prepares, and the subconsultants own the documents that they prepare. The architect and the consultants retain the copyrights in their own documents, but each grants a "limited license" to the owner so that the owner can use the documents to build the project.

[A] The Owner/Architect and Architect/Consultant Agreements

The architect's base contract form for traditional projects states that:

> 1.3.2.1 Drawings, specifications and other documents, including those in electronic form, prepared by the Architect and the Architect's consultants are Instruments of Service for use solely with respect to this Project. The Architect and the Architect's consultants shall be deemed the authors and owners of their respective Instruments of Service and shall retain all common law, statutory and other reserved rights, including the copyrights.[5]

Similarly, the Architect/Consultant Agreement states:

> 8.1 Drawings, specifications, and other documents, including those in electronic form, prepared by the Consultant are Instruments of Service for use solely with respect to this Project. Unless otherwise provided, the Consultant shall be deemed the author and owner of the Consultant's Instruments of Service and shall retain all common law, statutory and other reserved rights, including the copyrights.[6]

Under the AIA documents, the architect grants the owner a "non-exclusive license" to reproduce the documents as long as the owner complies with the contract, including the obligation to pay promptly. The license terminates when the agreement is terminated. The owner may not transfer the license to anyone else without written agreement of the architect. Similarly, under the 1997 version of AIA's C141, Architect-Consultant Agreement, the consultant grants to the architect a "non-exclusive license" to reproduce the consultant's documents *provided* that the architect complies with all obligations, including prompt payment of all sums due. The implication is that non-payment terminates the license. The architect can include the consultant's documents in a non-exclusive license to the owner.[7]

[B] The AIA Design-Build Contracts

The AIA carries its philosophy from its traditional contract forms into its design-build forms. Paragraph 3.1 of the B901 Design-Builder/Architect Agreement

---

[5]AIA's B141 Owner/Architect Agreement (1997 edition).

[6]AIA's C141 Architect/Consultant Agreement (1997 edition).

[7]*See* ¶¶8.1-8.6 of the C141 Agreement (1997 edition).

(1996 edition) states that the drawings and specifications furnished by the architect are instruments of service and the architect retains all rights, including the copyrights. The Design-Builder is permitted to use the documents only for the particular project. If the parties only execute a preliminary Part One design-build agreement, the design-build contractor is prohibited from using the documents without the written permission of the architect, nor can they be used on other projects without permission and compensation to the architect.

One of the criticisms of design-build is the owner's loss of the entire team, builder and designer, if the prime design-build contract is terminated. Recognizing this, the AIA contract states that if the design-build contractor defaults, the architect can grant the owner a license to use the documents for the completion of the project on two conditions: (1) the owner must cure the design-builder's default, if any, in payment to the architect; and (2) the owner must indemnify the architect "with regard to claims arising from such reuse without the Architect's professional involvement." The Part Two design-build agreement contains the same wording.

[C] The Architect's and Engineer's Concern

Architects and engineers alike are concerned about six things here. First, architects see themselves as selling services, not products. They do not "sell designs," but rather professional consultation and advice, of which design is one aspect. Second, the architect's reputation is based in large part on his or her designs, and the unique style or details used on a particular project. If that project is reproduced many times, or altered, the value of that design diminishes, which is not in the client's or architect's best interest. Third, the quality of the built structure may vary based on modifications made in subsequent duplications of the design, without the architect's involvement. Yet the public will associate the design with the architect, even if an inferior copy is built somewhere. Fourth, there is financial value in the design. Clients pay architects to produce the design. If the architect retains the copyright, then a fee can be charged for each reuse of that design by the owner or others for other locations. By giving up the ownership of the copyrights, the architect forfeits the right to collect those fees. Fifth, and maybe most important, is the professional liability that attaches each time an architect's design is reused.

An architect designs a building to be built for a specific site, on unique soil conditions, under applicable codes and ordinances. "Site adapting" a design to another location without the architect's involvement, or altering the design without the architect's input or advice can result in errors, omissions and, eventually, claims that create liability for the original designer, even though the architect had no involvement. An architect's or engineer's liability increases with each reuse of their design. This author has represented design professionals on claims where a particular design (which contained an error) was reused 25 times, each time repeating the structural error. The fee charged to design that one building was minor compared to the cost to repair all 25 buildings.

The sixth and final factor, particularly applicable to design-build projects, is the architect's ability to prevent being fired by the design-build contractor after providing preliminary "concept" design work, sufficient to win the award and establish a price. By preventing use unless a Part Two agreement is executed, the architect assures itself that it will remain a part of the team into the later phases of the project. By retaining the copyrights, the designer keeps its hooks in the project and reduces the risk of being let go before the design work is completed and all fees are paid.

§16.04 DBIA'S POSITION ON OWNERSHIP OF DOCUMENTS

While AIA has the architect retain the ownership (keeping in mind who is drafting the contract forms), DBIA has the Design-Builder retain the ownership. The DBIA official position on ownership of contract documents is as follows:

> As to the ownership of the completed construction documents once the project is built by the design-builder, DBIA believes that the documents are instruments of service and that the design-builder should retain ownership thereof. The owner, however, is permitted to make and retain copies of the documents and use them in connection with the use and occupancy of the project. Unless the parties specifically agree to the contrary, with the design-builder being afforded appropriate consideration, the documents should not be used for alterations to either the project or on another project. Any reuse without verification or adaptation by the design-builder for the specific purpose intended must be at owner's risk and without liability to the design-builder.[8]

§16.05 THE AGC POSITION ON OWNERSHIP

The AGC Preliminary Design-Build Agreement,[9] provides that upon the making of payment as required by the agreement, the owner receives ownership of the documents, "except for copyrights." It is implied that the owner can then use such documents for the project for which they were prepared. The owner is restricted, however, from using, reproducing, or making "derivative works" from these documents *for other projects* without written authorization of the design-build contractor. Any use of the documents on other projects is agreed to be at the owner's sole risk. Further, the owner is required to defend, indemnify, and hold harmless the design-builder, its architect/engineer, subcontractors and consultants from any claims, damages, etc., including attorneys' fees, arising out of such use of the documents by the owner. The design-build contractor agrees in the preliminary agreement to obtain from its design subconsultants all of the rights that correspond to what the owner is given by the design-builder.

The agreement should be more clear on the issue of "Who owns the copyrights?" and on the owner's obligation to indemnify the design-builder and its consultants if the owner terminates the relationship at the preliminary stage but uses the preliminary documents to finish this project with another contractor or

---

[8] DBIA Manual of Policy Statements, Policy No. 41, Oct. 1998.

[9] AGC Doc. No. 400, ¶4.1 (1999 edition).

team. There is no incentive for the owner to stay with the team that prepared the preliminary designs, for which the owner pays the preliminary fee once the client receives ownership of the documents.

The AGC's cost-plus and lump sum design-build contracts (after the preliminary phase) both address ownership of documents in the same way. Again, "upon the making of payment," the owner receives the ownership rights (except for copyrights) in the documents, including electronic data, prepared or procured by the design-builder, its architect/engineer, subcontractors, and consultants. The agreement gets interesting as it deals with termination of the contract. If terminated *for cause—i.e.,* due to some breach by the design-build contractor—then upon payment for all work performed, the owner gets the right to use, reproduce, "and make derivative works from the Design-Build Documents to complete the Work."

If, however, the AGC agreement is terminated *for convenience* (rather than for cause), the owner gets the ownership rights (excluding the copyrights) upon payment of all sums due but again including the right to make derivative works to complete the project. Sums due upon this type of termination, however, include payment of "a premium" which is stated up front in the agreement at the time it is signed. This is a "buy out" clause, under which the owner has to pay not only for what has been done, but an additional fee for the rights to use the documents to complete the project. Under either type of termination, the owner is prohibited from using the documents for "other projects" without written consent of the design-builder. As with the Preliminary Agreement form, the owner agrees that use on other projects is at the owner's sole risk and that the owner will indemnify, defend, etc. the design-build contractor and its team. This clause should be made clearer that the intent is not just to cover reuse "on other projects" but use on "this project" after termination, and without the design-builder's involvement.

§16.06 FEDERAL COPYRIGHT LAW

Prior to 1976, the courts recognized an architect's right to "common law copyright" in its documents, protected under state law.[10] This created differences from state to state in an area where some uniformity was required so that people doing business in multiple states could know their rights. Federal law changed that in 1976 by doing away with state copyright laws and replacing them with a uniform federal law, the Copyright Act of 1976.[11] As a result, there is no longer any such thing as a "common law copyright," although the AIA documents still refer to "common law" rights of the architect.

Although the two concepts are often mixed together, the law recognizes that the ownership of a design professional's *documents* is separate and distinct from ownership of *the copyrights* to the design depicted in those documents. The federal copyright statutes provide that:

---

[10] Kurfiss v. Cowherd, 121 S.W.2d 282 (Mo. Ct. App. 1938).

[11] 17 U.S.C. §§ 101 *et seq.*

> Ownership of a copyright . . . is distinct from ownership of any material object in which the work is embodied. Transfer of ownership of any material object . . . does not itself convey any rights in the copyrighted work embodied in the object.[12]

In a 1993 Pennsylvania case, an architect sued for copyright infringement after his drawings were copied by another architect who took over the project. The Court held that a contract clause that transferred only the ownership *of drawings* to the defendant does not suffice to transfer *the copyrights.*[13] One should keep this case in mind when taking over a project from another firm that has defaulted or had a falling out with either the owner or a design-build contractor and be sure that the contract with the predecessor firm gave not just the documents, but the copyrights in those documents to the owner or contractor before beginning work on the completion of that project. Otherwise, a design firm can get into legal trouble using copyrighted designs without proper permission.

§ 16.07 COPYRIGHT REGISTRATION AND ARCHITECTURAL WORKS

It is well established that architectural and engineering drawings are protected by copyright. For many years the courts have extended copyright protection to architectural and technical drawings.[14] The courts have found protection for such drawings under the copyright laws as "pictorial, graphic, and sculptural works."[15] Such works are defined as specifically including "technical drawings," which would include architectural and engineering drawings.[16] In 1990, Congress expanded the protection of such drawings through passage of The Architectural Works Protection Act, Section 701 of Public Law 101-650. This law took copyright protection beyond the physical drawings and expanded it to protect the building itself.

[A] Protection of Architectural Works

One of the most significant amendments in the 1990 Act established "architectural works" as copyrightable subject matter.[17] An "Architectural Work" is defined as "the design of a building as embodied in any tangible medium of expression, including a building, architectural plans, or drawings." As a result, not only are the drawings protected from being copied, but the actual structure itself is protected from being reproduced, even by photographing and redrawing the plans and elevations without the use of the original drawings.

---

[12] 17 U.S.C.A. § 202.

[13] Cassway v. Chelsea Historic Properties I L.P., 26 U.S.P.Q.2d 1791 (E.D. Pa. 1993).

[14] *See, e.g.,* Eales v. Environmental Lifestyles, Inc., 958 F.2d 876, 880 (9th Cir. 1992); Imperial Homes Corp. v. Lamont, 458 F.2d 895, 898 (5th Cir. 1972) (under pre-1976 copyright law); Demetriades v. Kaufmann, 680 F. Supp. 658, 663 (S.D.N.Y. 1988).

[15] 17 U.S.C.A. § 102(a)(5).

[16] 17 U.S.C.A. § 101.

[17] 17 U.S.C.A. § 102(a)(8).

[B] Buildings Versus Drawings

For years the courts made a distinction between the actual copying of drawings versus the construction of a building modeled after an existing building. A leading case on this distinction is *Demetriades v. Kaufmann*,[18] a 1988 case in which a homebuilder sued for an injunction and copyright infringement. The court held that though plans were protected, the house itself was not protected absent a design patent. The court stated, "Construction of a building imitating that depicted in copyrighted architectural plans does not . . . constitute infringement of those plans." An injunction was granted only to prevent further copying of plans or reliance on them. That all changed after 1990, when copyright protection was expanded beyond the drawings, to cover the design and even the building itself.

Federal regulations clarify that certain architectural works are not copyrightable. These include: (1) structures other than buildings (such as bridges, dams, walkways); (2) standard features (such as windows, doors, etc.); and (3) pre-December 1, 1990 designs, where the design was published or the building constructed prior to December 1, 1990.[19]

[C] Registration of a Copyright

Registration of a copyright does not have to be done until *after* there has been an infringement, although some remedies are available only if the copyright is registered early. The process for registration is fairly simple. A copy of the drawings or specification is sent to the U.S. Copyright Office in Washington, DC with the appropriate form, plus the filing fee. The Copyright Office will mail back a stamped certificate of registration. The process can take several months, but for a higher filing fee, an expedited registration will be done. The forms and instructions are all available over the Internet at the home page for the Copyright Office, which is: <http://lcweb.loc.gov/copyright/>.

To register a claim to copyright in an architectural work, send the following material in the same envelope to Register of Copyrights, Library of Congress, Washington, DC 20559:

1. properly completed application Form VA;
2. nonreturnable copy of the work; and
3. nonrefundable filing fee of $20 in the form of a check, money order, or bank draft payable to the "Register of Copyrights."

A claim to copyright in an architectural work is treated as distinct from a claim in technical drawings of the work. If registration is sought for both an architectural work and technical drawings of the work, two separate applications must be submitted. For works such as tract housing, a single work is one house model with all accompanying floor plan options, elevations, and styles that are

[18] 680 F. Supp. 658 (S.D.N.Y. 1988).
[19] 37 C.F.R. § 202.11.

applicable to that particular model. For registration of written specifications, use form TX.

§ 16.08 COPYRIGHT INFRINGEMENT AND REMEDIES

In order to establish a case for infringement of a copyright, the plaintiff must prove "ownership" of the copyrighted material and unauthorized copying by the defendant.[20] Standard form design-build contracts, like the AIA, DBIA, AGC and other forms, often address this issue of ownership of the documents, but they do not always mention who owns the copyrights. In absence of such a clause, the courts have held that the architect owns the copyrights in his work. Under 17 U.S.C. section 201, copyrights are owned (or vested) initially in the author or authors of the work. The authors of a joint work are deemed to be co-owners of copyright in the work. In the case of a work performed by an employee, the employer owns the copyrights, under what is called the "work for hire" doctrine. But when work is done by an independent contractor, the "work for hire" doctrine does not apply and the independent contractor—*i.e.*, the architect or engineer—owns the copyrights.[21] To avoid any argument on this topic, design firms should make it clear that they are: (1) independent contractors; and (2) not co-authors of the design, but the sole author.

[A] The "Substantial Similarity" Test

"Copying" is rarely admitted by the defendant.[22] As a result, copying is generally established by proving that the defendant had access to the copyrighted material and that there is a "substantial similarity" between the two works.[23] However, where the defendant admits copying the protected work, the person claiming a copyright does not need to prove that the defendant had access to the work, nor substantial similarity.[24] The "substantial similarity" is often made obvious by examining the original drawings and comparing those to the infringing drawing prepared by the defendant, made even more apparent by examining a side-by-side comparison. [25] Subtle changes in two designs may not prohibit a copyright claim as long as there is a "substantial" similarity and proof of access to the original work.

---

[20] Demetriades v. Kaufmann, 680 F. Supp. 658 (S.D.N.Y. 1988) (architectural plans copied); *see* Allied Mktg. Group, Inc. v. CDL Mktg., Inc., 878 F.2d 806, 810 (5th Cir. 1989); Apple Barrel Productions, Inc. v. Beard, 730 F.2d 384, 387 (5th Cir. 1984).

[21] *See, e.g.* Bryce & Palazzola Architects & Assoc., Inc. v. A.M.E. Group, Inc., 865 F. Supp. 401 (E.D. Mich. 1994); Kunycia v. Melville Realty Co., Inc., 755 F. Supp. 566 (S.D.N.Y. 1990).

[22] *Demetriades*, 680 F. Supp. 658.

[23] *Allied Marketing*, 878 F.2d 806; *Apple Barrel*, 730 F.2d 38.4.

[24] *Kunycia*, 755 F. Supp. 566 (architectural plans copied by admission of defendant).

[25] Slight differences between a protected work and an accused work do not preclude a finding of infringement. *Kunycia*, 755 F. Supp. 566.

[B] The Copyright Notice

One of the best ways architects and engineers can protect their designs from unauthorized copying is to put a copyright notice on every drawing and specification that leaves their offices. The notice shows to all the world that they are claiming a copyright in the design and documents and prevents anyone from claiming they did not know otherwise (known as the "innocent infringer" defense). The issue of "notice" is covered by copyright statute 17 U.S.C.A. § 401 which states that whenever a copyrighted work is published by authority of the copyright owner, a notice of copyright may be placed on publicly distributed copies.[26] The notice is to consist of three elements:

1. the symbol ©, the word "copyright," or the abbreviation "Copr.";
2. the year of first publication of the work; and
3. the name of the owner of the copyright in the work.[27]

An example would be "Copyright 1999 Jones Design-Build, Inc."

[C] Registration of the Copyright

Copyright statute 17 U.S.C.A. § 411 states that "no action for infringement of the copyright in any work shall be instituted until the registration of the copyright claim has been made in accordance with this title." Courts have held that in copyright law, a copyright registration certificate is *prima facie* evidence of originality and authorship, and compliance with the statutory formalities. [28] If a copyright is awarded, it grants the owner the exclusive right to prevent the unauthorized reproduction or copying of the copyrighted work.[29]

[D] Damages for Infringement

Courts have held that once the plaintiff has demonstrated a *prima facie* claim for copyright infringement, irreparable harm is presumed.[30] Sections 504 and 505 of the Copyright Act address the types of damages available for copyright infringement to redress this irreparable harm. Section 17 U.S.C.A. § 504(a) provides that:

> an infringer of copyright is liable for either—
> (1) the copyright owner's actual damages and any additional profits of the infringer, as provided by subsection (b); or
> (2) statutory damages, as provided by subsection (c).

---

[26] This section of the statutes previously read that such notice "shall" be placed on the work. Public Law 100-568 amended this section in 1988 with the word "may," making notice permissive, not mandatory. It is still highly recommended, however, to use the notice.

[27] 17 U.S.C.A. § 401(b).

[28] Broadcast Music, Inc. v. Moor-Law, Inc., 484 F. Supp. 357, 362 (D.C. Del. 1980).

[29] 17 U.S.C.A. § 106.

[30] *Demetriades*, 680 F. Supp. 658.

Actual damages and profits under section 504(b) include the actual damages suffered by the plaintiff as a result of the infringement, and any profits of the infringer that are attributable to the infringement and are not taken into account in computing the actual damages. To establish the defendant's profits, the plaintiff only needs to present proof of defendant's gross revenue from use of the copyrighted work. The burden is on the infringer to prove his deductible expenses and the elements of profit attributable to factors other than the copyrighted work. In architect and engineering cases, the courts have held that the copyright owner's actual damages is determined based on the "fair value" of the drawings.[31] The term "fair value" has been defined as what a willing buyer would have been reasonably required to pay a willing seller for the architect's work.[32]

Statutory damages may be recovered at the option of the copyright owner instead of actual damages and profits.[33] Statutory damages shall be "in a sum of not less than $500 or more than $20,000 as the court considers just." When there has been a willful infringement, however, the court has the discretion to increase the award of statutory damages to a sum of not more than $100,000.[34] Statutory damages are permitted only if registration of the copyright is made within three months after the first publication of the work.[35] Under the Copyright Act an architect may sue for copyright infringement of plans for actual damages and the defendant's profit, plus court costs. The copyright statutes also give the court the discretion to award the recovery of costs and attorneys' fees to the prevailing party in an infringement case.[36] Injunctions are sometimes sought to prevent use of infringing documents. However, it has been held that the drawings may not be impounded pending litigation.[37]

[E] Willful Infringement

A willful copyright infringement under the Copyright Act may be found when the defendant had actual knowledge that it was infringing on the plaintiff's copyright.[38] A case of willful infringement may be found where the "defendant was or should have been aware" that its activities constituted infringement of copyright. [39] The plaintiff's burden of showing willfulness can be met by showing that the defendant should have known his conduct was infringing or acted in reckless disregard of the plaintiff's rights.

---

[31] Nucor Corp. v. Tennessee Forging Steel Serv., Inc., 513 F.2d 151, 153 (8th Cir. 1975); *Eales*, 958 F.2d 876.

[32] *Eales*, 958 F.2d 876.

[33] 17 U.S.C.A. § 504(c).

[34] 17 U.S.C.A. § 504(c)(2).

[35] 17 U.S.C.A. § 412.

[36] 17 U.S.C.A. § 505.

[37] Van Deurzen v. Sanders, 21 U.S.P.Q.2d 1480 (D. Kan. 1991).

[38] 17 U.S.C.A. § 504(c)(2). *See, e.g.*, Coleman v. Payne, 698 F. Supp. 704, 706-707 (W.D. Mich. 1988).

[39] Little Mole Music v. Spike Inv., Inc., 720 F. Supp. 751, 755 (W.D. Mo. 1989); Wow & Flutter Music v. Len's Tom Jones Tavern, Inc., 606 F. Supp. 554, 556 (W.D.N.Y. 1985); Fallaci v. New Gazette Literary Corp., 568 F. Supp. 1172, 1173 (S.D.N.Y. 1983).

[F] Criminal Liability

Federal copyright law makes it a crime for any person to "willfully and for purposes of commercial advantage or private financial gain" infringe a copyright. Upon conviction, the infringer may be imprisoned for up to five years for a first offense, 10 years for a second offense, and fined.[40] In addition, the defendant must forfeit all infringing copies, which may be destroyed upon court order. If a false copyright notice is placed on or removed from any article, with fraudulent intent, the defendant may be fined up to $2,500. Criminal liability is not covered by professional liability insurance, so design firms must use extra care in this area.

[G] Sample Cases

In *Eales v. Environmental Lifestyles, Inc.*,[41] an architect sued the homeowners, their builder, his company and superintendent for copyright infringement. The plans were reproduced with the title block removed. The court held that the architect-copyright owner was entitled to actual damages of $11,968 (market value of the plans), plus the builder's profit of $45,250. In a 1992 case, the architect was entitled to recover the infringing architect's profit of $3,000, but not actual damages or attorneys' fees.[42]

Recovery of the infringer's profits can be substantial. For example, in a 1995 Illinois case, an infringing homebuilder was held liable for $294,443 in profits, plus $177,409 in attorneys' fees for copyright infringement in building a $789,000 house based on plans made from a brochure obtained by homeowners during a walk-through of a house they liked in an adjacent subdivision.[43] The court held that the sales brochure was a "derivative work" of the copyrighted plans and that slight differences between the two designs would not preclude a finding of infringement.

In a 1997 design-build case, an architect who copied the original design-builder's plans at the owner's request was held liable for copyright infringement.[44] The plaintiff, Johnson, was licensed as both an architect and a builder in Michigan. He presented the owner with an AIA Owner/Architect contract which contained the typical AIA clause granting the architect the ownership of the documents, including the copyrights. While negotiations over the contract continued, Johnson prepared a program for the project and started work on the drawings. The owner asked Johnson to serve as the contractor, so he gave the owner an AIA Standard Form Agreement Between Owner and Design-Builder, which stated that the drawings and other documents are not the property of the owner and could not be used for completion of the project by others, except by written agreement of the parties. Johnson and the owner were unable to

---

[40] 17 U.S.C. § 506; 18 U.S.C. § 2319.

[41] Eales v. Environmental Lifestyles, Inc., 958 F.2d 876 (9th Cir. 1992).

[42] Pfanenstiel Architects, Inc. v. Choteau Petroleum Co., 978 F.2d 430 (8th Cir. 1992).

[43] Harris Custom Builders, Inc. v. Hoffmeyer, 874 F. Supp. 899 (N.D. Ill. 1995).

[44] Johnson v. Jones, 149 F.3d 494 (6th Cir. 1998).

work out all the terms of their agreement and so the owner terminated Johnson's contract and then proceeded to hire a different architect and builder to take over the project. The owner refused to pay Johnson's invoice for $19,966 for services performed prior to termination. The new architect expressed concerns about using Johnson's drawings since they contained Johnson's copyright notice and would not proceed until that issue was resolved. After the owner's lawyer assured the new architect that there were no copyright problems, the architect traced some of Johnson's drawings, and merely replaced Johnson's name and seal with his own on other sheets for submission to the city. Johnson filed suit for copyright and trademark infringement after seeing the plans he created being used on the jobsite. The court held that the new architect did not have a license to use Johnson's drawings, nor permission to represent them as his own work product. The new architect was found liable for violation of Johnson's copyrights, plus violation of the Lanham Act for "false designation of origin" of the plans. The court ordered the infringing architect to pay as damages his entire $16,560 fee earned on the project, plus Johnson's attorneys' fees stating that "an architect who steals another architect's drawings and uses them as his own will not incur the same expenses that he would if he were forced to start from scratch." The court denied Johnson's claim for his lost profit as the builder of $259,970, stating that such damages were too speculative.

In a surprising outcome in another case, a New York court held that an architect's preliminary drawings were not protected by copyright and were not infringed upon since they were only concepts and ideas, not subject to copyright protection.[45] The architect had been retained to help find an appropriate solution to the hospital's expansion needs. The architect prepared drawings and sketches for a concept that included building a platform over FDR Drive on the East River in New York City. The hospital hired the architect and another architectural firm to further develop the concept, but the two architects found that they could not work together. As a result, the hospital terminated the contracts and provided competing firms with copies of the design concept. While several firms competed for the work, the original architect registered his copyrights in his initial report. When the hospital selected a new firm that made tracings of the original design concepts, the original architect filed suit under the Copyright Act and the Lanham Act. The court rejected the lawsuit finding that while there were many similarities between the original concept and the new schematic drawings, there was no infringement because the Copyright Act does not protect ideas and concepts, but only the expression of those ideas and concepts. The court did not rule out the possibility of protection for preliminary drawings in other cases, but found that here the original drawings were only generalized ideas and concepts and, in some cases, existing conditions, neither of which were entitled to protection.

[45]Eli Attia v. Society of the New York Hosp., 201 F.3d 50 (2d Cir. 1999).

§16.09 OWNERSHIP OF TECHNICAL PROPOSALS

In two-step procurement, where design-build teams submit a technical proposal as part of the selection process, owners get to see up to five different designs and innovative technical approaches to the same project. Even though the contract is awarded to only one of the five prequalified teams, there may be ideas or design details that the owner liked from another team's submission that the owner wants to incorporate into the final design. Is this fair to the teams who were not selected? Does the payment or acceptance of a minor stipend grant the owner unlimited rights to use the designs and ideas that were submitted by the teams? This issue is a very common concern among design-build teams who do not want to see their design used by the owner unless they are awarded the contract for several reasons, such as:

- They were not selected! If the owner likes the design of another team, then the contract should have been awarded to that team.
- The use of stipends, often less than one-tenth of one percent of the project cost, should not be used as a source of funding cheap design competitions in which the owner can obtain the rights to five designs by paying a fraction of the cost incurred to produce those designs. This is a windfall to the owner and devalues the benefit derived from the teams' efforts.
- The teams who are not selected, and who came up with innovative ideas, should be free to market those designs to other owners on other projects, where they might be able to win the award of a contract. Prisons, parking garages, treatment plants, schools, and other projects may be appropriate for use on other projects with similar programs.
- Exposure to liability.
- Windfall to the selected team. The team that was chosen gets the benefit of using innovative ideas that were the result of hard efforts by a competitor.

Some legislation has been proposed to give the public agency ownership of the designs in return for the acceptance of a stipend or honorarium. For example, Missouri's Senate Bill No. 772 (2000), contained this provision:

> Upon payment of the stipend to any unsuccessful design-build proposer, the state shall acquire a non-exclusive right to use the design submitted by the proposer, and the proposer shall have no further liability for its use by the state in any manner. If the design-build proposer desires to retain all rights and interest in the design proposed, the proposer shall forfeit the stipend.

In other words, the state gets the rights to five designs by paying a stipend to the proposers, and the design-builders give up a license in their copyrights. The release from liability granted by this bill may give the design-build team protection from claims by the state agency, but would not protect the team from a claim by some third party, such as a building user, since the proposed wording does not indemnify or defend the design-build team from third-party claims.

[A] DBIA's Position

DBIA's official Policy Statement No. 42, "Who Owns the Proposal," recommends that ownership of unsuccessful proposals and the drawings, specifications, and information in them should remain with the design-build team that submitted the proposal. DBIA states that "unless and until a proposal is accepted, the drawings, specifications and other information in the proposal shall remain the property of the proposer. The owner shall maintain the confidentiality of information contained in the proposals and shall not disclose the proposals or information contained therein to any other proposer or to the public."[46]

As with the proposed Missouri legislation, however, DBIA's official position recognizes that "in the event those proposers are compensated by a reasonable honorarium, the honorarium shall be considered for the purchase of the proposal and the information contained therein."

[B] The AIA/AGC Position

The AIA and AGC take the opposite view in a publication endorsed by DBIA.[47] In the AIA/AGC Guidelines, it is stated that "[d]esign work generated for a proposal should be the property of each of the individual design-build teams unless and until the proposal is accepted by the owner. An owner has no need for the design work and the design-build team may have a proprietary interest in all or part of the design." Considering the payment of a minor stipend as a "purchase" of the unsuccessful proposal, with unlimited rights to use the information, without paying fair value, is the wrong direction for design-build to go. Stipends of less than one-tenth of one percent, which are arbitrarily set by selection committees and the staff of public agencies, come nowhere near the value of a talented design and construction team's efforts that go into such technical proposals. Putting teams in the awkward position of choosing between a minor stipend, which might defray printing and postage costs at best, and no stipend in order to protect the design is unfair to the competing teams and represents a windfall to public owners. DBIA's statement is premised on the payment of a "reasonable stipend" but most stipends, while maybe reasonable to help defray expenses, are not "reasonable" as compensation for valuable designs and design ideas.

§ 16.10 OWNERSHIP OF PRELIMINARY DESIGN DOCUMENTS

Design-build introduces a new reason why architects and engineers want to retain the copyrights and ownership of drawings and designs. There is a fear that a contractor might hire an architect or engineer to prepare schematic drawings, sufficient to develop pricing and win the award of the design-build contract, but then fire the designers and attempt to finish the project without the assistance of a

[46] DBIA Manual of Policies, Policy No. 42, Oct. 1998.

[47] *Recommended Guidelines for Procurement of Design-Build in the Public Sector* (1995).

design professional in order to save fees. This fear has resulted in licensing regulations in some states that require the design-build contractor to inform the owner as to who the architect will be for the project, keep the architect involved throughout the project, and replace the architect immediately with another if terminated by the contractor.[48] One of the best ways to keep the contractor from firing the architect or engineer and attempting to use the design to complete the project is for the design firm to retain the copyrights in the design and ownership of the documents. With this legal right, the designer can prevent its design from being used or copies being made without its professional involvement.

The design-build contractor, however, has a competing interest. Having paid an architect to prepare the design that wins the award, the contractor may be eternally locked to a specific architect, even if the relationship breaks down and the need to replace the architect with another firm arises. If the contractor cannot continue to use the design that has been paid for, and have other professionals take over the remaining services, the project comes to a halt. This result is not fair to the contractor, nor to the owner, who may face legal action for copyright infringement if it attempts to use the design after terminating the contract with the architect. There is a need for a balancing of these interests so that the architect is protected, the contractor not handcuffed, and the owner not penalized due to disputes that may arise among the design-build team. DBIA's Policy Statement on ownership of the design states that:

> if an owner and design-builder proceed into schematic design and design development in good faith, but are unable to reach agreement as to price, the owner should have the option to shift to another proposed design-builder for the project. This option should include having the right to use some or all of the original design work developed during this early period of design-builder involvement.

However, DBIA recognizes that the team's "greatest intellectual effort" is usually made during the early design phase. Therefore, DBIA does not believe that an owner should be able to take advantage of this expertise in the early stages and then "shop" the work to other contractors in an attempt to get a lower price. DBIA's official policy is that if an owner and design-builder cannot reach agreement on price after preliminary design work is done, and the owner wants to change contractors, then in exchange for the right to use the design, the owner should agree to compensate the design-builder "with a premium over its initial compensation." In addition, the owner should indemnify the design-build contractor and its design partners or consultants against any liability relating to the project and the design, since these parties will not be able to complete their work, yet face some risk by having their incomplete designs used by others.[49]

---

[48] *See, e.g.*, 69 Ill. Adm. Code § 1150.85; S.D. Code § 20:38:22:04.

[49] DBIA Manual of Policy Statements, Policy No. 41, Oct. 1998.

§ 16.11 OWNERSHIP OF FINAL DESIGN DOCUMENTS

More and more owners want to own the documents and the designs. They do not want to have to go back to the original architect, engineer, or contractor each time they make changes to the building, add an addition, or build a duplicate structure or building. Today's owners want not only the ownership of the documents, but they want them on CAD diskette version, which is easily modified. Most owners recognize the risk that this exposes the designer to and, in return for the convenience of owning the design, are willing to release, indemnify, and defend the architect and its consultants from any reuse or modification to the design done without the architect's professional involvement.

What many design professionals fail to consider, however, is that their standard details, wall sections, specification sections, etc. are going to be used again by their firm on other projects. By giving the client exclusive copyrights in this information, they may be prevented from using their own designs on other projects. Whatever rights are granted to the client, an architect must be sure that he or she retains, at a minimum, the right to use standard details and specification sections on other projects without reproducing the entire design in whole; and that the client agrees to release, indemnify, and defend the architect, its consultants, and its respective officers and employees from any claims, damages, losses, costs or expense (including reasonable attorneys' fees and costs of defense) that arise out of the reuse of its documents or designs. It is also recommended that architects retain a hard copy of whatever documents are given to the client in electronic version so that they can verify in the future what information was on the diskette at the time it was transferred to the client. An architect also should remove its firm's name, logo, seals and other indications of authorship so that a future alteration of the design is not attributed to its firm.

§ 16.12 BID SHOPPING

One risk in design-build is that the owner will take the design-build contractor's design and use it as a basis to get bids from other contractors. For example, in an Oregon court case,[50] the owner hired a design-build contractor (BSI) to prepare a design that met the owner's financial requirements, with the understanding that the parties would negotiate a final price. After the design was prepared, the owner objected to BSI's $245,000 cost estimate and took the design to several other contractors, one of whom was hired to build it for $200,000. BSI sued for breach of contract and for fraud. The court of appeals upheld the jury's $51,000 award of punitive damages for fraud. Design-build contractors should consider including specific contract language that addresses use of the drawings by the owner to prevent the owner from "shopping" the design to other contractors. Retaining the copyrights is a key step toward preventing the use of the drawings by the owner.

---

[50] Building Structures, Inc. v. Young, 883 P.2d 1308 (Ore. App. 1994).

CHAPTER 17

DESIGN-BUILD PROJECT TYPES AND PROJECT STUDIES

Not too long ago, use of design-build was often limited to parking garages and jails, or warehouses—simple projects that the clients thought did not require complex designs. Design-build has come a long way, with its use covering every imaginable project type from city halls to stadiums, from light rail to high rise, from power plants to chicken plants, from hotels to highways, from arenas to aquariums. This chapter gives a sampling of projects done using design-build in the public and private sectors.

§17.01 PUBLIC SECTOR WORK

Public sector work has had some of the greatest growth in design-build. "The general trend for design-build now is in the public sector," according to Jerry Novacek, P.E., an associate with Zweig White Associates, Inc. Novacek says there was a five-to-six-year cycle where design-build took off in the private sector and that is now what is happening in the public sector. More and more legislation is being enacted each year to permit public bodies to do design-build. *See* Appendix E. That, coupled with success stories coming from many public users and a breaking down of old paradigms in government agencies, is resulting in more use of design-build on public projects of all types.

§17.02 TRANSPORTATION PROJECTS

One of the biggest growth areas for design-build has been in transportation projects. In 1998, the Transportation Equity Act for the 21st Century (TEA-21) was signed into law. The new law was based on the Intermodal Surface Transportation Efficiency Act of 1991 (ISTEA). TEA-21 is the largest transportation bill in history with spending projected to reach almost $200 billion by the year 2003. TEA-21 endorses design-build for its speed, economy, and efficiency.[1] As a result of these and other new transportation acts, highway departments have become one of the most aggressive users of design-build. It was reported in 1999 that the Federal Highway Administration had approved or was considering use of design-build on over 100 projects in 21 states. One U.S. Department of Transportation official was reported to have said: "Design-build is the future. There's no going back—people now know that roads and transit systems can be built fast and they're not willing to put up with the old, slow method—not with the traffic congestion we have today."[2] Ron Oakley, president of Fluor Daniel, Inc. said that over five years time, design-build grew from nothing to as much as half of all $100-plus million highway projects, predicting that over $6 billion of major transportation projects would be awarded using design-build over 14 months in 2001-2002.[3]

[1] Pub. L. No. 105-178, § 1307 (1998), codified at 23 U.S.C. § 112 (f)(2)(D).

[2] *Commercial Outlook*, Design-Build, Dec. 1999, p. 47.

[3] *Market Forces Dictate Use of Design-Build on Transportation Projects*, Dateline, DBIA, Mar. 2000, p. 3.

In October 1999, the American Road and Transportation Builders Association (ARTBA) adopted a policy that supports use of design-build provided that the public agency uses a two-step procurement and pays a fee to losing—but responsive—proposers. ARTBA's policy states that competitive bidding is preferred for public works, and that design-build should be used only on well-defined projects.[4]

[A] Light Rail Projects

Design-build has been used, and is being used, on several light rail projects, with the prime contractor usually responsible for everything except for the light rail vehicles themselves. In New Jersey, the Hudson/Bergen Light Rail and South Jersey Light Rail projects are both being done using design-build. In September 1999, the Minnesota Department of Transportation (Mn/DOT) published a solicitation for Letters of Interest from design-build firms for the construction of an 11.5 mile light rail project in Minneapolis and Bloomington, Minnesota, running past the HHH Metrodome through the Minneapolis airport to the Mall of America in Bloomington. The project included not only the rail system, but all civil works, trackwork, utility relocation, vehicle service facility and 14 surface rail stations. According to the request for proposals (RFP), the project has a two-year schedule for a test train segment, then phased construction with final completion in late 2004/early 2005. For this project, Mn/DOT conducted a two-phase selection process for the design-build team, with phase one used to develop a short list of teams with light rail and tunnel experience.

Colorado's DOT solicited proposals in March 2000 from design-build teams to improve 17 miles of two interstate highways (I-25 and I-225), plus add 19 miles of new double-track Light Rail Transit, including 13 new stations. One design-build team was to be selected for the entire project with all elements under a single contract except for the light rail vehicle procurement, collection system, and some minor portions of the project. Teams are being selected on a number of factors including the ability to meet or beat the project schedule and the budget.[5]

On the west coast, the Los Angeles to Pasadena Metro Blue Line project sought design-build proposals in February 2000 for a 13.9 mile light rail transit system, using a single contract for all system and facility elements other than the light rail vehicles. This project was specially authorized through California's Senate Bill No. 1847.[6] Farther north in California, San Francisco's Bay Area Rapid Transit District (BART) has undertaken a $750 million design-build expansion project, part of one of the largest public works programs in the United States, to connect the existing 95-mile, two-way BART system to San Francisco

---

[4] *ARTBA Accepts Design-Build*, Engineering News Record, Oct. 18, 1999, p. 19.

[5] *Request for Letters of Interest for Design-Build Proposals*, Engineering News Record, Mar. 6, 2000, p. 55.

[6] Enacted as Pub. Cont. Code §§ 132410(a)(4) and 132420(d). *See* Appendix E.

International Airport.[7] The rapid rail extension project adds a total of 8.7 miles of track to BART, including 6.5 miles of subway, plus four train stations. James R. Van Epps, BART's executive manager for the West Bay extension, says: "The design-build aspect has enabled us to [construct] about 43 percent of the project in less time than it would have taken us to design it." Van Epps estimates that savings in financing, administration costs, and inflation will run into the tens of millions of dollars. Two different teams are building the project, with team one consisting of Tutor-Saliba Corp. and Slattery Skanska, Inc., and team two made up of Sverdrup Civil, Inc. and Conco Cement Co. (a designer-led team, in which Sverdrup has 51 percent control). Two-thirds of the way through the project, Tutor-Saliba reported no claims, with the estimated completion costs increased by only 1.3. percent.

Seattle's Sound Transit, the public agency in charge of the city's $3.9 billion commuter rail line is using a design-build procurement for a $700 million contract to design and build five miles of tunnels and deep stations that will take trains beneath Portage Bay in north Seattle, Washington. The agency short listed three teams for a project planned for completion in 2006.[8]

Even the international market is using design-build for light rail, including a 16.8 kilometer, 35-station light rail system in Barcelona, Spain. The $200 million design-build-operate contract was awarded to Group TramMet, a consortium of companies led by Paris-based Alstom.[9]

[B] Heavy Rail Projects

In October 1998, the cities of Long Beach and Los Angeles, California (acting through the Alameda Corridor Transportation Authority) awarded a $712 million design-build contract to the Tutor-Saliba Team, a joint venture of Tutor-Saliba Corp., O&G Industries, Inc., Parsons Transportation Group and HNTB Design-Build, Inc., to build a 10-mile-long, double track segment of a 20-mile rail system. The 10-mile section, known as the Mid-Corridor portion, is to be built some 30 feet below street level. The Transportation Authority (known as ACTA) issued an RFP on January 21, 1998 and received three proposals. The Tutor-Saliba Team was selected using a "lowest ultimate cost" approach permitted by Los Angeles city charter. "This project involves everything that you dream about when you begin your engineering career. The trench, which will be 30 feet deep by 51 feet wide, will stretch for 10 miles," said George Hale, HNTB's project manager. "It's an incredible engineering feat." The project is scheduled for completion in April 2002 and carries liquidated damages of $125,000 to $200,000 for each day past the contract completion date.[10]

---

[7] *Subway by the Bay Hits Full Speed*, Design-Build, Feb. 2000, p. 15.

[8] *Seattle Short-Lists Three for $700 Million Commuter Rail*, Design-Build, Feb. 2000, p. 7.

[9] *Light-Rail Projects Rolling in Spain and Israel*, Engineering News Record, May 8, 2000, p. 9.

[10] *By Design*, Los Angeles Lawyer, July-August 1999, p. 32; *California Firm Willing to "Tutor" the State on Design-Build Potential*, Design-Build, Feb. 2000, p. 16.

In Puerto Rico, the $1.7 billion urban transit system, Tren Urbano, is the largest to date in the Federal Transit Administration's demonstration program to adopt a turnkey, design-build approach. The civil contracts for design and construction were divided into seven separate design-build contracts. Puerto Rico's largest contractor, Redondo Construction Co., and the Spanish firm Entrecanales, have four segments of the project. The project used a form of bridging, with the owner's consultants, Daniel, Mann, Johnson, & Mendenhall (Los Angeles) and Frederic R. Harris (New York) providing design to 30 percent, then managing and administering the seven contracts under a separate $170 million consultant contract. The preliminary design made assumptions about volumes and types of materials. The contracts include unit prices for differing site conditions actually encountered. Breaking up the job into seven contracts allowed the owner to start early without waiting for design documents and right-of-way for all sections of the project, which would have delayed the project a year.[11]

[C] Highways

Highway construction seems to be leading the public sector's use of design-build, with state after state announcing new projects to be constructed using that method. Design-build has been used in 23 states on highway projects and on more than 100 federal-aid projects.[12]

The largest design-build project in the United States in the 1990s has been the I-15 highway project in Salt Lake City, Utah. The $1.59 billion project was the largest project under a federal program that evaluated and tested innovative contracting schemes. The joint venture of Sverdrup and Kiewit was one of three teams that competed for the contract, which involved tearing out and reconstructing 16 miles of highway and 137 bridges in 4.5 years to help facilitate Salt Lake's hosting of the 2002 Winter Olympics. The two unsuccessful teams were paid a stipend of $950,000 each, but it is reported that teams spent as much as $3 million each in developing their designs and proposals for the project. The project is ahead of schedule and on track to be finished in time for the Olympics. Local polls show that 80 percent of the public thinks the contractor is doing a good job, a rating far above the 30 percent rating that Utah DOT officials consider good for public projects.

On the heels of the successful I-15 project, the State of Utah Department of Transportation (UDOT) announced plans for and solicited proposals from design-build teams for a $350 million project for 18 kilometers of parkway linking to I-15. The project includes pedestrian, bicycle and equestrian trails parallel to the highway, as well as maintenance obligations. Letters of interest from design-build teams were received in December 1999 for this project, with a planned contract award in late 2000 and project completion in fall of 2003. Utah officials are so pleased with their design-build projects that when the Utah Transit Authority

---

[11] *People Get Ready,* Design-Build, Feb. 2000, p. 27.

[12] *Commercial Outlook*, Design-Build, Dec. 1999, p. 49.

sought state permission to use design-build for its projects, state legislators asked why a project would be done any other way.[13]

The Texas Turnpike Authority, a division of the state Department of Transportation, announced a 91-mile, $1 billion design-build "option" project in 2000 for State Highway 130, a six-lane roadway. The Department solicited proposals from teams to develop, design, construct and potentially to maintain, repair, operate and/or participate in the financing of the project.[14] Whoever is selected to develop the project will also pick the project delivery system.

After Colorado passed House Bill 1324 permitting the state Department of Transportation (CDOT) to use design-build, the CDOT went to work planning to expedite highway and light rail construction on the Southeast Corridor, I-25 and I-225. "Design-build is used by many states to accelerate large transportation construction projects," said CDOT executive director Tom Norton. "Instead of going through two separate, time-consuming efforts to first hire a design firm and then a construction company, design-build consolidates the process. Based on experiences in other states, design-build can reduce construction time by as much as 40 percent and reduce project costs. In a project as large as the Southeast Corridor, it isn't necessary to delay the start of work until plans are 100 percent complete."[15] CDOT has used design-build on five projects already, including: (1) a $5.1 million overpass at I-70 and I-225, completed in 280 working days; (2) a $25.9 million 12-mile stretch of I-70 east of Aurora; (3) a $26.3 million 17-mile restoration project on I-25; (4) a $20.6 million 11.4-mile section of I-70 east and west of Agate; and (5) a $9.2 million 8-mile reconstruction of U.S. 287, near Lamar.

In early 1999, the Arizona DOT awarded a $79.7 million design-build contract to widen I-17 in Phoenix. The project includes widening seven miles of I-17, including two bridges, 200,000 feet of freeway barrier and a state-of-the-art electronic freeway management system, including cameras and loop detectors. Work was planned for 21 months and was awarded to a joint venture of Granite Construction Co. and Sundt Corp. The state of Arizona's Department of Transportation is also authorized to use design-build on two projects in excess of $40 million per year under legislation enacted in 2000.[16]

The new 27-mile Eastern Toll Road project in Orange County, California was awarded the 1999 American Consulting Engineers Council Award. This design-build project included 69 bridges, 10 toll plazas, and a 50-acre freshwater wetland. It was one of the largest design-build projects in the nation. Using fast-track design and construction, the team of CH2H Hill and Silverado Constructors opened the road to traffic one year ahead of schedule and under budget. Another toll road project, this one in Texas, is being built as a bypass around Laredo, Texas

---

[13] According to Tom Warne, Executive Director of Utah Department of Transportation, quoted in *Commercial Outlook*, Design-Build, Dec. 1999, p. 48.

[14] *TTA Seeks Qualification Submittals on Turnpike Project*, DBIA's Dateline Newsletter, May 2000.

[15] *CDOT to Use Design-Build in Southeast Corridor*, Colorado Construction, July 1999, p. 12.

[16] *Arizona Passes Design-Build Legislation*, DBIA's Dateline Newsletter, May 2000.

using design-build. The $50.1 million project includes a 22-mile toll road, which is being privately funded. It is being built by Granite Construction of Watsonville, California with an 18-month schedule. Austin-based Carter & Burgess, Inc. is the design subcontractor.[17]

New Jersey used design-build for the first time on a 3.6 mile, four-lane roadway project, the Hightstown Bypass on Route 133 in East Windsor Township. The contractor, Goodkind & O'Dea, an affiliate of Dewberry & Davis, completed the project two years ahead of schedule and $10 million under budget.[18]

The city of Orlando, Florida hired a design-build team of Martin K. Eby Constrution Co., Inc. and HNTB Corporation for a one-mile urban four-lane divided roadway project, called the John Young Parkway Extension. The project team exceeded city minority business enterprises (MBE) and women business enterprises (WBE) goals and began construction within seven days after the Notice to Proceed, prior to execution of fully developed plans. The entire main lane roadway opened in seven months, and was ahead of schedule. The project received praise from city officials and received AGC's Marvin M. Black Excellence-in-Partnering Award as well as the APWA's Project of the Year Award.[19]

Several states use private developer contracts when there is no statutory authority for design-build. In New Mexico, the state legislature has authorized design-build for all state agencies except for the highway department.[20] Due to the tight restrictions, the state Department of Transportation (DOT) came up with an innovative way to get some of the advantages of design-build by hiring a private developer to widen state Highway 44 on a $295 million project between Bernalillo and Bloomfield. The state awarded a contract to Mesa PDC, a subsidiary of Koch Industries of Wichita, Kansas, as developer, who then retained Flatiron Structures of Longmont, Colorado and CH2MHill of Denver to design the project. With Mesa's assistance, the state awarded separate construction contracts to four contractors to perform the construction work. The developer's $46 million contract included design, inspection, and management and general liability insurance for the four contractors. The state will also pay the developer another $62 million just to warrant the roadway for 20 years and the bridges for 10 years.[21]

Texas is another state whose DOT uses private developers to get around the lack of statutory authority for design-build procurement.

[D] Bridges

The state of Maine, in conjunction with the Federal Highway Administration, used design-build on the Sagadahoc Bridge, a $46.6 million project built

---

[17] *Private Texas Toll Road Goes Design-Build*, Engineering News Record, June 21, 1999, p.26.

[18] DBIA's Dateline newsletter, January 2000.

[19] DBIA's Project Spotlight, Dateline newsletter, January 2000.

[20] N.M. Stat. § 13-1-119.1. See Appendix E.

[21] *New Mexico Road Project Tries Something New, Something Old*, Engineering News Record, Mar. 13, 2000, p. 22.

using concrete segmental and cast-in-place box superstructures on eight-foot-diameter concrete drilled shafts. The contract was awarded to Flatiron-Figg after a weighted two-phase selection process that involved a 19-member scoring committee. *See* §14.12[D] above. The project required U.S. Coast Guard permits for navigation clearances and safety, plus a Corps of Engineers Wetlands permit and consultation from marine fisheries regarding threatened and endangered species. The state required a 10-year warranty on the bridge deck joints and the waterproofing membrane, and a five-year warranty on many other components. The contract included $1,500 per day of liquidated damages, but also gave an incentive bonus of $3,000 per day early, with a cap of $1 million.

[E] Aviation Projects

The Aviation Investment and Reform Act for the 21st Century,[22] passed by Congress in 2000, is a $40 billion aviation bill that authorizes a pilot program to allow up to seven design-build projects funded by federal grants. Section 139 of the Act allows the head of the Federal Aviation Administration to approve an application to use design-build if several conditions are met, including a selection process "as open, fair and objective as the competitive bid system" and at least three bids per project.[23]

Work to upgrade the existing Los Angeles International Airport (LAX) involves a $112 million design-build program based on preliminary bridging designs by Ted Tokio Tanaka Architects. Proposers were given extensive performance specifications developed by the Tanaka firm. The city decided to depart from normal contracting procedures to meet the project's tight schedule, reduce overall change orders, and minimize interpretation problems. Phase one is a $75 million portion of the project.[24]

§17.03 INDUSTRIAL, POWER, AND PROCESS PLANTS

According to F.W. Dodge, design-build projects in this segment of the market will account for $11.7 billion in the United States, predicted to climb by 9 percent in 2000. An article in *Engineering News Record* said, "The hottest market now seems to be power."[25] Clarence Ray, CEO of Duke-Fluor Daniel, a joint venture in the power market, says, "Domestically, this is the largest market I've seen in 20 years."[26] Some design firms are taking advantage of the market as lead contractor.

In Washington, for example, Black & Veatch is the prime design/build contractor on a $30 million electric transmission project, which consists of adding

---

[22] H.R. 1000, Subtitle B, § 139, Design-Build Contracting (2000).

[23] *Id.* The pilot program expires Sept. 30, 2003.

[24] *LAX Chooses Design-Build for Latest Upgrade*, Design-Build, Dec. 1999.

[25] *Hot Markets Shift into Higher Gear*, Engineering News Record, Apr. 10, 2000, p. 56.

[26] *Id.*

several (19 miles) of 230-KV transmission lines, plus two new 230-KV substations and the upgrade of existing lines and substations. As of April 1999, it was the largest design-build contract ever for Black & Veatch and the largest EPC contract among all U.S. utilities. Using fast-track design and construction, the project was six months ahead of schedule and $1.5 million under budget. Due to a concern over a power outage, the contract contains a $100,000 per day liquidated damages clause, which caused some firms to think twice before bidding the project.[27] Black & Veatch's vice president of marketing and strategic planning, Cheryl Muench, says that the power market is so robust for the firm, "we had to turn away work because of a lack of manpower or equipment."[28]

Stone & Webster is the design-build contractor for a $10 million compost plant on the island of Nantucket, designed to have nominal capacity of 50 tons per day of municipal solid waste and five tons per day of sewage sludge. The plant will convert non-recyclable municipal solid waste and sewage to produce compost material for use in landscaping.

§17.04 WATER TREATMENT PLANTS

The Tolt River Filtration Plant outside of Seattle, Washington was awarded on a design-build basis to Camp Dresser & McKee, Inc. of Cambridge, Massachusetts following a change in state law in 1995 authorizing alternative public works contracting procedures. The contract was awarded on a design-build-operate basis, under which the contractor operates the facility for 25 years after completion. According to the director of Seattle Public Utilities, "if you have a 25-year contract, those who are building also have to operate it, and they build with operation in mind." The winning proposal by Camp Dresser of $101 million beat the city's estimate by 40 percent. Proposals were based on 60 percent technical criteria and 40 percent financial strength. To the city's surprise, all of the proposals received were at least 24 percent less than the city's estimate.[29] Following the success of the Tolt River project, Seattle Public Utilities solicited proposals for a $110 million design-build-operate contract for the city's Cedar River water treatment plant. Together, the Tolt and Cedar plants will provide up to 370 million gallons of water per day.[30]

A $135 million contract to design, build and operate a 60 million gallon-per-day water treatment plant was awarded by Tampa Bay Water in April 2000. The contract covers $79 million in capital costs and $56 million in operation and maintenance fees, and is estimated to save 21 percent over a traditional design-bid-build contract process. The contract was awarded to a team led by USFilter,

---

[27] *Utility Saves Money and Time Using EPC Contract Approach*, Engineering News Record, April 12, 1999, p. 20.

[28] *Hot Markets Shift Into Higher Gear*, Engineering News Record, Apr. 10, 2000, p. 56.

[29] *A Watershed for Seattle*, Design-Build, Oct. 1999, p. 27.

[30] *Seattle's Savings at Tolt Plant Push Another Design-Build Job*, Engineering News Record, May 1, 2000, p. 15.

part of the French utility conglomerate Vivendi, with design support from Camp Dresser & McKee, Inc. and construction teammate Clark Construction Co., of Lansing, Michigan.[31]

In Puerto Rico, the island's first use of public design-build is a $305 million "Superaqueduct," a 50-mile long aqueduct that will supply water to the island at a rate of 100 mgd. The owner is the Puerto Rico Aqueduct and Sewer Authority (PRASA) and the design-build team is Thames-Dick Superaqueduct Partners, a joint venture of Dick Corporation of Pittsburgh and Thames Water PLC of the United Kingdom.[32]

§17.05 PARKING GARAGES

Public owners have used design-build on parking garages for many years, some starting out with that project type as a test of the procurement system, since garages are seen as fairly simple design projects. The trend has continued and today a large percentage of public parking structures are built using design-build. Some examples include a $9.4 million contract awarded to McCarthy by the state of California in 1999 for a six-level structure with a 385-day completion schedule. Another is an $11.7 million, five-level parking structure for the Jacksonville, Florida Transportation Authority. The Authority hired The Haskell Company to design and build the garage with completion scheduled for July 2000.

The University of Kansas obtained special legislation in 1977 to use design-build on a new parking garage at the school's medical center. A rival school, the University of Missouri, has built several parking garages using design-build. In January 2000, the university solicited design-build proposals for a budgeted $11.6 million, 1,410 car parking garage and pedestrian bridge on its Kansas City campus, to be completed in 540 calendar days or less. *See* §14.12[B], above.

In September 2000, the Denver Zoological Foundation sought qualifications statements from design-build teams for a $16 million, 764-minimum parking space, 3-level garage. This is one of many major design-build projects in Denver. *See* §§17.02[C], 17.06, 17.07, 17.08, 17.09[B], and 17.12.

§17.06 GOVERNMENT BUILDINGS

Justice projects are also among the fastest growing market segments. The Scott M. Matheson Courts Complex in Salt Lake City, Utah is a great example of design-build well done. The project was awarded to the team of Big-D Construction and MHTN Architects of Utah, with the San Francisco office of HOK Architects selected after a national design competition. The complex consists of 37 courts including the State Supreme Court, appellate courts, juvenile courts and 28 district courts, expandable on site to a total of 50 courts. It also has underground parking for 750 cars. The $68 million project had no change orders during con-

[31] *Tampa Bay Taps Design-Builders*, Engineering News Record, May 1, 2000, p. 14.
[32] *Puerto Rico's Superaqueduct Back on Track for Spring*, Design-Build, Feb. 2000, p. 7.

struction. The 120-foot rotunda is the focal point of the building, whose classical style was made more affordable by the team's selection of fiberglass columns. *See* § 14.14, above.

The state of California hired Colorado-based Hensel Phelps Construction and design teammate Fentress Bradburn Architects, of Denver, to design and build 360,000 square feet of office space in Sacramento to house several thousand state employees. The $68 million project is known as the Capitol Area East End Office Complex and will house staff from the Departments of Education, General Services and Health Services. A separate $225.6 million design-build contract was awarded to The Clark Construction Group, Inc. and Gruen Associates for a four-building, mixed use office development as part of the complex, to include 1.1 million square feet of new building space, as well as 630,000 square feet of parking facilities. The total project encompasses four city blocks.

Hensel Phelps was also the design-build contractor for the city of Oakland's Administration Buildings project, consisting of the new Dalziel Building, the historic Broadway Building, and the 6.3 acre Frank H. Ogawa Plaza. The project received DBIA's Design Excellence Award in October 1999.

The U.S. State Department hired the team of J.A. Jones, Inc. and HOK to rebuild two African embassies damaged by terrorist bombings in 1998. The $64.5 million design-build contract includes new office buildings and compounds for sites in Dar es-Salaam, Tanzania and Nairobi, Kenya incorporating the State Department's latest security standards.

In Denver, a new $124 million city and county office building is being built using design-build, with the contract awarded after a design competition. The unsuccessful proposers received a stipend of $100,000 each for their efforts. Finally, the city of West Sacramento, California is also using design-build procurement for construction of its new city hall and civic center, a project estimated at 74,000 square feet and $12 million in cost.

§ 17.07 ENTERTAINMENT AND SPORTS

The New Mile High Stadium in Denver, Colorado is a $366 million design-build project led by a joint venture of Turner Construction and HNTB Design-Build, Inc. The facility will have 76,125 seats and can expand to 80,000 when the open end of the horseshoe is filled in with seating. The new stadium is located on an 83-acre site between the existing Mile High Stadium and the old McNichols Arena, both of which were to be demolished during construction of the new stadium. The city of Denver had used the site as a city dump for years, requiring environmental cleanup and removal and backfill of 15 feet of material over the entire site. The Denver Broncos need the stadium to be open for the team's August 2001 preseason games, so the project has a very tight schedule. "The only way to do this job is design-build," said Turner's Rod Michalka. "We couldn't touch it otherwise." The project team's initial $6 million in fees was totally at risk during the initial design and planning phases until a November 1998 public vote. To cover the risk, the team bought a special insurance policy which would have assured the fee in the event the vote failed. A Super Bowl appearance for the Broncos in January 1998, plus John Elway's agreement to return for his final season in 1998 helped the underwriters get comfortable enough to issue the unique policy. At peak, the construction work will cost $18 million per month in construction costs.

Another NFL design-build project is the $12 million Atlanta Falcons training and headquarters complex in suburban Atlanta, Georgia. The project was awarded to Clayco Construction Co. of St. Louis, Missouri, the firm's second NFL training facility. The design-build project was on a fast track schedule to assure that the team could use the facility for the 2000 football season.

Even professional golfers have used design-build to meet tight schedules. When the Las Vegas LPGA Tour planners needed their project done in time for a golf tournament, they turned to design-build. The Desert Inn Clubhouse and Conference Center, a 60,000 square foot facility, was led by Clark & Sullivan Constructors, Inc., teaming with Tate & Snyder Architects for the design.

The Sevier County Baseball Stadium in Tennessee is a $13.2 million minor league baseball/multipurpose stadium designed and built by the team of Denark-Smith, Inc. and its partners, HNTB Sports Architecture and Barber & McMurray, Inc. The project was reportedly delivered two weeks ahead of schedule and on budget.[33]

Design-build has been used at the college sports level as well. It was reported that Eastern Michigan University in Ypsilanti, Michigan was ready to abandon plans for a new arena due to cost estimates using traditional design-bid-build procurement when a board member suggested use of design-build. The university eventually awarded a $29.6 million design-build contract to Barton Malow Design for the new 204,000 square foot Convocation Center and Arena. The contract was awarded after a design competition among three firms, with the two unsuccessful teams being paid $10,000 each for their effort. The university was so impressed with the new arena's savings in cost and time that it then used design-build on two other projects.[34]

The Family Arena, a 10,000 seat sports arena in St. Charles, Missouri, was a design-build project that is home for minor league sports. J.S. Alberici Construction Co. and Hastings + Chivetta Architects teamed up to design and build the 192,855 square foot structural steel facility. Enough material to make a layer 10 feet deep was brought in to compact the base before raising it out of the flood plain to the 100-year flood level. Two 17-foot deep steel trusses span between columns, with super long 185 foot steel joists spanning between the trusses. The project was finished in September 1999, three weeks ahead of a 75-week schedule, despite 15 days lost to weather. The original bid of $31 million was trimmed to $27 million.[35]

[33] *Denark-Smith/HNTB Deliver Baseball Stadium*, DBIA's Dateline Newsletter, May 2000.

[34] Building Design & Construction, Jan. 2000.

[35] St. Louis Construction News & Review, Jan.-Feb. 2000, pp. 20-21.

§ 17.08 MUSEUMS AND OTHER ATTRACTIONS

Denver, Colorado has had several of its most prominent projects of the late 1990s built using design-build. One of those projects is Ocean Journey, a $93 million aquarium project which opened in 1999. The project had several challenges for the design-build team which consisted of Denver-based Hensel Phelps Construction Co., as prime contractor, and two design firms, RNL Design and Anderson Mason Dale. It had to blend in with the red brick historic buildings in the area. The final design was a large red brick building with a serpentine glass curtain of 650 panes, evoking images of waves and ocean movement. The foundation is a square brick box, which saved costs in construction and permitted easier installation of exhibits and life support systems. The waving glass structure sets over the square base and, with the help of numerous skylights, gives form and light to the aquarium. Though the designers were given freedom to produce an attactive public building, Hensel Phelps had control of the purse strings and helped to advise the design team on cost control issues. "For a unique one-of-a-kind building that has cost constraints, it was so valuable to have a creative design team and implementers who would build it, and who all were involved from day one," according to Nancy Yelverton, executive vice president of Ocean Journey. The project came in on time and within budget, seven years after Hensel Phelps was hired as the design-builder in 1992. A key design decision was not to put the water tanks and life support systems below ground, based on problems that other projects had faced in the area. As a result, the second floor of the building has to keep one million gallons of water suspended on a 35,000 square foot floor plate, weighing 12 million pounds, made of 2,500 cubic yards of reinforced concrete.[36] Every detail of the building was hashed out between Hensel Phelps and the subcontractors. One such detail was the skylights, which needed to let in light, while not heating up the exhibits or the water too much, yet permit sufficient UV radiation to sustain plant and animal life. The result was more than 184,000 pounds of acrylic.

Another aquarium project done using design-build is the $50 million Long Island Aquarium in Bay Shore, New York project, awarded to the design-build team of Turner Construction Co. and Chermayeff Sollogub and Poole, a 110,000 square foot facility that includes a 260-seat IMAX theater and habitat areas, with completion scheduled for Summer 2002.[37]

In Orlando, Florida, the new $275 million Opryland Hotel is a design-build project awarded to the joint venture of Perini Building Co. and Suitt Construction Co. as general contractor, with Hnedak Bobo Group as designer. The hotel will have 1,406 guest rooms and suites, plus a separate 727,000 square foot convention and expo center with 80,000 square feet of meeting rooms, exhibition space, and two ballrooms. The complex also includes three glazed glass atriums, three restaurants, 15,000 square feet of retail space, as well as two swimming pools, a spa and fitness center. Completion is scheduled for February 2002.

---

[36] *Aquarium Reaches Destination*, The Denver Business Journal, Oct. 1, 1999, p. 8B.

[37] *Contracts,* Design-Build, Feb. 2000, p. 72.

§17.09 CORRECTIONS AND JUSTICE FACILITIES

[A] Federal Prisons

Another leader in use of design-build is correctional facilities. Tougher crime legislation, "three strikes" laws, and mandatory sentencing guidelines all mean that more people are going to prison for longer periods of time. This creates a market for housing of prisoners at the state and federal levels. The Federal Bureau of Prisons (FBOP) is reported to need 2,600 new beds per year to keep up with demand. The average facility has 1,000 beds, with a cost of $80,000 per bed, meaning $200 million worth of new federal prison work per year is needed. Conservative estimates of national prison population growth are 4 percent per year, meaning a market of $2.88 billion per year through 2006, with half of that work projected to be done design-build.[38] The FBOP did its first design-build project in 1997 in Arkansas. Three major 1,000 bed projects in Florida, Virginia, and Kentucky were built in 1998 using design-build, at roughly $80 million each.

[B] State and Local Agencies

In Aurora, Colorado, the $48 million Ridge View Youth Services Center was awarded under a design-build-operate agreement with the state. The general contractor, GE Johnson Construction Co. of Colorado Springs, has teamed with KMD of San Francisco, California as architect and Rite of Passage of Minden, Nevada as facility operator.

The state of Nevada's first design-build project was a 96-bed medium security juvenile facility in Las Vegas, built by Clark & Sullivan Constructors, Inc. who teamed with Sheehan-Van Woert-Bigotti Architects and Correctional Services Corporation, the contract operator.

The state of Missouri built a $48 million 1,200 bed state prison in Cameron, Missouri in 1995 using the design-build team of J.E. Dunn Construction Company and HNTB after a contested two-phase selection procurement process that was done without any special legislative authority. *See* Chapter 14, § 14.12[A]. The protesting low bidder dropped its lawsuit and injunction after the court ordered a bond posted for nearly $2 million while the case was litigated and the project put on hold. Following that experience, the state built a minimum security correctional facility in the Kansas City area using design-build but only after passing special legislation to authorize that single project.[39]

The city and county of San Francisco, Bureau of Architecture awarded an $85 million contract to the design-build team of Morse Diesel International and Kaplin McLaughlin Diaz for County Jail No. 3. The project includes an administrative services building, two housing buildings and vehicle and general storage facilities.

---

[38] Thomas M. Wierdsma, P.E., *Prisons—Watchtower on a Growing Market*, DBIA/AIA Conference, Chicago, Illinois (Oct. 1998).

[39] R.S. Mo. § 221.500 (1997).

§17.10 EDUCATIONAL FACILITIES

The University of Nevada, Reno Fire Science Academy/Crisis and Emergency Management Institute is an award-winning example of good design-build. The multi-million dollar project built by Clark & Sullivan Constructors, Inc. and was named the Most Outstanding Public Building Project of 1999 by *Southwest Contractor Magazine*. According to the prime design-build contractor, the project would have taken six years under the University's and State Public Works procurement system; however, design-build cut the project time to just 13 months. The project was developed through a finance/leaseback program.

Clayco Construction Co. of Fenton, Missouri is the design-builder for a $5 million, 30,000 square foot office/retail building in St. Louis that will serve as the main metro St. Louis campus for Sanford-Brown College.

§17.11 MANUFACTURING PLANTS

The single-point responsibility aspect of design-build has been called a plus by one owner because there is: "Just one chicken to choke!" St. Louis-based McCarthy took chicken choking to a whole new level in May 1998 when it completed a new state-of-the-art poultry processing plant for Foster Farms, capable of processing 672,000 chickens per week, with a capacity of 140 chickens per minute. The $22 million design-build project has over 100,000 square feet of process area and 10,000 square feet of administration space and was completed in less than nine months, which was still two weeks ahead of schedule and $2.2 million under budget. The contract provided for liquidated damages for late completion, increasing the pressure on McCarthy.

The site selected by the owner in the Columbia River Valley had annual rainfall of 80 inches per year. Soil borings revealed the site was river silt and sand, requiring 6,000 loads of sand dredged from the Columbia River to surcharge the 110,000 square foot building footprint, compacting the sandy soil about 18 inches. The imported material was left on site for four months, then rolled and reused to construct a large parking lot adjacent to the plant. To meet the rigorous schedule and avoid wet weather conditions, all areas around the plant were covered with ATB, an asphalt material, early on to ease movement of material during the rain and to keep mud out of the plant. All buildings and equipment had to meet U.S. Department of Agriculture specifications, including no standing water on floors in the process area. The concrete floors were poured with a one-quarter inch per foot slope. According to design partner Al Coltrane, of Vaughn, Coltrane & Associates, "This job was one of the most efficiently executed I've ever worked on."[40] Precast concrete walls and roof were completed just days before the rain began and the concrete floors were poured inside the dry enclosure.

---

[40] DBIA's Dateline Newsletter, Nov. 1999.

The speed at which design-build projects is sometimes amazing. Two manufacturing plants are good examples. V.I.P. Structures completed a 300,000 square foot manufacturing and warehouse facility for American Woodmark Corporation in Gas City, Indiana in only 110 days.[41] Even faster was the 150,000 square foot office/distribution facility at Mitchell International Airport completed by CON/STEEL Alliance in just 100 days, despite restrictions on use of a lifting crane for the concrete tilt-up panels in the flight path of the airport runway.

In Leesburg, Alabama, the team of BE&K and Suitt Construction was awarded a $46 million design-build contract for a 175,000 square foot auto parts plant. The project was scheduled for completion mid-March 2001, with production start-up set for November 2001.[42]

§17.12 PRIVATE OFFICE BUILDINGS AND HIGH-RISE RESIDENTIAL PROJECTS

Kiewit Construction is the design-build contractor on a campus for Level 3 Communications, Inc. in Broomfield, Colorado, an 800,000 square foot, $152 million project on a 46-acre site that includes four separate office buildings, two parking structures with space for 1,450 cars and on-grade parking for another 1,200 vehicles. The project was planned for a 27-month completion schedule using fast-track techniques. The architect is Holland-Basham Architects of Omaha, Nebraska.

A $31 million building in downtown Denver, Colorado was built by The Neenan Company, the largest design-build contractor in that state. The project was designed using 3D CAD modeling, with the subcontractor trades electronically transferring their drawings to Neenan. Though Neenan has some 40 architects on staff, it routinely pairs with an outside design firm, here teaming with the architectural firm of Shears + Leese Associated Architects, who built the 3D model which is used on a project Web site to check actual progress via an on-site camera.[43]

The Beck Group of Dallas, Texas was the contractor on a 60,000 square foot, two-story Revenue Management Call Center for Southwestern Bell in Ft. Worth, Texas. Beck not only designed and built the project, but provided the financing and purchased the land, completing the project in 10 months using fast-track scheduling and design-build MEP subcontractors with call center experience and tilt-wall design.[44]

The $30 million John Deere Credit facility involves two three-story buildings, each joined in the center by a common link, built in Johnston, Iowa, a suburb of Des Moines. The facility will house 800 employees. The design-build contractor was Ryan Companies US, Inc. of Minneapolis. The 240,000 square

---

[41] DBIA's Dateline Newsletter, Jan. 2000.

[42] *Contracts,* Engineering News Record, Sept. 11, p.80.

[43] *Getting the Big Picture*, Colorado Construction, p. 19, June 1999.

[44] DBIA's Dateline Newsletter, Jan. 2000.

foot facility was designed with a structural steel frame, with an exterior of 60 percent brick and 40 percent precast concrete, with over 500 precast panels. The fast-track project schedule required that the contractor complete half of the structure first so that the owner could begin moving in while the second half was still under construction. Using fast-track construction, Ryan started on drilling the over 240 piers before the architect had finished designing the building. In addition, structural steel was ordered prior to all the steel detailing being completely designed. Ryan self-performed all of the concrete work on the job, including the walls, footings and slabs in the buildings. Ryan also self-performs many of the interior finishes. The first meeting with the owner was in Fall 1998, and the project was scheduled for occupancy by May 2000. Communication between Ryan and its architect, OPN Architects, was aided because the companies had worked together on several projects in the past.[45]

Charles Pankow Builders, Ltd. is the design-build contractor for two projects valued at $137 million. In Daly City, Pankow is building an office building, parking structure, retail stores, and a 20-screen multiplex in its first phase. In San Francisco, California, Pankow is building a 39-story luxury apartment complex and it is reported to be the tallest concrete structure ever built in the highest-risk seismic Zone 4.[46] Another residential project is The Mansions on the Plaza, a gated residential community consisting of 222 apartment units by Clayco Construction Co. of St. Louis, Missouri and Humphreys & Partners Architects of Dallas, Texas. The firms teamed to design and build this five-story building on a 3.8 acre site in University City, Missouri.[47]

§17.13 NON-PROFIT ORGANIZATIONS

Design-build has been used in the non-profit sector, where most organizations operate on lean budgets, with minimal staff and no construction experience. Construction projects in this sector are usually the result of long term "capital campaigns," with fundraising from foundations, private donations and government grants. With non-profit dollars in tight competition among the thousand of charitable organizations, it is very important to non-profit organizations to get the most "bang for their buck" when making alterations, building additions, or new structures. Design-build may be an appropriate method in this sector. However, due to staff restrictions, non-profit organizations should consider hiring a consultant to advise them on preparing an RFP, budgeting, and administering a construction project. Many non-profits have contractors, architects, or lawyers on their board of directors who can give advice. However, due to concerns about conflicts of interest and time constraints, the role of project consultant and design-builder will usually be filled by others through competitive bidding or negotiation.

---

[45] *Fast Tracking For John Deere Credit,* Rent Right, Feb. 2000, p. RR6.

[46] DBIA's Dateline Newsletter, Jan. 2000.

[47] *Id.*

One such project, The Alfond Youth Recreation Center in Waterville, Maine, was done by the team of Cianbro/SMRT. The owner, the Maine Youth Alliance, hired a consultant (Wagner-Hohns-Inglis, Inc.) as its construction project manager. The consultant assisted the owner in the selection of the design-build team and monitored the team's performance during the design and construction period. Design-build reduced the total project time from 2.5 years to just 14 months. According to Mel Tillatson, P.E., of Wagner-Hohns-Inglis, Inc.: "Only with design-build could the project delivery time be reduced, given the limited construction season. Equally important, the costs were kept under control." The project was built for less than the $8 million budget.

CHAPTER 18

DESIGN-BUILD CONTRACTS

§18.01 TYPES OF CONTRACTS

The design-build contracts needed for a project typically include: a contract between the owner and the design-builder; subcontract(s) between the design-builder and the architect/engineer; and subcontracts or purchase orders between the design-builder and the various trade contractors and material suppliers. Standard contract forms have been developed by the national associations for architects, engineers and general contractors and are frequently used in the industry. Newer forms developed by the Design-Build Institute of America (DBIA) are also catching on with owners and design-builders. This chapter will discuss the various contract forms, highlighting issues for design professionals.

[A] Cost-Plus or Lump Sum

According to a recent survey, three-fourths of the design-build firms report that lump sum/fixed fee are the most common design-build contracts.[1] This is similar to traditional construction where surveys show that lump sum/fixed price contracts are still the most common.[2] One advantage of the cost-plus a fee type contract, with a guaranteed maximum price (GMP), is the ability to increase profits by sharing in savings.

[B] Industry Standard Forms

When the trade associations first started publishing form contracts for construction, it was a joint effort between architects and builders. The early contracts published by the American Institute of Architects (AIA) in the late 1800s were the joint work of the AIA and the National Association of Builders. For decades, the AIA and the Associated General Contractors (AGC) have jointly authored the AIA's A201 General Conditions, which were published by the AIA but endorsed by the AGC for use by their members. But the associations have realized that there is money to be made in the form contract business and, as a result, all of the major industry associations now publish their own contract forms.[3] This is certainly true with design-build forms and design professionals have a wide variety of contracts from which to choose when doing a design-build project today.

§18.02 THE AMERICAN INSTITUTE OF ARCHITECTS (AIA)

Until 1978, the AIA's Code of Ethics prohibited its members from engaging in design-build, due to a perceived conflict of interest where the architect had a fi-

---

[1] *1999 Design-Build Survey*, Zweig White & Assoc., Inc., p. 26.

[2] For industrial and non-residential work, contractors surveyed by CFMA in 1999 said that they used contracts in these percentages: Fixed Price (55%), Cost Plus (GMP) (27%), Cost Plus (7%), Time & Materials (4%), Construction Mgmt. (5%); and Unit Price 1%. Construction Industry Annual Financial Survey, CFMA (1999).

[3] AGC has even started publishing their own form of General Conditions, AGC Doc. No. 2000, first released in 1998. AGC's goal is to become the owner's preferred contract form of choice.

nancial interest in the project. *See* § 2.06[B]. The AIA Board rescinded the prohibition and in 1985, the AIA published a series of design-build contract forms. The documents are structured for two phases of services, beginning with a Part 1 Agreement, for preliminary design and budgeting. The next phase, the Part 2 Agreement, is for final design and construction. The AIA design-build documents published are as follows:

- AIA Doc. No. A191 Agreement Between Owner and Design-Builder (Parts 1 and 2);
- AIA Doc. No. A491 Agreement Between Design-Builder and Contractor (AIA's "subcontract" for construction); and
- AIA Doc. No. B901 Agreement Between Design-Builder and Architect.

The AIA updated its design-build contract forms in 1996. The new edition incorporates mediation into the dispute resolution process, among other changes in both format and content. The AIA plans to update its design-build series in 2000, with publication of the new forms in 2001. According to a 1999 survey, the AIA forms are the most widely used, with almost three-fourths of design firms responding that they use AIA's design-build forms over the others.[4] This is no doubt due to the architects' familiarity with, and trust of, the AIA documents that they use on traditional projects. Under the AIA series, however, the architect is a separate entity from the design-builder, and serves as a subcontractor. Some firms find that the AIA forms are too restrictive in that regard and do not reflect the way they do business.

The AIA was the only publisher of a Purchase Order form,[5] however, since 1991 AIA no longer prints this form. AGC has filled this void and now publishes a Purchase Order form for use on design-build projects as mentioned in § 18.03.

§ 18.03 THE ASSOCIATED GENERAL CONTRACTORS OF AMERICA (AGC)

AGC was the first to begin publishing design-build contract forms beginning in the early 1980s. Many design-builders find the AGC forms more practical to work with than the AIA's forms. These documents, like the AIA series, begin with a preliminary agreement and then lead in to a formal consruction agreement. The AGC revised their forms in 1999 and released 17 forms, which include prime contracts, subcontracts, bonds, change orders, and purchase orders. AGC's documents committee obtained input from many industry groups, including sureties, insurance, specialty contractors and an owner group called the Private Industry Advisory Council, which included representatives from companies like Monsanto, Disney, banks, pharmaceutical companies, and others. The negotiations did not always go smoothly, but

[4] *1999 Design-Build Survey*, Zweig White & Associates, Inc., p. 25.

[5] AIA Document No. G605 (1983 edition).

the forms show that careful thought was put into even the smallest detail. According to Bill Ernstrom, AGC's Documents Committee chairman, there was a six-hour discussion with the owners Advisory Council and AGC on whether the "cost of work" would include correction of defects if not due to the Design-Builder's negligence. Here are the AGC forms available through the 1999 edition:

- AGC Doc. No. 400 Preliminary Design-Build Agreement;
- AGC Doc. No. 405 Design-Build Guidelines (1991 edition);
- AGC Doc. No. 410 Design-Build Agreement and General Conditions Between Owner and Design-Builder (where basis of compensation is cost of work plus fee with a guaranteed maximum price option);
- AGC Doc. No. 415 Design-Build Agreement and General Conditions Between Owner and Design-Builder (where basis of compensation is a lump sum);
- AGC Doc. No. 420 Agreement Between Design-Builder and Architect/Engineer for Design-Build Projects;
- AGC Doc. No. 450 Agreement Between Design-Builder and Subcontractor;
- AGC Doc. No. 455 Agreement Between Design-Builder and Subcontractor (with contingent payment clause);
- AGC Doc. No. 460 Agreement Between Design-Builder and Subcontractor (where subcontractor provides a guaranteed maximum price);
- AGC Doc. No. 465 Agreement Between Design-Builder and Subcontractor (where subcontractor provides a guaranteed maximum price plus contingent payment clause);
- AGC Doc. No. 470 Design-Build Performance Bond (including design services);
- AGC Doc. No. 471 Design-Build Performance Bond (excluding design services);
- AGC Doc. No. 472 Design-Build Payment Bond (including design services);
- AGC Doc. No. 473 Design-Build Payment Bond (excluding design services);
- AGC Doc. No. 491 Application for Payment (Cost Plus);
- AGC Doc. No. 492 Application for Payment (Lump Sum);
- AGC Doc. No. 495 Change Order (Cost Plus); and
- AGC Doc. No. 496 Change Order (Lump Sum).

The AGC documents are very user-friendly, easy to read, with lots of "fill-ins" throughout the documents. Of particular interest to design professionals is AGC's subcontract between Design-Builder and Architect/Engineer, Document No. 420. A set of suggested supplementary conditions to that contract is included in this book as Appendix C, drafted with the design professional's viewpoint in mind.

§18.04 THE ENGINEERS' JOINT CONTRACT DOCUMENTS COMMITTEE (EJCDC)

The three professional engineering societies—National Society of Professional Engineers (NSPE), American Consulting Engineers Council (ACEC)—and American Society of Civil Engineers (ASCE), comprise the engineers' joint contract documents committee (EJCDC). In 1995, the EJCDC released a comprehensive nine-document set of design-build contract forms and guides. These include:

- Doc. No. 1910-42 User's Guide;
- Doc. No. 1910-40 General Conditions of Contract Between Owner and Design-Builder;
- Doc. No. 1910-40-A Owner-Design-Builder Contract (Lump Sum);
- Doc. No. 1910-40-B Owner-Design-Builder Contract (Cost Plus);
- Doc. No. 1910-41 Subcontract Between Design-Builder and Engineer;
- Doc. No. 1910-43 Agreement Between Owner and Owner's Consultant for Professional Services;
- Doc. No. 1910-48 General Conditions for Subcontract;
- Doc. No. 1910-48-A Subcontract (Lump Sum); and
- Doc. No. 1910-48-B Subcontract (Cost Plus).

The EJCDC documents are available in hard copy and on disk with a variety of word processing formats for WordPerfect and for Microsoft Word for DOS, Windows or Macintosh applications. The 1998 Zweig White survey showed that EJCDC documents are used by about 12 percent of the firms doing design-build.[6]

§18.05 THE DESIGN-BUILD INSTITUTE OF AMERICA (DBIA)

In response to the emergence of design-build project delivery, the Design-Build Institute of America (DBIA) was founded in 1993. Just five years later, in late October 1998, DBIA released its first set series of contract forms, with additional forms released in October 1999 at the DBIA annual meeting. The DBIA forms were developed by a 10-member committee led by attorneys Michael C. Loulakis and Bennett Greenberg, chair and vice chair respectively of the DBIA drafting committee. The committee took great steps to tailor the forms to the design-build type of project delivery. The document series is well written, user friendly and incorporate some of the more current concepts in construction contracting. DBIA publishes these 10 contracts:

---

[6] *1998 Design-Build Survey*, Zweig White & Assoc., Inc., p. 107B.

- No. 501 Contract for Design-Build Consultant Services;
- No. 520 Standard Form of Preliminary Agreement Between Owner and Design-Builder (1998).
- No. 525 Standard Form of Agreement Between Owner and Design-Builder, Lump Sum (1998);
- No. 530 Standard Form of Agreement Between Owner and Design-Builder, Cost Plus with an Option for a Guaranteed Maximum Price (1998);
- No. 535 Standard Form of General Conditions (1998);
- No. 540 Standard Form of Agreement Between Design-Builder and Designer (1999);
- No. 550 Standard Form of Agreement Between Design-Builder and General Contractor, Cost Plus with an Option for a Guaranteed Maximum Price (1999);
- No. 555 Standard Form of Agreement Between Design-Builder and General Contractor, Lump Sum (1999);
- No. 560 Standard Form of Agreement Between Design-Builder and Design-Build Subcontractor, Cost Plus with an Option for a Guaranteed Maximum Price (1999); and
- No. 565 Standard Form of Agreement Between Design-Builder and Subcontractor, Lump Sum (1999).

A 1999 survey found that only 17 percent of design-build firms used the DBIA forms, no doubt due to the fact that they were new in 1998, and the entire family of contracts was not available until October 1999. DBIA publishes a Manual of Practice (non-member price $125.00) which is a two-volume binder set, which includes the 10 contract forms with instructions, procedural guidance, sample formats, code of professional ethics, risk management guidelines, and regulatory data. DBIA also publishes guidebooks on public RFQ/RFP processes, both for the public owner and the design-builder, and a number of surveys of state procurement and licensing laws. To order publications or to get membership information, call (202) 682-0110, or write to DBIA at 1010 Massachusetts Ave., N.W., Suite 350, Washington, DC 20001. The DBIA Web site is at <http://www.dbia.org>.

Of particular interest to design professionals is DBIA's Doc. No. 540, Agreement Between Design-Builder and Designer. A set of suggested supplementary conditions, drafted from the architect/engineer's viewpoint is included as Appendix B to this book.

§18.06 THE FEDERATION INTERNATIONALE DES INGENIEURS-COUNSEILS (FIDIC) CONTRACTS

The Federation Internationale des Ingenieurs-Counseils (FIDIC) is an international association of consulting engineers based in Lausanne, Switzerland. FIDIC was founded in 1913 by three associations of consulting engineers in

Europe and now has membership in more than 60 countries.[7] FIDIC publishes several form contracts for use on international design-build projects, including Engineer-Procure-Construct (EPC) projects. The FIDIC forms include:

- The Orange Book. Conditions of Contract for Design-Build and Turnkey (1995 edition).
- The Red Book. Conditions of Contract for Works of Civil Engineering Construction (1987 edition). This is the fourth edition of this contract form, intended for traditional civil engineering projects, where the owner furnishes the design to the contractor.
- The Yellow Book. Conditions of Contract for Electrical and Mechanical Works (1998 edition). This is the third edition of this contract for traditional engineering projects.
- The Silver Book. Conditions of Contract for EPC/Turnkey Projects (1999 edition). Intended for use on process or power plant, factories or infrastructure projects done on a turnkey basis in which the contractor takes total responsibility for the design and construction of the project with little involvement from the owner, providing a fully equipped facility ready for operation (at the "turn of the key").
- Conditions of Contract for Construction, for Building and Engineering Works Designed by the Employer (1999 edition). Intended for traditional projects in which the design is furnished by the owner, though there may be certain portions of the project that are contractor-designed.
- Conditions of Contract for Plant and Design-Build (1999 edition). Intended for electrical or mechanical plant, and for design-build projects, both building or engineering works. The contractor furnishes the design in accordance with the owner's requirements.
- Conditions of Contract for Plant and Design-Build for Electrical and Mechanical Plant, and for Building and Engineering Works Designed by the Contractor (1998 edition).
- Short Form of Contract (1999 edition). This form is recommended for smaller, traditional projects in which design is provided by the owner and for repetitive work or work of short duration.

The FIDIC contract forms include forms of Letter of Tender, Contract Agreement, General Conditions, Dispute Adjudication Agreement, and guidelines for use. The FIDIC forms contain many of the same topics dealt with in design-

---

[7] The FIDIC can be reached by phone at 41 21 654 41 11 or by fax 41 21 653 32. On the Internet, FIDIC's Web site is at <http://www.fidic.com/>. FIDIC contracts are also available from the bookshop of the American Council of Engineering Consultants (ACEC) at <http://www.acec.org/>.

build forms published in the United States, but address issues unique to international projects. Design-build contracts are dealt with in very straight-forward language. For example, the EPC form states simply that: *"The Contractor shall design, execute and complete the Works in accordance with the Contract, and shall remedy any defects in the Works. When completed, the Works shall be fit for the purposes for which the Works are intended as defined in the Contract."*

§18.07 CUSTOM DESIGN-BUILD CONTRACTS

Many large design-build contractors have developed their own contract forms, either out of dissatisfaction with the available form contracts, or due to a desire to control risks on their own terms. These forms vary greatly. Many owners, especially public owners, have developed their own contracts for design-build. Some are only slightly modified traditional construction contracts, while others are detailed and customized for design and construction in one contract. One should be very cautious in reviewing these forms, which may not be well thought out or may contain hidden traps.

§18.08 LIMITATION OF LIABILITY CLAUSES

"A topic that was added to the AIA contracts in 1997 is picked up in the FIDIC forms as well, and that is the waiver by each party to lost profit and indirect or consequential loss or damage suffered in connection with the contract. Unlike the AIA forms, however, the FIDIC contract also limits the contractor's liability to the owner (other than indemnity, fraud, reckless, or delibirate acts) to a sum stated in the Particular Conditions or, if not stated there, to the Contract Sum." These types of limitations of liability have been used by the airlines industry (look at your ticket next time you fly) to limit their liability for lost luggage, by parking garages (read your parking stub), by home security companies, Internet service providers, even by Kodak ® on their boxes of film (limiting their liability to a replacement roll of film). The construction industry has caught on that these clauses can be used to help allocate risk.

[A] Design Contracts

Limitation of liability clauses are becoming more common in architectural and engineering contracts where the fee charged is disproportionate to the financial risk of liability. These clauses typically limit the liability of the professional to its fee, to an agreed lump sum, or to available insurance proceeds in an effort to cap liability between contracting parties. Were fees charged based on the exposure to liability that flows from the providing of professional services, the public would in many cases be unable to obtain services at affordable prices. By use of these limiting clauses, architects and engineers are able to perform the services in reliance on a known limit of their legal exposure. A typical limitation clause reads like this:

> **Limitation of Liability**. To the fullest extent permitted by law, and notwithstanding any other provision of this Agreement, *the total liability, in the aggregate, of Architect*, its officers, directors, employees, agents, and consultants, and any of them, to Owner and anyone claiming by, through or under Owner, for any and all claims, losses, liabilities, costs or damages whatsoever arising out of, resulting from or in any way related to the Project or this Agreement from any cause, including but not limited to the negligence, professional errors or omissions, strict liability, breach of contract, or warranty (express or implied) of Architect, its officers, directors, employees, agents or Consultants or any of them, *shall not exceed the total compensation actually received by Architect under this Agreement, or the sum of $50,000, whichever is greater.*

Such limitation clauses are often used where the type of service desired is not adequately covered by liability insurance—*e.g.*, environmental hazards or asbestos abatement. If not for such contractual limitations on the amount of risk involved, the public could not obtain certain high-risk services from qualified firms. Limiting liability is a fair and equitable way to approach project risks and to give the public the benefit of reasonable fees.

AIA and EJCDC both publish suggested limitation of liability language for use in appropriate contracts. According to a 1986 study by the *Professional Services Management Journal*, 33 percent of the work performed by architectural and engineering firms surveyed was covered by a contractual limitation of liability. Presumably, that figure is higher now, especially with the publication of standard contract language by the professional associations. There is a developing body of case law around the country upholding limitation of liability clauses in architectural and engineering contracts as valid and enforceable.[15] These cases have generally upheld limitations clauses, finding that businesspersons can agree by contract to allocate risks in a variety of ways, and that agreeing to limit one party's liability is one way to do that.

[B] Expression of Legal Consideration

There is one court case in which a limitation of liability was rejected in a contract for home inspection. In *Schaffer v. Property Evaluations, Inc.*,[16] the court

[15] *See, e.g.*, Florida Power & Light Co. v. Mid-Valley, Inc., 763 F.2d 1316 (11th Cir. 1985); Georgetown Steel Corp. v. Union Carbide Corp., 806 F. Supp. 74 (D.S.C. 1992); Wausau Paper Mills Co. v. Chas. T. Main, Inc., 789 F. Supp. 968 (W.D. Wis. 1992); Central Hudson Gas & Elec. Corp. v. Combustion Eng'g, Inc., 1989 WL 85868 (S.D.N.Y. 1989); Long Island Lighting Co. v. IMO Delaval, Inc., 668 F. Supp. 237 (S.D.N.Y. 1987); William Graham, Inc. v. City of Cave City, 709 S.W.2d 94 (Ark. 1986); Markborough California, Inc. v. Superior Court, 227 Cal. App. 3d 705, 277 Cal. Rptr. 919 (1991); FMC Corp. v. Continental Grain Co., 355 So. 2d 953 (La. App. 1977).

[16] 854 S.W.2d 493 (Mo. App. E.D. 1993).

refused to enforce a limitation clause. It held that damages for breach of a home inspection contract were not limited to a $153 inspection fee, despite language in the form contract that the inspection company's liability for any client post-inspection claims was limited to maximum of inspection fee paid. Though the homeowners were college graduates who read the clause, and even discussed it with the contractor, the court held there was no indication that any terms of contract were bargained for. This case suggests that separate "consideration" must be given for such a limitation in Missouri indicating that adequate consideration has been given for the limitation. The case has been criticized, however, leading one court to remark that: "So far as we can determine, *Schaffer* is leading a parade of one; in any event, we decline to adopt its reasoning."[17]

[C] Design-Build Contracts

There have been several court cases dealing with contractual limitations on liability either between the owner and the prime design-build contractor or between the prime contractor and its design subconsultants. The results have been mixed depending largely on the clarity of the wording in the contract clause. For example, in a design-build case involving an electrical generating facility, the contract provided for liquidated damages for failure to meet specific performance requirements by a certain date, but limited the contractor's liability to the amount of the contract. The clause stated, "In no event shall the liquidated damages payable pursuant to this Section 8.2 (when added to any liquidated damages payable pursuant to Section 7.2) exceed the Contract Price," and that "[c]ontractor shall not be otherwise liable for any incidental or consequential damages or for lost profits as a result of such failure." The project failed to meet certain performance criteria for a period of 56 days. Applying a contract formula to the 56-day delay would have resulted in liquidated damages of $167 million. However, the contract set a maximum amount of liquidated damages at the full contract price, approximately $32 million. Though the clause was challenged as being an unreasonable penalty, the liquidated damages were upheld by the court, finding that the contract was fairly negotiated and unambiguous.[18]

By contrast, however, in *CRS Sirrine, Inc. v. Dravo Corp.*,[19] the project engineer and contractor entered into a joint venture agreement for a $100 million design-build power plant for the U.S. Navy in Virginia. The joint venture agreement stated that the engineer had no liability for design errors it detected if it notified the contractor prior to any costs being expended on that work; if costs were expended in reliance on defective plans, then the engineer was not liable for the first $750,000 in such costs, but would have unlimited liability above that.

[17] Moler v. Melzer, 942 P.2d 643 (Kan. App. 1997).

[18] CIT Group/Equip. Fin., Inc. v. ACEC Maine, Inc., 782 F. Supp. 159 (D. Me. 1992).

[19] 445 S.E.2d 782 (Ga. Ct. App. 1994).

The contract then contained the following clause, which said:

> 9.3 Notwithstanding any of the forgoing, Sirrine shall have no risk, liability, or accumulation will occur of error and omission charges for construction material quantity variations if the actual quantities are different from those in the bid to the Navy.

The court held that section 9.3 was ambiguous as a matter of law and that "it was not the intent of the parties that the section relieve Sirrine of all responsibility for damages resulting from increases in construction material quantities caused by any errors or omissions in Sirrine's work." The court found that the engineer was responsible for errors and omissions in its work and that even the $750,000 exemption did not apply because the engineer failed to give its contractor partner notice of the errors, as required by the contract clause.

§18.09 WAIVERS OF CONSEQUENTIAL DAMAGES

When a construction contract is breached, there may be direct damages in the form of repair costs, for example, to a roof that leaked. But if that roof leak damaged computer equipment in the building, causing a shutdown of a manufacturer, the consequences of the construction defect can have results far beyond repairing the roof, such as layoffs of workers, delay in shipment of orders, loss of customers, etc. Often, the "consequential damages" from a construction defect can far exceed the cost of the repair of the construction defect, thereby exposing the contractor to unlimited damages. As a result, the AIA's 1997 A201 General Conditions contains a waiver of consequential damages clause barring the owner from making claims beyond the direct damage caused by the breach of contract.[20] The standard design-build contracts published by AGC, DBIA and FIDIC also include waivers of consequential damages. AGC's wording waives claims by the owner for "loss of use of the Property, all rental expenses incurred, loss of services of employees, or loss of reputation."[21] DBIA's wording is set out in all capital letters and its scope is broader, including a waiver of "any consequential losses or damages, whether arising in contract, warranty, tort (including negligence), strict liability or otherwise, including but not limited to losses of use, profits, business, reputation or financing."[22]

Unlike the AIA forms, the FIDIC contract also limits the contractor's liability to the owner (other than indemnity, fraud, reckless or delibirate acts) to a sum stated in the Particular Conditions or, if not stated there, to the Contract Sum.

[20] 4.3.10, AIA General Conditions of the Contract for Construction, Doc. No. A201 (1997 edition).

[21] 11.8, AGC Standard Form of Design-Build Agreement, Doc. No. 410 (1999 edition).

[22] 10.5, DBIA Standard Form of General Conditions, Doc. No. 535 (1998 edition).

§ 18.10 OWNERSHIP AND COPYRIGHTS IN DESIGNS AND DOCUMENTS

A common source of disputes in contract negotiations for design-build projects involves ownership of drawings and copyrights. *See* Chapter 16 for a comparison of the AIA, AGC, DBIA, and FIDIC documents and the associations' respective positions on this topic.

§ 18.11 IMPLIED WARRANTIES

Some courts have held that persons who design and construct can be held to an implied warranty that the building will be fit for the owner's intended purpose.[23] The Uniform Commercial Code (UCC) applies certain warranties as a matter of law to the sale of goods and products. Some have suggested that design-build might open up a contractor to claims of implied warranties under the UCC just like the maker of products. A preventative measure to avoid this would be to disclaim any such warranties in the design-build contract, just as manufacturers give limited warranties and disclaim all others. AGC's standard design-build contract is the only form that has just such a disclaimer.[24] This type of disclaimer is not contained in AIA or DBIA design-build forms but should be considered as a revision or additional condition.

§ 18.12 SHARING IN THE SAVINGS

Whether leading the design-build team as prime contractor, or by joint venturing, or even acting as a subcontractor, there are opportunities for sharing in the savings that are not available in traditional construction delivery. Much of design-build is contracted for under a "cost-plus" type of contract with a guaranteed maximum price (GMP). Neither the AIA, AGC, nor the EJCDC design-build contracts provide for a savings split in the "cost-plus" form. By contrast, the DBIA Doc. No. 550 Cost Plus contract has an entire Section 6.6 entitled "Savings," which gives the parties an option to choose to split the savings on a pure percentage basis, or based on a specific dollar amount.

§ 18.13 INCENTIVE BONUSES IN DESIGN-BUILD CONTRACTS

Bonuses for early completion can be another source of increased profit. Some design-build contracts provide for liquidated damages for late completion, and

---

[23] *Robertson Lumber Co. v. Stephens Farmers Cooperative Elavator Co.,* 143 N.W.2d 622 (Minn. 1966); *Kennedy v. Bowling,* 4 S.W.2d 438 (Mo. banc 1928).

[24] Par. 3.8.2, AGC Standard Form of Design-Build Agreement and General Conditions, Doc. No. 410 (1999 edition).

for a bonus for early completion. The bonuses can be huge in some contracts. *See* Chapter 9, § 9.09, for examples of some projects where bonuses were paid on design-build projects. DBIA's design-build prime contracts contain an "Early Completion Bonus" with a blank space to be filled in with a dollar amount for each day that Substantial Completion is attained earlier than an agreed date. DBIA's standard forms are the only ones that contain such a clause.

§ 18.14 LIQUIDATED DAMAGES

The standard forms vary in the way that they treat damages for late completion. As you might expect, neither the AIA nor the AGC design-build contracts address the topic of liquidated damages for late completion. The engineers, on the other hand, address damages for late completion in both the forms published by EJCDC and the international counterpart published by FIDIC. EJCDC's Doc. No. 1910-40-B "Cost Plus" design-build contract form contains a lengthy liquidated damages clause that has two blanks to fill in, with one for delays in Substantial Completion, and a second amount for failure to achieve final completion. Under the FIDIC design-build forms, if the contractor does not complete the work by the agreed date, it must pay delay damages to the owner in the sum stated in the Particular Conditions for each day between the contract completion date and the actual date stated in the Taking-Over Certificate, not to exceed a maximum amount stated in the Particular Conditions.

DBIA deals with liquidated damages for late completion just before the bonus clause for early completion, discussed in § 18.11, above. If the owner incurs a loss for late completion, there should be a savings for early completion, according to that form.[25] Be aware that when using the DBIA forms, par. 11.7.2 of DBIA's subcontract between design-builder and designer, DBIA Doc. No. 540 (1999 edition) passes on to the designer any liquidated damages that the owner assesses against the design-builder and that are due to the designer.

§ 18.15 INDEMNITY CLAUSES

Indemnity is the one common theme in the standard industry design-build forms. All of the contracts require the design-build contractor to indemnify the owner for claims of bodily injury and property damage caused by the negligence of the contractor. These clauses are generally written around the types of claims that are covered by a contractor's general liability policy of insurance. There are also standard clauses requiring indemnity by the design-builder for patent and copyright infringement actions. Indemnity is also addressed in the subcontracts with trade contractors and design professionals. For example, there are several indemnity clauses sprinkled throughout the DBIA subcontract between design-

[25] *See* DBIA Doc. No. 525, par. 5.4 (1998 edition).

builder and designer, DBIA Doc. No. 540 (1999 edition). *See* Chapter 5, § 5.05[M], above.

The international forms published by FIDIC include a broad indemnity clause much like that found in U.S. design-build contracts, covering claims for bodily injury, death, and property damage.[26] While the contractor is not required to defend the owner under the FIDIC forms, the obligation to indemnify includes the owner's legal fees and expenses. Unlike the U.S. counterparts, the FIDIC form requires that the owner indemnify the contractor for similar claims or damages caused by the owner, its personnel, or agents as well as for certain uninsured losses to the contractor.

§ 18.16 INTERNATIONAL CONTRACT CLAUSES

For international design-build contracting, the FIDIC forms include some clauses and terms that are not found in standard U.S. contract forms. Unlike the U.S. forms and terminology, the FIDIC forms refer to the project owner as the "Employer." The contractor's invoice, or application for payment, is called the "Statement," with the final invoice called the "Final Statement." In the United States, owners publish a Request for Proposals (RFP) seeking interested contractors and the contractor submits a bid or proposal. Outside the United States, the owner prepares "tender documents," which FIDIC recommends be prepared by qualified engineers and reviewed by a lawyer. The tender documents set out the requirements of the project, with Instructions to Tenderers (instead of Instructions to Bidders). In response, the contractor submits a Letter of Tender, a signed offer to perform the work. The "Base Date" is the date for the start of the bidding period, within which contractors may submit their Tender. FIDIC publishes a form of Tender Security, which is the equivalent of a bid bond, to be submitted with the tender.[27]

This term "variation" is used instead of "change order" to mean any change in the owner's requirements for the project. The contractor must comply with any variation issued by the owner unless it cannot obtain the materials requested, or it will reduce safety or have an adverse impact on performance requirements of the contract. When the work is completed, the owner issues a Performance Certificate to the contractor certifying acceptable performance, as well as a Taking-Over Certificate stating the date on which the work (or sections of it) was completed except for minor work and defects, and when the owner was able to take over and use those portions of the project.

In the United States we take for granted the type of currency used for payment, unlike in the international marketplace where Euros, Francs, or Pesos are

[26] Par. 17.1, FIDIC EPC/Turnkey Contract (1999 edition).

[27] FIDIC also publishes a guidebook on tendering, called FIDIC Tendering Procedure, 2nd Edition, Doc. No. W-0064 (1996 edition).

used. The FIDIC forms require that payment is to be made in the currency specified in the Particular Conditions. If more than one currency is stated, then the fixed rate of exchange to be used is to be stated in the contract and any change orders (called "variations") must break down the expected currency proportions for each change.

FIDIC forms cover the rights in relics or "articles of antiquity" found on the job site. This topic is not covered by U.S. form construction contracts, but is needed in international work, which is more likely to encounter buried relics on a construction site. Under the FIDIC contracts, all items of geological or archaeological interest found on the site are to be placed under the owner's care. Any delay caused by such discovery is the basis for a claim for additional time and/or money. As in the clauses in U.S. contracts, that deal with hazardous materials, the contractor is required to give the owner prompt notice after discovery. The owner will then issue instructions for how to deal with it.

The FIDIC forms specifically cover the risks of wars and terrorist acts since these occur more commonly on foreign soil than in the United States. These forms also provide for a time extension, as well as compensation for delay or damage to the project caused by war, hostilities, invasion, rebellion, terrorism, revolution, civil war, riot, disorder, and similar violent acts. Further, any circumstances that make it impossible or unlawful for either party to fulfill its obligations is grounds to terminate the contract and release such party from liability.

CHAPTER 19

COST ESTIMATING FOR DESIGN-BUILD FIRMS

§19.01 NEED FOR ACCURATE ESTIMATES

In a perceptive comment from a Louisiana case, the judge noted:

> There seems to be no reported case in which an architect is alleged to have *over*-estimated the cost of a building project. Indeed, no case reveals that an architect has [ever] correctly estimated the cost of a project although tradition assures that this does occur.[1]

The quote recognizes the reputation that architects have as cost estimators—not a good one. Many architects start right off sketching plans, elevations, and designing the building without any concept of whether what they are doing can be built within the budget. Using generic square foot pricing from R.S. Means or other resource books, the designers keep track of the budget by tracking the square footage only, not considering how fancy details, fixtures, and selection of materials will affect the unit costs. This causes delays and embarrassment when the bids come in too high. Site costs are especially tough to estimate.

The better architects are those who can truly understand the relationship between design and cost. Architects engaged in design-build need accurate estimates, since cost overruns are paid out of the profit margins. When the architect acts as a subcontractor to a contractor, there can be litigation if the contractor locks itself into a fixed price based on a bad estimate, which was based on defective plans. Design professionals looking to be prime contractors should consider hiring experienced estimators from a construction company.

§19.02 HOW CONTRACTORS PUT ESTIMATES TOGETHER

Before deciding to become a design-build contractor, design professionals should go to a contractor's office and watch what actually happens on bid day, especially during the last hour before submission of a bid. Though owners allow several weeks to prepare bids, most often the price quotes from subcontractors for major components are received in the last few minutes before the bid deadline. Multiple fax machines are running, the phone lines are tied up, and the project estimators are trying to make sure that they have everything covered, and that the proposals they are getting have no exclusions that will leave a hole in the bid. Contractors sometimes get no bids at all for portions of the project and have to do their own internal estimate or use a "dummy number" as a guess of what that portion might cost, then try to find a subcontractor to work for that price after the contract award. The reason these prices are put together at the last minute is that a subcontractor or major material supplier does not want his price released too far in advance for fear that it will be exposed to his competitors. Bids may come in from subcontractors that have been asked to bid, as well as unsolicited bids from companies the contractor does not know. This requires quick decisions. The result

[1] Williams Eng'g, Inc. v. Goodyear, 496 So. 2d 1012, 1014 (La. 1986).

is a hectic and stressful final hour at the contractor's office, before locking in a price and furnishing a bid bond to guarantee the price.

To insure the bid is not submitted late, a blank bid form is often given to an employee of the contractor who waits by the phone at the location where bids are to be submitted. The hallway outside the bid location may be lined with contractors on their mobile phones talking to the office only to fill in the final number minutes before the deadline for submission of bids, after the final calculations are done. This often results in bid errors, transposed numbers, a zero left off here or there, or even a portion of work left out entirely.

A contractor may not know of his mistake until the bids are opened and the low bidder is found to be 10 percent or more below the rest of the bids, which are tightly grouped together. Then comes the process of checking, and re-checking, the estimate sheets to make sure everything is correct. We have all heard of a bidder who "left $100,000 on the table," meaning that the low bid was that much below the rest of the pack, maybe due to an error. Depending on the law in the particular state, a contractor may not be able to get out of an erroneous bid, and may lose his bid bond. To avoid having to pay the surety for the cost of the bond penalty, the contractor may accept the contract, even with the mistake, hoping to make it up by change orders during the course of the job or by squeezing subcontractors to lower their prices. For this reason, it is questionable whether it is really in the owner's best interest to always take the low bid, especially in the face of a claimed bid error.

§ 19.03 BETTER COST CONTROL POSSIBLE THROUGH DESIGN-BUILD

Design-build provides for better estimating, with the design and construction team working closely together for months during the proposal stage, pricing as the design is developed and, thereby avoiding the last-minute rush that accompanies the traditional design-bid-build process and often causes bid mistakes. The team must stay abreast of actual costs on a constant basis so that adjustments can be made if projections indicate that costs are to be exceeded in some categories. Two-thirds of contractors surveyed in 1999 reported that management and project managers review the estimated cost to complete/percent complete on jobs monthly, with 18 percent reporting that costs were reviewed more frequently than monthly.[2] Forty-seven percent of respondents reported that project management understands the estimated cost to complete the process well and takes it seriously.

When working with a fixed budget, some design-build firms start the design work with the fixed costs first. Larry Maxfield, AIA says his firm starts with the "known costs" first, and then works on the building design. "We know we have to have site work, storm sewer, landscaping, paving, curbs, apron for driveway and utilities. These things are fixed, so we do a cost estimate on that first to see what is left to put into the building. Then we start to design the building," says Max-

[2] CFMA Construction Industry Annual Financial Survey (1999).

field, an architect-turned-builder. He says he literally keeps both cost estimates posted in front of him during the design process, since if there is a cost overrun on the building, those costs must come out of the site costs, and vice versa. It is then a balancing task through completion of the building design to keep the site and building costs in check to total up within the owner's budget.

§19.04 BIDDING METHODS

Brad Buchanan, AIA, former chair of the AIA's design-build PIA subcommittee, says that his firm uses two different types of bid systems. They give a "soft bid" based on 30 percent design after they have completed a site plan, building plans and elevations, very few building sections, and a schematic structural, mechanical and electrical plan. "Architectural plans are very rough at this stage, but that is the easy stuff to estimate. More difficult are the systems (structural, MEP) as well as the site work. We have found where we get nicked on costs are on site work like utilities and excavation, so we prepare a detailed site plan early on." The soft bid price is then used to get the owner's approval to proceed with the project. Then, the firm goes forward with detailed contruction documents where the soft bid price can be refined based on hard bids from subcontractors.

§19.05 LIABILITY WHEN ESTIMATES ARE BASED ON INCOMPLETE DESIGNS

In design-build competitions, the estimating is usually done based on preliminary and incomplete plans with outline specifications. Yet design-build contractors lock in multi-million dollar contracts on such preliminary designs, using experience and contingencies to fill in the missing design information. When jobs are significantly underbid, however, all the teaming relationships tend to break down as the prime contractor looks to recover its losses from its design partner. In two such cases, design firms have been held liable to design-build contractors that relied on defective plans when entering into construction contracts with owners. In a 1997 Washington case,[3] an architect who was hired by a design-build contractor was held liable for design defects that caused the contractor to incur extra work and interest on borrowed funds during a 32-day project delay. The contractor based its guaranteed maximum price on the architect's incomplete design. After a 14-week jury trial, the contractor was awarded $820,372 for extra work, plus $332,919 for interest paid to finance the work. The court of appeals rejected a "betterment" argument by the architect, holding that the contractor would not have undertaken the project had it known the true cost of the work.

Similarly, in a 1996 federal case decided under Missouri law, the court held that an engineering firm was liable to a design-build contractor who relied on the engineer's designs in preparing estimates for its bid.[4] The engineer argued that it

[3] Skidmore, Owings & Merrill v. Intrawest I Ltd. Partnership, 1997 WL 563159 (Wash. App. Div.).
[4] C.L. Maddox, Inc. v. Benham Group, Inc., 88 F.3d 592 (8th Cir. 1996).

did not warrant the accuracy of bidding information it prepared. The court held, however, that there was an "implied warranty" that the engineer's work would be properly performed and that the engineer could be liable to the design-build contractor for breach of that warranty. The contractor was awarded $5 million in damages at trial, though that sum was reduced on appeal to $2.7 million, still an amount that would exceed the insurance limits carried by most architectural and engineering firms. Some insurers are now offering insurance coverage to design-build contractors to cover cost overruns and estimating errors to protect against these very risks. The premiums are extraordinarily high and would not be feasible on most design-build projects.

§ 19.06 KNOWING THE MARKETPLACE

While it may seem apparent, there are surprising variances in the cost of labor across the United States. Figure 19-1 shows a comparison of eight construction trades among eight large U.S. cities, based on hourly union pay scales in December 1999. The variations in wages among the same trade are amazing. A 1999 study found that 50 percent of contractors responding reported that they operated as an open shop, while 35 percent were union contractors.[5] If a project is located in a strong union town, this will affect the wages that will be paid for trade labor. Wages range, for example, from a low of $18.75 for bricklayers in Dallas to a high of $47.36 per hour in New York City. Using square foot pricing for certain building types can be extremely misleading if market adjustments are not made and project specific building conditions are not factored in.

FIGURE 19-1. Hourly Union Pay Scales.

	Atlanta	*Boston*	*Chicago*	*Dallas*	*Kansas City*	*Los Angeles*	*New Orleans*	*New York*
Bricklayers	$20.12	$40.45	$35.71	$18.75	$29.05	$32.78	$19.25	$47.36
Carpenters	$19.94	$36.40	$35.47	$18.13	$29.03	$30.21	$19.56	$50.96
Masons	$18.84	$41.51	$35.69	$19.25	$27.03	$32.21	$15.90	$47.39
Laborers	$12.47	$29.50	$30.10	$12.81	$21.90	$27.47	$13.00	$38.73
Painters	$20.02	$37.44	$34.51	N/A	$26.83	$29.03	$15.60	$38.66
Plumber	$27.54	$41.95	$38.95	$22.65	$34.36	$36.65	$22.01	$56.64
Roofers	$16.95	$34.11	$33.72	$22.50	$27.59	$28.43	$13.96	$42.24
Truck Drivers	$12.25	$30.38	$29.19	N/A	$25.17	$31.88	$13.01	$42.31

Source: Engineering News Record, Dec. 20, 1999.

[5] CFMA's Construction Industry Annual Financial Survey (1999).

Material prices can also vary greatly depending on where the project will be built. For example, Figure 19-2 is a sampling of lightweight 8" × 8" × 16" masonry block prices across the country as of March 6, 2000 (per 100 block):[6]

Atlanta	*Baltimore*	*Boston*	*Chicago*	*Dallas*	*Denver*	*N.Y.*	*Phila.*	*Pitts*	*Seattle*
$ 118.50	$ 92	$ 196	$ 82	$ 110	$ 118	$ 113	$ 94	$ 135	$ 132

FIGURE 19-2. Masonry Block Prices (per 100).

So if a design-build contractor is based in Chicago, where block prices are $82 per 100, pricing a job in Boston, where the price is $196.00 per 100, you had better do some market research in the locale where the project is being built and adjust your bid price or you will find that you have underbid the cost of masonry block by 140 percent. The hourly wage for union masons is also 16 percent higher in Boston than Chicago, adding to the overall labor and material difference. *See* Figure 19-1.

Mistakes in knowing local market conditions will not be grounds for a change order or claim of a bid mistake. Such errors of judgment are usually not recognized by the courts as sufficient basis to grant relief. For example, in a 1982 case, the low bid on a highway project was 11 percent lower than the next lowest bid. After investigation, the Colorado-based contractor determined that a $620,000 error had been made because the bidder failed to include certain labor costs required by a local labor agreement applicable to the Missouri project. The court held that a mistaken bidder may be relieved of its obligation if the mistake is: (a) not obvious; (b) substantial; and either (c) the result of clerical or computational error; or (d) a misconstruction of the specifications. However, rescission of the bid would not be allowed for a mistake of judgment or lack of investigative effort. The court found that failing to research local labor conditions was not a math error, but a mistake of judgment and the court denied relief to the bidder.[7] This example shows the benefit of teaming with a local contractor and researching local labor conditions prior to bidding.

§19.07 SUBCONTRACTOR PRICING

Larry Maxfield, AIA recommends that to control costs, a design-build contractor should get "committed costs" as early as possible—that is, hard bids from subcontractors that can be plugged into the design-builder's estimate. One way to get these fixed prices for trade subcontracts early is to have the subcontractors design-build their part of the project. But, how does one get the subcontractors to do this without a developed design? "We give the subs a well-defined scope of work,

[6] *Construction Economics*, Engineering News Record, Mar. 6, 2000, pp. 29–30.

[7] State *ex rel.* Missouri State Hwy. Comm'n v. Hensel Phelps Const., 634 S.W.2d 168 (Mo. 1982).

which they then use to price out their part, based on a schematic design and some design assumptions," says Maxfield.

Maxfield also recommends getting hard bids from the specialty subcontractors for mechanical, electrical, plumbing, fire sprinkler, and any other component the design-build contractor is not familiar with. He says:

> We figure all our own demolition, interior finishes, doors, carpentry, cabinetry, footings, foundations, concrete flatwork, paving, etc. We have good unit prices from our own data base so we know what these things cost. Where the quantities can be easily taken off the plans, we can do a good job estimating the cost. But for the unknowns and specialty work, like MEP work, sprinklers, even roofing, we get those prices from the trade subcontractors. What an air-handling unit will cost, how much duct work of what sizes, etc. are better left to the trades to estimate.

§ 19.08 BID MISTAKES ON PUBLIC PROJECTS

Many lawsuits are filed over mistakes in bids and whether the bidder is bound to an erroneous bid or can be relieved from its bid. The issue is whether a public owner should reap a "windfall" by holding a contractor to an erroneous bid, or whether allowing relief due to mistakes would taint the competitive bidding process and encourage corruption. On federal government projects, when the contracting officer suspects a bid mistake, he is required to ask the bidder to verify the bid.[8] Bid spreads of 10 percent or more might indicate that an error has been made by the low bidder.[9] If the contractor is not allowed to withdraw the bid, he is faced with defaulting and losing his bid bond, proceeding to do the work at a loss, or worse yet, proceeding with the work and being unable to finish, resulting in termination. These are all factors that both the owner and the contrator must consider when a bid mistake is made.

[A] Case Law on Bid Mistakes and "Clerical Errors"

For many years courts have taken different views on bid mistakes and whether a contractor is entitled to relief. Very generally, most states and federal procurement agencies allow withdrawal or modification of a bid in which a "clerical or non-judgmental mistake" has been made.[10] A "clerical bid error" can have significant cost consequences. A clerical mistake is generally:

1. a math error such as extending a unit quantity to only 100 units instead of the intended 1,000 units;
2. a transpositional error—*e.g.*, a bid item intended to be $92,300.00 is recorded at $29,300.00; or
3. an inadvertent failure to include all items of work—*e.g.*, leaving out two needed HVAC units.

---

[8] 48 C.F.R. § 14.406-3(g)(1).

[9] One case even found that a general contractor had an obligation to notify its subcontractor of a suspected error in the subcontractor's bid. Sulzer Bingham Pump v. Lockheed Missiles, 947 F.2d 1362 (9th Cir. 1991).

[10] Aydin Corp. v. United States, 669 F.2d 681, 685 (Ct. Cl. 1982).

Proving a clerical error usually requires reference to bid preparation documents to show the intended bid and to explain the mistake. Keep all bid papers, worksheets, computer spreadsheets, and other records of the bid process. These documents may help prove an excusable bid error.

Relief will not always be granted, however, depending on the jurisdiction and governing law. For example, in a 1979 case, a contractor submitted a bid that was 20 percent lower than the next bid, and 60 percent lower than the engineer's estimate. All parties suspected an error and the low bidder soon determined that a clerical error had been made in transferring a figure from the adding maching tape to the bid estimate sheet. The contractor asked to withdraw its bid on the basis of a gross clerical error. However, the state Supreme Court refused to relieve the bidder from its bid, stating that if contractors could withdraw bid on public projects due to clerical errors, it would "seriously undermine and make the requirement or system of sealed bids a mockery; it could likewise open wide the door to fraud and collusion between contractors and/or between contractors and the Public Authority."[11] The court concluded that in absence of fraud, a unilateral mistake does not excuse nonperformance of a contract.

On the other hand, some courts have not only granted relief to bidders, but allowed them to modify their bid and remain low. In *Dick Corp. v. Assoc. Electric,*[12] the bidder made a $1 million error, submitting a bid for $13.6 million which should have been for $14.6 million. The other five bids ranged from $15.6 to $18.5 million. The contractor wrote a line item for field overhead and general conditions as $100,000, which should have been $1,100,000. The owner sent notice of award at $13.6 million, but the bidder sent a notice of mistake and intent to withdraw without forfeiting the 10 percent bid bond. The court held that no other bidder would be adversely affected if the bid was withdrawn, that it was a good faith mistake, and that the bid could be *reformed* to correct the mistake to $14.6 million. Under this result, the owner paid what it should for the construction, not an erroneous price, the contractor got what it intended, and the next low bidder was still $1 million off the mark. However, a contractor should not rely on the fact that it may be able to withdraw or reform a bid. The outcome varies depending on the state law and the facts applicable. Raising bids after a public bid opening could be abused by some contractors to increase profit, and courts will be reluctant to allow bid reformation.

[B] Bidder Mistake Statutes

As a result of the split in court cases and the uncertainty for bidders seeking relief, laws have been passed in several states to give bidders relief from estimating errors if certain procedures are followed in a timely manner. The Kansas Legislature passed just such a law in 1995 when it passed an act titled: "State Government and Subdivisions: Bidding Process, Mistakes, Corrections." Under

[11] Triple A Contractors, Inc. v. Rural Water Dist. No. 4, 226 Kan. 626, 603 P.2d 184 (1979). *See also* Anco Const. Co., Ltd. v. City of Wichita, 660 P.2d 560 (Kan. 1983), in which the Kansas Supreme Court denied relief to another contractor who inadvertently omitted $95,794 from its bid due to a clerical error. The court stated that relief is not available where the bid is not withdrawn until *after* the bid opening process has begun.

[12] 475 F. Supp. 15 (W.D. Mo. 1979).

the statute,[13] a bidder may correct mistakes or withdraw a bid *before* the bid opening. If there is an error in judgment, the bid cannot be withdrawn *after* it has been opened. If the awarding authority suspects a non-judgmental error, it may request the bidder to verify its number. If so, the bidder may, within two business days, either verify the bid as submitted, or withdraw the bid. A "non-judgmental" mistake is one that is evident on the face of the bid, or established by "clear and convincing" evidence. If the bid is withdrawn, the bidder is barred from doing any work on the project, even as a subcontractor. A low bidder can sue for an injunction to prevent enforcement of an erroneous bid under these criteria.

[C] Procedures When Errors Are Discovered

If an error is suspected by the owner or design criteria consultant on a design-build prjoect, the prudent thing to do is to notify the bidder and ask that the bid be either confirmed or withdrawn. If an architect is the low and successful bidder, the prudent thing to do once the error is discovered is to immediately contact the owner or owner's representative by telephone and by fax. A certified letter should be sent explaining the error in detail, how it happened, including supporting documents that show the nature of the clerical error. The bidder should ask permission to withdraw the bid. If only permission to correct the bid is requested, but denied, the bidder may be stuck with having the bid accepted and then facing the tough choice between losing their bid bond and performing at a loss.

§ 19.09 RESOURCES FOR ESTIMATING

Computers have linked the world together and they have changed the way contractors estimate projects. Now quotes on materials and equipment can be obtained over the Internet. One source, Contractors E Source, promotes getting price quotes over the Web, resulting in multiple quotes by submitting an on-line RFQ form, linking contractors with suppliers nationwide. The RFQ form is available at <http://www.contractorsesource.com/>.

Until a company develops its own internal database of construction costs from projects it builds as a design-build contractor, it will need to use some of the published cost guidebooks. The one most familiar to architects is *Means Building Construction Cost Data* published by R.S. Means Company, Inc. The Means Cost Data 2000, for example, has separate books for building costs, mechanical, repair and remodeling and electrical costs. The cost is very inexpensive at $85.95 for the softbound version of Means' *Building Construction Cost Data, 58th Annual Edition*. The book has over 700 pages of comprehensive unit price, square foot prices and other data including crew sizes, equipment tables, estimating tips by CSI division, and a City Cost Index which permits adjustments to the national average prices in 689 cities and 930 zip codes based on research by the Means' staff. The Means data is also available on interactive CD-ROM and is linked to the Means Web site for quarterly updates and building product information.[14]

---

[13] K.S.A. §§ 75-6901 *et seq.*

[14] For more information, *see* the R.S. Means Company's Web site at <http://www.rsmeans.com>.

CHAPTER 20
PROJECT MANAGEMENT

§20.01 COMPLEXITY OF PROJECT MANAGEMENT

In 1894, Theodore M. Clark, FAIA, wrote:

> Among all the business relations which men enter into, there are none, perhaps, more complex than those which are involved in the construction of a building, by the cooperation of a multitude of contractors, journeymen, and dealers in materials, under the supervision of an architect, for the owner of the land on which the building is erected, who is also the employer of the architect; and it speaks more for the general honesty and good faith with which such operations are carried on than for the prudence of the persons who engage in them that there are hardly any two classes of men whose legal status, in regard to other people, is so undefined as that of architects and builders.[1]

A 1981 court shared a similar view of construction, stating that:

> . . . except in the middle of a battlefield, nowhere must men coordinate the movement of other men and all materials in the midst of such chaos and with such limited certainty of present facts and future occurrences as in a huge construction project. . . . Even the most painstaking planning frequently turns out to be mere conjecture and accommodation to changes must necessarily be of the rough, quick and ad hoc sort, analogous to ever-changing commands on the battlefield.[2]

These observations point out what every contractor already knows—that project management on a large construction project is a difficult task. It requires skills learned, good judgment, and the right type of personality. It is questionable whether every architect or engineer can, or should, become a general contractor overnight to lead a design-build team. Design professionals would be concerned about contractors who decided that they can be designers for similar reasons. Even though contractors are not licensed in most states, there are many skills that contractors acquire, either through formal education or on-the-job training, that are not covered in traditional architectural or engineering studies. Project management is one of those skills acquired on the job rather than in a classroom.

Architects and engineers need to think differently about project management to succeed in design-build. Dave Shelton, AIA, started his own design-build firm in 1999. He says: "Most architects lack the knowledge to properly administer the construction contract. Many architects cannot communicate properly with contractors, which drives the two apart. They lose sight of the goal, which is to protect the owner. Instead, they try to protect themselves—try to survive the project." Successful design-build contractors tend to put the owner's needs first and try to "think like an owner," keeping value, cost and schedule as the primary goals, rather than just getting through the project.

Following are some of the skills an architect will need to acquire, either by training or by hiring skilled staff in order to run a construction company.

---

[1] Architect, Owner and Builder Before the Law, Macmillan & Co., 1894.

[2] Blake Constr. Co. v. C.J. Coakly Co., 431 A.2d 569 (D.C. 1981).

§ 20.02 PROJECT ADMINISTRATION SKILLS

Design-build compresses the time schedule greatly compared to traditional procurement. Decisions are made quickly, in-house or with a phone call to make a change in the field. This means that an architect or engineer will need to develop a procedure to manage changes, by developing a record of the decision-making process with less paper, in a shorter time than on traditional projects. Use of a project Web site is the solution chosen by many design-builders, who need one place to efficiently handle requests for information (RFIs), change orders, ASIs, etc. A team must identify a team leader who will manage the paper, whether in hard copy or electronically. If done electronically, some consultants recommend that this be used as the exclusive communication devise—no faxes or letters, so everything is concentrated in one place for the whole team to see, avoiding the delay of mail.

Coordinating a dozen or more trade subcontractors, plus managing a labor staff, and keeping the job on schedule requires special skills. It may be necessary to hire an experienced project manager from a construction company, rather than attempt to develop those skills in-house. A good project manager is hard to find, but even harder to develop from scratch. Many design firms hire these managers from construction companies. Architect Larry Maxfield recommends hiring a carpenter as a superintendent, then training him or her to be a project manager. He prefers to find younger staff, who have not developed a "hard bid" mentality, which he says too often means: "Tough on subs, tough on design, tough on everybody. That philosophy does not work with design-build." By catching them early, before they develop bad habits, one can train staff to adopt the firm's attitude toward clients, design and subcontractors. "This doesn't exclude older superintendents, but you need to know their personality," says Maxfield.

Most small architect-led design-build firms subcontract out 100 percent of the construction work. Some will self-perform only a very small job of $100,000 or less, with a short time schedule, such as 30 to 60 days. In that case, one may hire a working-superintendent. It can be very expensive on a small job to have a superintendent on site who is not also performing labor due to the overhead expense. A working superintendent, such as a carpenter, can manage the project and help perform labor, thereby avoiding the job cost of a project manager.

If a design firm does not have an experienced project manager on staff, some contractors and subcontractors will offer to superintend the project or parts of it for a fee. This is obviously more dangerous than having one's trusted and trained employees on the site. One architect-contractor says that his firm uses the trade subcontractors sometimes to superintend parts of the job. On a $12 million shopping center, before construction began on the building, the architect hired the excavator's superintendent to supervise the grading as well as the site utilities, curbs, and footings by paying the excavator an extra fee. "They already had a guy on the jobsite to do the grading. For an extra fee, he supervised the other site activities so we did not have to have a non-working superintendent on site as an overhead cost."

Some general contractors will offer to "loan" a project manager for a fee, though this is less likely when work is up and project managers scarce. If there is a fee in it, and risk is low, architects may find contractors willing to loan them an employee for a job in which they do not have the risk of cost, time, or quality. More like an employment agency, they merely provide a skilled worker to handle a position that needs to be filled. This may help a contractor to keep a staff member busy rather than laying him or her off during a slow period.

§20.03 BIDDING AND ESTIMATING SKILLS

Most contractors will agree that the most valuable employee of any construction company is the estimator. The company obtains work on competitively bid work by being the lowest bidder; the company makes or loses money based on the accuracy of the estimator's work. So good estimators not only bring in business, but make profit for the company. Of course, a good estimate depends on a good design.

One of the products being touted as specifically created for design-build contractors is PerSpective, a performance-based master spec system developed jointly by The Design-Build Institute of America (DBIA) and the Construction Specification Institute (CSI).[3] Another product, developed to interface with Microsoft Excel, is Vertigraph, Inc.'s BidPoint XL software, designed for estimators and quantity surveyors. BidPoint XL is a digitizer interface that enables the estimator to click on Microsoft Excel workbook cell and electronically measure areas and lengths and count items in seconds using the digitizer cursor. In addition to the measurements, a graphic representation of the digitized measurement is also saved to the Excel cell.[4] Another digitized takeoff and estimating software for earthwork is Roadwork and Trenchwork by Quest Solutions. The systems calculate and fill quantities, volumes, end areas, topsoil stripping, and mass diagrams. The visual templates feature allows users to create cross-sections of individual stations.[5]

As with most functions within a contractor's office, there is computer software for estimating. One of the fastest growing cost estimating systems is D4Cost-2000, a software, database and project case study library of real buildings that have actually been constructed. It includes actual costs, broken down into the 16 CSI Masterformat sections, plus the date and place of the construction. By using regional modifiers and cost escalation adjustments, the estimator can use actual building costs for a similar building based upon national averages.[6] Another reliable industry source is McGraw-Hill, which publishes the Dodge Unit Cost Guide in hard copy and on CD-ROM version, with over 14,000 unit component costs for all 16 CSI specification divisions. The data includes crew composition, crew output per day, man-hours per unit (*e.g.*, linear foot) and, of course, cost including overhead and profit. The publication includes local

[3] This product can be reviewed at <http://perspectivenet.com/>.

[4] For more information on this product, check out Vertigraph's Web site at <www.vertigraph.com>.

[5] For more information, check out Quest Solutions' Web site at <www.questsolutions.com>.

[6] This product can be reviewed at the manufacturer's Web site <http://www.d4cost.com>.

multipliers for locations throughout the United States, plus a Project Estimate Worksheet compatible with the unit cost data.[7]

Some design-build firms include the material quantities and bills of materials right on the design drawings. Since most CAD systems are capable of doing material takeoffs or can be integrated with estimating software, this saves time and probably gives a more accurate count of materials than other methods.

§ 20.04 PROJECT PLANNING

One senior vice president of a large engineering firm, experienced in design-build recommends analysis of project risks up front, with proper allocations as to the best party to handle that task and risk. Some of the issues to be addressed before the bid or proposal are due include:

- Obtaining all major permits and clearances.
- Obtaining final geotechnical and foundations reports.
- Acquiring rights of way.
- Negotiations with utility companies, railroads, and departments of transportation.
- Quality control and quality assurance.
- Force majeure events that will permit time extensions and cost increase.
- Differing site conditions.
- Hazardous materials.[8]

Sequencing of construction is an important factor, especially when there are temporary structures to be built, sheet-piling, cofferdams, tower cranes, scaffolds, etc. A layoff area for materials that are delivered to the site needs to be accessible and large enough to handle storage of anticipated rebar, steel, equipment and other deliveries. By early involvement of the contractor, the designer can plan the design around quicker ways to construct. Many design-build jobs use tilt-up or precast panel construction as a way to speed construction of the building shell. Consider where the construction crane will be positioned on site. Will it have to reach over an area that has already been completed? Is it quicker to prefabricate brick panels and hang them on the building than to have a crew of masons on site, subject to weather conditions? By thinking through the construction process during the design phase, months can be taken off the construction process.

§20.05 UNDERSTANDING THE CONTRACT

It is shocking how many project managers and design professionals have never read their contract. They operate on a general sense of what is in these con-

[7] For more information, visit <http://www.mcgraw-hill.com> or the order line at <http://www.mhorder.com/dodgeunit.html>.

[8] Darius Irani, as quoted in Contractor's Business Management Report, Jan. 1999.

tracts without knowing the specifics. Design-build is different enough, as is each project, that the key personnel should be given a copy of the contract and a sheet to sign acknowledging that they read it. Time limits on claims, performance requirements, notices, milestones, etc. are different in each contract and must be understood. Set aside time in the schedule to let the project manger familiarize himself or herself with the contract documents.

§20.06 SCHEDULING SKILLS

Most architects do not receive any formal education in project scheduling. Though bar charts are fairly easy to understand, a detailed critical path method (CPM) schedule takes some training to develop and understand. In order to complete the job on time and make money the design-builder must know where he or she is in terms of schedule, what functions need to happen and in what order. Many contractors hire outside scheduling consultants to prepare their schedules. An architect-builder should be sure to involve its trade subcontractors in the scheduling process, since they know better than the architect how much time they need to perform their portion of the work.

The end date for the project is often set in stone for the owner, with financing or production of some product tied to a completion date. Design-build teams need to understand that no matter what happens in the preconstruction phase, that end date does not change, so changes in design or materials may be needed to make up for time spent on design, site selection, or other preconstruction tasks. Design-build requires flexibility in the schedule to react to changes. Use the "float" time in some activities to plan better ways to perform upcoming tasks.

An architect needs to have someone with experience prepare the schedule with as much detail as can be afforded. If needed, the architect can hire a scheduling consultant to prepare the schedule. It is important to get input from key subcontractors rather than imposing an unrealistic schedule on them and having conflicts when it is not met. A copy should be posted on the wall of the job trailer so that every sub can see it on a daily basis. The schedule should be discussed at each weekly progress meeting to be sure all trades know what is coming, what is holding work up, and what is needed to get their part done on time, get them off site and get them paid.

§20.07 CHANGE ORDER REVIEW PROCEDURES

Every contract and subcontract addresses change orders. They are the most common source of disputes, however. One should look at the time requirements of the prime contract, the design contract, and subcontracts to make sure they mesh. If the prime contract requires claims for time or money to be made in 21 days from an occurrence, one needs to be sure that the subs have to submit their claims within 14 days from the occurrence, so there is a week to review the sub's claim, get more information, and properly document the situation before submitting to the owner. Failing to make a timely claim for a change order can result in a waiver or even a valid claim.

§ 20.08 WORKING DIRECTLY WITH THE SUBCONTRACTORS

AIA standard contracts prevent the architect from having much direct contact with the trade subcontractors. For example, paragraph 4.2.4 of the 1997 AIA General Conditions, Document No. A201, states that "[c]ommunication by and with Subcontractors and material suppliers shall be through the Contractor." Design-build, however, requires direct contact between the design professional and the subcontractors to help with design and constructability input and, most importantly, problem solving.

[A] Developing a Subcontractor List

As a designer-led design-build firm, a project will largely succeed or fail due to the strength of the subcontractors involved. One architect whose firm has been the lead on many design-build projects says, "You are only as strong as your weakest subcontractor." Most architects, and even many contractors, will subcontract 100 percent of the construction work to trade subcontractors on a design-build project. Some of the subcontractors may be designing their own portions of the project. But whether the quality of construction meets the owner's expectations will depend on the quality of subcontractors who are hired. As a result, an architect will need to develop a subcontractor list early on and update it as he comes across trade subcontractors who do the type of work with the attitude he wants on his projects.

The architect can begin to develop the list by looking through the directory of the local contractor's association, the phone book, lists from prior projects he has designed and even more informal methods. One architect-led design-build firm admits that early on "[w]e would drive through construction sites looking to see who the subs were on various projects—a process we called 'trolling for subs.' It became a rather odd hobby until we developed a list we could rely on." The same firm claims to have two such lists: an A-list, with subcontractors who do good work, on time with the least trouble; and a B-list, with those subcontractors that they never will work with again. "You will spend 90 percent of your fee on the two problem subcontractors," said the architect owner of the design-build firm. Maintaining a list of reliable subcontractors is a way to avoid that problem.

[B] Have a Field Office

Architect/contractor Larry Maxfield, AIA recommends that whoever is in charge of project management should be located in the field, not in the home office. This will mean the added overhead of a job trailer, or some office space within the facility, but by being on site, the project manager can see in minutes any problem that arises. "You can see it, touch it, get muddy," says Maxfield. "More important," he says, "you can relate daily to the subcontractors and the workers. You are not just coming on site occasionally with a white hard hat and clean shoes. They need to know you are part of the team. Architects need to come down from the mountaintop and get to know the people in the field." A field office gives the design firm that presence on site and allows better monitoring of work, prompt solutions to field problems, and more relationships with the trade subcontractors.

[C] Listen to the Subcontractors' Ideas

Design-build contractor Dave Shelton, AIA, gave an example of a project in which the project designer produced an attractive design for a paving pattern which the trade subcontractors said would be costly and difficult to do. So Shelton asked the subs "OK, what would you do?" He then took the subcontractors' design to the project architect who at first resisted, but after Shelton began to mediate between the designer and the craftsmen, the project architect later agreed to changes. "I was trying to balance the designer's goals with the subcontractor's practical concerns. Through a give and take process, we ended up with a design that was buildable, affordable, and kept us on schedule. Everybody was happy."

This example points out a tendency among designers to feel that if they draw it, it is then set in stone and cannot be changed. Whether due to egos or real concern about design integrity, there has to be more openness to change, and greater flexibility to react to site conditions, cost, and schedule in order to succeed in design-build. Architects need to solicit and incorporate the input of the trade subcontractors and craftsmen, whose experience and practical approaches will help design-build projects stay on schedule and within budget.

[D] Treat Everyone Fairly

Every design issue can involve many trades. As a result, Larry Maxfield says, "treating the subcontractors fairly can permeate the whole team. If you treat them right, people will work with their hearts, not with their backs." He also warns that one can develop a reputation very quickly among the subcontracting community. Starting out as an architect or engineer-led design-build contractor may raise some suspicion among the subcontractors. Most have had more than one bad experience where they feel they were mistreated by a design professional. This can result is some subs not bidding an architect's work, or some charging a "premium" in their bid to cover a perceived risk. According to Maxfield: "You have to walk your talk. Back up what you say. Treat people fairly."

Architects should begin with subcontractors they know from prior traditional projects. Most architects know subcontractors in their community from past projects and from industry associations. These subcontractors should be used first and the architect should build its reputation through them. Most contractors will admit that if an architect helps them, the subs will help the architect, but if they are treated unfairly, forget about it.

[E] Coordinate Work Packages and Scopes of Work

When entering into subcontracts, it is essential to make sure that there are no gaps, and that the entire project is covered by the various subcontract scopes. Problems often arise when work is covered by two different trades, such as the electrical hookup to a piece of mechanical equipment, or the roof penetration by a plumbing stack or vent. One architect who leads design-build projects says that he meets with each sub before signing the subcontract and goes over the scope of work in detail looking for not only gaps, but items that are covered in more than one person's bid. "If I find an item of work that I know is in someone else's scope, I tell the sub that it does not have to be in his subcontract too and ask what deduct he will give me to take it out of his work."

§20.09 KICK-OFF AND PROGRESS MEETINGS

To make sure that everyone—designers, contractor and key consultants, and subcontractors— understands their roles, a design-builder should start off with an initial meeting. It is important to go over the project scope, deadlines, unique site or public constraints. All procedures should be reviewed for responsibility, communication, and owner requirements. Everyone should be encouraged to offer suggestions for improvements and to deal with problems that will arise during construction. One design-build team sets aside every Friday afternoon starting with a pizza lunch at the contractor's office, followed by a mandatory meeting of the project managers from the architect and its key consultants, plus the contractor and key subs. They get to talk out problems face-to-face, discuss upcoming work, sketch out details and, most importantly, develop the "team" relationship needed for a successful project.

§ 20.10 SHOP DRAWING REVIEW PROCESS

With design-build there are many ways to speed up the shop drawing process. One practice that is becoming more common is to have the structural engineer prepare the shop drawings for the precast, trusses, or steel. Since the person who would check the shop drawings is the one preparing them, one step in the process can be eliminated. Chuck Williams, P.E., of HDR's design-build division, says: "Take off the stamp that says REVISE AND RESUBMIT. If you have a problem with a submittal, get with the subcontractor involved and work it out. You don't need to take the time to go through a second submittal process." Small changes like this can take days or weeks off the time to fabricate a building component.

§ 20.11 MANAGEMENT SOFTWARE

Software is available from numerous vendors to help with project management. These new tools allow contractors to manage project cash flow, pinpoint cost and schedule variances, know when materials are needed to keep the project moving, as well as have communication with the entire project team through a project Web site. Software even alerts the contractor to project deadlines and warns when thresholds are about to be exceeded, even alerting those responsible. One such product is Primavera Project Planner® (P3e™).[9]

Project Web sites can be set up with other software products as well, without hiring a Web consultant. Meridan Project Systems sells a product called "Prolog Web Site" which helps contractors set up their own project-specific Web site without knowing anything about HTML or the Internet.[10] Use of this or similar systems during the design phase helps the design team and construction team to review documents, redline drawings, have "discussions," and interact without being in the same office. During construction, the Web site becomes the hub of business, managing paperwork and processes, such as RFIs, submittals, change orders, scheduling, and punch list. Connections to weather sites and posting of progress photos give the entire team current data on progress. More elaborate Web sites have digital video cameras mounted on site, providing the team and the owner with "real-time" images of the project's progress and weather conditions.

The Neenan Company, Colorado's largest design-build contractor, has a dedicated Web site for its $31 million office building in lower downtown Denver. The Web site is maintained at the job trailer, which permits problems to be identified and changes made at the most convenient location. "Everyone has accessibility to the 3D design via a private Internet site," says Neenan's director of information systems, George Dial. "The fact that all the designers generate their piece in 3D and then transfer it to the Web site ensures that all of the various pieces fit together logically."[11] Neenan placed a camera (a web cam) on top of the building across the street from the project, which takes pictures of the building's progress every two minutes and updates the Web site's image. The 3D model of the building has been placed against the view from the camera angle so that actual construction can be compared to the model. When the project is completed, Neenan will have a time-lapse video of the building's construction to use for marketing efforts.

§20.12 ISO 9000 CERTIFICATION

For those design-build firms looking for the best practices in project management, there is the prospect of obtaining ISO 9000 Certification from the International Organization for Standardization (ISO). ISO is a Geneva, Switzerland-based federation of the national standard organizations of 91 countries. The idea behind its ISO certification program is to promote standardization in procedures for the international exchange of goods and services. ISO certification is more common in manufacturing industries such as the automotive industry, but it is catching on in construction. Many U.S. contractors choose to become certified under ISO 9001, a

[9] This product can be reviewed at <http://www.primavera.com/p3e/>.

[10] This product can be reviewed at <http://www.mps.com/>.

[11] *Getting the Big Picture*, Colorado Construction, June 1999, p. 119.

standard that includes design.[12] In order to become registered or certified under ISO, contractors must be audited by a third-party registrar to make sure they comply with all of the elements of the standard. To obtain the ISO 9001 certification, there are 20 elements covering such things as management responsibility, quality system, contract review, design control, documentation and data control, and purchasing. U.S. registrars must be accredited by the Registrar Accreditation Board of the American Society for Quality, based in Milwaukee. Some firms hire a consultant to guide them through the process of ISO certification. The process involves considerable investment of staff time. One contractor estimated its investment in ISO certification at $100,000, another between $500,000 and $700,000 in employee time and direct expenses.[13] Many overseas owners are ISO certified and look for contractors who have the ISO certification to do their work. It is one way to prequalify contractors in a manner that owners can relate to. In the United Kingdom, it is estimated that 80 percent to 90 percent of contractors are ISO 9000 registered. Being ISO certified is one way a firm can set itself apart from the crowd as design-build grows and becomes more competitive.

[12] *ISO Registration Creeps Slowly into Construction*, Engineering News Record, Dec. 13, 1999, P. 32.

[13] *Id.*

CHAPTER 21

DESIGN-BUILD RISKS AND LIABILITIES

§21.01 INCREASED RISK IN DESIGN-BUILD: MYTH OR REALITY?

Many architects shy away from design-build due to a perception that it is a higher risk than traditional work. Certainly the risk of design errors is always present, whether in traditional or design-build projects. However, those involved in design-build say that they are less at risk because they can catch and correct those errors more easily, and with less friction than on traditional projects. Architects Clifford Bedar, AIA and Michael Behm, AIA say that increased risk is an illusion, not a reality. "The irony in the liability debate is that no matter how much we try to contractually isolate ourselves from the builder, our clients always perceive architect and contractor as a team. When there is a problem, it's everyone's problem. When we finally understand our client's perspective, we will understand the concept of a single point of responsibility."[1] Design-build, especially when designer-led, gives more control over risk to the design professional.

[A] Less Risk Because of More Control

A survey of design-build firms in 1999 found that two-thirds of those responding felt that it was less likely they would get involved in a dispute when doing a design-build project, and nearly all (97 percent) of the firms said they had never been involved in a lawsuit concerning a design-build project.[2] While there are some different risks, designer-led design-build actually offers ways to reduce risks, not available on traditional projects. Brad Buchanan, AIA, partner in the Denver, Colorado design-build firm of Buchanan Yonushewski Group, LLC sees a way to decrease risk through design-build by better communication. He says, "Communication between the architect and the construction team becomes more difficult as the project increases in complexity. Because we are both the architect and the contractor, we don't have that problem. There is no finger-pointing between the architect and contractor. We can solve any problem that arises at the site."[3] Buchanan can solve the problem because he is in control of the design and construction. There is no RFI, no change order, no delay claim, or dispute. He just fixes the problem and moves on. That level of total control reduces the risk of claims.

[B] Insurance Claims Reduced

A report by CNA/Schinnerer in 1996 showed that about 55 percent of the claims against design-build firms are brought by owners, just as 57 percent of the claims against pure design firms are brought by the owner. This is not unusual, nor unexpected, since the person most likely to have a complaint or suffer damage due to design error is the client. More interesting was the statistic that there

---

[1] *Architects' Attitudes Toward Design-Build Are Ill-Conceived and Self-Defeating*, Architectural Record, February 1999, p. 22.

[2] *1999 Design-Build Survey*, Zweig White & Assoc., Inc., p. 29.

[3] *How to Increase Your Fees in a Tough Market*, Architectural Record, Nov. 1999, p. 62.

are about half as many claims against design-build firms as there are against traditional design firms. In 1992, for example, CNA-insured design-build firms experienced 12.9 claims per 100 insured firms, while in that same year, insured A-E firms experienced 22.6 claims per 100 firms.[4] Why the drop in claims for design-build firms? According to a CNA spokesperson, the reason for such measurable differences in success are:

1. Strong underwriting;
2. Restricted coverage; and
3. Practical incentives to identify problems early and resolve them quickly—such as maximizing profit and less owner intrusion.

According to one industry analyst, "fifteen years of experience analyzing and underwriting design-build exposures convinced CNA/Schinnerer that its long-standing approach to design-build projects overestimated their risk."[5] As a result, CNA reportedly has plans to revise its policy to automatically cover insureds who perform occasional design-build services. The cost to repair faulty workmanship will still be excluded, but only if it is not performed in accordance with the design for the project, or was performed directly by the insured architect itself.

[C] Managing the Risks

There are basically five ways to manage risk on a construction project: (1) shift it by contract; (2) insure it; (3) share it; (4) avoid it; or (5) accept it. Architects and engineers have been urged for many years by legal counsel, insurers, and professional societies to do the first four. With design-build, architects and engineers must begin to accept more risks. Design professionals in design-build see fewer claims because errors are caught during bidding or in the field and resolved without the normal posturing and adversarial nature that attaches to traditional construction. Design errors are not exploited by the contractor with "premium" costs and large mark-ups. Instead, the design-build team works directly together to resolve the problem, often without involving the owner and without cost to the owner. It may be that the design professional pays some portion of the corrective work, but as prime on the job, this is paid out of profits and contingencies in the design-build contract, rather than out of slim design fees.

§ 21.02 NEGLIGENCE AND PROFESSIONAL STANDARD OF CARE

Most design professional lawsuits involve some allegation of "malpractice," a term that has no real legal meaning. This is really an action for professional

---

[4] These statistics are a dramatic drop from claims statistics in the mid-1980s, when claims were running around 44 per 100 firms. Those statistics peaked in 1984 and have been gradually dropping ever since, with the most recent statistics showing claims down to about 22 per 100 firms.

[5] A. Abramowitz, *Professional Liability Insurance in the Design-Build Setting*, The Construction Lawyer, Forum Committee on the Construction Industry (ABA Press, Aug. 1995).

negligence. Negligence cases have four elements that need to be proved to establish fault. If any one of the elements cannot be proved, then there is usually no liability. The elements are:

1. A *duty* owed by the defendant to the plaintiff;
2. A *breach* (or violation) of that duty by the defendant;
3. A *causal connection* between the negligent conduct and the resulting harm; and
4. Actual *loss or damage* suffered by the plaintiff.

When the plaintiff proves all four elements of his or her case, it is known as a "prima facie" case of negligence. The defendant must then come forward with evidence to refute one or more elements of the plaintiff's case to avoid a finding of liability and damages. We have all heard of cases where the jury found someone at fault, but awarded only $1 in damages. Merely being negligent does not mean that the plaintiff gets a huge award. The plaintiff must also prove the amount of damages, and persuade the judge or jury that they are reasonable.

[A] The Professional Standard of Care

The accepted definitions of negligence all involve conduct which falls below the "standard of care" established by law for the protection of others against harm. In everyday life, people are held to the "reasonable person" standard of care. In other words, we must all act like a reasonable person would act if faced with the same situation. If we act unreasonably, our conduct falls below the standard of care and we are negligent. For everyday events, like driving a car, cleaning the walk, or throwing a ball, a jury of our peers is capable of judging the conduct based on their own life's experiences. However, in cases of professional negligence involving architects, engineers or other trained professionals, like doctors, nurses or lawyers, the jury needs to consider how others trained in *the same* profession would have acted in the defendant's situation. This is established through expert testimony by other professionals.

When the design professional agrees to perform services on a project, the professional is assuming obligations to properly perform the undertaking. A failure to properly perform the obligations can lead to liability. Before there can be liability for any error, there must be a duty owed to the injured or damaged party. The legal foundation for a duty may be grounded either in the contract obligations owed to those parties benefitting from the contract, or in tort law which imposes duties on individuals and entities for the protection of society in general. Tort law may give the design professional additional obligations not only to the parties to the contract, but in some cases to individuals or entities not part of the contract relationship, such as contractors, sureties, lenders and the public.

Architects, engineers, and land surveyors are required by law to perform their services in a manner that is consistent with the way other professionals would in the same situation. This is known as the "professional standard of care." When a design professional's conduct falls below the standard of care, the law calls this "negligence." Design professionals are required to perform their services in a non-negligent manner, not necessarily perfect, but not negligent. The courts have generally grouped design professionals with doctors, lawyers, and others who deal in "inexact sciences."[6] Perfection is not required of these professionals, nor is any guarantee of results.

A frequently quoted case is the federal Court of Appeals' decision in *Aetna Insurance Co. v. Hellmuth Obata & Kassabaum, Inc.*,[7] In that case, the court stated:

> an architect is not a guarantor or an insurer but as a member of a learned and skilled profession he is under a duty to exercise the ordinary, reasonable technical skill, ability and competence that is required of an architect in a similar situation. . . . The standards of reasonable care, which apply to the conduct of architects, are the same as those applying to lawyers, doctors, engineers, and like professional men engaged in furnishing skilled services for compensation. . . .

As a subcontractor on a design-build team, the architect is held to this traditional standard. However, as prime contractor, an architect might be required by the owner to guarantee the results, thereby raising the standard and increasing the risk.

[B] Licensing Board Regulations

Some state licensing boards have incorporated the professional standard of care into state licensing regulations. For example, § 4 C.S.R. 30-2.010(2) of the Missouri board regulations states that when practicing, design professionals:

> shall act with reasonable care and competence, and shall apply the technical knowledge and skill which are ordinarily applied by registered architects, professional engineers or land surveyors of good standing, practicing in Missouri.

This is a codification of the standard of care as established by Missouri courts. It reinforces the concept that professional conduct is judged by ordinary practice by other professionals in the state. Missouri courts have refused, however, to allow a lawsuit to be maintained against a design professional based merely on a violation of the licensing laws.[8] The courts' rationale is that these laws are designed to protect the public at large, not any individual, and the licensing laws have their own procedures for punishing those who violate them without

---

[6] *See* Chubb Group of Ins. Cos. v. C.F. Murphy Assocs., Inc., 656 S.W.2d 766 (Mo. App. W.D. 1983) (labeling architecture an "inexact science"); City of Mounds View v. Walijarvi, 263 N.W.2d 420 (Minn. 1978) ("architects, doctors, engineers, attorneys, and others deal in somewhat inexact sciences"). *But see* Tamarac Dev. Co., Inc. v. Delamater, Freund & Assocs., P.A., 675 P.2d 361, 364 (Kan. 1984) (express warranty by engineer gave rise to a holding that the owner could expect an exact result).

[7] 392 F.2d 472 (8th Cir. 1968).

[8] Burns v. Black & Veatch, Inc., 854 S.W.2d 450 (Mo. App. W.D. 1993); Business Men's Assurance Co. v. Graham, 891 S.W.2d 438 (Mo. App. W.D. 1994); Gipson v. Slagle, 820 S.W.2d 595 (Mo. App. 1991).

creating grounds for a lawsuit as well. While there are no licenses for "design-builders", a design firm will be subject to standards imposed by architect/engineer regulations like this one regardless of the method of project delivery.

[C] AIA Code of Ethics

The 1997 AIA Code states in Rule 1.101 that "[i]n practicing architecture, Members shall demonstrate a consistent pattern of reasonable care and competence, and shall apply the technical knowledge and skill which is ordinarily applied by architects of good standing practicing in the same locality." AIA members can be disciplined for violation of the Code, again whether doing traditional or design-build work.

[D] Raising the Standard of Care

While ordinary professional services are judged by the prevailing standard in the profession, a design professional by agreement may accept a higher standard of care. For instance, if a design-build contract says that the services are to be performed to the "highest level of care" or that the architect shall warrant a specific result, the standard is raised above what the law otherwise requires. In these cases, evidence of what the typical architect or engineer would have done is no defense, because the bar has been raised to require proof of what *the best* architects and engineers would have done using the "highest" degree of care. Caution must be exercised against making any such agreement because most professional liability policies exclude coverage for promises of higher than normal performance. See § 13.17. The standard of care for contractors is different than that for design professionals. Contractors warrant that their work will comply with the plans and specifications. If the specifications require a specific performance result, an output, or production, the contractor must meet that performance criteria. If the criteria are not met, then the contractor has breached its warranty to deliver an exact result.

[E] The DBIA Position on Standard of Care

DBIA's official policy on "Design-Build Standard of Care" states that "if performance requirements contractually exist and can be objectively measured, the design-builder should be obligated to meet such performance requirements, *regardless of the ordinary industry standard of care*."[9] For construction contractors this will come as no revelation. But for design professionals entering into design-build, whether as prime, lead contractor or as a subcontractor, the concept of being held to a performance standard without regard to the ordinary standard of care is a radical departure from what the AIA, NSPE, the insurance industry, and legal advisors recommend.

[9] DBIA Manual of Policies, Policy No. 45, Oct. 1998.

DBIA's standard form of subcontract between design-builder and designer includes an insurable standard of care in paragraph 2.2.1, consistent with state law in most jurisdictions holding the designer to that level of care and skill "ordinarily used by members of the design profession practicing under similar conditions at the same time and locality of the Project." However, the clause goes on to add: "Notwithstanding the preceding sentence, if the Design-Build Agreement contains specifically identified performance standards for aspects of the Services (which shall be set forth in an exhibit to the Design-Build Agreement entitled Performance Standard Requirements), Designer agrees that all Services shall be performed to achieve such standards [*i.e.*, regardless of the standard of care]."[10] This may create a "warranty" or guarantee not covered by professional liability insurance. Design professions must use care to either avoid such clauses or to modify them to bring the risk back within insurance coverage. *See* Appendix B for an example.

DBIA acknowledges in its Policy Statement on standard of care that "the owner is best served through the broad use of performance-based specifications. This enables the owner to define its needs and permits the design-builder to do what it does best—develop solutions to those needs. The result is consistent with the design-build philosophy of single point responsibility when performance can be objectively measured."

The reality here is that owners want, and expect, single-point responsibility from design-build. If a design firm steps into the lead role, it must be willing to take on risks that are not always covered by insurance, including express warranties. This is the risk that is passed on to contractors every day by design professionals and now that the shoe is on the other foot, design professionals must be willing to either accept that risk or not participate in design-build.

§21.03 THE ECONOMIC LOSS DOCTRINE

When we talk about claims by third parties, we must break them into two groups: (1) those for damages to persons or property; and (2) "economic loss." If a roof leaks due to faulty design or construction, and someone slips and falls, or water damages stored goods, there is personal injury or property damage. Anyone who suffers such damage can sue even though they had no contract with the architect or builder if their injury is reasonably forseeable. However, if the damage is merely the cost to repair the roof, or lost rent or other "economic" injury, the ability to sue may be limited by state law. "Economic damages" are defined as including inadequate value, costs of repair, replacement of the defective product, loss of use of the defective product, cost of replacing the product, lost profits to plaintiff's business, or cost of renting a temporary replacement.[11] More broadly, they cover anything *other than* damage to persons or their property.

---

[10] ¶2.2.1, DBIA Standard Form of Agreement Between Design-Builder and Designer, Doc. No. 540 (1999 edition).

[11] Professional Lens Plan, Inc. v. Polaris Leasing Corp., 675 P.2d 887 (Kan. 1984); Elite Professionals, Inc. v. Carrier Corp., 827 P.2d 1195 (Kan. App. 1992).

There is no question that a contractor may sue the architect/engineer for personal injury or property damages. However, there is split authority around the country as to whether or not a contractor can maintain an action against an architect/engineer *for pure economic loss* when the contractor has no contract with the architect/engineer. In what has become known as the "economic loss" doctrine, many states have held that a contractor or subcontractor has no claim for pure economic loss against the architect/engineer. Instead, the contractor must pursue his remedy for economic loss against the owner or other party with whom he has a contract. The courts in several states have recognized the economic loss doctrine as a defense to third-party claims against architects/engineers.[12] In these cases, the courts have held that generally, in the absence of a direct contract, no cause of action exists in tort to recover economic damages against design professionals involved in the drafting of plans or specifications. In most traditional projects, there is no direct contact between the contractor and the design professional and, therefore, the economic loss doctrine applies.

Design-build, of course, changes the contractual relationships and when the design professional has a direct contract with the contractor, there would be no "economic loss" defense. So, in this sense, design-build increases the liability of architects and engineers by exposing them to more suits by contractors and subcontractors with whom they have direct contracts.

[A] Some Design-Build Court Cases

In a 1998 Colorado design-build case, the owner of a restaurant sued the contractor for negligence, seeking damages for mental anguish, anxiety, and inconvenience when the design left insufficient room for a drive-through window. The court held that these types of damages are not recoverable in contract disputes and that the economic loss doctrine prohibited negligence claims based on a breach of contract duties. As a result, the owner could not recover for these "non-economic" damages.[13]

In another design-build case, *Dakota Gasification Co. v. Pascoe Building Systems*,[14] Kaiser Company was hired as a "turnkey" contractor to design and build a synthetic natural gas production plant in North Dakota, to be one of the largest synthetic fuel plants in the world. Kaiser subcontracted to Lotepro Corporation to provide the labor, material, and equipment. A lower-tier subcontractor, Pascoe Building Systems, was to provide a preengineered metal building to

[12] *See, e.g.*, Malta Constr. Co. v. Henningson, Durham & Richardson, Inc., 716 F.2d 1466 (N.D. Ga. 1989); 2314 Lincoln Park W. Condominium Ass'n v. Mann, Gin, Ebel & Frazier, Ltd., 555 N.E.2d 346 (Ill. 1990); Briar Contracting Corp. v. City of New York, 550 N.Y.S.2d 717 (1989); Craft Floor Covering, Inc. v. Parma Community Gen. Hosp. Assoc., 560 N.E.2d 206 (Ohio 1990); Linde Enters., Inc. v. Hazelton City Auth., 602 A.2d 897 (Pa. Super. Ct. 1992); Sensenbrenner v. Rust, Orling & Neale Architects, Inc., 374 S.E.2d 55 (Va. 1988).

[13] Terrones v. Tapai, 967 P.2d 216 (Colo. App. 1998).

[14] 91 F.3d 1094 (8th Cir. 1996).

enclose the plant. The owner defaulted on construction loans guaranteed by the U.S. Government, so the Department of Energy took possession of the plant, later selling the $3 billion plant to Dakota Gasification for less than $100 million, plus a percentage of the plant's profits. The purchase contract said that Dakota took the plant assets "as is" and without warranty. Eight years after the project was completed, part of the roof collapsed under the weight of ice and snow. Dakota sued Kaiser, Lotepro, Pascoe, and others for damages. The trial court entered summary judgment against Dakota on all claims based on the "economic loss" doctrine, which barred any tort claims when the only damage was to the building product itself, on which Dakota had waived any warranty. On appeal, the court held that the sale of the metal building and its components was a "sale of goods" and, therefore, the economic loss doctrine barred Dakota's claims for the defects.

§21.04 EXPRESS WARRANTIES

Architects and engineers are told not to warrant anything because they cannot guarantee a perfect result. One architect says: "When my clients ask for a warranty, I tell them that I will only warrant two things: The skylights *will leak*; and the concrete *will crack*. Beyond that, I won't guarantee anything." As stated above, the courts do not hold architects and engineers to a standard of perfection and recognize that they are only required to exercise reasonable care and skill. Professional liability insurers exclude coverage for express warranties or guarantees. (*See* Chapter 13 § 13.17). However, the owner will expect the same warranty from an architect as contractor as he or she would get from any other contractor, whether design-build or traditional. A one-year warranty is a standard part of every construction contract.[15] As a result, an architect needs to either qualify any warranties as applying only to the construction work, or be prepared to self-insure the risk of a claim that is brought under the contract warranty clause.

The AGC standard design-build contract[16] has a good example of a typical warranty clause. Paragraph 3.8.1 states:

> The Design-Builder warrants that all materials and equipment furnished under the Construction Phase of this Agreement will be new unless otherwise specified, of good quality, in conformance with the Contract Documents, and free from defective workmanship and materials. Warranties shall commence on the Date of Substantial Completion of the Work or of a designated portion. The Design-Builder agrees to correct all construction performed under this Agreement which is defective in workmanship or materials within a period of one year from the Date of Substantial Completion or for such longer periods of time as may be set forth with respect to specific warranties required by the Contract Documents.

There is usually no insurance, whether one is a contractor-led or a designer-led design-builder, for correcting one's own defective workmanship. Owners

---

[15] *See, e.g.*, AIA's A201 General Conditions, ¶12.2.2 (1997 edition).

[16] AGC Doc. No. 410 (1999 edition).

looking for single-point responsibility will expect an architect to agree to this same warranty without limitation. So it is important to understand that with design-build comes more risks like this that will be covered out of company (or subcontractor's) profits, not by insurance. Refusing to give a warranty and assume risk will give construction-led teams an advantage over those who will not warrant the work. An architect needs to be sure that whatever warranties are contained in their prime contract also are included in the subcontracts, so that the architect can turn any warranty claim over to the appropriate trade contractor to resolve, a way of transferring the risk.

§21.05 THE IMPLIED WARRANTY OF FITNESS FOR A PARTICULAR PURPOSE

Some courts have held that persons who design and construct can be held to an implied warranty that the building will be fit for the owner's intended purpose. At least one court case has held that a design-build contractor can be held liable for breach of an implied warranty of fitness for a specific purpose. In *Robertson Lumber Co. v. Stephens Farmers Cooperative Elevator Co.*,[17] the owner contracted with a lumber company to design and build a grain storage building. The evidence was that the contractor expressly guaranteed that the building would safely store 100,000 bushels of wheat for a period of one year (with a total weight of some six million pounds). However, the building collapsed after being loaded with wheat. The court found that the contractor was held to an implied warranty of fitness for the purpose intended, since: (1) the contractor held itself out, expressly or by implication, as competent to undertake the contract; and the owner (2) had no particular expertise in the kind of work contemplated; (3) furnished no plans, design, specification, details or blueprints; and (4) tacitly or specifically indicated his reliance on the experience and skill of the contractor, after making known to him the specific purposes for which the building was intended.

Similarly, in *Kennedy v. Bowling*[18] a contractor was hired to design and build a four-story warehouse to store chemicals, with a very fast track of only four months to construct the building. The contractor represented to the owner that he could store things to the ceiling if needed, saying, "When we get through, you can put anything you want in it." Famous last words and, of course, the wood floor joists could not withstand the weight and had to be reinforced with steel by the owner. The contractor hired the architect directly to design the warehouse and the owner had given the contractor a specific floor load of at least 250 pounds per square foot. The jury awarded the owner its costs of repair under a theory of "implied warranty for a particular purpose." That verdict was upheld on appeal with the court stating that:

[17] 143 N.W.2d 622 (Minn. 1966).

[18] 4 S.W.2d 438 (Mo. banc 1928).

> If [the contractor] represented themselves as experienced and skillful in the construction of building and in the designs for such buildings, and the plaintiffs [owners] relied upon their judgment, and defendants undertook the construction on the building, then the implied warranty arose that in design and workmanship the building should be reasonably fit for the purpose intended.[19]

In yet another case in 1978, a contractor designed and built the shell for a house. The foundation settled and cracked and the builder refused to fix it. The court awarded the owners $19,000 for breach of an implied warranty that the plan was fit for use in the construction of a house. The court explained that it was not holding the contractor liable as a "builder/vendor," a seller of a product, but for the sale of a design not fit for use for a particular purpose.[20] These cases show that, at least in some jurisdictions, a design-builder may be held liable to impliedly warrant the fitness of the building.

§21.06 UNIFORM COMMERCIAL CODE (UCC) WARRANTIES

The UCC has been adopted in every state except Louisiana. This code applies certain warranties as a matter of law to the sale of goods and products. Probably the most widely litigated UCC warranty is that contained in section 2-314, the implied warranty of merchantability. This warranty is usually not applicable to a company providing professional services.[21] It has been applied in a construction setting, however, to suppliers of materials and equipment that prove to be defective. There is no reported court case as yet holding design-build contractors to a UCC warranty. However, at least one commentator has speculated that a design-build contractor might fit the definition of a "merchant" under UCC section 2-104(1), which provides in part that:

> Merchant means a person who deals in goods of the kind or otherwise by his occupation holds himself out as having knowledge or skill peculiar to the practice or goods involved in the transaction or to whom such knowledge or skill may be attributed by his employment of an agent or broker or other intermediary who by his occupation holds himself out as having such knowledge or skill.[22]

Yet another law review article has argued that because the UCC applies to transactions in goods, it is not possible to apply the UCC to a design-builder

[19] *Id.* at 445.

[20] O'Dell v. Custom Builders Corp., 560 S.W.2d 862 (Mo. banc 1978).

[21] Although one federal court of appeals has held that "[t]he person who designs the structure is responsible for insufficiencies in the effectiveness for the purpose intended." Kelley v. Bank Bldg. & Equip. Co., 453 F.2d 774, 779 (10th Cir. 1972), citing U.S. v. Spearin, 248 U.S 132 (1918).

[22] H. Block, *As the Walls Came Tumbling Down: Architects' Expanded Liability Under Design-Build/Construction Contracting*, 17 John Marshall L. Rev. 1, 45 (1984).

of buildings, which are not "products."[23] One Illinois case has held that a multi-level parking garage is not a "product."[24] It would not seem that the UCC and laws dealing with sales of "goods" should apply to a completed building. However, a preventative measure would be to disclaim any such warranties in the design-build contract, just as manufacturers give limited warranties and disclaim all others. AGC's standard design-build contract has such a disclaimer which says:

> To the extent products, equipment, systems or materials incorporated in the Work are specified and purchased by the Owner, they shall be covered exclusively by the warranty of the manufacturer. There are no warranties which extend beyond the description on the face of any such warranty. To the extent products, equipment, systems or materials incorporated in the Work are specified by the Owner but purchased by the Design-Builder and are inconsistent with selection criteria that otherwise would have been followed by the Design-Builder, the Design-Builder shall assist the Owner in pursuing warranty claims. **ALL OTHER WARRANTIES EXPRESSED OR IMPLIED INCLUDING THE WARRANTY OF MERCHANTABILITY AND THE WARRANTY OF FITNESS FOR A PARTICULAR PURPOSE ARE EXPRESSLY DISCLAIMED.**[25]

This author is not aware of any case yet that has applied such a disclaimer to a design-build contract, but it is likely that the courts would uphold contract language that was negotiated and not in violation of public policy.

§21.07 STRICT LIABILITY FOR DESIGN-BUILD CONTRACTORS

[A] Strict Liability in History

The concept of a "master builder" who both designed and constructed, has been traced as far back as 1800 B.C. Strict liability for a master builder has its roots in the ancient Babylonian Code of Hammurabi, which imposed the same liability on the master builder as that suffered by his client. The ancient Code stated that:

> If a builder has built a house for a man, and his work is not strong, and if the house he has built falls in and kills the householder, that builder shall be slain.
>
> If the child of the householder be killed, the child of that builder shall be slain.
>
> If the slave of the householder be killed, he shall give slave for slave to the householder.

---

[23] B. Miller, *The Architect in the Design-Build Model: Designing and Building the Case for Strict Liability in Tort*, 33 Case W. Res. L. Rev. 116, 140 n. 154 (1982).

[24] Lowrie v. City of Evanston, 365 N.E.2d 923 (Ill. App. Ct. 1977).

[25] ¶3.8.2, AGC Standard Form of Design-Build Agreement and General Conditions, Doc. No. 410 (1999 edition).

> If the goods have been destroyed, he shall replace all that has been destroyed; and because the house that he built was not made, and it has been fallen in, he shall restore the fallen house out of his personal property.
>
> If a builder has built a house for a man and his work is not done properly, and a wall shifts; then that builder shall make that wall good with his own silver.[26]

There was no requirement of proof that the "standard of care" was violated, no expert witnesses, just automatic liability imposed on the builder. If you built it and it fell down, you were liable automatically. This is the same basic premise behind the modern-day concept of strict liability. Remnants of strict liability were written into French law in the 1800s which stated:

> If the edifice constructed at a fixed price perishes, totally or in part, through vices of construction, even by the vice of the soil on which it is built, the architect and the contractor are responsible for it during ten years.[27]

This legal standard is still part of our laws in America, though not normally applied to design and construction contracts.

[B] The Modern Day View of Strict Liability

There is a famous old English case, *Winterbottom v. Wright,*[28] which held that a person injured by a defective product could not sue the manufacturer in the absence of a contractual relationship between the two. This rule of law prevailed during the period of the Industrial Revolution until the early part of the twentieth century, when courts began to realize that this was no longer acceptable. The 1916 landmark case of *MacPherson v. Buick Motor Co.*[29] changed this when the New York Court of Appeals held that a negligent manufacturer could be held liable without a direct contract if it sold defective goods or goods that were inherently dangerous. This concept of "strict liability" in tort was written into a model law in 1965 by the American Law Institute, in a restatement of the case law called The Restatement of the Law of Torts. Section 402A of the Restatement set out the concept of strict liability for manufacturers and that section has been adopted now by most state courts as the law of the land.

[C] The Restatement (Second) of Torts, Section 402A

The theory of law embodied in section 402A makes the seller of a defective product liable if the product poses an unreasonable risk to the user, consumer, or his property. The intent of the law is to provide a means for an injured party to re-

[26] H. Block, *As the Walls Came Tumbling Down,* note 22, above.

[27] The French Civil Code, Article 1792, as quoted by T.M.Clark, FAIA, *Architect, Owner and Builder Before the Law*, Macmillan & Co., (1894), p. 57.

[28] 152 Eng. Rep. 402 (1842).

[29] 217 N.Y. 382, 111 N.E. 1050 (1916).

cover from the manufacturer with whom the party has no contract. However, the law was directed toward manufacturers who mass produce products for use by consumers. This liability applies even though "the seller has exercised all possible care in the preparation and sale of his product." This is a stark contrast to standard architect liability, based on the exercise of reasonable care and skill.

[D] Should This Law Apply to Design-Build?

There has been some speculation that design-build will expose architects and engineers to claims of "strict liability." The argument is that because of the packaged services provided by design-builders, the services are more like those of a manufacturer and, therefore, a strict liability standard should apply when a building defect causes injury.[30] This is a flawed theory, however, because it fails to recognize the distinct differences between manufacturing and construction. First of all, buildings are not "products," as has been recognized by the courts.[31] Manufacturers of products that are reproduced over and over have the opportunity to do better quality control, and have the chance to make samples and test them before putting them into the stream of commerce. Buildings, on the other hand, are normally built one time only, on a unique site, with special soil conditions. Aside from the manufactured housing industry or motor homes, there is no repetition, no chance to build a sample product and test it before putting it on the market. Applying this theory to design-build would be a mistake for the courts and would likely result in the swift termination of this innovative mode of construction, with the numbers of architects, engineers, and contractors fleeing from design-build as quickly as they entered it.

There are ample remedies available to the public through contract, negligence, and express/implied warranty theories and there is no public policy reason to expand strict liability outside of the manufacturing setting and apply it to design-build. Nonetheless, clever lawyers will try to advance whatever theories they believe will get their client the most relief. What they may fail to consider, however, is that there are normally exclusions in professional liability policies for manufacturing of products by the insured and a claim of strict liability is likely not covered by an architect's or engineer's insurance. Professional liability coverage is designed to cover negligent acts, errors or omissions, not strict liability. This may leave the plaintiff with a clever new legal theory, but no way to collect on a judgment that is based on strict liability.

[E] Court Cases on Strict Liability in Construction

Architects and engineers are generally not liable in the absence of proof of their negligence. As stated by one court:

---

[30] B. Miller, *The Architect in the Design-Build Model,* note 23, above.

[31] Lowrie v. City of Evanston, 365 N.E.2d 923 (Ill. Ct. App. 1997). The court rejected a strict liability theory being applied to the design and construction of a parking garage under § 402A.

> Professional services do not ordinarily lend themselves to the doctrine of tort liability without fault because they lack the elements which give rise to the doctrine [of strict liability]. There is no mass production of goods or a large body of distant consumers whom it would be unfair to require to trace the article they used along the channel of trade to the original manufacturer.[32]

The Missouri Court of Appeals held in the Kemper Arena roof collapse case that neither the architect nor the contractor could be sued under a strict liability theory. However, the court stated that the theory could be applied to the manufacturers of building components used to build the arena.[33] In an Illinois case, the court held that design-build contracts focus more on a "final product", not pure services. Therefore, remedies should be more "product oriented."[34] Much law is yet to be written by the courts on this aspect of design-build.

§21.08 LIABILITY OF DESIGN-BUILD TEAM FOR DEFAULT

One of the drawbacks to design-build is that when the owner terminates the prime contract, it loses the entire team—designer, builder, and all. Unlike the traditional method, where defaulting a construction contractor does not replace the project architect, one can stay under contract to assist the owner in working with the surety and replacement contractor. So one would think that firing a design-build contractor would result in a significant cost increase to the owner. When the School Board of Broward County, Florida terminated its design-build contractor in March 1999, the Board feared the worst. However, the replacement contractor's cost was actually slightly less than the original contract, despite design problems. The school designed by the replacement contractor's architect did not match the foundation of the old one. As a result, the original foundation was demolished and rebuilt, not only due to the new design, but due to the liability of taking over another contractor's work. The School Board approved the demolition, since it knew that if the new contractor had to build on what was already in place, it would require the Board to hold it harmless for any problems connected with the foundation.[35]

That case may be the exception rather than the rule, but it points out how design-build can result in innovative cost-savings ideas that help lessen a bad situation.

§21.09 THE *SPEARIN* DOCTRINE AND IMPLIED WARRANTIES OF PLANS

Though design professionals are not held to a standard of perfection, and do not warrant their plans as perfect, the law is different concerning owners and con-

---

[32] LaRossa v. Scientific Design Co., 402 F.2d 937, 942 (3d Cir. 1968).

[33] Chubb Group of Ins. Co. v. C.F. Murphy, 656 S.W.2d 766 (Mo. App. W.D. 1983); *See also* Counts v. MK-Ferguson Co., 862 F.2d 1338 (8th Cir. 1988); Commercial Dist. Center v. St. Regis Paper Co., 689 S.W.2d 664 (Mo. App. W.D. 1985).

[34] Kishwaukee Community Health Servs. Ctr. v. Hospital Bldg. & Equip. Co., 638 F. Supp. 1492 (N.D. Ill. 1986).

[35] *Board Replaces Design-Builder*, Engineering News Record, Apr. 19, 1999, p. 24.

tractors. Federal court cases have held that on public projects, when a contractor bids on a set of plans and specifications, the government warrants the adequacy of the design specifications, and if such specifications are deficient, the government bears the risk, not the contractor.[36] These cases rely on the "*Spearin* doctrine" as derived from the U.S. Supreme Court's landmark decision in *U.S. v. Spearin*. The *Spearin* doctrine holds, generally, that when the owner furnishes detailed plans and specifications to a contractor, the owner is deemed to *impliedly warrant* that those plans and specifications are accurate and suitable for their intended use. The *Spearin* case held that "if the contractor is bound to build according to plans and specifications prepared by the owner, the contractor will not be responsible for the consequences of defects in the plans and specifications."[37] The *Spearin* doctrine is widely recognized by state courts as well as federal courts.[38]

This implied warranty gives rise to a large number of claims by contractors, especially on public, hard bid construction work. In design-build, however, when there is no design furnished by the owner, not only is the owner's implied warranty eliminated, but owners often get a broad warranty from the contractor in a reversal of the normal roles, which covers all design work as well as construction warranties.

The *Spearin* doctrine may still have some application to public design-build projects. When bridging documents or detailed RFPs are given to design-build teams, there may still be an implied warranty made by the owner. In one federal contract appeals board case, the design-build contractor was entitled to an equitable adjustment from the owner, the U.S. Air Force, for costs it incurred relying on erroneous drawings included in the government's RFP. The government argued that a design-build contractor cannot rely on information in the RFP, but the appeals board held that there was no distinction between standards applicable to traditional contractors and those applicable to design-build contractors.[39]

§21.10 PROMISSORY ESTOPPEL AND BID PRICES

A risk for design firms going into design-build is the new experience of dealing with dozens of subcontractors and suppliers. That risk starts when getting subcontractor bids based on preliminary designs, and then holding those subs to that bid price after the firm is awarded the contract by the owner. When a subcontractor gives his bid to a general contractor, which is used in computing the contractor's bid to the owner, there is normally no written agreement between the contractor and subcontractor until after the bid is accepted by the owner and the prime contract is awarded. Yet the contractor relies on that number in computing

---

[36] Big Chief Drilling Co. v. U.S., 26 Cl. Ct. 1276, 1293 (1992).

[37] United States v. Spearin, 248 U.S 132, 136 (1918).

[38] *See, e.g.*, Miller v. City of Broken Arrow, 660 F.2d 450, 457 (10th Cir. 1981); Housing Authority v. E.W. Johnson Constr. Co., Inc., 573 S.W.2d 316 (Ark. 1978); Sanders Company Plumbing & Heating, Inc. v. City of Independence, 694 S.W.2d 841 (Mo. App. 1985). *See generally The Owner's Warranty of the Plans and Specifications for a Construction Project*, Pub. Contract L.J., Vol. 14, No. 2, p. 240 (Feb. 1984).

[39] In re M.A. Mortenson Co., ASBCA No. 39978, 93-3 BCA ¶26,189 (1993).

the overall contract price. If a subcontractor backs out and refuses to honor its price, it leaves the contractor at risk if the contractor cannot find another sub who will perform for the same price. When design firms act as subcontractors to general contractors, they are asked to quote a fee to the contractor, which is relied upon by the contractor in fixing a lump sum or GMP price with the owner. If your firm makes a mistake in setting the fee and attempts to drop out of the project, the contractor would have the same rights as it would against any other subcontractor to hold the firm liable for the increased price of getting another design firm involved. Of course, this risk is much less in design-build, since the design firm has more invested in the proposal stage generally than a subcontractor who has merely furnished a quote. The design firm will have participated in design preparation in many cases and risks forfeiting any stipend or fee if it does not stay part of the team and honor its bid for design fees.

[A] Enforcement of Subcontractor Bids

There are numerous cases addressing the question of whether a general contractor can hold a subcontractor to a telephone bid. However, there is a split of authority on the issue. The minority view is that the subcontractor or supplier is not bound by an oral bid.[40] However, this view is not shared by most courts. The majority view, based on the legal theory of "promissory estoppel," is that the subcontractor is bound if the general contractor notifies the sub of acceptance within a reasonable time after being awarded the prime contract.[41] The leading case on this issue is *Drennan v. Star Paving Co.*,[42] which is cited in almost every case on the issue of enforcing subcontractor bids. In *Drennan*, a contractor was awarded a contract based on his low bid. The contractor's bid included a paving subcontractor's phone bid. The sub then refused to perform claiming a bid error. The court held that although there was no bilateral contract binding on the parties, nor any consideration to support a contract, the subcontractor would still be liable for damages under the principal of "detrimental reliance." The court stated: "Reasonable reliance serves to hold the offeror in lieu of the consideration ordinarily required to make the offer binding."

---

[40] *See, e.g.*, Anderson Constr. Co. v. Lyon Metal Prods., 370 So. 2d 935 (Miss. 1979) (UCC statute of frauds); Home Elec. Co. of Lenoir, Inc. v. Hall & Underdown Heating & Air Conditioning Co., 358 S.E.2d 539, *aff'd*, 366 S.E.2d 441 (N.C. 1987); Line Dickson Co. v. Union Oil Co., 635 P.2d 103 (Wash. banc 1981) (UCC statute of frauds precludes enforcement).

[41] *See, e.g.*, Allen M. Campbell v. Virginia Metal Indus., Inc., 708 F.2d 930 (4th Cir. 1983); Harry Harris, Inc. v. Quality Constr. Co., 593 S.W.2d 872 (Ky. Ct. App. 1979); Meade Constr. Co. v. Mansfield Commercial Elec., 579 S.W.2d 105 (Ky. 1979); Alaska Bussell Elec. Co. v. Vern Hickel Constr. Co., 688 P.2d 57 6 (Alaska 1984); Lyon Metal Prod., Inc. v. Hagerman Constr. Co., 391 N.E.2d 1152, 1155 (Ind. Ct. App. 1979); Arango Constr. Co. v. Success Roofing Inc., 730 P.2d 720 (Wash. Ct. App. 1st Div. 1986) (refusing to follow the *Line Dickson* case). *See generally The Construction Industry Bidding Cases: Application of Traditional Contract, Promissory Estoppel, and other Theories to the Relations Between General Contractors and Subcontractors*, 13 John Marshall L. Rev. 565 (1980).

[42] 333 P.2d 757 (Cal. banc 1958).

For example, in a 1994 case, the general contractor relied on and used the low bid of a roofing subcontractor to obtain a renovation contract. Following award of the prime contract, the contractor sent a written subcontract to the roofer. The roofer informed the contractor that it was not going to perform because it was not certified to apply the specified roofing material. As a result, the contractor hired another roofer to do the work at a cost $18,565 higher than the roofer's bid. The court held that the contractor was entitled to recover this $18,565 in damages because it had relied on the roofer's bid to its detriment and that the oral bid was enforceable.[43]

[B] Prime Not Bound to Accept Sub's Bid

Interestingly, although subcontractors may be responsible to general contractors that rely on their bid, general contractors are rarely obligated to enter into subcontracts with subcontractors whose prices they use.[44] The law is different for general contractors than for subcontractors. As stated above, a sub or supplier who quotes a price to a general contractor may be bound to that price even before the offer is accepted. The sub cannot back out of its bid even if there is an error, since the general contractor has relied on the quote in making a bid to the owner. On the other hand, a general contractor is generally *not bound* to a subcontractor merely because he used the sub's bid in making a quote to the owner. The courts say that the mere use of a bid does not equal acceptance of the offer. This has resulted in the industry problem of "bid shopping" under which a general contractor uses sub A's price to get the job, then "shops" the price around to subs B and C after the owner awards the contract to get a better price and maximize the profit margin. Laws have been passed in some jurisdictions that require a general contractor to list his subs when bidding and prevent any changes in the list absent proper justification.

[C] Promissory Estoppel

Most of these subcontractor bid cases rely on Restatement (Second) of Contracts § 90 as setting forth the law of promissory estoppel. The three elements for promissory estoppel are: (1) a promise; (2) the promisee's reliance; and (3) an injustice which can be avoided only by enforcement of the promise. Courts have held that contractors who rely on sub prices meet all three elements.[45] However, courts do not approve of bid shopping. In one federal case, the court said that "a contractor should not be free to delay acceptance after he has been awarded the

---

[43] Branco Enterprises, Inc. v. Delta Roofing, Inc., 886 S.W.2d 157 (Mo. App. S.D. 1994).

[44] *See* Williams v. Favret, 161 F.2d 822 (5th Cir. 1947); Southern Cal. Acoustics Co., Inc. v. C.V. Holder, Inc., 456 P.2d 975 (Cal. 1969); Mitchell v. Siquieros, 582 P.2d 1074, 1077 (Idaho 1978) (citing nine other cases for this proposition). *See also* Electrical Constr. v. Maeda Pac. Corp., 764 F.2d 619, 621 (9th Cir. 1985).

[45] Debron Corp. v. National Homes Constr. Corp., 493 F.2d 352 (8th Cir. 1974).

general contract in hopes of getting a better offer." The court said that if there had been such delay, or the contractor attempted to reopen bargaining with the sub, the result would have been different.[46]

[D] Changing Prime Contractors May Waive Bid Prices

As discussed in Chapter 13, many design firms cannot get bonding for design-build projects, so they either pair up with a bonded contractor through a joint venture or let the contractor be the prime and the design firm act as a subcontractor. A design firm should check out any bonding requirements well in advance of getting subcontractor prices. If a subcontractor gives a price to the design firm as prime, but the contract is later held by a general contractor, the price may be no good. That was the result in a 1988 Missouri case, *A.L. Huber & Son, Inc. v. Jim Robertson Plumbing, Inc.*[47] In that case, the court found that a plumbing subcontractor *was not* bound by its bid under an estoppel theory. The facts are unique to that case. In *Huber*, the subcontractor made its bid to "Clevenger Homes" as general contractor. When Clevenger couldn't get a bond, it entered into a joint venture with Huber, where Huber was to act as the general contractor. Huber sued to enforce the bid claiming "promissory estoppel". The court found that there was no promissory estoppel nor "detrimental reliance" since the subcontractor's bid was made to Clevenger, not Huber. As a result, the subcontractor could not have foreseen that either Huber or the joint venture would rely on the bid. One should keep this in mind when forming joint ventures and in getting subcontractor pricing. Be sure that subcontractors and suppliers give their bids to the company that will hold the prime contract.

§ 21.11 DEALING WITH MULTIPLE SUBCONTRACTORS

Coordinating a group of 20 subcontractors, spread out over a several acre site can be the biggest risk and challenge for a design-build contractor. Architects usually work with just two or three consultants with whom they have a relationship, who work at their own offices and do not need to be monitored closely. But when architects become contractors, they may be dealing with dozens of low-bidding subcontractors with whom they have never worked before, and who need careful supervision to make sure the work is done right and stays on schedule. The risks begin when getting bids, as addressed in § 21.10. The next risk is when the architect gives the owner its list of proposed subcontractors.

[A] Owner Objection to Proposed Subs and Suppliers

A problem can arise when the owner rejects a proposed subcontractor or supplier, on whose price the design-builder has relied. A common provision in

---

[46] *See also* Drennan v. Star Paving, 333 P.2d 757, 760 (Cal. banc 1958).
[47] 760 S.W.2d 496 (Mo. App. 1988).

public and in some private construction contracts is one allowing the owner to object to any subcontractor hired by the contractor and forcing the contractor to replace any such subcontractor upon request. The owner pays the difference in price if the replacement subcontractor costs more than the one the contractor wanted originally to use. *See, e.g.*, AIA A201 General Conditions, ¶5.2 (1997 edition). If that is not in the proposed design-build contract, a clause like this one should be added:

> If the Owner has reasonable objection to a proposed Subcontractor or Design Consultant, the Design-Builder shall propose another to whom the Owner has no reasonable objection. The Contract Sum and Contract Time shall be increased or decreased by the difference, if any, occasioned by such substitution and a Change Order shall be issued before commencement of the affected work.

The problem with replacing subcontractors on a design-build job is magnified, since the design-builder may have delegated the design of a part of the project to that subcontractor, using its price and design ideas in preparing the proposal. If a contractor has to replace the subcontractor, he may have to pay a fee for the design work the subcontractor performed and get the copyrights to whatever designs the subcontractor prepared so that he can use them. The replacement subcontractor may have to have the design completely redrawn, however, so that the plans can be sealed by a new engineer. State licensing laws prohibit one engineer sealing drawings not done by that engineer or under his or her direct supervision. This adds to the cost of replacement of a subcontractor.

If a design firm is acting as a subcontractor on a design-build project, a clause letting the owner object to proposed subcontractors literally could give the owner the right to have the design firm replaced. Since most design-build selections are based in large part on qualifications of the entire team, especially the design team, this risk seems to be a small one. Nonetheless, one should watch for those clauses and modify them accordingly to exempt the primary design subcontractors, assuring that the owner cannot cut out a design firm whose efforts and qualifications helped to obtain the design-build contract.

[B] Subcontractor Delays and Defaults

As mentioned above, architects are used to working with their favorite subconsultants. Most design firms have a "stable" of consulting engineers that they have worked with over the years and with whom they have a relationship. However, when acting as a general contractor, an architect will be getting bids from dozens of subcontractors and suppliers with whom there is no relationship. Due to price constraints, the architect will often have to accept the low bidder to meet budget constraints. There is always the risk that the subcontractor will: (1) neglect the job for one with his "favorite" general contractor; (2) default in performance, and delay the job; (3) perform defective work; (4) not pay his bills; or

(5) go broke and fail to finish the work. These risks are not covered by any insurance policy and will require careful screening of subs, and possibly bonding of subs the architect is unfamiliar with (though this will increase their prices and the architect's total cost). Subcontractors delays and defaults may be the most risky aspects of becoming a contractor.

One of the biggest problems in construction is subcontractors who do not show up for work. "Architects aren't used to dealing with these day-to-day phone calls to subcontractors," says Larry Maxfield, AIA. Unbeknownst to the prime contractor, a subcontractor may have bid on too much work, resulting in stretching the capabilities of the company over too many jobs at the same time. An architect will need to watch for patterns, such as a sub who shows up only part days and leaves to finish other jobs, or who does not show up until noon, or who is missing from the site even on good weather days. This may mean that the sub is spreading one crew over two or three jobs, responding to whomever yells the loudest or who pays the most.

A design-builder should be prepared to take action by: (1) having a written subcontract that requires the subs to follow the schedule; (2) giving each subcontractor a copy of the project schedule; (3) having subcontracts that include short time notice provisions allowing you to supplement or replace a subcontractor who fails to show up according to schedule. Some subcontracts require two separate seven-day notices before a subcontractor can be terminated or supplemented. Most contractors find this unworkable. If the sub shows up on day seven, then does not come back for five days, the notice process has to start over. Waiting for 14 days is too long when one trade may be holding up other trades and delaying critical path work. Many contractors have short two-day, or three-day notices, that allow the contractor to supplement, replace, or terminate any subcontractor who is not performing on schedule or consistently failing to follow the contract documents. Standard form subcontracts may need to be modified to tighten up default provisions and reduce delays.

[C] Liens by Unpaid Subs, Suppliers, and Lower-Tier Entities

Although architects and engineers have lien rights in many states, they are not used nearly as often as they are by construction subcontractors and suppliers. Lien waivers become an important tool to assure that subs have paid their bills. If a lien is filed by a lower-tier sub or supplier, the prime contractor will usually be required to remove the lien (by paying it or "bonding it" off). If the lien is not removed, the contract or lien statutes may require that the contractor defend the owner against a lien claim. The owner will often hold retainage to assure that the lien is paid and the suit properly defended. Says one architect-contractor, "I refuse to pay a single subcontractor unless he walks into my office and signs a lien waiver. No exceptions." Each pay period, this firm has the subcontractors send an authorized representative to pick up the check, but not until the lien waiver is executed.

As a "prime contractor," the design-builder is required in some states to give the owner a specific written notice, or file a notice with the county or the court in order to have any lien rights for itself.[48] Giving this notice is a condition precedent to the creation, existence, or validity of any mechanic's lien in favor of a prime contractor. One should check the law of the state in which the project is located for specific details, since the requirements and timing for filing or giving lien notices varies greatly. Failure to comply can result in a waiver of the design-builder's lien rights.

§21.12 JOB SITE SAFETY

Site safety is addressed in more detail in Chapter 22. However, there may actually be *less* liability when the architect or engineer is the prime contractor under design-build. The reason that so many claims are brought against design professionals by injured workers is that the worker cannot sue the general contractor under state workers' compensation laws. Under state law, for example, if an employee of a subcontractor is injured, he can recover under the workers' compensation laws. These claims are usually insured and covered by the subcontractor's insurance policy. The law then provides that the general contractor is immune from further suit by the worker or his family as a "statutory employer." Since the worker cannot sue the contractor, he often sues the architect or engineer. Suits increased to the point that the legislatures in several states passed laws providing immunity to the design professional from such claims except when: (1) liability for site safety is specifically assumed by contract; or (2) the accident was the result of a design error.[49]

However, when acting as the prime design-build contractor the architect/engineer would be entitled to the same immunity from claims that a general contractor has. Legislation may need to be modified to clear up the conflict between the design professional exceptions and the general contractor immunity. This still leaves open the possibility of an OSHA claim and fines for unsafe conditions on the job site. Such fines are not covered by either professional liability or general liability insurance. Recent cases have split on whether an engineer or construction management can be liable under OSHA laws.[50] However, there is no question that a design professional acting as the prime contractor can be liable under OSHA, thereby increasing the risk for site safety.

§21.13 UTILITY RELOCATIONS/DIFFERING SITE CONDITIONS

Construction can come to a complete halt if site conditions are encountered that were not anticipated. More and more, owners attempt to place total responsi-

---

[48] *See, e.g.*, R.S. Mo. § 429.012.

[49] *See, e.g.*, R.S. Mo. 287.150.5; K.S.A. § 44-501; Okla. St. § 12.

[50] Secretary of Labor v. Foit Albert Assoc., 1997 WL 197016, where AIA contracts were used, design firm held not liable for safety. However, in Secretary of Labor v. CH2M Hill Central, Inc., 1997 WL 197011, where the engineer acted as construction manager and had authority to direct the work of the contractor, review safety programs and approve them, the engineer was initially held liable under OSHA regulations. That decision was later reversed in CH2M Hill, Inc. v. Herman, 192 F.3d 711, 713 (7th Cir. 1999).

bility for site conditions on the contractor, using disclaimers and statements that require the contractor to acknowledge that the contractor has examined the site, satisfied himself of the conditions both above and below the ground, and agrees to waive any claim for actual conditions that might be encountered. Such conditions can be the discovery of a buried utility that was not shown on existing surveys, or the discovery of rock at a higher elevation than expected based on soil borings. Since owners look to the design-build contractor for acceptance of all risks, some owners will expect the contractor to bear those types of risks. The problem with this is that contractors, whether design-build or traditional, try to price their jobs competitively, based on foreseeable costs. If the borings show rock at a specific elevation, or no utility lines in the way of the construction, the contractor does not put contingent money into the contract for that. If, however, the owner puts contract language in the agreement that bars any claim for unknown or differing site conditions, then contractors have to put sufficient contingency figures in the bid to cover the potential costs of an unknown condition. If the conditions are as expected, the contractor reaps a windfall by keeping the contingency.

Even in design-build, owners need to understand that some risks must be shared. The better approach seems to be to permit the contractor to rely on surveys or soils reports and for the owner to pay for verified conditions that come up that could not have been expected. Many contracts contain a "rock clause" which includes a unit price for rock removal. This way the owner only pays for whatever rock is encountered—no more, no less—and the contractor does not have to put a large contingency in its price, nor "bet the farm" by taking all the risk of the unknown. It is, after all, the owner's rock that is being removed and the owner's utilities that have to be relocated, and there is no benefit to the contractor in doing this work without compensation.

One should be careful of contract clauses, however, that place total risk on the contractor with no recourse for differing conditions that might be encountered. For example, in a 1993 federal court case, the bid documents stated that:

> LOCAL CONDITIONS. Each bidder shall visit the site of the work and thoroughly inform himself relative to construction hazards and procedure, labor, and all other conditions and factors, local and otherwise, which would affect the prosecution and completion of the work and the cost thereof. . . . It must be understood and agreed that all such factors have been properly investigated and considered in the preparation of every proposal submitted, as there will be no subsequent financial adjustment, to any contract awarded thereunder, which is based on the lack of such prior information or its effect on the cost of the work.
>
> SUBSURFACE CONDITIONS. The determination of the character of subsurface materials . . . shall be each bidder's responsibility. Borings, field testing, and laboratory tests have been performed for the project design. This information has been bound separately and is issued with these specifications. . . . Logs of test borings may not be indicative of all subsurface conditions that may be encountered.[51]

[51]Green Const. Co. v. Kansas Power & Light Co., 1 F.3d 1005 (10th Cir. 1993).

The court held that under this contract, the risk of uncertainty as to site and soil conditions was placed on the contractor. The contractor's claim for over $1.9 million due to differing soil conditions was denied by the court with the court holding that "[w]hen a contract contains a site inspection clause, it places a duty on the contractor to exercise professional skill in inspecting the site and estimating the cost of work. . . . Thus, [the contractor] is not entitled to additional compensation merely because the project was more expensive due to unexpected soil moisture."[52] This size of cost overrun would break many contractors, despite the risk assumed by contract. This is a risk that is best shared with the owner.

§21.14 INCREASED ADA LIABILITY FOR THOSE WHO "DESIGN AND CONSTRUCT"

One area where risk is definitely greater with design-build is exposure to claims under the Americans with Disabilities Act (ADA).[53] Since the enactment of the ADA, there have been several court cases on the issue of whether architects and contractors can be held directly liable for violations of this civil rights law. Architects and contractors have defended themselves by claiming that they: (1) are not liable under the ADA because they do not typically fall within the classes of potential parties mentioned in section 302 of the Act; and, (2) are not liable under the ADA because they do not typically handle both the "design and construction" of the construction project. Of course, design-build would impact this second defense.

There has been a split in the courts on the issue of architect and contractor liability under the ADA. The cases focus on two provisions of the ADA that may create liability for architects and contractors.[54] First, section 302(a) of the ADA provides:

> *(a) General rule.* No individual shall be discriminated against on the basis of disability in the full and equal enjoyment of the goods, services, facilities, privileges, advantages, or accommodations of any place of public accommodation by any person who owns, leases (or leases to), or operates a place of public accommodation. 42 U.S.C. §12182 (a).

Architects and contractors do not typically "own, lease or operate" places of public accommodation, but rather, their client owns and operates the buildings they design and construct. As a result, some courts have concluded that section 302(a) of the ADA, standing alone, does not provide the necessary grounds for ADA liability against an architect or contractor.[55]

---

[52] *See also* Air Cooling & Energy, Inc. v. Midwestern Const. Co., 602 S.W.2d 926 (Mo. Ct. App. W.D. 1980) (upholding a subsurface conditions disclaimer and barring a claim for differing site conditions).

[53] Thanks to Eddie James, Esq., for his assistance in researching and writing this section.

[54] *See* Johanson v. Huizenga Holdings, Inc., 963 F. Supp. 1175 (S.D. Fla. 1997).

[55] *See* Paralyzed Veterans of America v. Ellerbe Becket, 945 F. Supp. 1, 2 (D.D.C. 1996).

Rather than relying on section 302(a) of the ADA, the Department of Justice and some disability rights groups have focused on another provision in the ADA as the basis for the liability for architects and contractors. Section 303(a) of the ADA provides:

> *(a) Application of term.* Except as provided in subsection (b), as applied to public accommodations and commercial facilities, discrimination for purposes of section 302(a) includes—(1) a failure to *design and construct facilities* for first occupancy later than 30 months after the date of enactment of this Act [enacted July 26, 1990] that are readily accessible to and usable by individuals with disabilities, except where an entity can demonstrate that it is structurally impracticable to meet the requirements of such subsection in accordance with standards set forth or incorporated by reference in regulations issued under this title. 42 U.S.C. §12183 (a).

Those courts finding architects or contractors liable under the ADA have relied on the "design and construct" language of section 303(a) to find ADA obligations.[56] The two sections are linked together. Section 303, by its terms, only defines discrimination "for purposes of section 302 (a)." Liability under section 302, however, is limited to owners, lessors, lessees, or operators. Therefore, it has been argued that architects and contractors cannot be held liable under the ADA since they do not fall into one of the section 302 categories. This argument has been a successful defense in several ADA cases.[57] If this defense is not accepted by the court, then those involved in design-build may have more exposure than their peers who do not both "design and construct."

The second defense to ADA liability most commonly used by architects and contractors is based upon the "design and construct" language of section 303. In a 1996 ADA case, the court dismissed a suit against an architectural firm, stating that "the phrase 'design and construct' is distinctly conjunctive" and that the architectural firm could not be liable under the ADA since it only *designed* the pro-

[56] *See, e.g., Johanson*, 963 F. Supp. at 1177-1178; United States v. Ellerbe Becket, Inc., 976 F. Supp. 1262, 1266 (D. Minn. 1997).

[57] *See Paralyzed Veterans*, 945 F. Supp.; United States v. Days Inns of Am., Inc., 22 F. Supp. 2d 612, 615 (E.D. Ky. 1998) (holding that it is "apparent . . . that Congress intended to limit liability under § 303 to owners, operators, lessors, and lessees."); United States v. Days Inns of America, Inc., 8 Am. Disab. Cas. (BNA) 491 (E.D. Ca. 1998) ("reading most consistent with Congressional intent and the text and design of Title III is that § 302(a) describes the parties liable for both § 302 and § 303."). *But see* United States v. Days Inns of America, 151 F.3d 822, 825 (8th Cir. 1998), *cert. denied*, 119 S. Ct. 1249, 143 L. Ed. 2d 346, 67 U.S.L.W. 3585 (1999), which rejected this defense. The Eighth Circuit Court of Appeals focused on the difference between the "public accommodation" language of section 302 and the "commercial facilities" language of section 303 and held that the defense articulated above would "leave no entity liable for violations of the new construction accessibility standards for buildings which are commercial facilities only." 151 F.3d at 822. The defense, if allowed, would create a gap in ADA coverage and "would render meaningless section 303's inclusion of commercial facilities." *Id.* at 825.

ject at issue and did not "design *and* construct" the project.[58] While this case was good news for traditional design practitioners, it pointed out increased exposure for design-build firms. In another ADA case, the Eighth Circuit Court of Appeals stated: "we apply conjunctively the 'design and construct' language of section 303. Therefore, in order for [the defendant] to be liable under section 303, [the defendant] must have been responsible for the design and construction of the hotel."[59] The court added that "any party substantially participating in the design and construction of a facility shares the responsibility if the facility violates ADA's standards" and "[w]e conclude that to bear responsibility for an inaccessible facility under section 303, a party must possess a significant degree of control over the final design and construction of a facility."[60]

Architects, engineers and contractors engaged in design-build may still be able to argue that they do not "own, lease or operate" the building and, therefore, are not directly liable under the ADA. But, if they cannot get past that issue, they will be less likely to be successful asserting the "design and construct" defense. Design-build firms that provide turnkey services to own and operate a facility will have liability under sections 302 and 303 of the ADA since none of the exceptions mentioned above would apply to them as compared to architects and contractors practicing in a more traditional role.

---

[58] *Paralyzed Veterans*, 945 F. Supp. at 2.

[59] U.S. v. Days Inns of Am., 151 F.3d 822, 825 n. 2 (8th Cir. 1998) (citing *Paralyzed Veterans*, 945 F. Supp. at 2.

[60] *Days Inns of America*, 151 F.3d at 826.

CHAPTER 22

JOB SITE SAFETY

§22.01 SHIFTING OF TRADITIONAL RESPONSIBILITY

With the title of lead contractor comes a new responsibility, one that design professionals traditionally attempt to stay far away from—construction site safety. Most design professionals have had no training in job site safety either in school or in practice. Site safety is placed solely on the contractor by standard American Institute of Architects (AIA) construction contracts. As a result, the responsibilities of a contractor dealing with trenching, scaffolding, fall protection, and safety data sheets are completely foreign to many architects. Becoming a prime design-build contractor means that design professionals will have to learn about their new responsibilities for site safety and develop training programs for staff to help promote safety on the construction site. This is what the Occupational Safety and Health Administration (OSHA) would like to see happen regardless of design-build, according to H. Berrien Zettler, deputy director of OSHA's construction division. "OSHA believes much could be done to improve safety and health on the worksite if we could get designers and engineers and architects to pay attention from the beginning and design into blue-prints measures that would lead to a safer workplace, to think of the construction process and design for that as well as for end use."[1]

The mantra of design-build is "faster-cheaper-better." An important consideration is not sacrificing job site safety when building faster. Design-build contractors boast of being innovative problem solvers and there is no reason those innovative ideas cannot be used to make job sites safer for workers as well. Safety should never be compromised in order to complete a job faster. This chapter will discuss what design professionals need to know to manage site safety and control risk in this area.

§22.02 CONSTRUCTION—THE MOST HAZARDOUS INDUSTRY

Design professionals going into design-build as prime contractor have to realize that they are stepping into a high risk business in terms of worker safety. The U.S. Department of Labor issues annual statistics on workplace injuries and deaths in the country. The construction industry consistently tops the list as the most hazardous industry, with over 517,000 workers injured and 1,171 deaths in 1998.[2] Compare this to annual workplace injuries for architects, which total about 150 per year.[3] The construction industry accounts for 21.5 percent of all

[1] *Undeserved Attention? Designers Say OSHA Is Unfairly Expanding Safety Responsibility*, Engineering News Record, June 21, 1999, p. 28.

[2] U.S. Department of Labor, Bureau of Labor Statistics, *Incidence Rates of Non-Fatal Occupational Injuries and Fatal Occupational Injuries by Industry*, 1998.

[3] Typical injuries for architects and engineers are sprains (43.4%), bruises (8.1%), and carpal tunnel syndrome (5.5%). Injuries for contractors are usually more severe involving cuts and punctures (12.5%), fractures (10.3%), back pain (6.5%) and only 0.6% with carpal tunnel syndrome. U.S. Department of Labor, *Occupational Injuries and Illnesses: Count, Rates, and Characteristics*, 1997 (published Oct. 1999).

work-related fatalities, but only 5 percent of the nation's work force, showing that construction workers face a disproportionate risk of death on the job when compared to other industries. Falls are the number one cause of fatal accidents on job sites, accounting for one-third of all construction deaths. Electricians, roofers, plumbers, iron workers, and masons are at the highest risk according to Department of Labor statistics. Contractors often celebrate a job that is free of serious injury with a "topping off" ceremony, by placing an evergreen tree atop the structure. Others post signs at the site boasting the number of days free of deaths or serious injury. With the industry's high accident rate, an injury-free work site is something to brag about. It also helps remind employees and subcontractors of the importance of site safety.

One of the suspected causes of construction injuries is a high rate of drug and alcohol use among construction workers. A report by the U.S. Department of Health and Human Services in 1996 showed that construction supervisors, laborers, plumbers, and painters rank among the occupations with the highest levels of illicit drug and heavy alcohol use.[4] The study of about 29,000 people annually over a three-year period showed a decrease in use of drugs and alcohol, but still found that 11.6 percent of construction workers used illicit drugs within the previous month, well above the average for all industries. As for alcohol, the study showed that laborers, roofers, masons, and tilesetters each reported 20 percent were heavy alcohol users (five or more drinks, on five or more occasions in last 30 days). What these studies show is that the most dangerous industry in America employs workers who may be under the influence of drugs or alcohol while on the job. This is a dangerous and often fatal mix. These statistics also reinforce the need for a thorough site safety program to help prevent accidents and injuries.

§ 22.03 THE OCCUPATIONAL SAFETY AND HEALTH ADMINISTRATION (OSHA)

OSHA is the federal agency charged with regulation of industry in the United States and preventing workplace injuries. The Occupational Health and Safety Act of 1970 was passed to ensure that workers in America had safe working conditions. While Congress knows that it cannot ensure a risk-free workplace, the OSHA law attempts to eliminate or greatly reduce injuries and deaths on the job. The construction industry is among the most highly regulated industries in the country. Regulations range from covering how many hours a worker can use a certain tool, to how much a laborer can lift without mechanical assistance, to the bracing of trenches. Some experts estimate that regulations add 20 percent to 35 percent to the cost of construction.

OSHA regulations place responsibility for site safety on the prime contractor, whether it is a design firm, a design-build firm, or a traditional contractor. While OSHA has had mixed results when attempting to cite architects and engi-

[4] *Drug Problems Persist*, Engineering News Record, May 13, 1996, p. 10.

neers who perform only traditional design services, it has cited such firms when they undertake design-build. For example, one engineering firm was cited in connection with a December 1998 explosion at a fiber-optic cable plant in St. Cloud, Minnesota even though the accident involved a subcontractor. OSHA cited the engineering firm that was responsible for construction under a turnkey contract. "The more you take responsibility like that, your liabilities increase," said a spokesman for the cited firm.[5]

The first place to start learning about OSHA requirements is to get a copy of the agency's regulations. Many of the OSHA regulations are available on its Web site at <http://www.osha.gov> and can be downloaded from there. A hard copy of the OSHA construction industry regulations, 29 CFR Part 1926, can be ordered direct from the Government Printing Office at (202) 512-1800 or by mail at: Superintendent of Documents, U.S. Government Printing Office, P.O. Box 371954, Pittsburgh, PA 15250-7954.

The OSHA regulations can also be purchased from publishers such as Construction Book Express, (800)253-0541, or Building News, (888)264-2665. The Associated General Contractors of America (AGC) sells the complete version of OSHA regulations, 29 C.F.R. Part 1926, as their publication No. 125 (1999), available through local chapters for $35.20 (members) or $52.80 (non-members), or by calling the AGC at (800)AGC-1767, Publications Department. AGC's Publications and Videos catalog has an entire section on safety materials dealing with OSHA compliance, including manuals, training courses and dozens of video tapes on safety precautions and procedures. There is also a product called "OSHA Compliance Software," a package that can be installed on a PC using standard Windows to print off OSHA-required reports, OSHA 200 Log and Summary, 52 safety training talks, plus loss control programs to track a company's accident frequency rate, lost days, OSHA lost days, etc. The software sells for about $500 through some of the publishers mentioned above. Design firms getting into design-build as prime contractor need to educate their staff about these safety regulations.

§22.04 COMMON RISKS

Injuries can come in an almost infinite variety of sources on a construction site and can include nail-gun misfirings, eye injuries, back strain, hearing loss, and musculoskeletal injuries such as tendonitis, carpal and tarsal tunnel syndromes, and hand-arm vibration syndrome. In addition, exposure to environmental toxins and irritants such as wood or silica dust, asbestos, paint fumes, formaldehyde, or solvents can cause injury. The majority of construction fatalities are the result of falls, cave-ins or deadly contact, such as electrocution, falling objects, or toxic fumes. Cave-ins from unshored or improperly shored trenches and excavations are a frequent cause of injury and death.

---

[5] *Undeserved Attention?* note 1, above, at, pp. 28-29.

§ 22.05 ARCHITECT AND ENGINEER IMMUNITY

As a result of the high number of claims made against design professionals by injured construction workers, many state legislatures have passed laws granting limited immunity to architects and engineers from such claims. A typical statute is this one from Missouri that states:

> No construction design professional who is retained to perform professional services on a construction project or any employee of a construction design professional who is assisting or representing the construction design professional in the performance of professional services on the site of the construction project shall be liable for any injuries resulting from the employer's failure to comply with safety standards on a construction project for which compensation is recoverable under the workers' compensation law, unless responsibility for safety practices is specifically assumed by contract. The immunity provided by this subsection to any construction design professional shall not apply to the negligent preparation of design plans or specifications.[6]

Similar laws have been passed in Kansas, South Dakota, Oklahoma, and several other states.[7] These statutes grant immunity unless the accident is due to a design error or the responsibility for safety has been assumed by the design firm by a contract. South Dakota's statute also permits liability if the design professional assumes safety responsibilities "by that person's actual conduct." Therefore, when acting in the traditional role of project architect, there is immunity from liability for job site accidents. Even when acting as a subcontractor to a design-build contractor the immunity would still apply in most cases. However, when a design firm acts as the prime design-build contractor, whether by holding the contract with the owner alone or through a joint venture with a contractor, the statutes would not normally apply, since the prime contract will always place responsibility for site safety on the prime contractor, as would OSHA's regulations.

§ 22.06 WORKERS' COMPENSATION AND THE EXPERIENCE MODIFICATION RATING (EMR)

Workers compensation insurance rates reached a crisis stage in the United States when rates doubled between 1980 and 1987, and continued to climb thereafter. As one might expect, a high-risk industry like construction has high rates of workers' compensation insurance. Construction contractors, depending on their experience modification rate (EMR), pay anywhere from 5 percent to 20 percent of their direct labor cost for workers' compensation insurance. This is usually more than a contractor's profit percentage. By implementing an effective safety program to reduce job site accidents, a contractor can influence its workers' compensation premium cost, resulting in safer job sites, lower insurance rates and increased profits.

---

[6] Mo. Rev. Stat. § 287.150.5.

[7] Kan. Stat. Ann. § 44-501(f); S.D. Cod. Laws § 36-18A-68; 85 Okla. Stat. § 12.

An EMR is a rating system used to figure workers' compensation insurance rates for various trades and industries. It is a multiplier used in computing the insurance premium using the past experience of the company to forecast future benefit payment to employees who may have insurance claims. The EMR is the ratio of actual losses to expected losses over a moving three-year period. The period ends one year before the EMR becomes effective, and each year thereafter the loss experience for the earliest year is dropped off, with the newest year added. For example, the 1999 EMR for a company would reflect loss experience for 1995, 1996 and 1997, so the data is not the most current reflection of the company's safety practices. By lowering the frequency and severity of construction accidents, a contractor can lower its EMR which, in turn, will lower the company's workers' compensation insurance premiums.

For a larger contractor, with more employees, one would expect more accidents and more workers' compensation claims simply due to the number of workers as compared to a smaller contractor. However, the workers' compensation insurers expect that accident claims should result in a fairly stable distribution for the number and severity of claims per number of employees. Insurers say that the most important figure is not the size of the claim, but the frequency of claims. For example, a contractor with one claim of $30,000 is a better risk than another contractor experiencing 15 accidents costing $2,000 each, though both total $30,000 for the year.

[A] How the EMR Is Computed

An EMR of 1.0 reflects the expected average for a particular trade or industry. A contractor with an EMR of 1.34 is one whose experience is 34 percent greater than average. An EMR of 0.9 would show a contractor with 10 percent less accidents and injuries than average. Studies have shown that typical EMR values for construction contractors range from 0.3 to 2.0. A contractor with a high EMR may be paying twice the premium cost for workers' compensation insurance than a competitor. This will add greatly to labor costs and eat into annual profits. The lower the EMR value, the more it reflects an individual contractor's ability to prevent accidents through workplace safety programs. Most contractors in a 1999 survey (78 percent) listed an experience modifier for workers' compensation claims between 0.50 and 0.99.[8]

[B] How the Workers' Compensation Insurance Premium Is Computed

A company's workers' compensation insurance premium (WCIP) is figured based on three factors: (1) the company's EMR; (2) manual rate; and (3) payroll units. The formula used is as follows:

$$\text{WCIP} = \text{EMR} \times \text{Manual Rate} \times \text{Payroll Units}$$

---

[8] CFMA's *Construction Industry Annual Financial Survey* (1999).

The manual rate is an insurance premium rate based on the type of work performed. Jobs are grouped into categories, called classification codes, which are typically four-digit numbers. In figuring a company's premium, each classification code has a premium rate based on worker accident claim experience in that type of work in that state for a given year. The rate also includes overhead expenses and is expressed in "dollars per one hundred dollars of direct payroll." The last part of the formula, Payroll Units, is derived by dividing an employer's total annual direct labor cost by 100. The difference in cost between two competing contractors can be dramatic. Consider this example:

Contractor A has an EMR of 0.6, while contractor B has an EMR of 1.4. The premium might be calculated like this:

Contractor A (EMR 0.6)	Contractor B (EMR 1.4)
$0.6 \times \$15.00 \times \frac{\$10{,}000{,}000}{\$100}$	$1.4 \times \$15.00 \times \frac{\$10{,}000{,}000}{\$100}$
Insurance Premium: $900,000	Insurance Premium: $2,100,000

The safety record of Contractor A resulted in a savings of $1.2 million per year. If the cost of labor and materials is roughly equal, the workers' compensation premium could make the difference between the low bidder and the high bidder and between a year-end bonus or a loss. As a result, there is an increased emphasis on site safety in construction with many companies implementing a "zero injury" goal. A safe job site can be a more profitable job site.

§ 22.07 OSHA FORM NO. 200

OSHA requires employers to record and report accident information on Form No. 200, the Log and Summary of Occupational Injuries and Illnesses. Each reportable case is to be recorded on the log within six workdays after the occurrence and the log must be maintained and kept for a period of five years following the calendar year covered by the log. Form No. 200 is completed annually and is required by federal regulations.[9] OSHA Form No. 200 includes:

- Number of fatalities;
- Number of injuries and illness involving lost workdays;
- Number of injuries and illness involving restricted workdays;
- Number of days away from work;
- Number of days of restricted work activity; and
- Number of injuries and illness without lost workdays.

[9] 29 C.F.R. Part 1904.

Some say that Form 200 figures provide a better picture of a contractor's safety record, since it is more current and accurate as to the contractor's recent safety record. A company's incidence rate can be computed quickly and easily by using a simple formula. Count the number of OSHA recordable cases for the year from the OSHA Form 200 log. Then, using payroll or other time records, figure the number of hours all employees actually worked. "Hours worked" should not include non-work time even though paid, such as vacation, sick leave, holidays, etc. When actual hours worked are not available for employees paid on salary, hours should be estimated on the basis of scheduled hours or an eight-hour workday. With this information, a company can then compute its incidence rate using this formula:

$$\frac{\text{(a) Number of injuries and illnesses} \times 200{,}000}{\text{(b) Employee hours worked}} = \text{incidence rate}$$

The 200,000 figure represents the equivalent of 100 employees working 40 hours per week, 50 weeks per year, and provides the standard base for the incidence rates.[10] An example might show the following for a company that had 11 recordable injuries during the year. If the total hours worked by all employees was 130,000, the incidence rate of 16.9 would be figured like this:

$$\frac{11 \times 200{,}000}{130{,}000} = 16.9$$

This means that the company had an incidence rate of 16.9 injuries per 100 full-time employees. To compute the incidence rate for the company's more serious, non-fatal cases—*i.e.*, those involving missed work days—assume that of the 11 injuries, 5 involved days away from work. Therefore, it experienced a rate of 7.7 days-away-from-work injuries per 100 full-time workers using the same formula:

$$\frac{5 \times 200{,}000}{130{,}000} = 7.7$$

Owners are taking a more active role in ensuring that the contractor they select has a low EMR and that its incident and frequency rating are also low. Safety has become a category by which many owners select their general contractors. Owners, general contractors, and even subcontractors are looking at OSHA 200 forms, incident rates, and EMRs to get a picture of what type of safety program a contractor or subcontractor has. According to Carl Heinlein, director of safety and health services for AGC, "On some jobs, you can't get on site unless your ex-

[10] Appendix C, *Occupational Injuries and Illnesses: Counts, Rates, and Characteristics*, 1997, U.S. Department of Labor, Bureau of Labor Statistics, Oct. 1999.

perience modification rating is typically around a rating of one or just above that, and some owners are looking for a number that's even lower."[11] A safe job site means fewer claims against the owner and more attention to detail.

§ 22.08 MATERIAL SAFETY DATA SHEETS (MSDS)

Also known as "haz com" for hazardous materials communications, OSHA regulations 29 C.F.R. §§ 1926.59 and 1910.1200 spell out the requirements for handling, storage, and transportation of toxic and hazardous substances in the workplace. This includes many common chemicals used in construction. OSHA first applied the haz com regulations to construction in March 1989. Failing to have Material Safety Data Sheets (MSDS) for each chemical at the work site is one of the most frequent reasons for an OSHA citation. In one study, it was reported that 30 percent of all OSHA citations issued in one year were due to violation of hazardous materials communication regulations, accounting for over 21,000 citations.[12] This accounted for more citations than scaffolds and electrical safety violations.

How does one know if there is a product on site that requires an MSDS? If it is tagged or labeled with any key words such as "danger," "caution," "flammable," "warning," etc., this is a signal to get an MSDS from the supplier, though it is the responsibility of suppliers and distributors to provide MSDS's for the products they are selling. If a firm does not get an MSDS with the shipment, it is the firm's responsibility to ask for one, and the request should be in writing. The firm should designate one person to coordinate the compliance efforts for all of the job sites. He or she can become familiar with what documentation should be on each job site, check periodically to see that the people in the field are complying, and help educate the staff on their obligations. Under the HazCom regulations, a company is required to have a written program on hazard communications. This is very important during an OSHA inspection since it is likely the first thing that an OSHA field inspector will ask to see when on the project site. There are many publications available from OSHA, the AGC, the Associated Builders and Contractors, Inc. (ABC) and other publishers on compliance with haz com regulations. It is important that the company is not left open to an OSHA fine for failure to have the right paperwork on site. Don't be one of next year's 21,000 haz com citations

§ 22.09 LESS LIABILITY THROUGH DESIGNER-LED DESIGN-BUILD?

Insurance statistics show that about 8 percent of claims made against architect and engineers on traditional projects are by injured construction workers.[13] With over 500,000 workers injured and over 1,000 killed on construction sites every

---

[11] *Jobsite Safety Begins with Attitude*, Colorado Construction, Dec. 1999, p. 40.

[12] *Penalty Shift Shakes OSHA's Image*, Engineering News Record, Nov. 25, 1991, p. 8.

[13] CNA Insurance Claims Statistics July 2000.

year, each of these accidents present a potential claim against the design professional by a worker or his family. As stated by one court, "[w]henever accidental death occurs, it is human nature to place blame."[14] Since the worker cannot sue his employer, due to the immunity granted contractors under workers' compensation laws, the worker or his survivors often strike out at others who participated in the project, blaming the design professional for either causing or not preventing the accident. The project architect and engineer are prime targets for these costly lawsuits, which often seek damages in the hundreds of thousands or millions of dollars, depending on the severity of the injury involved. These suits are covered by professional liability insurance, but not without spending the design firms' deductible (which can range from $5,000 to $1 million), plus taking hundreds of hours of time to defend.

While acting as a prime contractor on a design-build project does expose the A-E firm to OSHA liability, it actually results in *less liability* to injured workers. Under state workers' compensation laws, the employer (usually a subcontractor) as well as the prime contractor (called the "statutory employer") is immune from claims by the injured worker or his or her family due to a job site accident. The worker must generally seek his or her recovery from workers' compensation insurance as the sole remedy, and cannot sue the contractor for additional damages. As a result, when the A-E firm (architect/engineer) is the prime contractor, the workers' compensation laws protect the firm from claims, resulting in *less liability*, fewer lawsuits and no deductible payments.

§22.10 LIABILITY OF CONTRACTOR FOR DESIGN CONSULTANTS' SAFETY

Though most court cases dealing with site safety deal with injured construction workers, there are claims from time to time by design professionals injured on a job site. In one such case, *Aubrecht v. Acme Electric Corp.*,[15] an architect was injured when a scaffold he was using to inspect an interior wall soffit collapsed. The architect sued the general contractor who owned and had assembled the scaffolding. The court held that the architect was within the class of persons protected by New York's Scaffolding Act. As a result, the court denied the contractor's motion for summary judgment.

§22.11 SHORING AND SCAFFOLDING

As mentioned above, trench collapses result in many deaths each year and high dollar OSHA citations. Shoring failures and scaffolding falls account for a large percentage of serious construction accidents. Failure to follow OSHA regulations on trench shoring can result in accidents, injuries and substantial OSHA fines. For example, in 1991, OSHA cited a Birmingham, Alabama general con-

[14] CH2M Hill, Inc. v. Herman, 192 F.3d 711, 713 (7th Cir. 1999).

[15] 692 N.Y.S.2d 544 (App. Div. 1999).

tractor for 10 separate safety violations, each of which carried a $50,000 fine, for a total of a proposed $580,000 fine for an unsafe trench that was 12 feet deep for an underground utility installation, even though no injuries occurred.[16] A federal appeals court in 1999 upheld a $70,000 fine against a contractor whose workers did not have a safe means of egress from a trench even though nobody was injured.[17] OSHA regulations require that "a stairway, ladder ramp or other safe means of egress shall be located no more than 25 ft. . . . of lateral travel for employees."[18] In that case, the court found that the workers placed a ladder in the trench only after the OSHA inspector began videotaping the scene. In some cases, even criminal charges have resulted from trench collapses. In 1987, a Texas contractor pleaded no contest to a charge of criminally negligent homicide in connection with a trench cave-in that killed one worker. The company was fined $20,000 on the criminal charge. While all OSHA regulations are important to know, due to the frequency of incidents, someone in a design-build firm should be familiar with, and keep updated, sections 1926.650 to 1926.652 of the OSHA regulations entitled, Subpart P—Excavations.

Another high risk area on construction sites, and the source of numerous injuries each year, is scaffolding accidents. OSHA regulations cover this topic in 29 C.F.R. §§ 1926.450 to 1926.452, titled Subpart L—Scaffolds.

§ 22.12 OSHA INSPECTIONS

There are two types of OSHA inspections—programmed and unprogrammed. Programmed inspections are scheduled in advance by OSHA, while unprogrammed are unplanned, resulting when OSHA learns that a workplace may be in violation of safety regulations and standards. Unprogrammed inspections are usually given priority and are conducted as a result of: (1) some report of an imminent danger that threatens the employees at a particular site; (2) a fatality or catastrophe on the job site; (3) a formal complaint made to OSHA by an employee about a dangerous condition, or a referral from some other safety or health agency; or (4) a report in the news media about an accident. Programmed inspections are those planned by OSHA for a particular industry or employer. The inspections are normally scheduled in cycles established for each fiscal year. Under special emphasis programs, the agency will give particular emphasis to workplaces within a certain industry which it feels may pose a higher than average risk to employees. The construction industry has been designated as just such an industry.

In January 2000, OSHA announced that it planned to target the construction industry for increased inspections. The agency's budget for 2001 included $1 million to develop a survey that will indentify contractors with above-average injury rates. Those contractors with higher rates will be subject to "wall-to-wall" inspec-

[16] *OSHA Seeks $580,000 Over Injury-Free Site*, Engineering News Record, Sep. 16, 1991, p. 14.

[17] Dakota Underground, Inc. v. Secretary of Labor, 200 F.3d 564 (8th Cir. 2000).

[18] 29 C.F.R. §1926.651(c)(2).

tions once the agency analyzes the survey data, which will be collected in 2002. Prior to the year 2000, construction contractors were not included in OSHA's targeting programs because collecting data was difficult given the mobile nature of the industry. However, OSHA's $426 million budget for 2001 was 11.6 percent above the more than $381 million Congress appropriated for OSHA in 2000. The increased budget included hiring 63 additional inspectors.[19]

OSHA exempts some workplaces from scheduled inspections under its voluntary protection programs (VPP). The agency developed these programs in 1982 to encourage employers to develop their own comprehensive safety programs. To qualify, applicants must have lower than average accident rates, written safety programs, injury and exposure records available for OSHA review, inform their employees about the program and their rights under OSHA laws, and notify the agency annually of their injury rates. The program frees OSHA inspectors to visit job sites that are less likely to meet safety requirements.

One experimental safety program in Colorado for residential contractors, known as "Homesafe," requires contractors participating to voluntarily comply with certain conditions. In return, those contractors may operate without programmed OSHA inspections and can be cited or penalized only for serious hazards. They may also receive a discount on workers' compensation premiums by developing their own personalized safety programs. The goal of the program is to encourage both builders and OSHA to emphasize preventing serious injuries rather than reacting to citations for minor safety issues. During a three-year period under the program, one builder showed an 86 percent savings in workers' compensation claims.

§22.13 OSHA FINES AND CITATIONS

Following an OSHA inspection, a citation may be issued either immediately or delivered later by mail. The citation should describe the violations found during the inspection in detail, with reference to the particular standard alleged to have been violated, and it should set a deadline to correct (or "abate") the violation. Employers must post all such citations at or near locations where the violation occurred, and keep them posted for three working days or until the violations are corrected.[20] Failure to post the violation can result in civil penalties and a loss of the right to challenge the citation.[21]

OSHA violations fall into six different categories:

1. *De minimus*, conditions that violate a standard, but have no direct or immediate relationship to safety or health. OSHA inspectors are required to document such violations but not to include them on a citation.
2. *Other-than-serious*, conditions that probably would not cause serious injury or death, but would still affect safety on the site.

[19] *OSHA Will Target Contractor Injuries*, Kansas City Bus. J., Feb. 11, 2000, p. 23.

[20] 29 C.F.R. § 658(b); 29 C.F.R. §1903.16(b).

[21] 29 C.F.R. § 666(i).

3. *Serious*, conditions involving a substantial probability of death or serious physical harm.
4. *Repeat*, conditions substantially similar to ones noted in earlier citations involving the same employer.
5. *Willful*, violations involving intentional and knowing disregard of OSHA standards, in which the employer knew that a hazard existed and that it violated a standard, but failed to make a reasonable effort to correct it.
6. *Criminal/willful*, violations that were willful and resulted in the death of an employee.

OSHA may combine multiple instances of a violation of a standard in the same citation, so that a condition that was found at four different locations on the same site would result in one citation, with each instance listed. The penalties for an OSHA violation depend on the gravity of the alleged violation, the size of the business, and the employer's good faith in showing genuine efforts to comply with safety standards before the inspection, as well as during and after and, finally, the employer's history of previous violations. The size of the penalty can run into the tens of thousands of dollars.[22] These fines are not covered by any insurance policy and so a safety violation, as well as the legal fees incurred to defend against an OSHA charge, come right out of the firm's profits.[23] Again, a safe job site is a profitable job site.

OSHA has assessed sizeable fines against contractors in trenching accidents as part of a continuing crackdown on unsafe trenching practices. In 1997, a Florida trenching contractor was fined $448,000 and a Pennsylvania excavation contractor was assessed $230,400 in fines based on unsafe trenching operations.[24] The Atlanta West office of OSHA once proposed a $580,000 fine against a contractor for an unsafe trench that OSHA claimed endangered many workers, even though no injuries occurred.[25] In a Wisconsin crane collapse case, in which three workers were killed, OSHA assessed fines totaling $539,000 against three construction companies.[26] In 1989, OSHA cited Denver-based engineering firm CH2M Hill Central, Inc. for 46 willful safety violations and proposed a $460,000

---

[22] 29 C.F.R. § 666.

[23] The Federal government is the largest law firm in the nation. While the government uses its staff attorneys to handle its litigation, private contractors must retain their own legal counsel to fight an OSHA citation. Due to the cost of defending against an OSHA citation, a bill was introduced in Congress in 2000 that would help small contractors recover their legal fees when a safety violation is found to be without merit. The Fair Access to Indemnity and Reimbursement Act would allow construction companies with net worth not over $7 million and less than 100 employees to recover their attorneys' fees and other expenses if cleared of charges brought by OSHA or a labor violation filed by the National Labor Relations Board (NLRB). According to the Associated Builders and Contractors, many contractors settle meritless charges because it costs less than the legal fees.

[24] *Trench Crackdowns Continue*, Engineering News Record, Sept. 15, 1997, p. 16.

[25] *OSHA Seeks $580,000 Over Injury-Free Site*, Engineering News Record, Sept. 16, 1991, p. 14.

[26] Building Design & Construction, Feb. 2000, p. 9.

fine relating to a 1988 methane tunnel explosion that killed three supervisors. The engineer claimed that it should not be liable since the firm did not engage in construction work, exercise substantial supervision over the construction, or create the hazardous condition. The firm was later cleared of the charges.[27]

According to Berren Zettler, deputy director of construction of OSHA, a larger fine—sometimes as much as $100,000—may be assessed to a high profile contractor to get the attention of smaller contractors.[28] Zettler says that OSHA has traditionally had problems keeping tabs on smaller contractors who have a fatality rate that is disproportionately high. Short-term subcontractors, on and off the site in a short period, sometimes feel that it is unnecessary to promote safety on the site. But OSHA has gone beyond merely citing and fining general contractors for job site accidents and injuries. In two highly publicized court cases, OSHA went after engineering firms with citations and fines for unsafe site conditions. In *Secretary of Labor v. Foit Albert Assoc.*,[29] where AIA contracts were used, the OSHA Review Commission looked to the contract and to the nature of the site activities and held that the engineer had no duty or authority to direct the work. The fact that the engineer made its concerns known about safety violations on site did not expand its duties.

In the other case, *CH2M Hill, Inc. v. Herman*,[30] three workers died in an explosion during construction of a sewer system in Milwaukee. OSHA cited the engineering firm that had been hired by the city and imposed a series of fines. In the opinion issued by the Commission, where the engineer acted as construction manager on a wastewater treatment plant where three workers were killed in a methane gas explosion, the Commission found that the engineer had authority to direct the work of the contractor, review safety programs, and approve them.[30] The Commission found the level of supervision to be greater than a normal design professional and, as a result, the engineer was liable under OSHA regulations for fines totaling $200,000. In 1999, however, that decision was overturned, and the Seventh Circuit Court of Appeals ruled that the engineer did not exercise "substantial supervision" over construction.[31] As a result, there is still much confusion as to what amount of supervision will trigger OSHA liability for a design professional. The engineer's contract stated that its day-to-day inspections and construction management of the project would not cause the engineer "to be responsible for those duties and responsibilities which belong to the construction contractors and which include, but are not limited to, the contractors' responsibilities for the techniques and sequences of construction and the safety precautions incidental thereto." OSHA contended that the engineering firm functioned as an "employer" who was, therefore, responsible to "protect the

[27] *CH2M Hill Cleared in Tunnel Explosion Case*, Engineering News Record, Sept. 6, 1993, p. 6; CH2M Hill, Inc. v. Herman, 192 F. 3d 711 (7th Cir. 1999).

[28] *Jobsite Safety Begins with Attitude*, Colorado Construction, Dec. 1999, p. 40.

[29] 1997 WL 197016.

[30] 1997 WL 197011.

[31] CH2M Hill, Inc. v. Herman, 192 F.3d 711 (7th Cir. 1999).

employment and places of employment of each of his employees engaged in construction work" under 29 C.F.R. Section 1910.12(a). The Court of Appeals held that the contract language was controlling here and that "[t]o ignore the manner in which the parties distributed the burdens and benefits is contrary to our notion of contract law . . . [OSHA] cannot ignore the fact that parties to a contract bargained for one to maintain safety responsibilities and for the other to refrain from such responsibility." Finding no significant evidence that CH2M Hill exercised substantial supervision or control over safety activities, the court vacated the citation against the engineer.

When an architect or engineer goes from being a design firm to being a contractor, however, there is no need for OSHA to wrestle over whether the firm is a proper target for a safety citation. By taking on the construction responsibility as prime contractor or, possibly, even by a joint venture, the design professional opens its firm up to OSHA regulations and fines. These fines can run up to $7,000 for each violation or up to $70,000 if willful, and are often calculated based on each day the violation existed, which can run into the tens and hundreds of thousands of dollars.[32]

§ 22.14 CRIMINAL LIABILITY FOR UNSAFE CONDITIONS

Violation of OSHA regulations can result in criminal prosecutions against the design-build contractor if the violation is deemed to be willful. For example, in 1996 a company was indicted by a federal grand jury of two counts of violating 29 U.S.C. section 666(e), which makes it a criminal misdemeanor for an employer willfully to violate any OSHA regulation if that violation causes the death of an employee. The defendant was hired by the general contractor on a post office project to fabricate and erect the structural steel. Two iron workers were killed when over 60 structural steel members collapsed during construction.[33] Again, there is no insurance to assist in defending a criminal charge against a contractor.

§ 22.15 MULTI-EMPLOYER CITATION POLICY

Since 1975, federal courts have held that a general contractor, as well as its subcontractor, can be held liable for a safety hazard under what is called the "multi-employer doctrine."[34] The doctrine provides that "an employer at a multi-employer construction worksite is responsible . . . in the absence of exposure of its own employees, for any hazardous conditions which it creates or controls."[35] The doctrine is based on 29 U.S.C. § 654(a), which requires employers to provide to each employee a place of employment free from recognized hazards

[32] 29 U.S.C. § 666.

[33] U.S. v. Pitt-DesMoines, Inc., 970 F. Supp. 1346 (N.D. Ill. 1997).

[34] Brennan v. OSHRC, 513 F.2d 1032 (2d Cir. 1975).

[35] Anthony Crane Rental, Inc. v. Reich, 70 F.3d 1298, 1305 (D.C. Cir. 1995).

likely to cause death or serious physical harm, and to comply with all OSHA standards.

In a 1999 case, a general contractor was cited and fined for safety violations made by a subcontractor.[36] The contractor argued that the statute does not impose liability outside of the employer/employee relationship. However, OSHA argued that under its "multi-employer work site" doctrine, an employer who controls or creates a work site safety hazard may be liable even if the employees threatened by the hazard are solely employees of another contractor. The Court of Appeals found that the general contractor may be the only person on site with authority to compel compliance with OSHA safety standards and that the fine was proper even though the safety hazard was created by a subcontractor.

Whether OSHA cites a general contractor for a safety violation by a subcontractor depends on whether the contractor created, exposed, or corrected the problem and whether the contractor controls safety on the entire job site. In 1999, OSHA published a formal statement on its multi-employer citation policy.

[A] The 1999 OSHA Multi-Employer Citation Policy

In December 1999, OSHA issued Instruction No. CPL 2-0.124, the Multi-Employer Citation Policy, which suspended Chapter III.C.6 of the OSHA Field Inspection Reference Manual (CPL 2.103). The new Policy gives several examples of when multiple citations may be issued. OSHA defines a "Creating Employer" as one that caused a hazardous condition; an "Exposing Employer" is one whose own employees are exposed to a hazard; a "Correcting Employer" is one on the same job site as the exposing employer and who is responsible for correcting a hazard; and a "Controlling Employer" is one who has "general supervisory authority over the worksite, including the power to correct safety and health violations." The examples in the Policy show that OSHA expects the general contractor to make reasonable efforts to monitor safety compliance by subcontractors and that OSHA can cite a contractor for failure to correct safety violations not remedied by subcontractors. One example discusses a painting subcontractor who fails to comply with OSHA fall protection requirements, even though the general contractor points out the violation during each inspection. Since the general contractor is the "Controlling Employer" with general supervisory authority over the site, including the right of control over the painter, OSHA's Policy would be to cite the general contractor for failure to enforce safety compliance on the job site.

[B] Architects and Engineers as Controlling Employers

OSHA's 1999 Multi-Employer Citation Policy states that: "Architects, engineers, and other entities are controlling employers only if the breadth of their in-

[36] Universal Const. Co. v. Occupational Safety and Health Review Comm'n, 182 F.3d 726 (10th Cir. 1999).

volvement in a construction project is sufficient to bring them within the parameters discussed above." In an example given, the Policy discusses an architect with traditional duties with no authority over means of construction, but who does call safety hazards to the general contractor's attention during site inspections. OSHA says that in this role:

> [the architect's] responsibilities are very limited in light of numerous other administrative responsibilities necessary to complete the project. It is little more than a supplier of architectural services and conduit of information to [the Owner]. Its responsibilities are insufficient to confer control over the subcontractors and it did not exercise control over safety. The responsibilities is does have are insufficient to make it a controlling employer.

Merely pointing out safety violations did not make the architect a controlling employer in this example.

The Policy does not end there, however, and goes on to state that even if a party has no explicit contract rights with respect to safety, it can still be a controlling employer "if, in actual practice, it exercises broad control over subcontractors at the site." The Policy directs OSHA inspectors to consult the Regional Solicitor's office before issuing any citations in this situation.

§ 22.16 WHAT TO DO WHEN OSHA COMES CALLING

OSHA does not generally notify a contractor in advance of an inspection. In fact, federal law makes it a crime for any person to give advance notice of an inspection without the agency's authorization, with up to a $1,000 fine and a maximum jail term of six months.[37] So what should a firm expect when OSHA comes calling? A firm has certain rights if an OSHA inspector shows up on its job site. With limited resources, OSHA visits less than 10 percent of all workplaces each year, but the agency tends to focus on big employers. OSHA inspections are carried out by a Compliance Safety and Health Officer (CSHO).

[A] No Advance Warning Allowed

The purpose of not warning the employer is to preserve the element of surprise. As a result, the OSHA field inspector may show up on a job site unannounced with pen, paper, camera and tape measure. Under federal statutes, any person who gives advance notice of an OSHA inspection can be fined or imprisoned, or both.[38] The inspector may examine the equipment, materials, conditions and structures at the site, as well as question management and employees to determine whether the site is in compliance with the relevant safety and health standards. After talking to the superintendent and walking around the site, the

---

[37] 29 C.F.R. § 1903.6(c).

[38] 29 U.S.C. § 666.

inspector may take some notes and a few pictures, and then give the contractor an OSHA citation for unsafe practices. It is important for a firm to know its rights so that it knows its options.

[B] Warrant Can Be Demanded

By law, an employer has the right to insist that the OSHA inspector obtain a warrant prior to entering the job site and may even refuse entry without such a warrant. OSHA inspectors can also obtain a subpoena in order to obtain records. If the employer permits an inspection, but limits or interferes with the inspector, the OSHA inspector is directed to contact his supervisor as to whether the conduct constitutes a refusal to permit inspection. The OSHA Field Operations Manual states that in that event, the inspector is not to engage in argument, but is to leave the premises and immediately report the refusal to the OSHA supervisor. If OSHA is compelled to obtain a warrant, the inspection will take place after the warrant is served on the employer. Federal regulations permit OSHA to seek a warrant in advance of an inspection although the agency's policy is generally not to seek warrants without evidence that the employer is likely to refuse entry to the job site or business.[39] To obtain a warrant, OSHA must present a U.S. district judge or magistrate with evidence that a contractor is violating a safety standard. If a contractor refuses to permit an inspection even after being served with a warrant, OSHA can seek to have the contractor held in contempt of court in a criminal proceeding.

[C] Inspection Procedures

The OSHA manual for its field inspectors states that inspections are to be made during regular business hours. The inspector is supposed to contact the person in charge at the job site, usually the general contractor's superintendent, and present his credentials. If the person in charge is not present, the inspector is to attempt to contact the employer by telephone and request the presence of a management official. However, the inspection is not to be delayed to await the arrival of a supervisor and the OSHA manual says that the delay should not exceed one hour. On a multi-employer site, such as a construction site, the inspector may request the general contractor to identify all subcontractors on the site, with names of the individuals in charge of their operations. The inspector may then request that all such persons be contacted and asked to assemble in the general contractor's office or job trailer.

The inspector will conduct an "opening conference" with the contractor to show credentials and discuss the purpose of the inspection and to inspect records that the contractor is required to maintain. Among the records to be inspected are the material safety data sheets (MSDS). The inspector will also check for compliance with posting requirements. The inspector will then conduct a "walkaround,"

[39] 29 C.F.R. §1903.4.

in which he examines the actual working conditions at the site and evaluates the contractor's safety training and enforcement on the site. Violations of OSHA standards are to be noted on a worksheet. Conditions of the job site are to be noted by the inspector upon arrival, as well as any changes that occur during the time he is there. If there is some imminent danger, the inspector is to get to the location of the dangerous situation as quickly as possible. If there has been an accident or fatality, the inspector is to inform the employer that an investigation will be conducted and that extensive interviews of witnesses will be needed.

The employer (contractor) can designate a "walkaround representative" to accompany the OSHA inspector during the site inspection. The contractor's representative is not to interfere with a full and orderly inspection of the job site.[40] The inspector may use a tape recorder during the inspection to preserve statements of witnesses or employees. However, if the contractor objects to being taped, the inspector is to stop using a recorder.

Following the walkthrough inspection, the OSHA inspector holds a "closing conference" with the contractor to describe any apparent violations found during the inspection and discuss ways to abate the violations. The inspector is to give the contractor certain safety publications, including a pamphlet titled, "Employer Rights and Responsibilities Following an OSHA Inspection" and discuss the information and answer any questions the contractor might have, including the right to challenge any citations that might result from the inspection.

There are several instructional videos on how to handle an OSHA inspection. AGC sells one titled, "Ready or Not: OSHA's Inspection and Citation Process," priced in April 2000 at $115.50 for AGC members and $173.25 for non-members, available at (800)AGC-1767, AGC's Publications Department. Another video entitled, "Ready for Inspection: What to Do During an OSHA Inspection," sells for about $70 and includes a checklist of each step in responding to an OSHA inspection. The video is available through private publishers like Construction Book Express, which can be contacted at (800)253-0541. Educating your staff on how to handle an OSHA inspection is money well spent.

§22.17 POSTING REQUIREMENTS

OSHA regulations require the following documents to be posted at the work site:

1. OSHA poster informing employees of their rights and obligations under the OSHA Act;
2. OSHA Form No. 200, Log and Summary of Occupational Injuries and Illnesses during the month of February;

---

[40] 29 C.F.R. §1903.8(d).

3. Current citations, if any; and
4. Petitions for Modifications of Abatement Date (PMAs).[41]

An employer who violates any of the posting requirements can be assessed a civil penalty for each violation.[42] Compliance is relatively simple, but might be overlooked without proper employee training.

§22.18 SAFETY TRAINING AND PLANS

Most contractors have a designated safety officer charged with keeping the company in compliance with OSHA and state safety laws. This is good business for several reasons. First and foremost is the protection of workers on the job site. Each workplace injury takes a valuable employee off the job for an indefinite time. Second, each claim made against a company's workers' compensation policy increases its EMR ratio, which results in a potential increase in the company's insurance premiums. Any increase in overhead costs, such as insurance, ultimately eats into the company's profits. Third, OSHA investigations can result in substantial fines and citations that may expand beyond the area of the accident to include any other unsafe condition noticed by the OSHA inspectors during a site visit. These fines are not covered by any insurance policy carried by most contractors and, therefore, come straight out of the "bottom line" profit margins for the company. Fourth, a good safety program can boost company morale. Many contractors give awards to a project team with good safety records. A 1999 study reported that 64 percent of contractors use an awards program within their company.[43] The AGC both nationally and locally gives out safety awards at recognition dinners for contractor members with outstanding safety records. Employees can take pride in knowing that neither they, nor their co-workers, were injured on their job. Fifth, many owners are beginning to inquire during the selection process about a contractor's safety history. A high EMR ratio can tell an owner that this contractor does not perform its work safely, creating risk to the owner of claims by injured workers, plus indicating an overall lack of attention to detail on the job site. Last, accidents and injuries on the job site can cause project delays for which the owner is not required to give a time extension. This can result in a need to accelerate work to pick up on lost days or can result in liquidated damages being assessed by the owner for late completion, also cutting into potential profits.

Formal safety programs are more rare among design firms. After OSHA proposed almost $1.4 million in fines against design firm HDR, Inc. relating to exposure of workers to asbestos and other hazards during an airport construction

[41] 29 C.F.R. §§1903 and 1904. A Petition for Modification of Abatement Date (PMA) is a petition under 29 C.F.R. §1903.14(a) under which an employer seeks additional time to abate an unsafe condition noted by the OSHA inspector.

[42] 29 U.S.C. § 666.

[43] CFMA Construction Industry Annual Financial Survey (1999).

project, the firm reportedly settled the claim with OSHA for $700,000, agreed to hire a full-time safety director, and increased its yearly safety training budget for 1998 to $800,000.[44] Most design firms, however, have no such programs. By contrast, a 1999 study reported that 97 percent of contractors have formalized safety plans for their companies and that 83 percent use on-site safety orientation programs.[45] Design firms are way behind in safety education.

Safety programs vary and some contractors have used creative ways to stress safety and awareness. On Denver's $160 million Pepsi Center project, contractor Mortenson initiated a comprehensive "zero-injury" safety program nearly three months before ever breaking ground. The plan required each worker to be 100 percent tied-off when exposed to a fall of six feet or more. "If you didn't tie-off, you were asked to leave the site. We won't give you a second chance to have an accident," said James Mitchell, Mortenson's on-site safety director. The safety environment included early morning calisthenics, led by an exercise leader (job foreman). A crowd of nearly 260 employees warmed up every morning as a group, with exercises including head rolls, shoulder rolls, runner's stretches and a full minute of jumping jacks.[46] Incentives for good safety included $25 cash bonuses each month to Mortenson employees, and regular team barbecue lunches for the entire job site, hosted by Mortenson as a reward for maintaining a zero lost-time accident record.

A noted safety expert, Dr. E. Scott Geller, says that "behavior-based" safety programs are important for creating a safe culture on the job.[47] Field personnel have the greatest opportunity to make change because concern shown among peers is more likely to promote safe behavior and decrease at-risk behavoir. Paul Hamari, Safety Director at The Neenan Company, Colorado's largest design-build firm, says that companies should never expect that a specific method or procedure will work for them from a safety manual or from a safety class. He says that while theories can be shared, it is ultimately up to each individual company to devise specific methods and procedures that best fit its culture, and safety must be customized to fit each company. Hamari says, "In order to make safety fit your culture, everyone must be involved in the process. Educate all of your workers, from management down, on why certain safety concepts are being applied, why they are practical, how they are scientifically proven and how they will improve safety conditions . . . Get the employees involved."[48] Some ways to make a job site safe, according to Hamari, include:

- Have subcontractors submit a job hazard analysis, in which they provide a step-by-step description of tasks, identify potential hazards, and develop procedures to follow for safe performance.

---

[44] *Undeserved Attention?* note 1, above at p. 31.

[45] CFMA Construction Industry Annual Financial Survey (1999).

[46] *Managing the Creation of a Landmark*, Pepsi Center Profile, supplement to Colorado Construction (1999).

[47] Paul Hamari, *Safety Measures Key in Construction Biz*, Denver Bus. J., Oct. 25, 1999.

[48] *Id.*

- Hold daily (not just weekly) safety review meetings.
- Provide OSHA "competent person" training classes.
- Provide safety orientations and overview classes to subcontractors to create a common safety knowledge base for everyone at a job site.
- Promote weekly participation by subcontractor foremen to perform job site safety assessments, which would include a written report.
- Conduct monthly full-job site safety meetings to discuss corrective actions due to any safety incidents, as well as discuss current safety concerns of everyone on the job.
- Establish an incentive program in which safe behavior is recognized.

§22.19 LEARNING ABOUT SITE SAFETY

The intent of this chapter is to raise awareness of the risks inherent in construction as a general contractor. If an architect or engineer is starting a design-build firm or joint venturing on a design-build project, he or she should consider attending a full-day OSHA seminar or taking a short course on site safety through a local AGC chapter. Partnering with an experienced contractor who has a good safety record is probably the best way to get hands-on mentoring. Joining the AGC or the Associated Builders & Contractors (ABC), or a local contractors' organization is also a great way to get information on safety programs. A design professional will not likely find these same programs offered through the local AIA or NSPE chapters. There are dozens of books and video tapes available on site safety. A design professional should contact his or her local AGC chapter or call the AGC Publications Department at (800)AGC-1767 and ask for the AGC Publications and Videos catalog, which is full of safety publications, tapes, manuals, and forms.

CHAPTER 23

WHEN DESIGN GOES BAD: LESSONS LEARNED FOR LITIGATION

Legal action design-build teams by owners is nothing new. Of course, in the ancient times of the early master builders, justice was strict and swift under the ancient Babylonian Code of Hammurabi, which imposed the same liability on the master-builder as that suffered by his client. The Code provided that "if a builder has built a house for a man, and his work is not strong, and if the house he has built falls in and kills the householder, that builder shall be slain."[1] As centuries went by, the system of courts and civil suits replaced executioners as the way to resolve construction disputes. We still find that design-build has not always resulted in harmony among the team, nor with the owner. This chapter explores some of the actual court cases where design-build went bad.

§23.01 DESIGN AND ESTIMATING ERRORS BY DESIGN SUBCONTRACTOR

One of the most common sources of claims between design-build partners is when the design partner makes either design or quantity errors that are relied on by the contractor-partner in establishing a fixed price with the owner. If the errors result in extra costs, the parties may be able to resolve those disputes informally. However, case law shows that when the losses are substantial, the design-build partners may still end up in court. In design-build, however, the owner is not normally involved in the lawsuit as it is in traditional projects.

One example is the case of *C.L. Maddox, Inc. v. The Benham Group, Inc.*[2] In that case, Benham was a design subcontractor to Maddox, who held the prime design-build contract on a power plant in Joppa, Illinois. The job went badly and Maddox sued Benham for breach of contract and fraudulent and negligent misrepresentation, seeking $5.1 million in damages. The two parties had an oral subcontract for the proposal stage, during which Benham prepared plans and specifications, an equipment list, and quantity information for a $58,200 fee. Maddox relied on Benham's figures when bidding and was awarded the prime contract for $10.3 million. Maddox and Benham then entered into a $616,000 written subcontract for the remaining design work. According to Maddox, Benham's design work was late, its plans were incomplete and its estimations were wrong, resulting in $5.1 million in damages. As a result, Maddox defaulted on the prime contract and the owner called on the performance bond surety, USF&G, who paid $2.8 million to complete the work. After a five-week trial, the jury awarded Maddox $5 million in damages against Benham. The evidence showed that Benham "repeatedly assured Maddox . . . that it was well qualified to do the work and that it had the manpower and expertise to do the work. . . . Under Missouri law, these assurances created an implied warranty, allowing liability [against Benham] for the bidding errors."

---

[1] H. Block, *As the Walls Came Tumbling Down: Architects' Expanded Liability Under Design-Build/Construction Contracting*, 17 John Marshall L. Rev. 1 (1984).

[2] 88 F.3d 592 (8th Cir. 1996).

Practice Pointer. From this case, one can see that the preliminary design and quantity take-offs can have a dramatic effect on the project cost. The preliminary fee of $58,200 for a $10.3 million project is 0.5 percent. This may not have been sufficient, nor the time adequate, to develop designs and estimates to a level that is appropriate for a $10 million construction project. Nonetheless, if the design firm represents itself as being qualified to perform what is needed for the agreed fee, the courts will not relieve the firm of liability. Therefore, it is important that: (1) there is sufficient fee to cover the services; (2) there is adequate time to develop the design; (3) the most qualified member of the team does the material and equipment estimates; (4) if there is not sufficient time to develop what is needed, include adequate contingencies in the estimate to cover the unknown and undeveloped designs; and (5) contract provisions are included in which the parties agree to share these risks. This case also raises the issue of what level of professional liability insurance is appropriate on large-scale projects. Even though the $5 million jury award was reduced on appeal, this size loss would be enough to put most firms out of business without adequate insurance limits.

In a similar case, *CRS Sirrine, Inc. v. Dravo Corporation*,[3] the project engineer and contractor entered into a joint venture agreement for a $100 million design-build power plant for the U.S. Navy in Virginia. Under the agreement, Sirrine was to design and Dravo was to build and bond the job. The joint venture had $30 million in cost overruns and Dravo sued Sirrine for $12.5 million of those costs. The court held that Sirrine had breached its contract and was liable for $5.5 million in damages for providing insufficient and inaccurate information that Dravo relied upon in bidding. The primary error was Sirrine's failure to track quantities and notify Dravo that the estimated quantities would be exceeded. The court found that the actual quantities dramatically exceeded the estimate. The contract stated that both parties would indemnify each other for damages resulting from untimely, defective and non-conforming work, with liability limited to each party's respective scope of work. However, the subcontract contained certain additional clauses that limited Sirrine's liability for Sirrine's own errors and omissions. Under one clause, Sirrine had no liability for errors if Sirrine gave Dravo notice before any costs were incurred; under a companion clause, Sirrine had no liability for the first $750,000 in damages for errors if Sirrine gave Dravo notice after costs had been incurred, with full liability for such errors thereafter. The court found that the limitation clauses in the contract were ambiguous and that Sirrine did not provide the required notice and, therefore, the limitation did not apply. In a statement indicating just how complicated design-build litigation can become, the court noted that there were 2,500 exhibits listed by the parties and 130 witnesses planned to testify at trial.

Practice Pointer. As with the *Maddox* case, this case shows that it is important to be sure that there is sufficient time and fee to cover the preliminary design

[3] 445 S.E.2d 782 (Ga. App. 1994).

phase and, if not, the risks should be recognized and shared, plus adequate contingencies included in the bid price. If the parties agree on limitations on liability, these should be written in plain, unambiguous wording that allocates risk in a manner that can be understood by a court and applied to situations, whether due to errors in the preliminary phase or the construction document phase. A mediation clause might have resulted in a savings of significant legal fees that were incurred by the parties. This case also points out that insurance limits should increase on larger scale projects. When there are errors on a $100 million project, the damages will be proportionately greater than on a $1 million project.

Another example of inadequate design resulting in cost overruns is *Skidmore, Owings & Merrill v. Intrawest I Ltd. Partnership.*[4] In that case, a developer (Intrawest) hired Skidmore, Owings & Merrill (SOM) to design a high-rise complex with a tight budget and schedule. Based on SOM's design, the developer hired SDL Corporation to prepare the estimate and set a guaranteed maximum price (GMP) for the design-build contract. The evidence showed that the parties understood that SOM's design drawings had to be 90 to 100 percent complete at the bidding phase, and that after falling behind schedule, SOM requested an extension of time and warranted that its drawings would be at least 90 percent complete if given more time. However, the drawings were only 50 to 65 percent complete. The drawings were represented as complete and SDL prepared its estimate for the $29.5 million project based on those drawings. SDL testified that it would never knowingly establish a GMP from incomplete drawings. As a result of defects in the design drawings, the project was delayed by 32 days. SOM sued for unpaid fees and Intrawest countersued for cost overruns and interest on the construction loan during delays. After a 14-week jury trial, SOM received an award of $99,775 in fees which was overshadowed by Intrawest's award of $820,372 for contract changes and extra work, $252,088 in interest on borrowed funds, and another $332,919 in prejudgment interest. SOM admitted that its drawings were incomplete, but claimed that the extra work would have been required had the drawings been complete and that a judgment against the architect would result in a "windfall" to Intrawest. The evidence showed, however, that Intrawest would not have undertaken the project had it known the true extent of the cost. As a result, the award was upheld against the architect, SOM.

Practice Pointer. The clear problem in this case was not the estimate (which the court found was accurate based on SOM's designs), but the incomplete design drawings used to set the GMP for the design-build contract. A firm must not promise what cannot be delivered, on a schedule that is unrealistic. The case does not state why SOM did not deliver the 90 percent complete drawings, but understaffing or taking on too much work, or delays by subconsultants can all delay design work. If there is not sufficient time to prepare what is needed, a firm

[4] 1997 Wash. App. LEXIS 1505.

should: (1) get an adequate time extension; (2) be sure all parties understand what level of drawings can be prepared and that sufficient contingencies must be included; or (3) avoid the project if expectations are too high. Here, the actual added costs of $820,372 were only a 2.7 percent overrun. However, on a $29.5 million project, even small percentages of error result in large costs, plus delays. A limitation of liability clause with some agreed margin of error may be appropriate where the schedule is too tight to produce fully developed designs, with the team agreeing to share that risk.

§ 23.02 LIABILITY OF ARCHITECT/ENGINEER FOR CONSTRUCTION PHASE ERRORS BY CONTRACTOR

One of the earliest reported cases of architect liability is a French court case in 1844, in which a church was built in Saint Germain-en-Laye, France from the plans and under the superintendence of two architects. It seems that in a rush to complete the project, the carpenter used roof timbers that were not "well seasoned"—green lumber that failed. Due to using "green lumber," nine out of 29 tie-beams were discovered to be rotten. The church sued the carpenter and architect jointly to recover the cost of repairs. The court held that the carpenter was liable, but that the architect "who limited himself to preparing the plans, writing the specifications and directing the work, should not be held to the responsibility of Article 1792, when it is otherwise established that the plans are conformable to the rules of art, that the direction of the work has been conscientiously pursued, and that the verification of the materials well done."[5] The French Article 1792 imposed strict liability on the architect when it served as the prime contractor on projects. Here, however, the court recognized a type of "standard of care" defense, finding that the architects' direction of the work did not make them liable for the contractor's errors.

In *C.L. Maddox, Inc. v. The Benham Group, Inc.*,[6] the subcontract between the prime design-build contractor (Maddox) and its engineering subcontractor (Benham) stated, in part, that Benham would keep Maddox "informed of the progress and quality of the Work, and *shall endeavor to guard [Maddox] against defects* and deficiencies in the Work of [Maddox]." Maddox claimed $1.2 million of damages against its engineer even though those damages were caused by Maddox's own errors and those of the owner. The engineer's subcontract also contained standard language which stated that Benham did not have control of, and was not responsible for, construction means and methods, nor "for the acts or omissions of the contractor, its subcontractors or any other persons performing any of the Work, or for the failure of any of them to carry out the Work in accordance with the Construction Documents." The court held that this language was not sufficient to shift to Benham liability for Maddox's own acts or omissions,

[5] Ville de S. Germain-en-Laye v. Malpiece et Moutier, as reported by T.M. Clark, FAIA, in *Architect, Owner and Builder Before the Law*, Macmillan & Co. (1894), p. 58.

[6] 88 F.3d 592 (8th Cir. 1996).

and that Benham had "no duty under the contract to act as insurance against Maddox's own carelessness." As a result, the court reduced the $5 million jury award by $1.2 million.

Practice Pointer. Standard American Institute of Architects (AIA) and the Engineers' Joint Contract Documents Committee (EJCDC) contracts include the wording that the court accepted here that limits or excludes the design firm's liability for construction means, methods, procedures or for construction errors by the contractor or its subcontractors. Here, that wording saved the engineering firm $1.2 million in damages. One should be sure to use contracts that include appropriate risk allocation, with the design firm responsible for design errors, and the construction partner responsible for construction errors. One should avoid contracts that try to shift liability to your firm for work that you cannot control and do not perform.

§23.03 LIABILITY OF THE CONTRACTOR FOR THE DESIGN ERRORS BY THE ARCHITECT/ENGINEER

In *Rivnor Properties v. Herbert O'Donnell, Inc.*,[7] the court held that the design-build contractor was liable to the owner for both design and construction damages caused by design and construction defects in a glass curtain wall. The project was a three-story office building known as the "Honeywell Building" completed in 1981. The architect was hired directly by the contractor to provide plans and specifications only, with no construction supervision role. The parties used a standard AIA set of General Conditions (1976 edition), which referred to the architect's role during construction. However, the court ruled that having elected not to hire the architect for construction administration, the design-build contractor assumed all of those responsibilities, including periodic inspections to ensure that the building was free from defects.

Shortly before completion of the building, the owners complained about water leaks into the building from the glass curtain wall, which eventually resulted in excessive glass breakage and discoloration. Repair efforts were unsuccessful and the only solution was to dismantle and reassemble the window wall system. The owner sued the design-build contractor in 1985 for damages, including cost of repairs and diminution of value of the building. Other subcontractors and suppliers, their sureties and insurers were added as parties to the case.

After a one-year trial, the judge found the design-build contractor liable for all of the $1.99 million in damages, including $1.2 million in repairs, $200,000 in architect fees, $200,000 in attorneys' fees, $14,500 in expert fees, and $380,446 in rental income. The court found the architect 25 percent at fault, the glass subcontractor 35 percent at fault, and the drywall contractor 25 percent at fault. Presumably, the contractor was responsible for the remaining 15 percent of fault.

Prior to the trial court's ruling, the design-build contractor and its insurer

[7] 633 So. 2d 735 (La. App. 1994).

settled with the owners for $1,050,000 but assigned their rights and claims against the other defendants to the owner. The architect settled with the owner for its policy limits. Though the architect was found liable for 25 percent of the nearly $2 million in damages, the architect had only $250,000 in insurance. After claims expenses of $240,000 were deducted, the policy only had $10,000 left to pay damages. The owner contested the claims expenses, but the court held that the policy provision on expenses was valid and not against public policy. As to the remainder of the judgment not settled, the court held that the owner's only relief was against the design-build contractor with whom it had a contract. The owner was not permitted relief directly against the subcontractors other than as assignee of the contractor's rights.

Practice Pointer. This case has lessons for both the owner and the design-build contractor. The court did not permit the owner to maintain actions directly against the subcontractors since the owner had no contract with them. An owner would be better protected by requiring the design-build contractor to insert a provision in every subcontract stating that the owner was "an intended third-party beneficiary" of the subcontracts, so that the owner could pursue relief directly against any subcontractor if it so chose.

This case points out the drastic difference between a contractor's general liability policy and an architect's professional liability policy, under which attorneys' fees and expert fees are deducted from the policy limits. With only a $250,000 policy from the architect, a one-year trial and its preparation left only $10,000 to pay under the claim. This indicates that higher policy limits should be required and that parties involved in large, multi-party litigation need to recognize that the longer the case continues, the less coverage will be available under the architect's policy since legal costs eat away at the policy limits. One must find out what the available policy limits are prior to settling so there are no surprises.

With nearly $2 million in damages, but apparently only $1 million gained in settlement with the contractor and its insurers, and just $10,000 recovered from the architect, this case also points out some of the risks inherent in settling. Settlement here worked in favor of the contractor (who paid only $50,000 of the $1 million settlement) and the architect (who settled for only the remaining policy limits).

§ 23.04 INCENTIVE BONUS CLAUSES

It is not uncommon in fast-track design-build projects for the owner to give a bonus for early completion. In some contracts with specified performance requirements, there may even be a bonus for exceptional performance that exceeds the owner's performance requirements. In one 1994 case, however, the desire to maximize that bonus resulted in a lawsuit with allegations of fraud and deception. In *Combustion Engineering, Inc. v. Miller Hydro Group*,[8] Miller Hydro and Com-

[8] 13 F.3d 437 (1st Cir. 1994).

bustion Engineering entered into a design-build contract for a $24 million hydroelectric power plant in Maine with a capacity of 14 megawatts and 7,800 cubic feet of water per second (cfs). The contract contained a sliding-scale bonus for efficiency if the plant produced power in excess of 77,500 megawatt hours per year, with a corresponding penalty for lower performance.

In order to ensure the bonus, Combustion Engineering designed a much larger plant with a capacity of 18 to 19 megawatts and over 9,000 cfs, investing $1 million of its own money into the design in order to get an $8 million bonus. Miller refused to pay the bonus, claiming that the design of the larger plant was done secretly to get the bonus, without Miller's knowledge or consent and in violation of the license that Miller had obtained for the facility. Combustion Engineering filed a lien and sued to recover the $8 million incentive bonus, plus another $893,000 for an early completion bonus. Miller countersued for fraud, breach of contract, and violation of the federal racketeering statute (RICO) claiming that the over-design of the plant was done secretly and constituted a material change in the contract, done without Miller's consent. Combustion Engineering argued that the 7,800 cfs capacity was a "minimum" design requirement, which the court rejected, finding that the specified capacity was both a target and a maximum. Since there was no contract change to approve the increase, Combustion Engineering breached the contract by over-designing. As a result, there was no entitlement to a bonus. Miller lost on the counterclaim as well, since the court found that Miller had not suffered any damages because it did not have to pay out the bonus. The court told the parties that:

> This case could plausibly have been settled at the outset by, for example, payment of any remaining amounts due under the contract but without any bonus payment. Now, after wearisome and no doubt expensive litigation over an imperfect contract and imperfect conduct, neither side has gained what it sought at the outset. This may itself be a form of justice, but it could have been achieved at a lower price.

Practice Pointer. This case also points out how a mediation clause may have saved significant legal fees and expenses. More importantly, the case shows that contracts with performance criteria must be clear on what is required, whether the requirements are minimums or maximums, how the parties handle changes in those criteria, and whether bonuses can be achieved by exceeding the design criteria or not. Also, all licenses must be reviewed if they control the type of facility involved and restrict the size, output, or other component of the project. Design changes that may exceed the performance criteria or licensed limits should be discussed with the design-build team and the owner to avoid surprises and disputes.

§23.05 MISREPRESENTATION ABOUT PROJECT SCOPE

With no fully developed plans and specifications, it is important that the owner give a detailed program to the design-build team so that everyone knows the owner's expectations. At least one case shows, however, that even a 25-volume contract may leave some parties unclear about the scope and result in litigation over losses. There is no published result in this case, but we know from *Fluor Daniel International, Inc. v. General Electric Co.*,[9] that Fluor Daniel and its subsidiary, Fluor Daniel Arabia Ltd. sued General Electric (GE) and its subsidiary, Saudi American GE, for $60 million over a 380 megawatt, combined cycle power plant in Saudi Arabia. Under a 25-volume prime turnkey contract, GE was to be prime and Fluor the subcontractor to design and build the plant. Fluor Daniel sued GE for misrepresentation about the scope of the project, contending that it could not finish the job on time and it incurred substantial cost overruns. According to the lawsuit, GE pressured Fluor Daniel to accept the job under threat of canceling other contracts between the two. The central legal issue before the court was whether the two parties had to arbitrate this dispute, which the court found was mandatory under the contract. As a result, the outcome of the arbitration is not public. However, the case shows the importance of understanding the scope of work.

Practice Pointer. This case points out that even very large, sophisticated, and experienced companies like Fluor Daniel and GE, with huge staffs, can have disputes over scope of work. Here, even a 25-volume contract did not include sufficient detail, according to one of the parties, about the scope of work of the project. Design firms often include very brief scopes of work in traditional design contracts. This is risky enough on traditional projects, but with design-build it is critical that the contract contain a detailed scope of work so that the team understands what is required, and so that the owner knows what is and is not included in the contract price. One of the few grounds for a change order in design-build is a scope change. Without adequate detail, even those changes may be waived if one cannot point to something in the contract that supports the increase in cost based on a change in scope.

§ 23.06 DESIGN ERRORS BY DESIGN PARTNERS

In *17 Vista Fee Associates v. Teachers Insurance and Annuity Assoc.*,[10] a prime design-build contractor (17 Vista) was hired to design and build an office building in New York City, and to sell it upon completion. The contractor subcontracted the mechanical engineering, including a smoke purge system. The building was sold before completion based on a temporary certificate of occupancy. The owner could not get the permanent certificate of occupancy because the smoke purge system had not been approved by the fire department. Three years later, the system passed inspection after a second fan was installed. The buyer refused to pay the full purchase price and the contractor sued. When the buyer countersued for $452,000 in damages related to the smoke purge system, the

[9] 1999 U.S. Dist. LEXIS 12983 (S.D.N.Y. 1999).

[10] 259 A.D.2d 75, 693 N.Y.S.2d 554 (1999).

contractor joined with the engineer and sought indemnity for the engineer's negligence. On the eve of trial, the contractor and the buyer settled. The court then granted judgment to the engineer, finding that it was not liable for the contractor's own breach of contract with the buyer, plus there was no contract indemnity provision. The court of appeals reversed that decision, finding that the contractor was entitled to "common law indemnity" from the engineer since the contractor had "delegated full responsibility for the design of the smoke purge system" to the engineer, who had "exclusive responsibility" for the system.

Practice Pointer. Design errors, whether on traditional projects or on design-build projects, can result in liability to the design firm if the errors result in added costs or delays. Mistakes are made even by the best firms. As a result, one should carry professional liability insurance with adequate limits of coverage in light of the project size or unique risks. Be aware that even if there is no indemnity clause in the contract, a design firm can be liable for "common law" indemnity for added costs the owner or a design-build partner incurs. The case does not explain why it took three years to get the final Certificate of Occupancy, but a proactive effort to resolve problems early on can result in less costly solutions without expensive litigation.

§23.07 DELAY DAMAGES CAUSED BY DESIGN-BUILD CONTRACTORS

In *Royal Insurance Co. v. CNF Constructors, Inc.*,[11] Cogen Energy Technology, LP entered into a design-build contract with CNF under which CNF was to build a combined-cycle cogeneration facility including a steam turbine generator. Cogen was permitted to "go on-line" and sell energy before the plant was completed. The contract had several interim performance tests to determine that CNF was meeting performance criteria. Cogen sued after the steam turbine generator experienced a "catastrophic failure." Cogen also sought consequential damages from CNF as a result of a complete shutdown of the plant for three months while CNF repaired the generator. During this time, Cogen was unable to sell energy to its customer. CNF admitted that the generator failure was due to defects in its design or workmanship, but argued that it did not breach the agreement because the defect was repaired prior to the Final Acceptance date in the contract, and that Cogen's remedy was limited to the cost to replace and repair, and did not include consequential damages. Alternatively, CNF argued that the liquidated damages clause in the contract precluded recovery for consequential damages. The court held that "the liquidated damages provisions were included to cover the possibility of a delayed opening, not for an interim shutdown caused by the temporary failure of a component part." CNF next argued that Cogen carried insurance to protect itself from the risk of a shutdown during the project, showing an intent

[11] 1995 U.S. Dist. LEXIS 112 (S.D.N.Y.).

that it would bear the risk of such delays. The court held that it could not determine the parties' intent and allowed the case to proceed to trial.

Practice Pointer. This case shows that risks need to be considered whether they are delays in completion or damage to the project during construction. Insurance products can cover a wide variety of risks. A firm should discuss with its insurance agent what coverages make sense for each particular project. If the contract size and risk are large enough, the project may be appropriate for a special policy to cover liquidated damages or efficacy risks like performance output, delays, or damage during construction. Contract liquidated damages clauses should be clearly written to cover risks contemplated by the parties. A waiver of "consequential damages" is a standard provision in many design-build contracts. Such a clause may have provided great protection to the design-build contractor on this project, where the cost of fixing a construction defect is not as significant as the consequences to the owner, such as lost income from energy sales to customers.

§ 23.08 PAYMENT FOR PRELIMINARY DESIGN SERVICES

One concern among design-build firms is that the owner will terminate the contractual relationship after the preliminary design and estimating phase, then take the design and shop the price to other contractors. The AIA, AGC, DBIA, and EJCDC form design-build contracts all deal with this in different ways, but the general theme is that by controlling the ownership of the copyrights and the documents themselves, or by including a "buy-out" provision, the design-build firm can prevent the owner from using the proposed design unless the design-build firm consents, is paid, and is indemnified by the owner for use of the design without the design-build firm's involvement. Two cases, with different results, show how problems can arise from misuse of a design.

The first case is over 100 years old, but still addresses valid concerns today and teaches a valuable lesson. In *Walsh v. St. Louis Exposition & Music Hall Association*, the owner put out a request for proposals (RFP) on April 19, 1883 for architects to design and superintend the construction of a performing arts complex. The RFP had a program complete with space requirements for a 4,000 seat auditorium and a smaller 1,200 seat theater, with the site survey available at the owner's offices. Thomas Walsh, a St. Louis architect, submitted two designs and was selected as architect. However, the owner refused to enter into a contract with Walsh and hired others to build the project, in part, using Walsh's design. Walsh sued for his construction fee and won at trial. However, the case wound its way through several appeals until a final decision was rendered on November 17, 1890—over seven years after the RFP date—reversing the trial court's decision and holding in favor of the owner.[12]

When soliciting proposals, the owner offered a $500 stipend to those se-

---

[12] Walsh v. St. Louis Exposition & Music Hall Ass'n, 16 Mo. App. 502 (1885), *aff'd*, 2 S.W. 842 (Mo. 1887), *rev'd*, 14 S.W. 722 (Mo. 1890).

lected to compete. The owner then convened a selection committee, including "two experts" to review all the submissions and interview the proposers. At the interview, Walsh gave a performance bond under which he agreed to perform as architect and to superintend the construction for a fee of $30,000 at a construction cost not to exceed $398,000. The owner selected one of Walsh's designs, the "Dignus Laudie A" but passed a resolution requiring the plans to be modified "so as to suit the views of the board." The owner asked Walsh to make alterations to his plans, including making the building 13 feet wider and relocating the smaller theater. Walsh agreed and modified the plans, but insisted that the price be increased to $425,000 or he could not enter into the contract. The evidence was that experts felt that the building could not be built for the price quoted by Walsh and this caused concern by the owner that Walsh would not use suitable materials or workmanship. Negotiations went through three or four drafts before falling apart and neither party signed the contract. Though the owner sent Walsh the $500 stipend, he refused to accept the check. As a result, the owner hired another architect to draw new plans and to superintend a building very similar to Walsh's design, which was to cost $600,000. Walsh sued based, in part, on the use of his design and for the construction phase fee. However, the Missouri Supreme Court held that "the similarity between the building and plaintiff's design ceases to be of any value in light of the fact that defendant tendered to him the $500, and thereby became the owner of and entitled to use his plans."[13] The court said that the award to Walsh was conditional only, and did not entitle him to be hired to superintend the construction. Having tendered the $500 fee, the court held that he became the owner of the plans and was entitled to use them. Walsh's refusal of the $500 check was no defense.

The second case is *Building Structures, Inc. v. Young*, a 1994 Oregon case, in which the owner hired the design-build firm (BSI) to prepare a design to meet the owner's financial requirements, with the understanding that the parties would negotiate a final price. After the design was prepared, the owner objected to BSI's $245,000 cost estimate and took the design to other contractors, eventually contracting with one of them to build the project for $200,000. BSI sued the owner for breach of contract and fraud and was awarded $51,000 in damages on the fraud claim.[14]

Practice Pointer. These cases show the importance of retaining the ownership of the designs and the copyrights when submitting preliminary designs to an owner prior to being hired as the design-build contractor. DBIA's preliminary design-build agreement states that, "[a]ll drawings, specifications and other documents and electronic data furnished by Design-Builder to Owner under this Agreement ("Work Product") are deemed to be instruments of service and Design-Builder shall retain the ownership and property interests therein, including

[13] 14 S.W. at 725-726.

[14] Building Structures, Inc. v. Young, 883 P.2d 1308 (Or. Ct. App. 1994).

the copyright thereto."[15] The agreement then provides that if the owner does not hire the design-build contractor to complete the design and construction, the owner can use the design only on these four conditions: (1) the owner must pay all amounts due under the preliminary agreement; (2) use of the design is at the owner's sole risk; (3) the owner must defend, indemnify, and hold harmless the design-build contractor and its design consultants of any tier; and (4) the owner must pay an additional amount agreed upon up front by the parties. In addition to these conditions, one must read the RFP carefully to see if acceptance of a stipend conveys a license to the owner to use the designs submitted, and place a simple copyright notice on every drawing that is submitted to the owner that simply says: "Copyright 2001 XYZ Design-Build, LLC" with the year being the date of the submission and the name being that of the submitting firm.

§ 23.09 PERFORMANCE AND PRESCRIPTIVE REQUIREMENTS

An RFP for design-build work may contain a mixture of performance-based criteria as well as some detailed design requirements, known as "prescriptive" specifications. A design-builder must be careful about making assumptions that it has the discretion to vary from the prescriptive requirements. One government appeals board case held that the use of the term "design-build" did not relieve a contractor from compliance with prescriptive drawing requirements. In *Appeal of FSEC Incorporated*, the Navy awarded a contract to FSEC to build a painting facility in California.[16] The drawings showed two exhaust fans and two dust collectors in each of the two rooms. The specifications called for a minimum air handling capacity in cubic feet per minute. FSEC submitted shop drawings showing a single fan in each room, though the equipment would satisfy the minimum air handling capacity. The contracting officer directed FSEC to install two fans and two collectors, which FSEC said was a change, since the contract referred to the mechanical equipment as "design-build." On appeal, the board held that the design-build language did not supersede or conflict with the prescriptive requirement of two fans per room, holding that "[t]he prescriptive/design and performance labels do not alone create, limit or remove a contractor's obligations. Rather, the obligations imposed by the specifications determine the extent to which a particular specification is either performance or prescriptive, and it is not uncommon for a contract to contain both design and performance characteristics."

Practice Pointer. This case points out the need to follow prescriptive requirements in the owner's program or "bridge documents" where provided, or to request a variance in writing to deviate from those requirements where a firm believes that it can meet the owner's performance criteria with a different design.

---

[15] ¶4.1, DBIA's Preliminary Agreement Between Owner and Design-Builder, Doc. No. 520 (1998 edition).

[16] Appeal of FSEC Incorporated, ASBCA No. 49509 (July 28, 1999).

§23.10 UNCLEAR PAYMENT TERMS IN PRIME AND SUBCONTRACTS

Unclear payment provisions in prime design-build contracts and in subcontracts can lead to disputes, whether the subcontractor is a design consultant or a trade contractor. In a New York lawsuit, a design-build contractor hired an architect as a subcontractor, with the architect's fee to be a percentage of the "cost of construction".[17] The architect contended that its fee should be calculated on the entire cost of construction, including the design-build contractor's fee. The contractor argued that "cost of construction" meant only the reimbursable costs of labor and materials, excluding the fee. The court held, however, that although the contractor's fee was calculated on the reimbursable costs only, those provisions were not incorporated into the design subcontract and, because the contractor drafted the subcontract, any ambiguous term of that agreement would be construed against the contractor. The architect contended that the term "cost of construction" is a term of art, with an accepted meaning among architects, based on the AIA forms. Though the subcontract was not an AIA form, the court held that if the term had an accepted meaning among architects and the contractor wanted to deviate from that, it was the contractor's obligation to make that deviation clear.

Practice Pointer. If payment to the design consultant is a percentage-based fee, rather than a lump sum or other basis, this case points out the need to define just how that percentage will be calculated. If there is to be retainage held in the prime contract, one must make it clear whether or not retainage is to be held on subcontracts, and at what rate. If the payment to the subcontractors is to be contingent upon the owner making payment first under the prime contract, then this must be stated in clear and unambiguous language. The court's statement that any ambiguous terms in the subcontract would be construed against the contractor is consistent with general contract law that the party drafting the contract has the burden to make its intent clear. If its intent is not clear, the ambiguous term will be construed against the drafting party and in favor of the other party's interpretation, if reasonable. DBIA's subcontract between design-build contractor and architect/engineer does not contain a definition of "cost of the work" or "cost of construction," and leaves compensation to be defined in a separate exhibit.[18] AGC's subcontract with architect/engineer does not contemplate a percentage-based fee and does not define "cost of construction" either.[19] Even AIA's subcontract form does not define the term and so a modification would be

---

[17] Harza Northeast, Inc. v. Lehrer McGovern Bovis, Inc., 680 N.Y.S.2d 379 (App. Div. 1998).

[18] ¶6.1.2, DBIA's Agreement Between Design-Builder and Designer, Doc. No. 540 (1999 edition).

[19] ¶6.1, AGC's Agreement Between Design-Builder and Architect/Engineer, Doc. No. 420 (1999 edition). Note that ¶10.7 of the AGC form states that the agreement is to be considered "jointly drafted" and that "this Agreement shall be construed neither against nor in favor of either party, but shall be construed in a neutral manner."

needed whenever working with any of the major design-build forms if compensation is a percentage of the cost of construction.[20]

§23.11 IMPLIED WARRANTIES AND BUILDING CODES

When an owner furnishes its contractor with plans and specifications, the law is generally that the owner impliedly warrants the sufficiency of those documents and is responsible for errors contained in the design. *See* § 21.09 and the discussion of the *Spearin* doctrine. However, when the contractor furnishes the design, the law will imply a warranty from the contractor that its design will be sufficient and that it will meet the applicable building code requirements. That was the holding in a 1998 Texas case in which a contractor agreed to design and build an airplane hanger.[21] The contractor stopped construction in anticipation of a breach by the owner when the contractor discovered that the owner lacked sufficient funding. The owner claimed that it did not pay due to deficient work. The court awarded the contractor the amount due, less the cost to complete certain work and an offset for failure to comply with city fire codes. As to the code issues, there was testimony that the parties did not initially know what kind of fire protection would be required and that the cost of fire protection was not included in the contractor's bid, which was to be left for another day. The owner argued that the design-build contractor impliedly warranted to comply with the building codes so that the building is suitable for the purposes intended. The Court of Appeals agreed that every contract incorporates existing laws and that, in the absence of an agreement to the contrary, the burden to comply with building codes falls on the builder and an implied covenant is presumed. The court stated that "[c]ontractors, not owners, are in the best position to know about and comply with relevant building codes." According to the court, owners rely on builders to furnish structures that can be occupied and that an owner can sue its contractor for breach of contract, "where a contractor has failed to comply with building codes relevant to the intended use of the structure." However, in this case, the court ruled that the parties had made their own arrangements that modified the contractor's general duty.

Practice Pointer. Clearly when the design is furnished by the owner, the contractor relies on that design as being in compliance with the building codes. The AIA's 1997 A201 General Conditions state in paragraph 3.2.2 that "The Contractor is not required to ascertain that the Contract Documents are in accordance with applicable laws, statutes, ordinances, building codes, and rules and regulations, but any nonconformity discovered by or made known to the Contractor shall be reported promptly to the Architect." This is consistent with the *Spearin* Doctrine and the owner's implied warranty. However, where the contractor furnishes the design, the roles are reversed and the owner relies on the contractor to

---

[20] ¶9.1, AIA's Agreement Between Design-Builder and Architect, Doc. No. B901 (1996 edition).

[21] Tips v. Hartland Developers, Inc., 961 S.W.2d 618 (Tex. App. 1998).

provide a design that meets applicable codes. In some cases, like this one, there are design elements that are not fully worked out at the time the parties sign the contract. The parties in this case would have been better off if they had made specific provisions in the contract to deal with this topic, by either expressly omitting fire protection (to be added by change order later), or by including an allowance for fire protection (to be adjusted based on actual costs). Oral agreements are sometimes not admissible at trial if the contract is clear and unambiguous. One should not rely on oral statements which, while made in good faith, may later be either denied or contradicted when the relationship falls apart and the parties are in litigation.

§23.12 ACTION FOR FRAUD BASED ON MISREPRESENTATION

[A] Action May Not Be Allowed if Duty Breached Is Merely a Contract Duty

In a 1998 Virginia case, the public owner entered into an "Agreement for Design-Build Construction" for a new baseball stadium.[22] The design included 32 precast, post-tensioned concrete "bents" for its cantilevered roof and upper concourse seating. Each bent was to have hollow conduits containing steel tendons, and be grouted and tensioned for structural integrity. The design-build contractor submitted its applications for payment certifying that the work was "to the best of the Contractor's knowledge, information and belief" in accordance with the contract documents. More than 10 years after construction of the stadium, the owner discovered that many of the conduits contained no grout, or insufficient grout, resulting in corrosion of the steel tendons. Three conduits contained no steel tendons at all. The owner sued the design-build contractor for breach of contract and fraud. The trial court threw out the fraud action because the alleged misrepresentations were duties and obligations specifically required by contract and that, nothing established that the duty breached was separate and independent from the contract. The court summarized that the contractor's "failure to perform each and every one of these promises was a breach of its contract, not fraud." On appeal, the owner contended that the false applications for payment under oath were made to induce payments and that sealing the ends of the conduits with grout concealed the tube opening and went beyond the contractual duties. The Virginia Supreme Court disagreed with the owner and found that each alleged misrepresentation related to a duty or an obligation that was specifically required by the design-build contract. As a result, the contractor may have breached the contract, but its actions did not give rise to an action for fraud. The court held that cutting off and sealing the tubes "was part of the construction process and not an action undertaken by [the contractor] solely to hide its failure to inject grout into the conduits." The court concluded that nothing in the record suggested that the

[22] Richmond Metr. Auth. v. McDevitt St. Bovis, Inc., 507 S.E.2d 344 (Va. 1998).

contractor did not intend to fulfill its contractual duties at the time it entered into the design-build contract.

Practice Pointer. Allegations of fraud are often thrown about in lawsuits where one party feels that it was misled by another. This case shows that some courts will not allow more than an action for breach of contract where the duties breached are no more than what is required by the contract. Owners rely on design-build contractors to build in compliance with the plans and specifications and trust the contractor to self-inspect its work. Where the construction work is such that it is concealed from view in the construction process, such as reinforced concrete or grouted masonry block walls, or steel tendons like this stadium had, the owner may want to retain its own "special inspector" to ensure that the work is done properly. In many instances, special inspections are required by the building code and are to be provided by the owner, not the contractor, to keep the inspection process independent and free of conflicts of interest. The cost of such inspections needs to be figured into the construction budget at a frequency and unit price appropriate for the type of construction involved.

[B] Action Allowed If Fraud Induced a Party to Make the Contract

In a 1989 federal case, Aiken County, South Carolina hired an engineer to design a wastewater treatment plant and a general contractor to build the $20 million project to process 22 million gallons per day of effluent.[23] The contractor subcontracted the heat treatment system of the plant under a design-build subcontract to the BSP Division of Environtech Corporation, who was bonded by Travelers Indemnity Company. The heat exchanger failed after the plant went into operation, resulting in the blocking of pipes that carried the hot sludge from the reactor. The County became concerned about its ability to meet demands for wastewater treatment and so it filed suit against Environtech. After a 52-day bench trial, the court awarded the County $2.8 million in compensatory damages and $1 million in punitive damages against Envirotech. The general contractor was also held liable for the compensatory damages, but was held entitled to indemnity from the subcontractor. The design engineer was found not liable for the damages. The key subcontract contained a warranty clause under which Environtech warranted to the contractor and owner "that all of the Work will be of good quality . . . will be free from faults and defects . . . and will function as intended by the designer of the Work." The design specifications required the system to be operated continuously on a 24-hour per day basis, with not more than 15 percent of the total time required for maintenance and repairs. Evidence showed that the time spent cleaning the system exceeded the specified percentages and at times ran as high at 42 percent. The trial court held that the subcontractor was guilty of fraud in representing to the design engineer that it had tested the system and that it performed adequately. The evidence showed that

[23] Aiken County v. BSP Div. of Envirotech Corp., 866 F.2d 661 (4th Cir. 1989).

Envirotech had prior problems with irremediable plugging problems, but that the subcontractor had represented to the County's engineer that the equipment had been tested and had no problems. The court found that such representations were made intentionally so as to induce the County, which was unaware of their falsity, to refrain from rejecting Environtech's system and constituted actionable fraud.

Practice Pointer. It is hard to see how the County or the general contractor could have avoided this situtation. Ultimately, decisions made regarding design, construction, and selection of equipment or systems are based upon representations by others, including testing labs, manufacturers, and suppliers. If those representations prove to be false or misleading in order to induce the specification or purchase of that item, damages may be awarded for more than the cost of replacement. Insurance is normally not available for punitive damages and those costs can be sufficient to end a business either financially or by damage to reputation. While in this case the prime contractor was entitled to indemnity from the subcontractor, it is recommended that a firm require express indemnity from subcontractors upon whose representations it relies.

CHAPTER 24

ETHICAL AND CRIMINAL ISSUES IN DESIGN-BUILD AND CONSTRUCTION

§24.01 CONFLICTS OF INTEREST

Ethical questions are abundant in design-build. For example, an architect or engineer may be required to approve the contractor's pay application, but get paid from that same application. Can people really approve their own payment? Is there pressure on the design professional when acting as a subcontractor to approve work or payments going to the prime contractor? That is only the beginning. What about a design firm being required to certify completion on a job that has liquidated damages, knowing that the firm's design-build team will incur damages depending on the date certified. When material substitutions are proposed by a subcontractor or a joint venture-partner, the question is: What's best for the owner versus what's most profitable? Will an architect-as-prime fire a trade subcontractor whose performance is sub-standard when he knows that a replacement subcontractor will cost more, and that as prime contractor, the architect will have to eat that cost? These are just a few of the ethical questions that will confront design professionals in design-build.

Some architects and engineers feel that they cannot act in the client's best interest if they serve as both designer and builder and that it would compromise the checks and balances essential to the traditional three-party system of owner, design professional, and builder.[1] These feelings of unethical aspects of design-build have deep roots in the architectural profession and are embodied in the history of the American Institute of Architects (AIA).

[A] The AIA Code of Ethics (1909–1977)

When the American Institute of Architects (AIA) adopted its first Code of Ethics in 1909, the Code barred AIA members from engaging, directly or indirectly, in construction, which effectively prohibited its members from participating in design-build. This ethical restraint lasted up until the AIA's 1977 Code of Ethics, Rule 404, which stated:

> Members may not engage in building contracting where compensation, direct or indirect, is derived from profit on labor and materials furnished in the building process except as participating owners. Members may engage in construction management as professionals for professional compensation only.

This ethical rule was first passed at a time when the AIA felt that there were conflicts between the architect's obligation to the client and a potential competing self-interest in economic gain as a contractor. Those ethical issues have not gone away, but this rule has, after nearly 70 years. Architects are no longer prohibited by their professional association from engaging in design-build, whether as prime or as subcontractor.

---

[1] *Architects' Attitudes Toward Design-Build Are Ill-Conceived and Self-Defeating*, Architectural Record, Feb. 1999, p. 22.

[B] The AIA 1975 Policy Statement on Design-Build-Bid

In a policy statement on design-build-bid adopted by the AIA Board of Directors in March 1975, the AIA stated that the ethical conflict was in the architect's incentive to profit from construction, which might not be resolved even with a disclosure to the client. The 1975 policy statement said that "[w]hen the architect is employed directly as a professional consultant to the design/construct entity . . . . [t]he architect, in order to avoid any possibility of a claimed conflict of interest . . . should see that his role and extent of services with respect to the project are disclosed to the owner." However, the AIA Board warned that:

> If the participation of the architect is not such that he has only one client (the design/construct entity) to which he can give full professional loyalty, several ethical concerns come to the forefront. . . . Competition of the architect's financial interest with that of his client arising from the architect's ability to make profit-and-loss decisions during the construction process or from the architect's ability to derive compensation, directly or indirectly, out of labor and materials, points plainly to a relationship which cannot qualify under [of the AIA's Standards of Ethical Practice, 1974]. In rare instances, it might be conceivable that an extremely knowledgeable owner, after full disclosure, could "waive" his protection, but even if there is an informed "waiver" the architect still must be certain that he has resolved any potential conflict "in the best interest of the client" as required by [the Standards of Ethical Practice].

The AIA's policy statement even inferred that clients (*i.e.*, the owners) were the unethical ones in design-build, not the architects, when it stated that:

> Design-Build-Bid . . . does not contemplate that there ever will arise a traditional Architect-Client relationship. The owner is not the client. The client is in fact a composite design/construct entity which is furnishing single-point responsibility for the project to an owner who has chosen not to act as a client has traditionally acted, in order to shift the burden for price and time on the project to the design/construct team.

[C] The Antitrust Lawsuit and the AIA Code (1978–1986)

The AIA was sued for antitrust law violations by a member whose membership was suspended by the AIA over ethical violations under the 1977 Code.[2] The ethical issue in this case had nothing directly to do with design-build, but caused the AIA to repeal its entire 1977 Code and replace it with a "voluntary" code in 1980. After several years of voluntary compliance, AIA members voted in 1984 to reinstate a mandatory code of ethics for AIA membership and in 1986 the AIA adopted a new code of ethics.

---

[2]Mardirosian v. American Inst. of Architects, 474 F. Supp. 628 (D.C.D.C. 1979).

[D] The AIA Code of Ethics and Professional Conduct (1987-1997)

The several versions of AIA's Code of Ethics adopted from 1987 through 1993 suggested that architects engaging in design-build disclose the nature of their relationship with the design-builder to the owner.[3] The current 1997 AIA Code removed this portion of the commentary and is entirely silent on design-build, though it still requires AIA members to avoid and disclose conflicts of interest. The current (1997) AIA Code states:

> E.S. 3.2 Conflict of Interest: Members should avoid conflicts of interest in their professional practices and fully disclose all unavoidable conflicts as they arise.
>
> Rule 3.201 A Member shall not render professional services if the Member's professional judgment could be affected by responsibilities to another project or person, or by the Member's own interests, unless all those who rely on the Member's judgment consent after full disclosure.

Engineers should refer to the National Society of Professional Engineers (NSPE) Code of Ethics for Engineers,[4] which requires disclosure of all known or potential conflicts of interest that "could influence or appear to influence" the engineer's judgment. While not mentioning design-build, a disclosure may be necessary when an engineer is placed in a situation where decisions may be influenced by the contractor.

§24.02 CONSENTING TO ILLEGAL AND CODE VIOLATIONS OR ACTIONS AGAINST ADVICE

As part of a design-build team, especially as a subcontractor, architects and engineers may be faced with situations where their client (the prime contractor) wants to take a shortcut, make a substitution or a design change that violates the building code, OSHA standards, the ADA, or otherwise adversely impacts the project against the architect's and engineer's professional advice. Since the contractor may hold controlling interest in the joint venture or may be the prime design-build contractor, the contractor's decision may govern what happens. So what can one do as a professional? What is required by ethical codes?

The 1997 AIA Code says:

> Rule 2.105— If, in the course of their work on a project, the Members become aware of a decision taken by their employer or client which violates any law or regulation and which will, in the Members' judgment, materially affect adversely the safety to the public of the finished project, the Members shall: (a) advise their employer or client against the decision, (b) refuse to consent to the decision, and (c) report the decision to the local building inspector or other public official charged with the enforcement of the applicable laws and regulations, unless the Members are able to cause the matter to be satisfactorily resolved by other means.

---

[3] *See* Commentary, AIA Rule 3.202, Conflict of Interest (1987, 1992 and 1993).

[4] Rule II.4.a (July 1996).

This is similar to provisions in the NSPE Code of Ethics for Engineers and state licensing laws.[5] As a result, if an architect is acting as a subcontractor or joint venture partner with a contractor who is cutting corners over the architect's objection, the architect has ethical responsibilities to report such actions to the codes officials. Not only should an architect be concerned about the ethical and licensing ramifications, but he or she should be concerned about his or her own liability for participation in such decisions by passive agreement, as well as losing insurance coverage by participating in a knowing violation of a building code.[6] Professional liability policies contain exclusions like this one:

> any dishonest, fraudulent, criminal, intentional or malicious act, error or omission, or those of a knowingly wrongful nature or the willful violation of any statute, regulation, ordinance, or administrative complaint, notice or instruction of any governmental body or agency, committed by you or at your direction.[7]

Being part of a design-build team does not mean architects shelve their ethical obligations. Often a contractor's or client's decision to violate a building code can be reversed after the architect points out to the client or partner the ethical requirements he or she has and what the results will be if the architect reports this action to codes officials. Since reporting one's teammate to code officials will naturally sour the relationship, it is best to go over this ethical rule during the teaming agreement stage so that there is no misunderstanding about not compromising one's ethics.

§ 24.03 MAKING DECISIONS ON PAY APPLICATIONS, COMPLETION DATES, OR CLAIMS

The typical standard form construction contracts used for over 100 years make the design professional the judge of the contractor's performance. Many require that claims by the contractor be submitted to the architect or engineer for a decision. The AIA's B901 Design-Builder/Architect Agreement (1996 edition) places the architect in the traditional role of rejecting work, approving the contractor's submittals, certifying pay applications, and rendering decisions on contractor claims. Even claims relating to an alleged design error or omission are to be submitted for a decision to the very person who designed it. Paragraph 4.4.1 of the AIA's A201 General Conditions *requires* that even claims relating to errors or omissions of the architect be referred initially to the architect for a decision. Owners who traditionally have relied on the architect/engineer as "neutral" to

---

[5] *See* NSPE Code of Ethics for Engineers, Rule II.1.a and Rule II.1.e. (July 1996); NCARB Legislative Guidelines and Model Law/Model Regulations, Rule 100.803 (C)(1999).

[6] *See also*, AIA's Rule 2.106, which states "Members shall not counsel or assist a client in conduct that the architect knows, or reasonably should know, is fraudulent or illegal." (1997 Code of Ethics).

[7] Zurich-American's exclusion IV.A from their Architects and Engineers Professional Liability Policy, form no. U-PL-528-A CW (8/96 edition).

make decisions on claims and change order requests need to understand how the roles have changed in design-build. Some owners will still use older, traditional form contracts for a design-build project, containing clauses that require the architect or engineer to rule on contractor claims, even when the design firm is a subcontractor to that same contractor. This is the type of ethical conflict that would require a disclosure to the owner.

Paragraph 1.2.1 of the AIA's standard subcontract between Design-Builder and Architect[8] states that during construction administration, the architect is to provide services set forth in AIA Document A201, General Conditions of the Contract for Construction. That document requires claims by the contractor to be submitted to the architect for a decision as a condition precedent to arbitration. This raises several issues. First, an ethical disclosure needs to be made, as stated in § 24.01 [D]. Second, one needs to consider whether one wants to be put into this role at all. It is fertile soil for disputes among the design-build team to ask the designer to approve or disapprove of claims for time or extra money, not to mention the ethical dilemma when such decisions may impact on the architect's net profit for the design-build project. This is where a "bridge" consultant can be useful to act as a true neutral, with no financial ties to the design-builder.

[A] Immunity of Architect/Engineer as Neutral

An "arbiter" is a person chosen to decide a controversy—a referee. Architects and engineers routinely fill this role as a neutral party deciding disputes between the owner and the contractor. Courts have held that design professionals are immune from civil liability for decisions in their "quasi-judicial" role as arbiter of owner-contractor disputes. This immunity applies to actions brought by either party over the A-E's performance of his or her duties as arbiter. The rationale for such immunity is that if the design professional's decision can be questioned in suits brought against him or her by either of the parties to a dispute, there is a real possibility that the decision might be governed more by fear of suit than by the A-E's "unfettered judgment" as to the merits of the matter in dispute.[9]

The AIA's B901 Design-Builder/Architect Agreement says that in making decisions, the architect "shall not show partiality to either" the design-builder or the contractor. The AIA's A201 General Conditions contain this same wording but add that the architect "shall not be liable for results of interpretations or decisions so rendered in good faith."[10] The courts have held, however, that the architect has no immunity when not wearing the "arbiter" hat. In one case, the court stated that:

> In the role of arbitrator, *and in that role alone*, goes the cloak of immunity. If the tortious conduct with which he is charged is connected with and arises out of his determination of an owner-contractor dispute, he is usually immune

---

[8]AIA Doc. B901 (1996).

[9]Lundgren v. Freeman, 307 F.2d 104, 117 (9th Cir. 1962). One of the earliest cases recognizing immunity for the design professional came in 1896 where a Massachusetts court recognized the English law immunity concept. Corey v. Eastman, 166 Mass. 279, 44 N.E. 217 (1896).

[10]*See* A201 ¶4.2.12 (1997 edition).

> against the charge. If, on the other hand, that conduct is remote from and in no way associated with the performance of his arbitrator's function, he is liable for it in accordance with the usual principles of tort law.[11]

The privilege of immunity is qualified, not absolute, and it will not protect a design professional against actions taken in bad faith with intent to injure the contractor. One court stated it this way: "We think [the design professional] should be protected when he acts in good faith, however erroneously, and that such protection is enough. If he acts fraudulently, or with willful and malicious intent to injure the contractor, he should be liable."[12] It is, therefore, essential to protect the immunity defense that architects and engineers act impartially. This is often a difficult task, especially when the decision may be adverse to the client—who pays the architect's fees—whether that is the owner, developer, or design-build prime contractor. Keeping an objective viewpoint is essential to maintaining the integrity of the arbiter process. Architects and engineers walk a fine line between their different roles, often unaware that they wear different hats. Design-build adds to the complications since the architect/engineer may have a financial interest in the outcome of the claim.

[B] Design-Build Court Cases

Whether the architect/engineer's decision on claims, when part of the design-build team, is required or binding may depend on whether a full disclosure is made. In one 1997 court case, the prime design-build contractor hired Skidmore Owings & Merrill (SOM) as the designer and hired SDL Corporation as the contractor.[13] A dispute arose after the project suffered delays and cost overruns. Part of the claims that the prime contractor brought against SOM were "pass-through" claims of the contractor, SDL. Due to a mixing of AIA and AGC forms on the same project, there were conflicting terms on submission of claims by SDL to the architect. Nonetheless, the court found that these claims were not submitted to SOM initially, even though relating to architectural error and, as a result, the court of appeals threw out the award against SOM of the pass-through claims. The irony here is that the architect escaped liability by not being asked to review the claims against it, claims that were found to have merit.

In another case, where there was apparently no disclosure that the architect was a "silent partner" in the design-build firm, the court held that the architect could not function as a project neutral. In *Professional Builders, Inc. v. Sedan*

---

[11]Craviolini v. Scholer & Fuller Associated Architects, 357 P.2d 611 (Ariz. 1960). The contractor in that case had sought $450,000 in actual and punitive damages alleging that the architect had willfully and maliciously interfered with contract relations between the contractor and owner, and had sought intentionally to bankrupt the contractor. The court found such conduct to be outside the defense of immunity claimed by the architect, saying that where there is no dispute, there is no arbitration, and no arbiter—therefore, no immunity.

[12]*Lundgren*, 307 F.3d at 118. *See also* E.C. Earnst, Inc. v. Manhattan Constr. Co. of Texas, 551 F.2d 1026 (5th Cir. 1977), *cert. denied*, 98 S. Ct. 1246 (1978).

[13]Skidmore, Owings & Merrill v. Intrawest I Ltd. Partnership, 1997 Wash. App. LEXIS 1505.

Floral, Inc.,[14] the architect listed in the construction contract was also the vice-president and 50 percent owner of the contractor. He certified the project as complete, though the owner felt there were substantial defects. The trial court held that the architect had a conflict of interest since his decisions were binding on the owner.

§24.04 NATIONAL SOCIETY OF PROFESSIONAL ENGINEERS (NSPE)

NSPE has had its own share of problems with the Justice Department and, in a 1978 case decided by the United States Supreme Court, the organization was found to have restrained trade through its own ethical code.[15] NSPE's former Canons of Ethics prohibited its members from engaging in competitive bidding, contending that its ban protected the public safety by "minimizing the risk that competition would produce inferior engineering work." Public safety issues aside, the Supreme Court held that the only legal issue raised in the case was whether the ban had an anticompetitive impact, which the Court found that it did. The Court held that NSPE's defense that the ban was intended to prevent inferior work and to ensure ethical behavior was simply not relevant. The Court stated: "we may assume that competition is not entirely conducive to ethical behavior, but that is not a reason, cognizable under [federal antitrust law], for doing away with competition."[16]

Even the revised Code of Ethics, without the competitive bidding restriction, raises several ethical issues for design-build situations. Following are some of those issues:

[A] Accepting Commissions from Contractors

NSPE's Code of Ethics has for years contained a rule that states that:

> Engineers shall not accept commissions or allowances, directly or indirectly, from contractors or other parties dealing with clients or employers of the Engineer in connection with work for which the Engineer is responsible.[17]

Most likely, the intent of the rule was to prevent engineers from accepting bribes or other payments that might influence the engineer's judgment, or give the appearance of impropriety rather than to bar design-build work altogether. However, it could be read broadly that when the engineer is responsible for design on a design-build project, such as through a joint venture or engineer-led team, with a direct client relationship with the project owner, the engineer could not accept any payment from the contractor-partner whatsoever.

[14] 819 P.2d 1254 (Kan. App. 1991).

[15] National Society of Professional Engineers v. United States, 435 U.S. 679 (1978).

[16] 435 U.S. at 696.

[17] Rule II.5.b, *Code of Ethics for Engineers*, NSPE (July 1996).

[B] Accepting Free Engineering or Discounts from Suppliers

NSPE's former Code of Ethics prohibited its members from offering free engineering services. However, after the 1978 antitrust case, the organization repealed that rule. Even today, however, NSPE's Ethical Rule III.5.a prohibits engineers from accepting financial or other consideration, "including free engineering designs" from material or equipment suppliers in return for specifying their product. While in the traditional practice, these kinds of restrictions are understandable so as not to improperly influence the designer's decision, taking these into a design-build scenario raises new issues. What if the team preparing the technical proposal (for a meager stipend) has to choose between two suppliers for the team, one of which offers to prepare the design for the proposal stage for free? Or what if one supplier offers a discount for using it on several projects, or for early payment (without retainage) upon delivery of the equipment? It is easy to see how the ethical rules can raise new issues on design-build projects.

[C] Contingent Fee Arrangements

NSPE's Rule III.6.a presents another ethical question. It states that:

> Engineers shall not request, propose, or accept a professional commission on a contingent basis under circumstances in which their professional judgment may be compromised.[18]

This raises the issue of whether an engineer could enter into a teaming agreement with a contractor in which the engineer's fee for the technical proposal work is contingent upon the team being selected by the owner for the project. Of course, the rule is conditioned on situations where "professional judgment may be compromised," and if an engineer can participate on a design-build team on a contingent basis, without compromising its judgment on design issues, then the rule may not apply.

Contingent fees might be more suspect if the design-build team is asked to prepare a feasibility study as part of the technical proposal. In a 1981 ruling by NSPE's Board of Ethical Review, a village solicited proposals for engineering services on a $500,000 project. Three firms submitted proposals, one of which (Firm A) included a feasibility study, general layout and cost estimate, while the other two firms gave more traditional qualifications statements about their firms. Firm A also proposed that its proposal be accepted on the condition that its fee was contingent upon the village securing financing to proceed with a contract, otherwise the engineer would receive no fee. The NSPE ethics board held that Firm A went beyond submission of its qualifications and sought to influence the village by volunteering technical data, including a cost estimate and layout. The board felt that by reaching engineering conclusions as to feasibility of the project, with a contingent

---

[18]Rule 6.a, *Code of Ethics for Engineers*, NSPE (July 1996).

fee, the firm could no longer be impartial as to the client's interests in proceeding or not proceeding with the project and that the firm's practice was unethical.[19]

§24.05 USING THE TECHNICAL PROPOSAL OF ANOTHER FIRM WITHOUT CONSENT

It is common for public owners on design-build competitions to state that all design submissions by the teams become the property of the owner, usually in return for payment of a stipend to the unsuccessful teams. *See* Chapter 16. So what if a team is selected, but the owner wants the team to incorporate some of the design ideas it liked from one or two of the teams not selected? Or what if one design-build teammate approaches the other for a project type on which the contractor has made submittals for other owners, using other design consultants, and suggests that the team simply use the prior technical proposal and "site adapt" it for this project? Lots of ethical, not to mention copyright, issues arise in this situation.

Professional codes of ethics require that members not make misleading statements about their qualifications, and accurately state their responsibilities in connection with work for which they are taking credit.[20] NSPE's Code goes further, requiring engineers to "whenever possible, name the person or persons who may be individually responsible for designs." In a 1983 NSPE ethics board case, Engineer B submitted a technical proposal to a county council, and the proposal was given to Engineer A by a staff member of the council.[21] Without Engineer B's consent, Engineer A used the technical proposal in developing its own competing proposal. NSPE's Board of Ethical Review found that Engineer A's conduct was unethical and that the engineer had an obligation: (1) to seek and obtain Engineer B's consent before using the proposal; (2) if granted consent, to identify Engineer B in all cases of use of Engineer B's proposal; and (3) to negotiate and pay Engineer B "fair and reasonable" compensation for using the proposal. While this last item may be pushing the ethics board's authority (as to whether a fee must be paid if items (1) and (2) are met), it is certainly good advice whenever one is asked to use all or part of someone else's proposal to make sure the original author has consented to this use in order to avoid any ethical as well as copyright issues. As counter to marketing instincts as it may seem, the ethics rules would require one to give credit to, and name, the original designer for the portions of the design that are used, even if that designer is a competitor.

§24.06 PARTICIPATING ON MORE THAN ONE TEAM FOR THE SAME PROJECT

Design firms and contractors try to put together the most qualified team to win design-build competitions. The credentials and experience of the team,

---

[19]Contingent Contract, Case No. 81-1, NSPE Board of Ethical Review.

[20]AIA Rule 4.201- (1997); NSPE Rule III. 9 - (1996).

[21]Using Technical Proposal of Another Without Consent, Case No. 83-3, NSPE Board of Ethical Review.

including special consultants, is critical in weighted-criteria selections based on qualifications. Teaming agreements should prohibit team members from participating on competing teams for the same project, or from jumping to another team when it looks like a competitor has a better chance of winning. Imagine the surprise if a firm's top technical consultant showed up at the interview session with a competing team, not to mention the confusion by the selection committee.

NSPE's Board of Ethical Review dealt with this issue in a 1980 case in which two firms, called Able and Carlson (not the true names), were competing for the same project, both making the short-list. The public agency notified Able that his firm lacked qualifications in some specialized aspects and suggested that Able joint venture with another firm with such capabilities. Able contacted Engineer Baker, who agreed to participate as a joint venture. The public agency made the same suggestion to Carlson, who also contacted Baker, who agreed to participate in a joint venture with Carlson, but without notifying Able of this. Citing ethical rules about disclosing conflicts of interest, the Board ruled that is was unethical for an engineer to agree to participate in a joint venture with more than one of several firms being considered without making a disclosure to all of the firms.[22]

This situation also raises ethical concerns about accepting compensation from more than one party on the same project and about rules that prohibit architects and engineers from revealing information they have been asked to maintain in confidence, such as the ideas, pricing or technical designs of competing teams if one is a member of more than one team.[23]

§ 24.07 CHANGING MEMBERS OF THE DESIGN-BUILD TEAM AFTER THE INTERVIEW

The team that is identified in the proposal and participates in the interview is the one that the owner expects it is getting. Many RFP's or contracts will require the design-build contractor to identify all major subcontractors, including MBE and WBE participants, and prohibit any changes without the owner's written consent. Absent such restrictions, contractors sometimes use a price from Subcontractor A when putting together a bid, but then "shop" that number to other subcontractors after award and may let the actual subcontract to Subcontractor B, for the same or a lesser price. This type of "bid peddling" or "bid shopping" is considered by many to be unethical, but it may not technically be illegal. As a result, one may have a contractor teammate who is used to changing its subcontractors after award. On design-build projects, this raises new ethical issues when the team has been selected not just on a low-bid price, but based on the qualifications of the team. Changing the team without the owner's consent would likely be deemed unethical and may even be grounds to revoke the award.

---

[22] Participation of Engineer with Competing Firms for Same Contract, Case No. 80-4, NSPE Board of Ethical Review.

[23] *See, e.g.*, AIA Rule 3.401 (1997) and NSPE Rules II.4.b and III.4.a & b (1996).

Changing team members during the selection process has been held to be ethical if done with the owner's consent and the same opportunity is given to other competing teams. In a 1978 NSPE ethics board case, a public utility authority solicited proposals for the design of a power plant and, using a qualifications-based selection process, short-listed seven firms, one of which was a joint venture. Following the initial interview, the public agency told Firm A that the selection panel felt that its proposal did not indicate sufficient experience or technical personnel. Firm A then arranged for other participation on its team and asked permission to revise its proposal, which was granted. The ethical issue was whether it was proper for Firm A to seek to alter its qualifications proposal in order to improve its position. Without addressing whether the public agency had violated the state mini-Brooks Act (*see* Chapter 14, § 14.05), the board found that it was not only ethical, but that Firm A "had no ethical choice but to upgrade its qualifications" or to eliminate itself from further consideration, citing rules that prohibit engineers from seeking work for which they are not qualified. The board found that if the same opportunity is given to all the competing firms, there is nothing unfair about letting Firm A amend the qualifications proposal.

§ 24.08 REPLACEMENT OF THE DESIGN HALF OF THE DESIGN-BUILD TEAM

Contractors often find that they have to fire a subcontractor and replace it if the sub is not performing as required. There are standard construction contract clauses dealing with termination, replacement, and backcharges for costs incurred due to a default. One should be careful about prime contractors who treat design firms as if they are just another subcontractor. If a design-build contractor decides to change design consultants, the replacement firm needs to be cautious about taking on a project currently under contract to another firm, before that contract is terminated. Replacing design consultants raises issues dealing with licensing, ethics, copyrights, and legal restrictions on interference with contracts.

[A] The AIA Code of Ethics

For years, the AIA Code of Ethics prohibited one architect from interviewing for work already under contract to another architect. Many architects think the rule is still in effect, though it was repealed in the late 1970s. Since at least 1963, the AIA maintained in one form or another an ethical rule which in the May 1975 Standards of Ethical Practice was called Standard 9. It read:

> An architect shall not attempt to obtain, offer to undertake or accept a commission for which the architect knows another legally qualified individual or firm has been selected or employed, until the architect has evidence that the latter's agreement has been terminated and that the architect gives the latter written notice that the architect is so doing.[24]

[24] *Mardirosian*, 474 F. Supp. at 634.

This standard was also contained in nearly identical wording in AIA's 1977 Code of Ethics and Professional Conduct as Rule 605. From 1963 through 1977, the AIA processed 79 cases of alleged violations of Standard 9 and imposed sanctions in 28 of those cases.[25] But in 1977, the AIA was sued by a former member, whose membership was suspended by the AIA for violation of the Standard 9. In 1979, the federal court ruled that the standard was unlawful and violated federal antitrust statutes.[26] This rule was repealed and is no longer part of the AIA's Code. However, there are other legal reasons not to pursue work that is under contract to another design firm. Under a legal theory known as "intentional interference with contract," it is improper to interfere with a current contractual relationship and thereby cause that relationship to be terminated. There have been several court cases where an architect has sued a successor architect who took over the project, knowing that the original architect had a contract. In some cases, punitive damages have been awarded against the intervening firm.[27] As a result, while it may no longer be an AIA ethical violation, it is good business to make sure that any existing architect or engineering contracts have been terminated prior to undertaking design work on a project where another firm was previously hired.

[B] NSPE Code of Ethics

The NSPE's engineering Code of Ethics contains this rule:

> Engineers in private practice shall not review the work of another engineer for the same client, except with the knowledge of such engineer, or unless the connection of such engineer with the work has been terminated.[28]

Engineers are often hired by attorneys as expert witnesses to review the work of another engineer and are sometimes hired as consultants either for "peer review" of a set of plans prior to bidding, or to help analyze a design problem. The ethical situation could, of course, also come up when a design-build contractor is not satisfied with its current engineer and begins talks with a successor

---

[25] *Id.*

[26] Reportedly, the AIA paid an out-of-court settlement of $700,000 in 1981 to resolve the lawsuit. Engineering News Record, Aug. 2, 1984, p.19. The *Mardirosian* case was not the end of the AIA's antitrust problems. After a five-year investigation, in October 1990 the AIA signed a "Consent Decree" with the Department of Justice to settle another suit which alleged further AIA violations of federal antitrust laws. United States v. The American Institute of Architects, Case No. 90-1567 (D.C.D.C. 1990).

[27] *See, e.g.*, Williams v. Chittenden Trust Co., 484 A.2d 911 (Vt. 1984); Chaves v. Johnson, 335 S.E.2d 97 (Va. 1985). In the *Miradirosian* case, the AIA argued that its ethical standard merely reflected the legal theory against interference with contractual relations. The court held that as to interference with an existing contract, an architect is "privileged" to interfere with a contract that, by its terms, is terminable at will. 474 F. Supp. at 651. However, the court found that the AIA standard was far broader than the business interference covered by the common law doctrine of interference with contract.

[28] Rule 7.a, *Code of Ethics for Engineers*, NSPE (July 1996).

engineer before terminating the first engineer's contract. As stated above, whether or not such rules are anti-competitive, it is prudent for an engineer to consider whether there could be a claim of interference with contract by the original engineer if he or she begins negotiations with, or solicits work from, the design-build contractor prior to the termination of the first engineer's contract. Since design-build jobs are usually on a fast track, contractors who sense trouble with a subcontractor may start to line up a replacement long before formally terminating the subcontractor—whether that sub is a trade contractor or a design professional—so that there is no delay in the transition. One should talk to an attorney before engaging in discussions about reviewing the work of, or replacing, an engineer currently under contract with someone else to avoid legal and ethical complaints.

[C] State Licensing Laws

In at least one state, Idaho, there is a licensing regulation which requires a successor engineering firm to notify the firm it is replacing on a project.[29] In October 1999, the Idaho licensing board suspended the license of one engineer and reprimanded another engineer for violating the statute when it took over a city engineering project after the relationship between the city and the original engineering firm became strained and hostile.[30] The firm for which the two engineers worked reportedly spent $213,000 in legal fees and countless hours of partners' time fighting the charges.

In several states, including Illinois and South Dakota, the state licensing laws dealing with design-build specifically require that the client be advised immediately when the project architect is being replaced, and that the client consent to the replacement firm.[31] Failure to follow these procedures could result in an ethics charge against the new design firm. The obvious intent is to prevent a design-build contractor from winning the contract on the strength of a qualified design firm as a team member, and then replacing that firm with a cheaper or less qualified firm. Of equal concern is that the contractor would hire the architect to do preliminary design work and then fire the firm before construction documents are prepared in order to save costs on design fees.

[D] Copyright Issues

If a design firm is asked to take over as design consultant on a design-build project, or even to start on a new project in which the prime contractor furnishes the preliminary design, the firm needs to be sure that the prime contractor has permission of the original architect to use the design. Due to strong copyright laws, an architect who copies the drawings of another firm, or even redraws and alters the design of another firm can be liable for copyright infringement. The

---

[29] Idaho Admin. Code § 10.01.02, Rule 009. 03.

[30] *Idaho Acts on "Failure to Inform,"* Engineering News Record, Feb. 21, 2000, p. 14.

[31] 69 Ill. Admin. Code § 1150.85; S.D. Code § 20:38:22:04.

damages can be quite high, including a forfeiture of all profits gained by the infringing party. Intentional infringement can even be ground for criminal action.[32] Some design-build contracts give the copyrights in the design to the prime contractor under certain conditions.[33] One needs to be sure that the prime contractor shows proof that it owns the copyrights to the drawings and to the design, or has a license to use it. If the contractor cannot show proof of ownership, but offers to defend and indemnify the firm from any copyright action that the firm might get dragged into, it is important to ask yourself whether: (1) the contractor has the resources to defend and indemnify the firm; (2) the indemnity will protect the firm for punitive damages in the event of an intentional copyright action (probably not); (3) the indemnity will help the firm defend against any criminal charges that could be brought (again, probably not); and (4) whether the indemnity will give the firm any help in defending a licensing board, AIA, or NSPE ethical complaint, or help the firm if its license is suspended or revoked (definitely not). The best course of action is, therefore, not to undertake any design work on a project where there has been a prior design firm involved unless:

1. The contractor shows proof that the other contract has been terminated;
2. The contractor shows proof that it owns the copyrights or has a license or consent to use the design and drawings to finish the project with someone else; and
3. State licensing laws are checked for any applicable procedures relating to replacment of one firm with another.

§ 24.09 CRIMINAL LIABILITY IN THE CONSTRUCTION INDUSTRY

It is rare for criminal charges to be filed against a contractor and even more rare for charges to be filed against a design professional. Certainly participation in bribery, kickbacks, bid rigging, or other activity can result in criminal charges. Less obvious are infractions of other laws. The violation of state lien laws and licensing laws are often misdemeanors, though it is rare for criminal prosecution. A willful violation of copyright laws can be criminal.[34] OSHA violations can also result in criminal charges in cases of willful neglect. *See* § 22.14.

§ 24.10 CRIMINAL CHARGES AGAINST ARCHITECTS AND ENGINEERS

There are only a few reported criminal cases involving architects and engineers. Those reported often fall into a few categories—bribing public officials and conspiring with contractors.

---

[32] *See* Chapter 16, §16.08 [F].

[33] *See* 3.1.3 of AGC's Standard Form of Design-Build Agreement, Doc. No. 415 (1999 edition), which gives the owner the ownership rights in the documents, except for the copyrights. If terminated, the owner can use the documents upon payment for all work performed.

[34] 17 U.S.C. § 506.

[A] Bribing Public Officials

State licensing laws for design professionals often contain a prohibition against obtaining contracts through bribery. The AIA and NSPE Codes of Ethics likewise prohibit bribery. The AIA Code states:

> Rule 2.102.— Members shall neither offer nor make any payment or gift to a public official with the intent of influencing the official's judgment in connection with an existing or prospective project in which the Members are interested.
>
> Rule 2.103.—Members serving in a public capacity shall not accept payments or gifts which are intended to influence their judgment.[35]

The AIA's official commentary notes that this does not prohibit campaign contributions made in conformity with applicable campaign financing laws. There have been a few reported criminal cases in which architects or engineers have been prosecuted for bribes or attempts to bribe government officials. For example, in a 1975 Kansas case, 18 architects and engineers were indicted for bribing a public employee to award an architectural contract for work on the University of Kansas Medical Center.[36] Like the AIA Code, Kansas licensing regulation K.A.R. § 66-6-4(n) prohibits design professionals from offering a political contribution in order to secure work. The October 1996 issue of *Architecture* magazine reported another case in which the chairman of an Oklahoma architecture/engineering firm received a 30-month deferred sentence and a $1,000 fine in connection with racketeering charges involving the award of public contracts through bribes. The sentence was apparently based upon lying before a grand jury investigation into the scheme. In 1993, an FAIA member had his membership suspended for one year after pleading guilty to willfully violating state campaign finance laws, by paying an employee a $6,000 advance from the firm's corporate account, $5,000 of which was then donated by the employee to a gubernatorial candidate.[37] While contractors may not have similar ethical or licensing restrictions, be careful of overzealous marketing by your contractor teammate on design-build projects that could involve illegal bribes.

[B] Conspiring with Contractors

Many state licensing regulations prohibit design professionals from accepting compensation from more than one party for services on the same project unless there is a full disclosure and agreement among all parties. In addition, board regulations prohibit design professionals from soliciting or accepting financial or other valuable consideration from material suppliers for specifying their prod-

[35]AIA *Code of Ethics and Professional Conduct* (1997). *See also* NSPE *Code of Ethics for Engineers*, Rule II.5.b. (1996).

[36]State v. Campbell, 539 P.2d 329 (Kan. 1975).

[37]AIA National Ethics Council, Decision 93-7; AIA Memo, Feb. 1994, p. 9.

[38]*See, e.g.*, Kan. Admin. Regs. § 66-6-4(k) and (l); 4 Mo. C.S.R. 30-2.010(8), (10) and (11).

ucts; or contractors on projects where the licensee is responsible.[38] However, there have been several cases in which architects and engineers either took or paid bribes and kick-backs in return for illegal activity with contractors.

In a 1997 federal court case, an architect was sentenced to two years in prison and ordered to pay restitution for conspiring with an asbestos abatement subcontractor to over-bill for work not performed on a school district project.[39] The federal government then sued the architect under the False Claims Act, recovering triple damages of over $480,000, plus $20,000 in penalties. In an unrelated case, *Architectural Record* reported that 31 architects, real estate brokers, and construction managers pleaded guilty in October 1999 to bribery charges relating to bid rigging on interior construction work in New York City.[40] The article also reported that in one instance, nine architects, property managers, and other project consultants admitted to taking bribes from a contractor in exchange for giving the company work on specific projects. All the architects who pleaded guilty were sentenced to probation and ordered to pay fines. On top of that, the New York licensing board started proceedings to revoke the architects' licenses due to felony convictions and the AIA National Ethics Council was looking into action for violation of AIA's Code of Ethics.

§24.11 BID RIGGING AND ANTITRUST LAWS

Any design professional considering entering the construction arena as a design-build contractor needs to know a little about bid-rigging laws and the harsh legal consequences that can result. Competitive bidding is used in public construction because it gives everyone an equal chance to bid, and it saves taxpayers money. It is used to obtain the best work at the lowest price because taxpayers' money is being used. When bidders or design professionals agree to rig bids, however, the public trust is violated and a federal crime is committed. The penalties can be severe.

The federal government advises their agencies to look for any of the following signs when reviewing bids:

1. industry price lists referred to by bidders;
2. change from competitive bidding to identical bidding;
3. simultaneous price increases ("follow-the-leader");
4. rotation of bidders with each taking turns as low bidders;
5. division of market by agency or geography;
6. collusive price estimating system;
7. joint bid by two competitors when one is clearly qualified to bid on his own; and
8. identical calculation or spelling errors.[41]

---

[39] United States v. Peters, 110 F.3d 616 (8th Cir. 1997)

[40] *Architects Plead Guilty in New York Bid-Rigging Scandals*, Architectural Record, Dec. 1999, p. 59.

[41] 48 C.F.R. § 3.303.

These are the warning signs that bidders have engaged in one or more forms of bid rigging or price fixing. Federal and state laws prohibit price-fixing schemes that violate the spirit of competitive bidding. Any federal agency that suspects bid rigging is required to report the matter to the Department of Justice.

[A] What Is Bid Rigging?

The courts define "bid rigging" as "any agreement between competitors pursuant to which contract offers are to be submitted to or withheld from a third party." A construction company, whether a corporation or partnership, may be held criminally liable for bid rigging conspiracies committed by its employees if those employees were acting within the scope of their authority or apparent authority, and were acting, at least in part, with intent to benefit the business. There are several forms of bid rigging. These include:

1. **Submitting Intentionally High Bids.** Contractors submit inflated "complimentary" bids, often called "comp bids," in order to ensure that the designated bid is the low bid. This is the most common form of bid rigging, and is sometimes called a "set-up job."
2. **Agreeing Not To Bid.** An agreement between contractors to refrain from bidding on certain contracts. This may involve an agreement among two or more bidders to "take turns" being the only or low bidder on a series of contracts.
3. **Trading Jobs.** Contractors agree to rig bids in exchange for a reciprocal act of bid rigging or promises to reciprocate on future contracts. In September 1999, a construction company pleaded guilty to charges that it rigged bids through agreements with suppliers and others to allocate contracts on bridge jobs. The company agreed to pay a $720,000 fine.[42]
4. **Dividing Up the Market.** Contractors agree to divide up the markets according to the geographic proximity of the different bidders to the projects.
5. **Withdrawing Low Bidder.** A contractor will persuade the low bidders to withdraw their bids prior to contract award or will attempt to identify the bidders that are likely to submit competitive bids and attempt to persuade them to refrain from bidding or to submit inflated complimentary bids.
6. **Conspiracy with Owner's Agent.** The owner's agent (which could be a design professional) is persuaded to disclose competing bids or other relevant information, such as evaluation criteria, in exchange for bribes or kickbacks.
7. **Subcontractor Collusion.** Subcontractors enter into collusive agreements with prime contractors by agreeing to inflate their bids to other

[42]*French Firm Says It Rigged Bids*, Engineering News Record, Oct. 4, 1999, P. 19.

prime contractors in exchange for payment or assurances of subcontract awards from the conspiring prime contractor.

Design professionals should watch for signs of these schemes, whether on public or private construction, and use extreme care not to engage in any form of price-fixing agreement with competitors, nor particpate in any such scheme with contractors or others. Knowingly participating with a contractor on a design-build project to rig bids can lead to criminal prosecution as a conspiracy.

§24.12 GOVERNING LAWS

There are many federal laws that apply to bid-rigging schemes. Following are the most frequent used to charge contractors and others of bid-rigging schemes.

[A] Antitrust Laws

The Sherman Act declares illegal any contract, combination, or conspiracy in restraint of trade or commerce.[43] Bid rigging agreements "restrain trade" by undermining the competitive bidding process and a conspiracy to submit collusive, non-competitive, rigged bids is a violation of the Sherman Act which provides for criminal prosecution of violators. Conviction is a felony, punishable by: (a) fine, not to exceed $10 million if a corporation, or $350,000 if any other person; or (b) imprisonment, not to exceed three years; or (c) both.

In addition to criminal liability, a design professional or contractor can be sued in civil court for bid rigging. The Clayton Act permits the federal government, private individuals, or groups to sue. The Act provides that "any person . . . injured in his business or property by reason of anything forbidden in the antitrust laws may sue."[44] Plaintiffs in these cases may recover "treble" (triple) damages, plus costs, attorneys' fees, and interest. State attorneys general may also sue for antitrust violations either on behalf of the state or on behalf of its citizens. Penalties can include forfeiture of corporate charter, personal liability of stockholders, officers and directors, debarment from state work, injunctions, fines, imprisonment, civil penalties and civil actions for damages.

An August 28, 2000 article in *Engineering News Record* reported that a contractor pleaded guilty to violating the Sherman Act on two wastewater treatment plants by rigging bids. The contractor agreed to pay a $30 million fine over five years. Charges against other co-conspirators were expected to follow.

[B] Racketeering Laws

The Racketeer Influenced and Corrupt Organization Act (RICO) was enacted in 1970 to curb organized crime.[45] However, the law is broad enough to im-

[43]15 U.S.C.A. §1.
[44]15 U.S.C.A. §15.
[45]18 U.S.C.A. §§1961-1968.

pose civil liability on others who have no ties to organized crime. "Racketeering activity" includes 27 specific acts covered by federal criminal statutes. Mail fraud and wire fraud are among the criminal acts that qualify for RICO prosecution and are probably the most common RICO violations. Contractors engaged in bid-rigging schemes that use the mail, interstate fax, or telephone can be liable under RICO. Violators can be fined up to $25,000, imprisoned for up to 20 years, or both. A person, like the project owner, who is injured from a pattern of racketeering activity may sue for three times the actual damages sustained, plus attorneys' fees. The law covers more than just bid rigging. In a 1989 case, a contractor who developed a scheme to avoid paying subcontractors by defrauding them by use of the mail was found liable under RICO for over $150,000 in damages plus over $50,000 in attorneys' fees.[46]

[C] Mail and Wire Fraud

Because bid rigging schemes generally involve the use of the mail or wires, contractors engaged in bid rigging are usually prosecuted under the federal mail and wire fraud statutes. The statutes prohibit the use of services provided by the U.S. Postal Service or interstate transmission by means of wire of any communication for the purpose of executing a scheme to "defraud, or for obtaining money or property by means of false or fraudulent pretenses."[47] Even microwave transmission of a telephone call can support a conviction for wire fraud.[48] Violation of the mail and wire fraud statutes may result in fines of not more than $1,000 or imprisonment for not more than five years, or both for each use of the mail and for each wire transmission in connection with the scheme to defraud.

[D] Bribery and Kickbacks

There is a general federal statute that prohibits bribery to public officials, congressmen, officers or employees of the U.S. government in order to influence any official act.[49] In addition to that law, Congress has enacted specific laws that relate to the award of contracts, including construction contracts.[50] These include the Anti-Kickback Act of 1986,[51] which prohibits payments to any prime contractor or subcontractor to induce the award of subcontracts on federal projects. Under the Kickback Act, it is illegal for any person to provide, attempt to provide, offer to provide, solicit, accept, or attempt to accept a "kickback." It is also illegal to include the amount of any kickback in the contract price charged by a subcontractor. A person who knowingly and willfully engages in kickback activity can

---

[46]Atlas Pile Driving Co. v. DiCon Financial Co., 998 F.2d 986 (8th Cir. 1989).

[47]18 U.S.C.A. §§ 1341, 1343.

[48]United States v. Bohr, 581 F.2d 1294, 1303 (8th Cir. 1978).

[49]18 U.S.C.A. § 201.

[50]It is also criminal to bribe an arbitrator, judge or juror and even a lawyer in order to influence judicial proceedings. *See,* e.g., Mo. Rev. Stat. §§ 570.150, 575.260 and 575.280.

[51]41 U.S.C.A. §§ *See,* e.g. 51-58.

be sentenced to as much as 10 years in prison, plus fined, or both. The federal government can also sue individuals in a civil action under the Kickback Act to recover civil penalties equal to twice the amount of the kickback, up to $10,000. The employer or contractor of the violator can also be held liable in a civil action for the amount of the kickback. Federal regulations require the government to use contract clauses designed to detect and prevent kickbacks.[52]

Another anti-bribery law is the Hobbs Act,[53] which makes it a felony for any person to obstruct, delay, or affect commerce by extortion or attempts or conspiracies to do so. Federal courts have held that demanding a kickback falls within the Hobbs Act's definition of extortion. In one federal case, an employee of a prime contractor was convicted for demanding a $1,000 kickback as a condition to awarding a subcontract, by convincing the subcontractor that he would be denied any chance to obtain the work unless he paid the kickback.[54] "Extortion" means obtaining property from another, by wrongful use of actual or threatened force, either physical or economic. Violation of the Hobbs Act is punishable by a fine of up to $10,000, or imprisonment of up to 20 years, or both.

The Copeland Act[55] makes it illegal to demand kickbacks from persons employed on a construction project financed in whole or in part by the federal government. Persons convicted can be fined up to $5,000, imprisoned not more than five years, or both.

[E] Foreign Corrupt Practices Act (FCPA)

Probably the most misunderstood law is the Foreign Corrupt Practices Act (FCPA),[56] which prohibits bribes in obtaining contracts on foreign projects. The FCPA prohibits payment of money or gifts to: (a) any foreign official; (b) a foreign political party; or (c) any other person on behalf of a foreign official, political party or political candidate, for purposes of influencing any act or decision of such foreign official or inducing that official to use his influence in governmental functions in order to obtain business for the paying party. There are certain exceptions under the FCPA. For example, payments to an official are not illegal if they are lawful under the laws of the foreign country and are used to expedite governmental functions. Also excluded are gifts, including travel and lodging expenses, provided to a foreign official for promotional reasons or for performance of a contract with a foreign government. In a 1991 Eighth Circuit case, a business executive was convicted for providing airline tickets to an official's cousin to be used on his honeymoon, in order to help get approval of foreign contracts. The court held that this type of travel expense was unrelated to the "promotion,

---

[52] 48 C.F.R. § 52.203-7.

[53] 18 U.S.C.A. §1951.

[54] United States v. Brecht, 540 F.2d 45 (2d Cir. 1976).

[55] 18 U.S.C. § 874.

[56] 15 U.S.C.A. §§ 78dd-1 and 78dd-2.

demonstration, or explanation of products or services," which is permitted by the Act.[57] If a design-build firm pursues work in a foreign country, the firm must be cautious of local customs and "expediting fees" that are solicited in order to obtain contracts.

[F] Other Related Criminal Laws

Program Fraud is covered by 18 U.S.C. § 666 which prohibits bribery involving any agency or a state or local government entity that receives federal benefits. Violators can be imprisoned not more than 10 years and/or fined up to $250,000. Money Laundering is barred by 18 U.S.C. § 1956. "Money laundering" is the knowing conduct of financial transactions involving the proceeds of unlawful activity, including money obtained through bribes and/or kickbacks. Money laundering is punishable by imprisonment up to 20 years and/or a fine of up to $500,000 or twice the value of the money involved, whichever is larger.

[G] Debarment and Suspension

On federal projects, the government can debar or suspend contractors who engage in bid rigging. "Debarment" is the exclusion of a contractor from government contracting and government-approved subcontracting for a specified period of time. A contractor may be debarred for a conviction of, or a civil judgment for:

(a) committing fraud or a criminal offense in connection with obtaining, attempting to obtain, or performing a public contract or subcontract;
(b) violating federal or state antitrust statutes relating to bid submission;
(c) committing forgery, bribery, falsification or destruction of records, or making false statements; or
(d) committing a crime indicating a lack of business integrity or business honesty.

Generally, the period of debarment will not exceed three years, although it can be extended in order to protect the government.

"Suspension" is the act of temporarily disqualifying a contractor from government contracting and government-approved subcontracting. This is often a preliminary step to debarment. Suspension can be imposed if a contractor is *merely suspected* of any of the four offenses listed above for debarment. Conviction or rendering of a civil judgment is not required. The process for debarment is found at 48 C.F.R. Subpart 9.4. Under the federal law, a contractor can be suspended for up to 18 months if he or she is suspected or indicted for certain crimes. If convicted, the contractor can be debarred for up to three years. Many states have their own debarment laws which, like the federal law, permit suspension or debarment from state contracts due to illegal activities. During the period in which a contractor is being considered for debarment, the state can suspend a contractor's right to be considered for state contracts.[58]

---

[57] United States v. Liebo, 923 F.2d 1308 (8th Cir. 1991).

[58] *See, e.g.*, Kan. Stat. Ann. § 75-37, 103.

[H] False Statements and False Claims

The False Statements Act prohibits both oral and written false statements and the knowing and willful falsification or concealment of a material fact by "trick, scheme, or device" within the jurisdiction of any department or agency of the United States.[59] To establish liability, the government does not have to show that it or anyone else actually relied on the false statement. However, the statement must be intentional and known to be false. A contractor may be prosecuted for a criminal false statement if it submits a collusive bid and also furnishes a certification that the bid was not the product of any collusive or anti-competitive activity. Conviction under the Act may result in fines of not more than $10,000 or imprisonment of not more than five years, or both.

The False Claims Act[60] prohibits the knowing submission of a false, fictitious, or fraudulent claim for payment to a government agency. The term "knowing" means not only "actual knowledge" but "deliberate ignorance of the truth." The statute also prohibits conspiracy to defraud the government by getting a false claim paid. Conviction under the False Claims Act may result in fines of not less than $5,000 and not more than $10,000, plus three times the amount of damages the government sustained because of the act, plus costs to recover the penalty or damages.

§24.13 STATE AND FEDERAL PROSECUTIONS

The "double jeopardy" clause of the U.S. Constitution provides that "no person will be subject for the same offense to be twice put in jeopardy of life or limb. . . ." However, a civil antitrust action is not barred by the double jeopardy clause if the defendant has previously been criminally prosecuted and acquitted under the Sherman Act. The double jeopardy issue also arises where the federal government and a state separately prosecute a contractor for the same offense. Generally, a bid-rigging defendant may be subject to both federal and state prosecution for the same act without any constitutional violation, although this is rare.[61]

§24.14 WITNESS IMMUNITY

Nearly 90 percent of all federal criminal cases involve guilty pleas and many of these involve some form of plea agreement. This is certainly true in bid rigging cases. The government can grant "use immunity" to witnesses who cooperate in prosecutions and provide testimony. Testimony of such witnesses cannot

[59] 18 U.S.C.A. §1001.

[60] 31 U.S.C.A. §§ 3729(a)(1) and (a)(3).

[61] The Missouri antitrust statute says, however, that the attorney general shall not commence prosecution against a defendant who, at the time, is under indictment for federal antitrust violations. R.S. Mo. § 416.051.

later be used against them. Federal prosecutors frequently negotiate with minor players in bid rigging cases and grant them immunity in lieu of prosecution if the witnesses will cooperate in the prosecution of co-conspirators. In other instances, under plea agreements, a defendant may be charged with a lesser offense if he will plead guilty and agree to cooperate with prosecutors in cases against co-conspirators. Such cooperation can result in a reduced sentence under the Federal Sentencing Guidelines.

§24.15 GENERAL GUIDANCE

Although there are some close calls, most bid-rigging convictions involve conduct that one would recognize as illegal even before reading this chapter. If a contractor, subcontractor, or another design professional makes an architect or an engineer a proposal to submit bids that does not feel right, the design professional should go with his or her "gut instinct" and talk to a lawyer before proceeding. "Comp bids," while they may have been common in years past, are illegal. Companies have gone bankrupt and contractors and design professionals have gone to jail for bid rigging. A design-build firm should get the word out to its project architects/engineers, project managers, estimators and executives that no form of bid rigging will be tolerated by the company and will be grounds for immediate termination. The risks and penalties are just too severe to treat this subject any less seriously.

CHAPTER 25

CHANGES AHEAD FOR DESIGN PROFESSIONALS

§25.01 OPPORTUNITIES AND CHALLENGES IN DESIGN-BUILD

"Master Builder" is what history called the architects and engineers of old. This title hardly describes the role that is filled by many design professionals today. In a 1999 policy statement issued by OSHA on site safety, the agency states that architects whose contracts include only administrative responsibilities are *"little more than a supplier of architectural services and conduit of information"* to the owner.[1] Is that how the public sees the architect's role? As nothing more than a supplier of information? As protecting and improving the health, safety, and welfare of our communities. As shapers of not only the built environment, but having influence over the quality and direction of life in our communities, as planners, visionaries, and artists. Lawsuits, insurance premiums, and perhaps some inherent personality traits have driven many design professionals into more limited roles of low risk and low pay. Much has been written elsewhere about the public's decreasing perception of the architect and how the leadership role once enjoyed has been lost to construction managers, contractors, and now design-builders. Design-build presents an opportunity for those bold enough to accept the challenge to take back entire control of the construction process.

Architectural Record's Chief Editor, Robert Ivy, FAIA wrote that not only can the architect take control through design-build, but architects can create good architecture in the process. He said:

> At the large scale, design-build can produce economic efficiency through industrialized production, with consequent savings of time and effort. The soul of the design-build machine is construction engineering and the mastery of process, with a cursory nod to its architectural great-uncle, the Bauhaus. When synthesized with architectural vision, good or even great architecture can occur from coordinated, cooperative efforts; more often, when the architect's role in the design-build team diminishes and the budget constricts, the results can be mundane.[2]

By taking on a greater role (not just a subcontractor) in design-build work an architect can produce great architecture and take back the leadership role lost long ago.

[A] Diminished Role Can Result in Diminished Compensation

It is one of the true ironies that an architect who over a year or two plans, designs, draws, negotiates contracts, administers contracts, performs inspections and observations and, often, gets sued by someone in the process often makes the same 6 percent fee (much of which is paid out to consulting engineers) that a realtor makes in a few months (or less) of showing the facility to buyers and closing the sale. Architects have tried numerous ways to make more money. One way was to cut back on services, with less construction administration. This backfired

[1] OSHA Directive, CPL 2-0.124, Multi-Employer Citation Policy, Example 11, Dec. 10, 1999.
[2] Robert Ivy, FAIA, *Hands-On Architecture*, Architectural Record, May 2000, p. 23.

when construction managers stepped in and took over a part of that market. Then, architects relied on technology, with CAD drafting and standardized documents such as MasterSpec and others to reduce the cost of preparing construction documents. But investment in expensive equipment and staff training added to the cost of the process. The next wave has been the new AIA contracts in 1997, which give more opportunity to charge for additional services for things like added site visits, repeat shop drawing reviews, etc. While profits have inched up slightly (some say due only to the increase in work brought on by a better economy), design firms have struggled to make significant headway on fees and profitability. Is design-build a way back into the master builder role and a potential for increased compensation? Many firms have found greater control, higher profits and even more overall satisfaction through design-build.

"The opportunity to make money is there if you perform well," says David Evans, CEO of David Evans and Associates. "Selling hours is something we have to get away from. We have to stop selling ourselves as hourly employees. Design-build provides one way out," he adds.[3]

[B] Return to Role of Master Builder or Accept New Role as Subcontractor?

Even leading architects like Frank Gehry, FAIA have said, "I'd like to see the architect become the master builder again."[4] Architects are longing for a way to return to a leadership position in the industry. Design-build presents an opportunity to do just that—as the leader of the design-build team—but also presents a potential for a major step back, if architects become settled into the role of subcontractors to general building contractors. John C. Gaunt, FAIA, Dean of the University of Kansas School of Architecture and former CEO of Ellerbe Becket says, "The architectural profession has agonized for many years over its diminishing role as leader in the complex process of bringing a building program to fruition. Architects have watched in dismay as entrepreneurs with management skills have developed alternative project delivery systems that take control, and profit, out of their hands. In roughly a quarter-century such alternative processes have dramatically changed architectural practice and brought into question its future well-being." What the next quarter-century or more holds is, of course, not known, but many believe that it will see a strong movement to design-build as the preferred method of project delivery. Whether design firms choose to lead or be led is yet to be seen. Architects Clifford Bedar, AIA and Michael Behm, AIA of HDR in Chicago urge the profession to stop the hand-wringing over lost ground and get with the program by leading design-build. But they note:

> We wonder if our message will continue to fall on deaf ears. Like the cobbler's children, the architectural profession has no shoes. Our clients hire us to be visionaries, to create opportunities where they perceive only obstacles.

---

[3] *Hot Markets Shift Into Higher Gear*, Engineering News Record, Apr. 10, 2000, p. 56.

[4] Engineering News Record, Feb. 28, 2000, p. 77.

> Yet when it comes to the health of our own profession, we react too late, when the opportunities have disappeared.[5]

[C] Control Is Key for Many

A recurring theme in researching for this book was the number of design professionals and contractors who said that "control" is what makes a leader, and that architects usually lack control of the project once the drawings leave their office. Most agreed that control is a key element to successful design-build. The prime contractor has the most control and subcontractors have the least control over the project. Architects and engineers seeking more control can find not only that, but greater financial rewards through architect-led design-build. As stated by Robert Ivy, FAIA:

> Most architects are generalists by nature and training, capable of picking up new problems and applying a critical method, with results tailored for individuals and communities. However, for those architects who love the role of architect as builder, who seek more control of the entire design and construction process, architect-led design-build may be the path to greater satisfaction and, perhaps, higher compensation. Rather than ignore this new opportunity, let's learn about it, prepare for it, change it if we have to, and employ design-build when it is to our benefit. That would be designing, and building, success.[6]

As architects and engineers become subcontractors to general contractors, they will find that their level of control diminishes in addition to their direct contact with the client and end-user, further diminishing the role and public perception of the design professional. "Control and who is in control are key to the process," says Larry Maxfield, an architect who became a design-build contractor. "The moment you lose control, you lose the ability to deliver the project on time and on budget." Maxfield candidly admits that "[d]esign-build projects where the architect retains control are often failures. Architects are usually more concerned about maintaining design integrity. They are less concerned with cost and schedule. To succeed in design-build, one has to be concerned about cost and schedule." With design-build, especially when designer-led, the architect or engineer is in total control of design, cost, and schedule.

Brad Buchanan, AIA of Denver, Colorado compares design-build to snowboarding versus snow skiing. "Snowboarding feels to me like what skiing was supposed to feel like." He says that leading the design-build process feels like what architecture is supposed to feel like. "I sleep well at night when I am in control. When I don't sleep well is when I have a contractor that I am not comfortable with building my project." Control is a common theme among architects

---

[5] *Architects' Attitudes Toward Design-Build Are Ill-Conceived and Self-Defeating*, Architectural Record, Feb. 1999, p. 22.

[6] Robert Ivy, FAIA, *Not to Be Ignored*, Architectural Record, Oct. 1998, p. 15.

who are doing design-build, which they feel puts them in total control—an aspect lacking in traditional architectural practice. If the prime design-build contract is with a contractor and the designers are subcontractors, however, the designers have virtually no control over what goes on in the field and what decisions are made. Someone else is calling the shots in terms of the ultimate design decisions, cost, and schedule. As a design professional considers the design-build market, aside from the legal, insurance and contractual issues, one must also think about how this process can produce a better building for the client, improve the process, and create good, or even great architecture.

As one architect observed:

> There is no "design-build style," no such thing as "design-builderism" . . . design-build has no necessary formal characteristics. The term can apply equally to a highly crafted apartment renovation by an architect or a repetitive tilt-up industrial park, even though the scales vary radically and the outcomes lie at stylistic and aesthetic extremes. Both involve interdisciplinary collaboration in design and construction, while their stark differences reflect how architecture has fractured in the last century.[7]

By taking a leading role, or even sharing that lead with a contractor, architects can regain control of the process and ensure that clients get not only a faster project, within the budget, but that design quality is not overlooked or suppressed by concerns over cost and schedule. The real challenge is to produce good architecture through design-build, and to show that with architects in the lead role, that aspect of the project will not take a back seat to pressure over time and cost. Whether contractors can give clients that same assurance is questionable. This is where design firms can sell their services to clients as superior to a contractor-led project.

[D] Committing to Final Designs Early Is Required

Think of the normal time that a design firm has to work out schematic and design development phases of a project, reviewing periodic design options with an owner. Now think about a design-build competition in which the owner wants five teams to prepare and submit designs and pricing in a period of 30 to 60 days. With the compressed schedules and fast-track nature of design-build, architects and engineers are going to need to learn to design more quickly. Design-build contractor Roger "Bo" Harris of Harris Construction Co. says that architects getting into design-build need to know that they have to commit more design early on than they are comfortable with. This is because of the pressure to set a price early and the need to lock in the design quickly to permit pricing. This is difficult for many architects who are used to months of sketching, designing and trying alternative elevations, site configurations, and plans. Many architects agree that

[7] Robert Ivy, *supra* note 2.

their downfall as professionals is spending too much of the fee during the design stage and not leaving enough for construction administration, nor any profit at the end. In design-build, architects need to learn to design quickly and stay cost conscious.

As one architect said, "Our job is to move from abstract to real. Most architects have a hard time making that transition. Some never get there." Architects enjoy design and most are drawn to the profession by their natural artistic abilities. The hardest part of design for some is going from the soft, freehand sketching phase into working drawings. "Some architects cannot seem to get out of the buff-paper phase of design," said one mechanical engineer, who added that his firm works for architectural firms of all types, from the AIA-award winners, to the smaller, less flashy firms. Asked which one he prefers, he stated without hesitation, "the less flashy firms." He noted, unfortunately, that "[some firms] continue to design, revise, redesign and revise right up to the date the bid documents are printed. The base plans we are given to do the mechanical work from do not match the final architectural plans as a result. So there are addenda to the bidders, errors and inconsistencies. We all lose money on those jobs." By contrast, he says other firms do the design once, give the engineers the base design, the plans go out to bid coordinated, and "we all make money. I'll take that anytime given the choice."

§25.02 WHAT LIES AHEAD FOR DESIGN PROFESSIONALS?

All signs indicate that design-build will continue to grow and then level out as the industry finds which project types best lend themselves to this building method. The better design-build teams will find ways to resolve conflicts quickly and cooperate in exchanging ideas, recognizing that there are multiple approaches to every design challenge. The better teams will be those that are innovative in use of materials and sequencing of construction to facilitate fast-tracking of design and construction, keeping the owner's perspective at the forefront of every decision: "How can we do this faster and cheaper without sacrificing quality?" If, however, designers fail to make the switch from the disclaimer-laden, risk-shifting environment that grew during the liability crisis of the 1980s, to an atmosphere of risk acceptance and risk sharing, then designer-led design-build will not gain acceptance among owners, nor among the profession.

[A] New Attitudes Needed

Design-build does not work with the old attitudes of "design-bid-build-sue." Larry Maxfield says putting a "prima donna architect with a hard bid contractor" is a recipe for disaster. "You may as well just put the lawyers on retainer from Day One," he says. After getting his Bachelor of Architecture degree in 1974, Maxfield went to work for a general contractor as one of five architects on staff, doing design work and project management. "I learned the hard way by seeing my design mistakes in the field. You learn fast when it is your company that has to correct those mistakes." He sees a big difference, however, in how design errors are handled at a design-build company. "Because we were on the same team, the superintendent covered for my mistakes. We were on the same mission, to deliver the project just as we said we would. There was no reason to make each other look bad to the client." What a refreshing outlook on the contractor-designer relationship.

Another architect involved in design-build sums it up this way, "[w]e do not allow any finger pointing in the office." Design-build stresses the importance of the team approach to design and construction. If there is a problem, it is a team problem and will take a team solution. "Single-point" responsibility has replaced "finger-point" responsibility. Contractor Jim Neenan, president of The Neenan Company of Denver, Colorado, says that his company has 40 in-house architects, a staff that would place his firm among the largest architectural firms in the state, let alone the largest design-build contractor. Neenan's company stresses working out problems in the field, but learning from the mistake so it is not repeated. "Our staff have to change job descriptions depending on what the situation requires. The architects have to help resolve construction problems and the construction staff have to help with design."

This collaborative effort by designers and builders will run counter to the way many projects are handled, but the change in attitudes from walking in one's partners shoes will have rewards for the participants and for the industry. Engineer June Nakamura, president of Engineering Solutions, Inc. of Aiea, Hawaii, teamed with a contractor for a sewer line rehabilitation pilot project. The team held many partnering sessions to instill and maintain trust and communication among the team members and the client. She found that working side by side with the contractor opened her eyes to what contractors go through. They worked out details and resolved problems as colleagues, not adversaries, with a handshake often as the most formal agreement between them. "That's probably what I liked most about design-build," she says. "I finally have the chance to experience some of the things that he did."[8]

§ 25.03 CHALLENGES FOR EDUCATORS

If the predictions about design-build come true, architects and engineers will need to learn different skills in order to succeed in construction. As DBIA's former President, Dr. James Broaddus, says, "People in school now will find that construction is a multidiscipline process. We need to educate them to address the design-build process." Dean John Gaunt, FAIA, of the University of Kansas School of Architecture and Urban Design agrees that if the design-build projections hold true, education of architects will need to change. Gaunt says that adding construction and design-build courses to an architectural curriculum is difficult, though much needed if the universities are to turn out architects ready for the changing marketplace. This may meet with friction at schools whose faculties have no direct experience in design-build. Gaunt observed that "[a]rchitec-

---

[8] *CE and Contractor Side by Side*, American Consulting Engineer, Jan./Feb. 2000, p. 10.

ture faculties are not on the whole experienced in project delivery or sympathetic to methodologies beyond the traditional. For many educators, the Education/ Practice impasse is more than enough challenge in itself. The further consideration of processes in a more broadly considered construction industry represents a subject not easily accommodated in already compressed architectural curricula."

[A] More Hands-On Construction Experience

Robert Ivy, FAIA, said, "[i]f you've never built, but merely thought about building—planned it, drawn it, scheduled it—the craving for actualization can resemble unfulfilled hunger."[9] He notes that "hands-on architecture" programs have spread at universities from Georgia Tech to Nova Scotia and reports that "[s]tudents are dying to do this all over the country." At the University of Kansas, Dean John Gaunt, FAIA, and his faculty have developed combined continuing education and graduate-level courses in architectural management, design-build, and other variations. One successful course the university has offered has the students obtain financing, design and build a house, and perform all of the construction themselves other than electrical, plumbing, and other work that requires licensed contractors. The student work is supervised by a faculty member who has his own construction company and the entire house is designed and built by students in the course of one semester. This hands-on training teaches the students more about construction than any textbook can offer. Students have to commit to finish the project, even if that means staying on after graduation to complete the house during the summer, since the university's commitment to complete the house is based on donated student labor. In addition to the "hands-on" program, the University's first course on designer-led design-build had nearly 50 graduate students enroll, all practicing architects. Full-day seminars put on by the University of Kansas have drawn 60 to 100 attendees. Gaunt hopes that such education will allow "young practitioners of the future to have an enhanced ability to be 'players' in the competitive construction field while sustaining the architect's central focus on design and quality."

In addition to the programs at the University of Kansas, Purdue University started offering two courses on design-build in 2000. "Fundamentals of Design-Build" and "Design-Build Practicum" offer Purdue students the basics of design-build project delivery, plus hands-on experience with RFPs, schematic design, performance-based specifications, space planning, preliminary estimating and schedule, code reviews, MEP and ADA requirements. The University of Colorado at Denver also offers courses in design-build. Students in that program helped a small municipality develop a design-build RFP, interview and select teams, and award and help administer a contract for an actual project. These types of practical skills courses will prepare architects and engineers for their new roles in design-build.

---

[9] Robert Ivy, *supra* note 2.

[B] More Construction Education Needed

Architect-contractor J. Angel Martinez, AIA, says that his decision to get into design-build in 1974 was not like coming to a fork in the road and taking the construction path. "I have always felt that architects should be builders. But I didn't get the needed training in school on how to do construction, so I had to learn that by actually getting involved in construction." While working part-time for an architectural firm during college, Martinez realized that he was missing some needed understanding about construction and engineering, so he switched to the architectural-engineering program at the University of Illinois. "That engineering education has been very useful to me as a contractor," he says.

Robert Ivy, FAIA, adds that education in construction makes architects better at what they do, especially if that education includes a hands-on component. He says, "It isn't necessary to have built a wall before you design one, but if you have—at school, for yourself, or for a client—you'll build a better wall. Hands down."[10]

[C] Degreed Programs in Design-Build

Design-build courses are being offered at a handful of universities, usually as an elective as part of an architecture or engineering program. The first full-fledged degreed program at a major university was announced in 1999 by Stanford University's Department of Civil and Environmental Engineering. Stanford is offering a Masters of Science in Design/Construction Integration.[11] When construction management grew in the 1980s, universities put together degrees in construction science and project management to meet that need. With design-build taking hold in the 1990s and into the new century, it is likely that more programs like that offered by Stanford will be added at universities across the country.

---

[10] Robert Ivy, *supra* note 2.

[11] For more information on the new degreed program, *see* Stanford's Web site at <http://www.stanford.edu/group/dci/>.

APPENDIXES

APPENDIX A

AMERICAN INSTITUTE OF ARCHITECTS POSITION STATEMENT REGARDING DESIGN-BUILD

STATEMENT

"Design-Build" is a method of project delivery in which one entity signs a single contract accepting full responsibility for both design and construction services of the building facility. This entity is any party that meets the requirements within the public owners' jurisdiction with respect to offering and performing such services.

Design-build is defined as the selection of the qualified design-build entity through a competitive process which may require evaluation of the concept design and project cost, along with other criteria.

When a public agency employs the design-build method, selection of the design-build entity should be based on qualifications-based selection procedures, which require consideration of competence, capability, and a negotiated price that is fair and reasonable to the public. If the design-build selection process is utilized, however, the following are recommended criteria to meet the publics' desire for the architect to design building facilities that are safe, functional, attractive, and cost effective.

PRESELECTION

Select a short list of design-build entities based on competence utilizing qualification factors that include: (1) the ability to satisfy the project design and construction requirements; (2) past performance; (3) relevant experience; and (4) financial strength.

SCOPE OF WORK DOCUMENTS

Issue project-specific comprehensive scope of work documents prepared by licensed architects and other qualified professionals who are retained for the duration of the project. Include in the scope of work documents, at a minimum: (1) a procedure that accommodates interaction between the architect of record and the user agency; (2) detailed program statements that describe space and equipment requirements as well as other pertinent criteria; (3) site information, including a site survey and soil boring report describing subsurface conditions; (4) outline specifications; (5) budget parameters; and (6) project schedule.

REQUEST FOR PROPOSAL

Provide pre-qualified design-build entities with a comprehensive request for proposal (RFP) that includes: (1) the project scope of work documents described above; (2) the objective evaluation criteria that will be used as the basis for selection; (3) amount of the stipend to compensate the finalists; and (4) contract forms of agreement for the project.

PROPOSAL EVALUATION

Ensure that the proposals are evaluated by a jury of qualified professionals (including those licensed professionals who prepared the scope of work documents) according to the predetermined objective functional and technical criteria identified in the RFP.

COMPENSATION

Provide proposal preparation compensation to the design-build entities commensurate with the level of information required when a facility design and cost proposal are required by the RFP.

PROJECT EXECUTION

Require the design-build entity to retain and use the design architect of record throughout the duration of the project to maintain design integrity, functional and technical responsiveness, and to conduct on-site construction observation.

Board Approval—December 1997

Appendix B

SUPPLEMENTARY CONDITIONS TO DBIA DOC. NO. 540 (1999 EDITION); STANDARD FORM OF AGREEMENT BETWEEN DESIGN-BUILDER AND DESIGNER

These Supplementary Conditions amend Doc. No. 540 published by the Design-Build Institute of America (DBIA), titled Standard Form of Agreement Between Design-Builder and Designer, 1999 Edition.

SC-1 Conflicts. Amend Section 1.4.1 by replacing the last sentence with this new one: "In the event of any inconsistency, conflict, or ambiguity between or among the Contract Documents, these Supplementary Conditions shall control. If there are other conflicts between the Contract Documents listed in part .1, .2 and .4 of Section 1.3.1, the documents shall take precedence in that order."

SC-2 Communication with Owner. Add to the end of Section 2.1.4 this new sentence: "Notwithstanding the preceding sentence, the Designer may communicate directly with the Owner on matters of public health, safety and welfare when required by law or by any code of ethics if Design-Builder takes any action against Designer's recommendation or advice that violates any law, code or ordinance or places the Owner or the public at risk."

SC-3 Standard of Care. Delete the last sentence of Section 2.2.1 and replace with the following: "This standard shall apply to all services performed by Designer including, but not limited to, efforts to achieve any Performance Standard Requirements of the Design-Build Agreement."

SC-4 Estimating Errors. Add the following new sentence to Section 2.6.6. "In performing design services, Designer is expected and entitled to rely on Design-Builder's estimating services required by Section 3.3.3. The Designer shall not be required to revise the design or the Contract Documents at Designer's own cost where such revisions are due to estimating errors, mistakes or miscalculations of market conditions by Design-Builder. The Design-Builder shall pay Designer as Additional Services the cost of design revisions needed to the extent caused by such estimating errors."

SC-5 Site Observations. Delete the first sentence of Section 2.7.6 and replace with the following: "Designer shall at appropriate intervals visit the Site to determine if the construction is proceeding in general in accordance with the Construction Documents. Design-Builder recognizes that absent full-time resident services by Designer, much of the Work will go unobserved by the Designer and that full-time resident services are not part of Designer's contract services."

SC-6 Certifications. Add the following two new sentences to Section 2.7.8: "At a minimum, the Designer has the right to: 1) approve the form of certificate in advance; and 2) issue only those certificates Designer can reasonably do based on its level of involvement during construction. No such certification shall require Designer to issue any warranty or guarantee of the design or construction work."

SC-7 Reliance on Furnished Information. Delete the second sentence of Section 3.3.1 and replace with the following: "Designer can rely on the completeness and accuracy of such data in performing its services. If there are any limitations on Designer's ability to rely on such data, those limitations shall be stated in the blank space immediately following this paragraph. In the absence of any such limitations, there shall be none inferred or implied.

____

____

____

____

SC-8 Ownership and Copyrights in Work Product. Amend Section 4.2.1 by adding the following new sentences: "The copyrights in Designer's Work Product, including any Architectural Works, shall remain with the Designer. Notwithstanding any provision of the Design-Build Agreement, both Design-Builder and Owner are prohibited from future reuse of the Work Product on this Project or another project without an agreement in writing acceptable to Designer. In the event that either the Owner or Design-Builder desire to use the designs or Work Product of the Designer for other projects or for completion of this Project after termination of this Agreement, the parties may agree in writing to a limited license for such use provided: 1) Designer is first paid in full for all services rendered and expenses incurred in connection with this Agreement; 2) the party seeking Designer's consent agrees that such use without Designer's involvement is at the user's sole risk and without liability to Designer and further agrees to release, indemnify, defend and hold harmless Designer, its subconsultants and their respective officers, directors, employees and agents from any claims, damages, losses, liabilities, cost or expense, including attorney's and expert fees, whatsoever arising in any manner from such use including (where permitted by law) those claims alleging the negligence of any such indemnified person or party; and 3) Designer is paid an additional sum of $____________. This Section shall survive any termination of the Design-Build Agreement or this Agreement."

SC-9 Disputes. Delete Section 11.3.2. No decision by Owner or result of proceedings involving Owner shall be conclusive on Design-Builder's liability to Designer for any time extension or additional compensation or other relief.

SC-10 Applications for Payment. In the first sentence of Section 6.2.1, delete the word "certified."

SC-11 No Contingent Payments. Sections 6.2.3, 6.3.1 and 6.5.1 are not to be interpreted as contingent "pay when paid" or "pay if paid" clauses. No payment to Designer hereunder is conditioned on the Owner having first paid the Design-Builder. If Owner fails to make payment to Design-Builder for Designer's services, then Design-Builder shall nonetheless make payment to Designer no later than 45 days after receipt of Designer's Application for Payment. Interest on amounts due and unpaid shall accrue at the rate of 1.5% per month or the highest legal rate permitted by law, whichever is less. Delete Sections 6.6.1 and 7.1.1 entirely.

SC-12 Insurance. The parties recognize that Professional Liability Insurance is typically written on a one-year, claims made basis and that Designer cannot guarantee that such insurance will be available at reasonable rates after Final Payment is made. Design-Builder shall not insist on a certificate of insurance beyond that which Designer can reasonably provide. Sections 6.5.1 and 9.1.5 shall not require Designer to provide any certificate of insurance for Professional Liability Insurance for any specific period not ordinarily issued by Designer's insurer. In addition, nothing in Section 9.1.6 shall require the Designer to name the Design-Builder or others as additional insureds on Designer's Professional Liability and Workers Compensation policies, which normally do not permit additional insureds.

Design-Builder shall name the Designer as an additional insured on the Design-Builder's Commercial General Liability policy and Comprehensive Automobile Liability policy of insurance and shall provide the Designer with a certificate of insurance evidencing such coverage.

SC-13 Release of Claims. Nothing in Section 6.5.1 shall be construed as a release of claims that first arise after the submission of the Designer's final Application for Payment.

SC-14 Indemnity. This Agreement is amended on this topic as follows:
A) Section 6.8.1. Designer's obligation to indemnify the Owner and Design-Builder for claims of payment or liens by subconsultants is expressly conditioned on Designer having first been paid for those services.
B) Section 10.1.2. Designer's obligation to indemnify the Owner and Design-Builder for copyright and patent claims shall not apply when the Owner or Design-Builder provide the design in which case the Designer shall be indemnified by the Design-Builder for all claims, damages, loss or expense, including attorney's and expert fees, arising out of the use of such design.
C) Section 10.2.1. Delete the words "defend," "claims," and "liabilities."
D) Section 11.3.4. Designer's obligation to pay damages under this Section is limited to those damages caused by the negligent acts, errors or omissions of Designer, its employees or consultants.

SC-15 Liquidated Damages. Any liability of Designer under Section 11.7.2 shall be limited to those damages assessed to and paid by Design-Builder to the extent actually caused by the negligent acts, errors or omissions of Designer, its employees or consultants.

SC-16 Responsibility. Nothing in this Agreement shall be construed as making Designer liable for Design-Builder's own defects or deficiencies in the work, nor shall this Agreement be construed to make Designer a Controlling Employer as defined by OSHA for purposes of site safety.

SC-17 Licensing. The Design-Builder is duly qualified, licensed (where required) and authorized by law to act as general contractor and to offer design-build services in the state where the Project is located.

SC-18 Arbitration. Add the following sentence to Section 11.5.2: "Arbitration hereunder shall be governed by the Federal Arbitration Act, 9 U.S.C. sec. 1, et seq."

SC-19 Severability. Add the following to Section 12.4.1: "If the contract between the Owner and Design-Builder is incorporated by reference into this Agreement, then to the extent that the Design-Build Agreement is declared invalid or unenforceable, this Agreement shall be deemed severable. Any provision or part of the Design-Build Agreement which is held to be void or unenforceable under any law, shall be deemed stricken and all remaining provisions of this Agreement shall continue to be valid and binding upon the parties, including the Designer's right to payment for professional services rendered and expenses incurred. Designer's right to payment is not contingent upon the validity of the Design-Build Agreement.

SC-20 Electronic Media. Design-Builder may request computer disks containing some or all of the work prepared by the Designer either as record drawings or for use by subcontractors in shop drawings. Due to the potential that the information set forth on electronic media (disk) can be modified by the Design-Builder, or others, unintentionally or otherwise, and that design changes may be made after the data is downloaded onto a disk, the Designer shall remove all references to its authorship, corporation name, and/or involvement from each electronic display. Design-Builder recognizes that use of such electronic media (disk) will be at Design-Builder's sole risk and without any liability, risk or legal exposure to the Designer, its employees, officers or consultants.

SC-21 Record Drawings. If Designer's Basic Services include, in addition to items mentioned herein, the preparation of a set of reproducible record drawings showing significant changes in the Work made during construction, then Designer may utilize marked-up prints, drawings and other data furnished by the Design-Builder upon which the Designer shall be entitled to rely, plus change order and supplemental drawings issued by Designer.

SC-22 Limitation on Services. Basic Services of the Designer include the following. Services beyond these limits shall be provided as Additional Services in accordance with Section 2.8.1:

.1 up to _____ (___) reviews of each Shop Drawing, Product Data item, Sample and similar submittal of the Design-Builder.

.2 up to _____ (___) visits to the site by the Designer over the duration of the Project during construction.

.3 up to _____ (___) inspections for any portion of the Work to determine whether such portion of the Work is substantially complete in accordance with the requirements of the Contract Documents.

.4 up to _____ (___) inspections for any portion of the Work to determine final completion.

In addition, the following Construction Phase Services shall be provided by the Designer as Additional Services in accordance with paragraph 2.8.1:

.5 review of a Design-Builder's submittal out of sequence from the submittal schedule agreed to by the Designer.

.6 responses to Design-Builder's requests for information where such information is available to the Design-Builder from a careful study and comparison of the Contract Documents, field conditions, other Owner-provided information, Design-Builder-prepared coordination drawings, or prior Project correspondence or documentation.

.7 Change Orders and Construction Change Directives requiring evaluation of proposals, including the preparation or revision of Designer's drawings or specifications.

.8 providing consultation concerning replacement of Work resulting from fire or other cause during construction.

SC-23 Mediation. Add the following to Section 11.4.3: "The mediation shall be held in ______________________, __________, unless another location is mutually agreed upon by Design-Builder and Designer."

SC-24 Termination. Add the following new Section 7.6: "Designer shall have no liability for drawings, documents, designs or services which are incomplete due to an early termination of this Agreement."

SC-25 Payment Bonds. If Design-Builder provides Owner with a payment bond for this Project, then Designer's services shall be considered to be labor and Designer's Work Product shall be considered to be material for purposes of such bond and Designer shall be an intended beneficiary and claimant under such bond.

SC-26 Post Completion Services. As part of Basic Services, the Designer shall perform a one-year warranty walk-through with the Design-Builder if required by Owner, within one year after Substantial Completion of the Project. The purpose of this walk-through is to ascertain any defects or failures of the Work which may be covered by any equipment or other warranties on the Project. If such defects, deficiencies or failures are noted, the Designer shall assist the Design-Builder in notifying the appropriate party of the nature of the problem and the applicable warranty.

If additional architectural or engineering services are necessary to correct defects resulting from Design-Builder error or defective materials, and not the result of Designer error, such services shall be considered as Additional Services and the Designer shall be compensated accordingly, provided such services have been previously authorized in writing by the Design-Builder.

SC-27 Savings Sharing. To the extent that the Design-Builder shares savings with the Owner on actual costs less than the Guaranteed Maximum Price or other limitation, it is agreed that Design-Builder's portion of the savings shall be shared between Design-Builder and Designer in the following manner:

Design-Builder: ________________

Designer: ____________________

SC-28 Limitation of Liability. Design-Builder agrees that to the fullest extent permitted by law, and notwithstanding any other provision of this Agreement, the total liability, in the aggregate, of Designer, its officers, directors, employees, agents and subconsultants, and any of them, to Owner, the Design-Builder, and anyone claiming by, through or under them, for any and all claims, damages, losses, liabilities, costs or damages whatsoever arising out of, resulting from or in any way related to the Project or this Agreement from any cause, including but not limited to the negligence, professional errors or omissions, strict liability, breach of contract, or warranty (express or implied) of Designer, its officers, directors, employees, agents or consultants or any of them, shall not exceed the sum of $____________ or Designer's total fee under this Agreement, whichever is greater. The parties agree that adequate consideration has been given for this limitation.

SC-29 Miscellaneous. In Section 7.2.1, delete the words "to ensure" and replace with the word "so".

SC-30 Other Modifications. Insert here any other changes to the DBIA Doc. No. 540 (1999 Edition) agreed to by the parties:

DESIGN-BUILDER:	DESIGNER:
____________________	____________________
By: ________________	By: ________________
Name: ______________	Name: ______________
Title: ______________	Title: ______________

Appendix C

SUPPLEMENTARY CONDITIONS TO AGC DOC. NO. 420 (1999 EDITION); STANDARD FORM OF AGREEMENT BETWEEN DESIGN-BUILDER AND ARCHITECT/ENGINEER

[Note to user. These Supplementary Conditions should be considered by design professionals who are asked to act as subcontractors to general contractors under the Associated General Contractors of America (AGC) design-build contract forms.

[NOTE: Replace page 13 of the AGC Doc. 420 Agreement (1999 edition) with this.]

SUPPLEMENTARY CONDITIONS OR SERVICES

The following provisions amend and modify Article 10 of the Standard Form of Agreement Between Design-Builder and Architect/Engineer (AGC Doc. 420, 1999 edition) between the parties. In the event of any conflict, inconsistency, or ambiguity between the terms and provisions of the following supplementary conditions and the Agreement, the following Supplementary Conditions shall govern and control.

10.10 The following revisions to this Agreement are to be incorporated:

a. 3.2 Basic Services not required under this Agreement include the following:

[list here any Basic Services not required]

b. 3.2.7 Add the following: The review of shop drawings and submittals shall not constitute approval of safety precautions or construction means, methods, techniques, sequences or procedures. The approval of a specific item shall not constitute approval of an assembly of which the item is a component.

c. 3.2.11 Add the following: Nothing in this Agreement shall be construed as making Architect/Engineer liable for Design-Builder's own defects or deficiencies in the work. Also, delete the words "unless they are specified by the Architect/Engineer" in line 17.

d. 3.2.12 Add the following: Nothing in this Agreement shall be construed to make Architect/Engineer a Controlling Employer as defined by OSHA for purposes of site safety.

e. 3.3 Delete subpart .3 and replace with "Aerial photographs of the site."

f. 3.3 Delete subpart .5 entirely. Such services shall be provided by Design-Builder through other consultants. Architect/Engineer shall be entitled to rely on the completeness and accuracy of all soils, subsurface and environmental data furnished.

g. 3.4 Delete the words: "The Architect/ Engineer warrants and represents that. "

h. 4.1.1 In the second sentence delete the words "to the same extent as the Design-Builder."

i. 4.3 Add a new Paragraph. 4.3 The Design-Builder is duly qualified, licensed (where required) and authorized by law to act as general contractor and to offer design/build services in the state where the Project is located.

j. 5.1 and 7.1 Delete in their entirety and replace with paragraph 10.17.

k. 6.3.1 Add to the end: "In any event, payment to the Architect/Engineer shall be made within forty-five (45) days from the date of Architect/Engineer's invoice whether the Owner has paid Design-Builder or not. Interest on amounts due and unpaid shall accrue at the rate of 1.5% per month or the highest legal rate permitted by law, whichever is less.

l. 6.3.4 Architect/Engineer's obligations under this paragraph are conditioned upon Architect/Engineer having been paid in full for all outstanding invoices. Design-Builder shall not pay any lien claim by a subconsultant of Architect/Engineer over Architect/Engineer's reasonable objection.

m. 7.2.4 Delete the second to last sentence.

7.2.5 Delete the last sentence. The parties recognize that Professional Liability Insurance is typically written on a one-year, claims made basis and that Architect/Engineer cannot guarantee that such insurance will be available at reasonable rates after Substantial Completion. Design-Builder shall not insist on a certificate of insurance beyond that which Architect/Engineer can reasonably provide.

n. 10.1 Delete in its entirety, including all subparts, and replace with paragraph 10.14 and 10.15.

o. 10.3. Add to the end: "Arbitration hereunder shall be governed by the Federal Arbitration Act, 9 U.S.C. sec. 1, et seq."

10.11 If the contract between the Owner and Design-Builder is incorporated by reference into this Agreement, then to the extent that the Owner/Design-Builder Agreement is declared invalid or unenforceable, this Agreement shall be deemed severable. Any provision or part of the Owner/Design-Builder Agreement, or any provision herein, which is held to be void or unenforceable under any law, shall be deemed stricken and all remaining provisions of this Agreement shall continue to be valid and binding upon the parties, including the Architect/ Engineer's right to payment for professional services rendered and expenses incurred. Architect/Engineer's right to payment is not contingent upon the validity of the Owner/Design-Builder Agreement, nor the Owner's payment to the Design-Builder, and payment by the Owner shall not be a condition precedent to Architect/Engineer's right to payment.

10.12 Architect/Engineer is duly licensed and authorized to provide the professional services required of it by this Agreement. All design services shall be performed by Architect/Engineer or other licensed professionals and furnished to Owner by Design-Builder.

10.13 Design-Builder agrees that to the fullest extent permitted by law, and notwithstanding any other provision of this Agreement, the total liability, in the aggregate, of Architect/Engineer, its officers, directors, employees, agents and subconsultants, and any of them, to Owner, the Design-Builder, and anyone claiming by, through or under them, for any and all claims, damages, losses, liabilities, costs or damages whatsoever arising out of, resulting from or in any way related to the Project or this Agreement from any cause, including but not limited to the negligence, professional errors or omissions, strict liability, breach of contract, or warranty (express or implied) of Architect/Engineer, its officers, directors, employees, agents or Consultants or any of them, shall not exceed the sum of $_____________ or Architect/Engineer's total fee under this Agreement, whichever is greater.

10.14 All documents, drawings and designs prepared by Architect/Engineer, including the copyrights therein, are and shall remain the property of the Architect/Engineer whether or not this Project is built. Design-Builder shall not reuse such documents or designs on other projects without Architect/Engineer's written consent and appropriate compensation. Design-Builder shall prohibit the Owner and any subcontractors from using or reproducing such documents or designs without Architect/Engineer's written consent. Design-Builder's right to use and reproduce the Architect/Engineer's documents and designs for this Project terminates if this Agreement is terminated by either party.

10.15 In the event that either the Owner or Design-Builder desire to use the designs or documents of the Architect/Engineer for other Projects or for completion of this Project after termination of this Agreement, the parties may agree in writing to a limited license for such use provided: 1) Architect/Engineer is first paid in full for all services rendered and expenses incurred in connection with this Agreement; 2) the party seeking Architect/Engineer's consent agrees that such use without Architect/Engineer's involvement is at the user's sole risk and without liability to Architect/Engineer and further agrees to release, indemnify, defend and hold harmless

Architect/Engineer, its subconsultants and their respective officers, directors, employees and agents from any claims, damages, losses or liabilities whatsoever arising in any manner from such use including (where permitted by law) those claims alleging the negligence of any such indemnified person or party; and 3) Architect/Engineer is paid an additional sum of $______________.

10.16 Design-Builder may request computer disks containing some or all of the Work prepared by the Architect/Engineer either as record drawings or for use by subcontractors in shop drawings. Due to the potential that the information set forth on electronic media (disk) can be modified by the Design-Builder, or others, unintentionally or otherwise, and that design changes may be made after the data is downloaded onto a disk, the Architect/Engineer shall remove all references to its authorship, corporation name, and/or involvement from each electronic display. Design-Builder recognizes that use of such electronic media (disk) will be at Design-Builder's sole risk and without any liability, risk or legal exposure to the Architect/Engineer, its employees, officers or consultants.

10.17 Architect/Engineer agrees to indemnify and hold harmless the Design-Builder from and against all losses, damages, and costs, including reasonable attorney's fees and expenses, but only to the extent caused by the negligent acts, errors or omissions of the Architect/Engineer, its subconsultants and employees and those for whom they are liable arising out of this Agreement.

10.18 Design-Builder shall name the Architect/Engineer as an additional insured on the Design-Builder's Commercial General Liability policy and Comprehensive Automobile Liability policy of insurance and shall provide the Architect/Engineer with a certificate of insurance evidencing such coverage.

10.19 If Architect/Engineer's Basic Services include, in addition to items mentioned herein, the preparation of a set of reproducible record drawings showing significant changes in the Work made during construction, then Architect/Engineer may utilize marked-up prints, drawings and other data furnished by the Design-Builder upon which the Architect/Engineer shall be entitled to rely, plus change order and supplemental drawings issued by Architect/Engineer.

10.20 Basic Services of the Architect/Engineer include the following. Services beyond these limits shall be provided as Additional Services in accordance with paragraph 3.3:

.1 up to _____ (___) reviews of each Shop Drawing, Product Data item, Sample and similar submittal of the Design-Builder.

.2 up to _____ (___) visits to the site by the Architect/Engineer over the duration of the Project during construction.

.3 up to _____ (___) inspections for any portion of the Work to determine whether such portion of the Work is substantially complete in accordance with the requirements of the Contract Documents.

.4 up to _____ (___) inspections for any portion of the Work to determine final completion.

In addition, the following Construction Phase Services shall be provided by the Architect/Engineer as Additional Services in accordance with paragraph 3.3:

.5 review of a Design-Builder's submittal out of sequence from the submittal schedule agreed to by the Architect/Engineer.

.6 responses to Design-Builder's requests for information where such information is available to the Design-Builder from a careful study and comparison of the Contract Documents, field conditions, other Owner-provided information, Design-Builder-prepared coordination drawings, or prior Project correspondence or documentation.

.7 Change Orders and Construction Change Directives requiring evaluation of proposals, including the preparation or revision of Architect/Engineer's drawings or specifications.

.8 providing consultation concerning replacement of Work resulting from fire or other cause during construction.

10.21 The mediation shall be held in ____________________, __________, unless another location is mutually agreed upon by Design-Builder, Owner and Architect/Engineer. This supersedes the provisions of Paragraph 9.1.

10.22 This Agreement may also be terminated by the Design-Builder without cause, for the Design-Builder's convenience or if Owner exercises its right to terminate for convenience, upon seven (7) days' written notice to the Architect/Engineer. In such event, Architect/Engineer shall be paid: .1 for all services performed and expenses incurred prior to termination; plus, .2 reasonable termination costs and expenses, plus fair and reasonable sums for overhead and profit on items .1 and .2 above. Architect/Engineer shall have no liability for drawings, documents or designs which are incomplete due to an early termination of this Agreement.

10.23 As part of Basic Services, the Architect/Engineer shall perform a one-year warranty walk-through with the Design-Builder if required by Owner, within one year after Substantial Completion of the Project. The purpose of this walk-through is to ascertain any defects or failures of the Work which may be covered by any equipment or other warranties on the Project. If such defects, deficiencies or failures are noted, the Architect/Engineer shall assist the Design-Builder in notifying the appropriate party of the nature of the problem and the applicable warranty. If additional architectural services are necessary to correct defects resulting from Design-Builder error or defective materials, and not the result of Architect/Engineer error, such services shall be considered as Additional Services and the Architect/Engineer shall be compensated accordingly, provided such services have been previously authorized in writing by the Design-Builder.

10.24 Design-Builders portion. To the extent that the Design-Builder shares savings with the Owner on actual costs less than the Guaranteed Maximum Price or other limitation, it is agreed that savings shall be shared between Design-Builder and Architect/Engineer in the following manner:

Design-Builder: ________

Architect/Engineer: _______

10.25 **Other Basic Services.** The following other services shall be included as part of Basic Services: (if none, write "none" below).

Article 11

SCHEDULE OF EXHIBITS

The attached Exhibits are part of this Agreement.

EXHIBIT A	The Standard Form of Design-Build Agreement and General Conditions Between Owner and Design-Builder, dated ________________, _______ pages. (Copy has been furnished to Architect/ Engineer for review: _____ YES; _____ NO).
EXHIBIT B	Owner's Program, Design-Builder's schedule and estimate and any other relevant information, dated __________________________, ______ pages.
EXHIBIT C	Direct Personnel Expense Rates, dated ______________________, ______ pages.
EXHIBIT D	Key Project Personnel, dated ____________________________, _____ pages.
EXHIBIT E	Reimbursable Expenses, dated ____________________________, ______ pages.

This Agreement entered into as of the date entered in Article 1.

DESIGN-BUILDER:	ARCHITECT/ENGINEER:
______________________________	______________________________
By: ___________________________	By: ___________________________
Name: _________________________	Name: _________________________
Title: __________________________	Title: __________________________

APPENDIX D

TEAMING AGREEMENT FOR DESIGN-BUILD PROJECTS

THIS AGREEMENT is entered into between ABC Architects, Inc. ("Architect") and XYZ Construction Co., Inc. ("Contractor") relating to certain design-build projects that the parties want to pursue jointly. For good and valuable consideration, the receipt and sufficiency of which is acknowledged, the parties agree as follows:

1. The parties desire to jointly pursue contracts with various owners, for design-build work for of the following building type: ______________________ ____________________________________ in the states of ________________, ________________ and ________________. It is the parties' intention to form a limited liability company ("L.L.C."), owned and managed by the two parties, if successful in being awarded a design/build contract as a result of this Teaming Agreement. The parties agree that this proposed L.L.C. is formed to pursue work only for the above building type and geographical region, but it may be expanded to pursue work with others and in other states or countries in the future, by written agreement.

2. The proposed ownership in the L.L.C. is Architect 51% and Contractor 49%, subject to requirements or limitations by state licensing boards, which the parties realize may require that two-thirds of the L.L.C.'s members be licensed professionals. In that case, the Architect's share of ownership may be held by some, or all, of the following individual architects or engineers:

____

____

____

3. The parties intend to subcontract with the L.L.C., if permitted by state law, and to receive normal compensation for providing design and construction services. Architect anticipates architectural fees in the range of ___% to ___% of construction cost, and Contractor anticipates fee in the range of ___% to ___% of the construction cost, all depending on the scope of the project, subject to further agreement on a project-by-project basis.

4. The L.L.C. will receive fees for providing Project Management services, insurance, financial management and billing, assuming the risk under the prime design/build contracts, and coordination of consultants and subcontractors.

5. It is intended that the profit (or loss) be shared by the parties in this manner:

Architect ___% and Contractor ___%, with distributions made upon completion of each project, unless an earlier distribution is agreed on by the parties.

6. For a period of ___ months [or years] from the date of this Agreement, neither party will independently pursue design/build projects of the above building type, in the states listed above individually or as part of another design/build organization, team or joint venture, unless agreed to in writing by both parties, or unless the L.L.C. is terminated, or unless one party is in substantial breach of its obligations. This Agreement may be renewed or modified by written modification, signed by both parties.

7. In the event either party becomes aware of a potential client or owner who is not interested in the design/build method of project delivery, then either party is free to pursue their own independent arrangements to provide their individual specialties to that client or owner, with no obligation to the other.

8. The parties agree to share equally the legal and accounting fees necessary to set up the L.L.C., including any related filing fees with Secretary of State or Licensing Board offices, or other regulatory agencies. Legal counsel for the L.L.C. shall be the law firm of Smith & Jones, P.C.; the accountant for the L.L.C. shall be Taylor & Johnson Accountants, L.L.C.

9. The parties shall work to put together an insurance and bonding program for the L.L.C., through their respective insurance and bond agents. Premiums attributed to the L.L.C. shall be an expense of the L.L.C.

10. The parties understand that marketing costs will be necessary in order to obtain work, including costs to prepare qualifications statements, technical proposals and designs, and cost proposals. The parties also understand that some out-of-town travel may be necessary. It is agreed that such costs shall be borne by the parties or shared in this manner:

[describe agreement here]

In the event that the parties are awarded a stipend or honorarium for their proposal, that money shall be shared as follows:

[describe agreement here]

11. While it is the intent of the parties to form an L.L.C., it is recognized that some clients will require payment and performance bonds from the design/build contractor. In that event, it is agreed that Contractor shall provide a dual obligee bond which shall run to the benefit of the L.L.C. and the Owner. If that is not acceptable to the Owner, or not permitted by the surety, then it is agreed that the Contractor shall hold the prime design/build contract with the Owner and Architect shall be a subcontractor. Fees shall be adjusted in that case such that Architect receives a fee within the range set out in Section 3, above, plus a sum equal to ___% of the projected profit. Contractor will receive a fee within the range set out in Section 3, above, plus a sum equal to ___% of the projected profit.

12. Nothing in this Teaming Agreement shall be construed as a joint venture or partnership of any form, and the parties shall not be held jointly or severally liable for any obligations.

13. Decisions regarding which projects to pursue, costs to be incurred, the content of the qualifications statement, technical proposals, cost proposals, consultants and subcontractors to be utilized shall be made jointly. If the parties cannot reach agreement, after discussion, then the President of the Architect shall make the final decisions relating to subconsultants, design, technical proposals and marketing; and the President of the Contractor shall make the decisions regarding suppliers and subcontractors. If the parties cannot reach agreement on a cost proposal, then the team shall not submit a cost proposal and costs incurred for that project or proposal shall be shared as set out in Section 10.

14. If either party decides that the team should not pursue a design/build project within the building type and geographic region covered by this Teaming Agreement, then it shall notify the other party in writing of its intention within five (5) days from receipt of a written request of intentions. In that case, the other party hereto shall be free to pursue such project with another contractor or architect, or as part of another team, without being in breach of this Teaming Agreement.

15. The Architect shall be the spokesperson and main client contact, unless the parties agree that the Contractor has the closer relationship or is more appropriate to be the main contact (such as when the Contractor is a prime contractor and Architect is a subcontractor).

16. The parties agree that information disclosed to the other of a financial or proprietary nature, or where marked "Confidential," shall be kept in strict confidence during the term of this Teaming Agreement and thereafter and shall not be disclosed unless required by legal process or court order.

17. This Teaming Agreement shall not be assigned to anyone, for any reason whatsoever.

18. Copyrights in all designs developed by Architect shall be owned by Architect and may be used by it, in whole or in part, without further consent of Contractor, subject to further restrictions placed on the parties by the client or owner. Contractor shall not use such designs other than for work of the L.L.C. without written consent of Architect.

19. Disputes between the parties shall be resolved first by the presidents of Architect and Contractor informally. Failing an agreement, the parties shall submit their disputes to mediation, with mediator's fees shared equally. If not settled through mediation, nor within 60 days after the first mediation session, disputes shall be resolved by binding arbitration, pursuant to the Federal Arbitration Act, 9 U.S.C. § 1, et seq., each party to bear their own legal and expert fees. Any decision rendered in arbitration may be enforced by a court of proper jurisdiction. Unless mutually agreed otherwise, any mediation, arbitration or court proceeding shall be held in Bates City, Michigan and the parties agree otherwise in writing.

20. This Agreement shall be governed and construed in accordance with Michigan law and with the Federal Arbitration Act. It may be amended only by written document signed by the parties hereto.

21. The persons signing below represent that they are authorized to bind their respective companies.

Agreed this ____ day of ______________, 20___.

ABC Architects, Inc.	XYZ Construction Co., Inc.
By: ______________________________	By: ______________________________
Printed Name: ____________________	Printed Name: ____________________
Title: ____________________________	Title: ____________________________

APPENDIX E

STATE DESIGN-BUILD STATUTES (INCLUDING ATTORNEY GENERAL OPINIONS

ALABAMA

None.

ALASKA

I. Public Procurement Laws.

Alaska Stat. § 36.30.200. Conditions for use.

(a) Except as otherwise provided in this chapter, or unless specifically exempted by law, an agency contract shall be awarded by competitive sealed proposals if it is not awarded by competitive sealed bidding.

(b) The commissioner may provide by regulation that it is either not practicable or not advantageous to the state to procure specified types of supplies, services, or construction by competitive sealed bidding that would otherwise be procured by that method. When the procurement officer determines in writing with particularity that the use of competitive sealed proposals is more advantageous to the state than competitive sealed bidding, a contract may be entered into by competitive sealed proposals.

(c) When the commissioner of transportation and public facilities determines that it is advantageous to the state, a procurement officer may issue a request for proposals requesting the submission of offers to provide construction in accordance with a design provided by the offeror. ***The request for proposals shall require that each proposal submitted contain a single price that includes the design/build.***

II. Public Procurement Law.

Alaska Stat. § 36.30.115. Subcontractors for Construction Contracts.

* * *

(g) The requirements of this section ***do not apply to a design-build*** construction contract.

Alaska Stat. § 36.30.210. Request for Proposals.

* * *

(b) An offeror for a construction contract shall submit evidence of the offeror's registration under AS 08.18. A request for sealed proposals for a construction contract ***except a design-build construction contract***, must require the offeror, no later than five working days after the proposal that is the most advantageous to the state is identified, to list subcontractors the offeror proposes to use in the performance of the construction contract. The list must include the information required under AS 36.30.115(a). The provisions of AS 36.30.115(b)–(g) that apply to a construction contractor or an apparent low bidder apply to offerors submitting competitive sealed proposals for construction contracts ***except design-build construction contracts***.

Alaska Stat. § 36.30.270. Architectural, engineering, and land surveying contracts.

(a) Notwithstanding conflicting provisions of AS 36.30.100–36.30.260, a procurement officer shall negotiate a contract for an agency with the most qualified and suitable firm or person of demonstrated competence for architectural, engineering, or land surveying services. The procurement officer shall award a contract for those services at fair and reasonable compensation as determined by the procurement officer, after consideration of the estimated value of the services to be rendered, and the scope, complexity, and professional nature of the services. When determining the most qualified and suitable firm or person, the procurement officer shall consider the

(1) proximity to the project site of the office of the firm or person unless federal law prohibits this factor from being considered in the awarding of the contract; and

(2) employment practices of the firm or person with regard to women and minorities.

(b) If negotiations with the most qualified and suitable firm or person under (a) of this section are not successful, the procurement officer shall negotiate a contract with other qualified firms or persons of demonstrated competence, in order of public ranking. The procurement officer may reject all or part of a proposal.

(c) This section does not apply to contracts awarded in a situation of public necessity if the procurement officer certifies in writing that a situation of public necessity exists.

(d) Notwithstanding the other provisions of this section, a procurement officer may include price as an added factor in selecting architectural, engineering, and land surveying services when, in the judgment of the procurement officer, the services required are repetitious in nature, and the scope, nature, and amount of services required are thoroughly defined by measurable and objective standards to reasonably enable firms or persons making proposals to compete with a clear understanding and interpretation of the services required. In order to include price as a factor in selection, a majority of the persons involved by the procurement officer in evaluation of the proposals must be registered in the state to perform architectural, engineering, or land surveying services.

(e) This section does not apply to a contract that incorporates both design services and construction.

Alaska Stat. § 36.30.990. Definitions.

* * *

(23) ***"design-build construction contract"*** means a contract to provide construction in accordance with a design provided by the contractor.

Alaska Stat. § 4, ch. 15, SLA 1998.

* * *

(d) The adoption by a municipality, when exercising its powers under AS 29.35.020(a), of an ordinance for procurement of a facility or operation on a ***design-build construction contract*** basis satisfies the procurement requirements under (b) of this section.

ARIZONA

I. Anti-Indemnity Statutes.

A.R.S. § 32-1159. Indemnity agreements in construction and architect-engineer contracts void; definitions.

A. A covenant, clause or understanding in, collateral to or affecting a construction contract or architect-engineer professional service contract that purports to indemnify, to hold harmless or to defend the promisee from or against liability for loss or damage resulting from the sole negligence of the promisee or the promisee's agents, employees or indemnitee is against the public policy of this state and is void.

B. Notwithstanding subsection A, a contractor who is responsible for the performance of a construction contract may fully indemnify a person for whose account the construction contract is not being performed and who, as an accommodation, enters into an agreement with the contractor that permits the contractor to enter on or adjacent to its property to perform the construction contract for others.

C. This section applies to all contracts entered into between private parties. This section does not apply to:

1. Agreements to which this state or a political subdivision of this state is a party, including intergovernmental agreements and agreements governed by sections 34-226 and 41-2586.

2. Agreements entered into by agricultural improvement districts under title 48, chapter 17.

D. In this section:

1. ***"Architect-engineer professional service contract" means a written or oral agreement relating to the design, design-build,*** construction administration, study, evaluation or other professional services furnished in connection with any actual or proposed construction, alteration, repair, maintenance, moving, demolition or excavation of any structure, street or roadway, appurtenance or other development or improvement to land.

2. "Construction contract" means a written or oral agreement relating to the construction, alteration, repair, maintenance, moving, demolition or excavation or other development or improvement to land.

II. Public Procurement Laws.

A.R.S. § 34-601. Definitions.

In this chapter, unless the context otherwise requires:

1. "Architect services" means those professional architect services that are within the scope of architectural practice as provided in title 32, chapter 1.

2. "Construction":

(a) Means the process of building, altering, repairing, improving or demolishing any public structure or building or other public improvements of any kind to any public real property.

(b) Does not include the routine operation, routine repair or routine maintenance of existing facilities, structures, buildings or real property.

3. "Construction-manager-at-risk" means a project delivery method in which:

(a) There is a separate contract for design services and a separate contract for construction services.

(b) The contract for construction services may be entered into at the same time as the contract for design services or at a later time.

(c) Design and construction of the project may be in sequential phases or concurrent phases.

(d) Finance services, maintenance services, operations services, preconstruction services and other related services may be included.

4. "Construction services" means either of the following for construction-manager-at-risk, ***design-build*** and job-order-contracting project delivery methods:

(a) Construction, excluding services, through the construction-manager-at-risk or job-order-contracting project delivery methods.

(b) A combination of construction and, as elected by the agent, one or more related services, such as finance services, maintenance services, operations services, design services and preconstruction services, as those services are authorized in the definitions of construction-manager-at-risk, ***design-build*** or job-order-contracting in this section.

5. "Contract" means all types of agent agreements, regardless of what they are called, for the procurement of construction services or design services.

6. "Contractor" means any person who has a contract with an agent for construction services or design services.

7. "Design-bid-build" means a project delivery method in which:

(a) There is a sequential award of two separate contracts.

(b) The first contract is for design services.

(c) The second contract is for construction.

(d) Design and construction of the project are in sequential phases.

(e) Finance services, maintenance services and operations services are not included.

8. ***"Design-build" means a project delivery method in which:***

(a) There is a single contract for design services and construction services.

(b) Design and construction of the project may be in sequential phases or concurrent phases.

(c) Finance services, maintenance services, operations services, design services, preconstruction services and other related services may be included.

9. "Design requirements":

(a) Means at a minimum the agent's written description of the project or service to be procured, including:

(i) The required features, functions, characteristics, qualities and properties.

(ii) The anticipated schedule, including start, duration and completion.

(iii) The estimated budgets applicable to the specific procurement for design and construction and, if applicable, for operation and maintenance.

(b) May include:

(i) Drawings and other documents illustrating the scale and relationship of the features, functions and characteristics of the project, which shall all be prepared by an architect or engineer, as appropriate, who is registered pursuant to section 32-121.

(ii) Additional design information or documents that the agent elects to include.

10. "Design services" means architect services, engineer services or landscape architect services.

11. "Engineer services" means those professional engineer services that are within the scope of engineering practice as provided in title 32, chapter 1.

12. "Finance services" means financing for a construction services project.

13. "Horizontal construction" means highways, roads, streets, bridges, canals, floodways, earthen dams and landfills.

14. "Job-order-contracting" means a project delivery method in which:

(a) The contract is a requirements contract for indefinite quantities of construction.

(b) The construction to be performed is specified in job orders issued during the contract.

(c) Finance services, maintenance services, operations services, preconstruction services, design services and other related services may be included.

15. "Landscape architect services" means those professional landscape architect services that are within the scope of landscape architectural practice as provided in title 32, chapter 1.

16. "Maintenance services" means routine maintenance, repair and replacement of existing facilities, structures, buildings or real property.

17. "Materials":

(a) Means all property, including equipment, supplies, printing, insurance and leases of property.

(b) Does not include land, a permanent interest in land or real property or leasing space.

18. "Operations services" means routine operation of existing facilities, structures, buildings or real property.

19. "Person" means any corporation, business, individual, union, committee, club, other organization or group of individuals.

20. "Preconstruction services" means advice during the design phase.

21. "Procurement":

(a) Means buying, purchasing, renting, leasing or otherwise acquiring any materials, services, construction or construction services.

(b) Includes all functions that pertain to obtaining any material, services or construction, including description of requirements, selection and solicitation of sources, preparation and award of contract and all phases of contract administration.

22. "Services":

(a) Means the furnishing of labor, time or effort by a contractor or subcontractor that does not involve the delivery of a specific end product other than required reports and performance.

(b) Does not include employment agreements or collective bargaining agreements.

23. "Subcontractor" means a person who contracts to perform work or render service to a contractor or to another subcontractor as a part of a contract with an agent.

A.R.S. § 34-602. Project delivery methods for design and construction services; report.

A. An agent may procure design services, construction and construction services, as applicable, under any of the following project delivery methods:

1. Design-bid-build.

2. Construction-manager-at-risk.

3. Design-build.

4. Job-order-contracting.

B. Except as provided in section 41-2501, subsection C and section 41-2535, subsection B, For the design-bid-build project delivery method, the agent shall procure:

1. Design services as provided in this chapter and chapter 1, article 1 of this title.

2. Construction by competitive sealed bidding pursuant to chapter 2, articles 1, 2 and 3 of this title and pursuant to rules adopted by the agent pursuant to section 41-2501, subsection C.

C. An agent shall procure construction services under the construction-manager-at-risk, ***design-build*** and job-order-contracting project delivery methods pursuant to this chapter.

D. An agent shall procure design services relating to construction-manager-at-risk construction services projects pursuant to section 34-102, 34-103 or 34-603, as applicable.

E. For job-order-contracting construction services projects, if the agent does not include design services in the job-order-contracting construction services contract, the agent shall procure any design services relating to the job-order-contracting construction services projects under the contract pursuant to section 34-102, 34-103 or 34-603, as applicable.

F. On or before January 15 of each year, any agent that uses construction-manager-at risk, ***design-build*** or job-order-contracting to construction services shall submit a report to the governor, the president of the senate, the speaker of the house of representatives, the secretary of state and the director of the department of library, archives on the total benefits associated with the use of the procurement methods. The report shall include a statement of costs incurred by the agent in securing the services, any overall administrative benefits realized through the use of the procurement methods, the number of projects begun in the preceding calendar year using the procurement methods and the cost and description of each project, any anticipated cost savings resulting from the use of that procurement method and any actual cost savings realized through the use of that procurement method. The report is a requirement for participation in construction-manager-at-risk, ***design-build*** or job-order-contracting procurement agreements.

A.R.S. § 34-603. Procurement of professional services and construction-manager-at-risk, design-build and job-order-contracting construction services; definition.

A. Except for engineer services an agent procures by direct selection or pursuant to section 34-102, subsections C and D, except for architect services, landscape architect services, assayer services, geologist services and land surveying services an agent procures by direct selection or pursuant to section 34-103, subsections C and D and except for services an agent procures pursuant to section 34-604, an agent shall procure the following services pursuant to this section:

1. Architect services.
2. Construction-manager-at-risk construction services.
3. ***Design-build*** construction services.
4. Engineer services.
5. Job-order-contracting construction services.
6. Landscape architect services.

7. Assayer services.
8. Geologist services.
9. Land surveying services.

B. An agent shall provide notice of each procurement of professional services or construction services specified in this section and shall award contracts on the basis of demonstrated competence and qualifications for the type of professional services or construction services pursuant to the procedures prescribed in this section.

C. In the procurement of professional services or construction services pursuant to this section, an agent shall:

1. Issue a request for qualifications for each contract and publish notice of the request for qualifications in the same manner as provided in section 28-6713, subsection A. The request for qualifications shall include the number of persons or firms to be included on the short list. In a request for qualifications for a horizontal construction project, at least three but not more than five persons or firms shall be on the short list. In all other requests for qualifications, three persons or firms shall be on the short list.

2. Initiate an appropriately qualified selection committee for each contract. If the agent is procuring professional services, the agent shall determine the number and qualifications of the selection committee members. A selection committee for the procurement of construction services shall not have more than seven members, except that, if the contract involves the agent and additional governmental or private participants, the number of members of the selection committee shall be increased by one for each additional participant, except that the maximum number of members of the selection committee is nine. the selection committee for construction services shall include at least one person who is a licensed contractor and one person who is an architect or an engineer who is registered pursuant to section 32-121. A contractor is not required to serve on the selection committee if the contract does not include construction and if the contract does not require delivery of a design or a set of construction documents. These members may be employees of the agent or outside consultant. Outside contractors, architects and engineers serving on a selection committee shall not receive compensation from the agent for performing this service, but the agent may elect to reimburse outside contractors, architects and engineers for travel, lodging and other expenses incurred in connection with service on a selection committee. A person who is a member of a selection committee shall not be a contractor under the contract or provide construction, materials or services under the contract. The selection committee shall:

(a) Evaluate the statements of qualifications and performance data that are submitted in response to the agent's request for qualifications for the proposed contract.

(b) If determined by the agent and included by the agent in the request for qualifications, conduct discussions with at least the number of persons or firms to be included on the short list as stated in the request for qualifications but not more than the number of persons or firms to be included on the short list plus two as specified in the request for qualifications regarding the contract and the relative methods of approach for furnishing the required professional services or construction services.

(c) In order of preference, based on criteria established and published by the selection committee and included in the request for qualifications, select a short list of persons or firms the selection committee deems to be the most qualified to provide the professional services or construction services. The number of persons or firms on the short list shall be the number of persons or firms specified in the request for qualifications, except that:

(i) If a smaller number of responsive and responsible persons or firms respond to the solicitation, the selection committee may proceed with the selection process with the remaining persons or firms if at least two persons or firms remain or the selection committee may readvertise pursuant to this subsection as the selection committee deems necessary or appropriate.

(ii) As to professional services only, if only one responsive and responsible person or firm responds to the solicitation, the agent may proceed with only one person or firm on the short list and may award the contract to a single person or firm if the agent determines in writing that the fee negotiated pursuant to subsection E of this section is fair and reasonable and that either other prospective persons or firms had reasonable opportunity to respond or there is not adequate time for a resolicitation.

(d) Base the selection and order of preference on demonstrated competence and qualifications only.

(e) Not request or consider fees, price, man-hours or any other cost information in the selection or order of preference.

(f) For a request for qualifications for construction services only, not include on the short list any person or firm that includes or employs any person or firm that has provided for compensation any services relating to the project covered by the request for qualifications.

D. An agent shall award a contract for professional services or for construction services to one of the persons or firms on the short list prepared pursuant to subsection C of this section as provided in subsection E or F of this section, except that, if fewer than the number of persons or firms on the short list respond to the request for proposals pursuant to subsection F of this section but at least two persons or firms on the short list submit responsive proposals or if one or more of the persons or firms on the short list drop out of the selection process pursuant to subsection E or F of this section:

1. If there are three or more remaining persons or firms, the agent shall proceed with the selection process.

2. If there are only two remaining persons or firms, as the agent deems necessary and appropriate, the agent may elect to proceed with the selection process with the two persons or firms or may elect to terminate the selection process and may elect to readvertise pursuant to subsection C of this section.

3. If there is only one remaining person or firm, the agent shall terminate the selection process and may elect to readvertise pursuant to subsection C of this section, except that as to professional services only, the agent may award the contract to a single person or firm if the agent determines in writing that the fee negotiated pursuant to subsection E of this section is fair and reasonable and that either other prospective persons or firms had reasonable opportunity to respond or there is not adequate time for a resolicitation.

E. An agent shall enter into negotiations for a contract with the highest qualified person or firm for the professional services or for the construction services. The negotiations shall include consideration of compensation and other contract terms that the agent determines to be fair and reasonable to the agent. In making this decision, the agent shall take into account the estimated value, the scope, the complexity and the nature of the professional services or construction services to be rendered. If the agent is not able to negotiate a satisfactory contract with the person or firm considered to be the most qualified at a price the agent determines to be fair and reasonable, the agent shall formally terminate negotiations with that person or firm. The agent may undertake negotiations with the next most qualified person or firm in sequence until an agreement is reached or a determination is made to reject all persons or firms on the short list. If a contract for construction services is entered into pursuant to this subsection, construction shall not commence until the agent and contractor agree in writing on a fixed price or a guaranteed maximum price for the construction to be commenced.

F. As an alternative to subsection E of this section, an agent ***may award design-build construction*** services or job-order-contracting construction services as follows:

1. The agent shall use the selection committee appointed for the contract pursuant to subsection c of this section.

2. The agent shall issue a request for proposals to the persons or firms on the short list developed pursuant to subsection c of this section.

3. For ***design-build construction services*** and job-order-contracting construction services, the request for proposals shall include:

(a) The agent's project schedule and project budget or life cycle budget for a procurement that includes maintenance services or operations services.

(b) A statement that the contract will be awarded to the offeror whose proposal receives the highest number of points under a scoring method.

(c) A description of the scoring method, including a list of the factors in the scoring method and the number of points allocated to each factor. The factors in the scoring method shall include:

(i) ***For design-build construction services*** only, demonstrated compliance with the design requirements.

(ii) Offeror qualifications.

(iii) Offeror financial capacity.

(iv) Compliance with the agent's project schedule.

(v) ***For design-build construction services*** only, if the request for proposals specifies that the agent will spend its project budget and not more than its project budget and is seeking the best proposal for the project budget, compliance of the offeror's price or life cycle price for procurements that include maintenance services, operations services or finance services with the agent's budget as prescribed in the request for proposals.

(vi) ***For design-build construction services*** if the request for proposals does not contain the specifications prescribed in item (v) and for job-order-contracting construction services, the price or life cycle price for procurements that include maintenance services, operations services or finance services.

(vii) A offeror quality management plan.

(viii) Other evaluation factors as determined by the agent, if any.

(d) ***For design-build construction services*** only, The design requirements.

(e) A requirement that each offeror submit separately a technical proposal and a price proposal and that the offeror's entire proposal be responsive to the requirements in the request for proposals. for ***design-build construction services***, the price in the price proposal shall be a fixed price or a guaranteed maximum price.

(f) A statement that in applying the scoring method the selection committee will separately evaluate the technical proposal and the price proposal and will evaluate and score the technical proposal before opening the price proposal.

(g) If the agent conducts discussions pursuant to paragraph 5 of this subsection, a statement that discussions will be held and a requirement that each offeror submit a preliminary technical proposal before the discussions are held.

4. If the agent determines to conduct discussions pursuant to paragraph 5 of this subsection, each offeror shall submit a preliminary technical proposal to the agent before those discussions are held.

5. If determined by the agent and included by the agent in the request for proposals, the selection committee shall conduct discussions with all persons or firms that submit preliminary technical proposals. Discussions shall be for the purpose of clarification to assure full understanding of, and responsiveness to, the solicitation requirements. offerors shall be accorded fair treatment with respect to any opportunity for discussion and for clarification by the owner. Revision of preliminary technical proposals shall be permitted after submission of preliminary technical proposals and before award for the purpose of obtaining best and final proposals. In conducting any discussions, information derived from proposals submitted by competing offerors shall not be disclosed to other competing offerors.

6. After completion of any discussions pursuant to paragraph 5 of this subsection or if no discussions are held, each offeror shall submit separately the offeror's final technical proposal and its price proposal.

7. Before opening any price proposal, the selection committee shall open the final technical proposals, evaluate the final technical proposals and score the final technical proposals using the scoring method in the request for proposals. No other factors or criteria may be used in the evaluation and scoring.

8. After completion of the evaluation and scoring of all final technical proposals, the selection committee shall open the price proposals, evaluate the price proposals, score the price proposals and complete the scoring of the entire proposals using the scoring method in the request for proposals. No other factors or criteria may be used in the evaluation and scoring.

9. The agent shall award the contract to the responsive and responsible offeror whose proposal receives the highest score under the method of scoring in the request for proposals. No other factors or criteria may be used in the evaluation.

10. The contract file shall contain the basis on which the award is made.

11. ***For design-build construction services*** only, the agent shall award a stipulated fee equal to a percentage, as prescribed in the request for proposals, of the agent's project budget, as prescribed in the request for proposals, but not less than two-tenths of one percent of the project budget to each short list offeror who provides a responsive, but unsuccessful, proposal. If the agent does not award a contract, all responsive short list offerors shall receive the stipulated fee. The agent shall pay the stipulated fee to each offeror within ninety days after the award of the initial contract or the decision not to award a contract. In consideration for paying the stipulated fee, the agent may use any ideas or information contained in the proposals in connection with any contract awarded for the project, or in connection with a subsequent procurement, without any obligation to pay any additional compensation to the unsuccessful offerors. Notwithstanding the other provisions of

this paragraph, an unsuccessful short list offeror may elect to waive the stipulated fee. If an unsuccessful short list offeror elects to waive the stipulated fee, the agent may not use ideas and information contained in the offeror's proposal, except that this restriction does not prevent the agent from using any idea or information if the idea or information is also included in a proposal of an offeror that accepts the stipulated fee.

G. Until award and execution of a contract by an agent, only the name of each person or firm on the short list developed pursuant to subsection C of this section shall be available to the public. All other information received by the agent in response to the request for qualifications or contained in the proposals shall be confidential in order to avoid disclosure of the contents that may be prejudicial to competing offerors during the selection process. The proposals shall be open to public inspection after the contract is awarded and the agent has executed the contract. To the extent that the offeror designates and the agent concurs, trade secrets and other proprietary data contained in a proposal remain confidential.

H. An agent may cancel a request for qualifications or a request for proposals or reject in whole or in part any or all proposals as specified in the solicitation if it is in the best interest of the agent. The agent shall make the reasons for cancellation or rejection part of the contract file.

I. Notwithstanding any other law:

1. The contractor for construction-manager-at-risk, design-build or job-order-contracting construction services is not required to be registered to perform design services pursuant to title 32, chapter 1 if the person or firm actually performing the design services on behalf of the contractor is appropriately registered.

2. The contractor for construction-manager-at-risk, design-build or job-order-contracting construction services is not required to be licensed to perform construction pursuant to title 32, chapter 10 if the firm actually performing the construction on behalf of the contractor is appropriately licensed.

J. For job-order-contracting construction services only:

1. The dollar amount of an individual job order shall not be more than the amount set by the agent in an action noticed pursuant to title 38, chapter 3, article 3.1 or a rule adopted pursuant to title 41, chapter 6 as the maximum amount of an individual job order. This maximum amount shall not be more than seven hundred fifty thousand dollars. Requirements shall not be artificially divided or fragmented in order to constitute a job order that satisfies this requirement.

2. If the contractor subcontracts or intends to subcontract part or all of the work under a job order and if the job order construction services contract includes descriptions of standard individual tasks, standard unit prices for standard individual tasks and pricing of job orders based on the number of units of standard individual tasks in the job order:

(a) The contractor has a duty to deliver promptly to each subcontractor invited to bid a coefficient to the contractor to do all or part of the work under one or more job orders:

(i) A copy of the descriptions of all standard individual tasks on which the subcontractor is invited to bid.

(ii) A copy of the standard unit prices for the individual tasks on which the subcontractor is invited to bid.

(b) If not previously delivered to the subcontractor, the contractor has a duty to deliver promptly the following to each subcontractor invited to or that has agreed to do any of the work included in any job order:

(i) A copy of the description of each standard individual task that is included in the job order and that the subcontractor is invited to perform.

(ii) The number of units of each standard individual task that is included in the job order and that the subcontractor is invited to perform.

(iii) The standard unit price for each standard individual task that is included in the job order and that the subcontractor is invited to perform.

K. Notwithstanding anything to the contrary in this chapter:

1. Each project for horizontal construction ***under a design-build construction services contract*** and each project for horizontal construction services under a construction-manager-at-risk construction services contract shall be a specific, single project with a minimum construction cost of ten million dollars. The estimated cost of the project shall not include the cost to procure any right-of-way or other cost of condemnation.

2. An agent shall not procure any horizontal construction using the construction-manager-at-risk, ***design-build*** or job-order-contracting method of project delivery after June 30, 2007. For purposes of this paragraph, an agent procures horizontal construction when the contract for the construction services is executed by the agent and the contractor for the construction-manager-at-risk, ***design-build*** or job-order-contracting construction services. If a contract is executed for construction services on or before June 30, 2007, construction services under the contract may be rendered in whole or in part after June 30, 2007.

3. ***Each project under a design-build construction services contract*** or a construction-manager-at-risk construction services contract shall be a specific, single project. For the purposes of this paragraph, "specific, single project" means a project that is constructed at a single location, at a common location or for a common purpose.

L. Notwithstanding anything to the contrary in this section or this title, an agent shall not:

1. Enter into a contract as contractor to provide construction-manager-at-risk construction services, ***design-build construction services*** or job-order-contracting construction services.

2. Contract with itself, with another agent, with this state or with any other governmental unit of this state or the federal government for the agent to provide construction-manager-at-risk construction services, ***design-build construction services*** or job-order-contracting construction services.

M. The prohibitions prescribed in subsection L of this section do not prohibit an agent from providing construction for itself as provided by law.

N. For the purposes of this section, "professional services" includes architect services, engineer services, landscape architect services, assayer services, geologist services and land surveying services and any combination of those services.

A.R.S. § 34-604. Emergency procurements.

Notwithstanding any other provision of this title, an agent may make or authorize others to make emergency procurements of architect services, construction-manager-at-risk construction services, design-bid-build construction services, ***design-build construction services***, engineer services, job-order-contracting construction services, landscape architect services, assayer services, geologist services or land surveying services if a threat to the public health, welfare or safety exists or if a situation exists that makes compliance with this title impracticable, unnecessary or contrary to the public interest, except that these emergency procurements shall be made with such competition as is practicable under the circumstances. A written determination of the basis for the emergency and for the selection of the particular contractor shall be included in the contract file.

A.R.S. § 34-605. Multiterm contracts for job-order-contracting construction services.

A. A contract for construction services using the job-order-contracting method may be entered into for a period of up to three years, as deemed to be in the best interest of the agent, if the term of the contract and conditions of renewal or extension, if any, are included in the solicitation and monies are available for the first fiscal period at the time of contracting. Payment and performance obligations for succeeding fiscal periods are subject to the availability and appropriation of monies.

B. Before the use of a multiterm contract, the agent shall determine in writing that:

1. Estimated requirements cover the period of the contract and are reasonable and continuing.

2. Such a contract will serve the best interests of the agent by encouraging effective competition or otherwise promoting economies in the agent's procurement.

C. If monies are not appropriated or otherwise made available to support the continuation of performance in a subsequent fiscal period, the contract shall be cancelled and the contractor may only be reimbursed for the reasonable value of any nonrecurring costs that are incurred but not amortized in the price of the construction services delivered under the contract or that are otherwise not recoverable. The cost of cancellation may be paid from any appropriations available for these purposes.

A.R.S. § 34-606. Bid security for design-build and job-order-contracting construction services.

A. As a guarantee that the contractor will enter into a contract, bid security is required ***for all design-build construction services*** and all job-order-contracting construction services awarded by an agent by competitive sealed proposals pursuant to section 34-603, subsection F if the agent estimates that the budget for construction, excluding the cost of any finance services, maintenance services, operations services, design services, preconstruction services or other related services, will be more than the amount prescribed in section 41-2535, subsection B. ***Each proposal for design-build construction services*** or job-order-contracting construction services shall be accompanied by a certified check, cashier's check or surety bond in an amount equal to ten percent of the agent's budget for construction, excluding any finance services, maintenance services, operations services, design services, preconstruction services or other related services, for the project as stated in the request for proposals, except that, for job-order-contracting construction services, the bid security amount shall be the amount determined by the agent and stated in the request for proposals and shall not be more than ten percent of the agent's estimated budget for construction during the first year of the job-order-contracting contract.

B. The agent shall return The certified check, cashier's check or surety bond to the contractors whose proposals are not accepted and to the successful contractor on the execution of satisfactory payment and performance bonds and the contract as provided in this chapter.

C. Notwithstanding any other statute, the surety bond shall be executed solely by a surety company or companies holding a certificate of authority to transact surety business in this state issued by the director of the department of insurance pursuant to title 20, chapter 2, article 1. The bond shall not be executed by an individual surety or sureties, even if the requirements of section 7-101 are satisfied.

D. The conditions and provisions of the surety bond regarding the surety's obligations shall follow the following form:

Now, therefore, if the obligee accepts the proposal of the principal and the principal enters into a contract with the obligee in accordance with the terms of the proposal and gives the bonds and certificates of insurance as specified in the standard specifications with good and sufficient surety for the faithful performance of the contract and for the prompt payment of labor and materials furnished in the prosecution of the contract, or in the event of the failure of the principal to enter into the contract and give the bonds and certificates of insurance, if the principal pays to the obligee the difference not to exceed the penalty of the bond between the amount specified in the proposal and any larger amount for which the obligee may contract in good faith with another party to perform the work covered by the proposal, this obligation is void. Otherwise it remains in full force and effect. provided, however, that this bond is executed pursuant to section 34-606, Arizona Revised Statutes, and all liabilities on this bond shall be determined in accordance with the provisions of the section to the extent as if it were copied at length in this agreement.

E. If the request for proposals requires security, noncompliance requires that the agent reject the proposal for noncompliance with the security requirement, unless the agent determines that the bid fails to comply in a nonsubstantial manner with the security requirements.

F. After the agent opens the proposals, the proposals are irrevocable for the period specified in the request for proposals, except as provided in section 34-603, subsection F. If a proposer is permitted to withdraw its proposal before award, no action may be had against the proposer or the bid security.

G. All bonds given by a contractor and surety pursuant to this section, regardless of their actual form, are deemed by law to be the form required and set forth in this section.

A.R.S. § 34-607. Contracts for construction-manager-at-risk, design-build and job-order-contracting construction services; payments to contractor; security; recovery of damages by contractor for delay; progress payments.

A. An agent shall enter into a contract with the selected person or firm for construction-manager-at-risk construction services, ***design-build construction services*** or job-order-contracting construction services.

B. The terms of a contract entered into pursuant to subsection A shall include the following items:

1. A surety company bond or bonds as required by this chapter.
2. The owner by mutual agreement may make progress payments on contracts of less than ninety days and shall make monthly progress payments on all other contracts as provided for in this paragraph. Payment to the contractor on the basis of a duly certified and approved estimate of the work performed during the preced-

ing calendar month under the contract may include payment for material and equipment, but to ensure the proper performance of the contract, the owner shall retain ten percent of the amount of each estimate until final completion and acceptance of all material, equipment and work covered by the contract. An estimate of the work submitted shall be deemed approved and certified for payment after seven days from the date of submission unless before that time the owner or owner's agent prepares and issues a specific written finding setting forth those items in detail in the estimate of the work that are not approved for payment under the contract. The owner may withhold an amount from the progress payment sufficient to pay the expenses the owner reasonably expects to incur in correcting the deficiency set forth in the written finding. The progress payments shall be paid on or before fourteen days after the estimate of the work is certified and approved. The estimate of the work shall be deemed received by the owner on submission to any person designated by the owner for the submission, review or approval of the estimate of the work.

3. When the contract is fifty percent completed, one-half of the amount retained including any securities substituted under paragraph 5 shall be paid to the contractor on the contractor's request provided the contractor is making satisfactory progress on the contract and there is no specific cause or claim requiring a greater amount to be retained. After the contract is fifty percent completed, no more than five percent of the amount of any subsequent progress payments made under the contract may be retained providing the contractor is making satisfactory progress on the project, except that if at any time the owner determines satisfactory progress is not being made ten percent retention shall be reinstated for all progress payments made under the contract after the determination.

4. On completion and acceptance of each separate building, public work or other division of the contract on which the price is stated separately in the contract, except as qualified in paragraph 5, payment may be made in full, including retained percentages, less authorized deductions. In preparing estimates, the material and equipment delivered on the site to be incorporated in the job shall be taken into consideration in determining the estimated value by the architect, engineer or other person, as specified in the contract.

5. Ten percent of all estimates shall be retained by the agent as a guarantee for complete performance of the contract, to be paid to the contractor within sixty days after completion or filing notice of completion of the contract. Retention of payments by an agent longer than sixty days after final completion and acceptance requires a specific written finding by the agent of the reasons justifying the delay in payment. No agent may retain any monies after sixty days that are in excess of the amount necessary to pay the expenses the agent reasonably expects to incur in order to pay or discharge the expenses determined by the agent in the finding justifying the retention of monies. In lieu of the retention provided in this section, the agent, at the option of the contractor, shall accept as a substitute an assignment of time certificates of deposit of banks licensed by this state, securities of or guaranteed by the United States of America, securities of this state, securities of counties, municipalities and school districts within this state or shares of savings and loan institutions authorized to transact business in this state, in an amount equal to ten percent of all estimates that are retained by the agent as a guarantee for complete performance of the contract. If the agent accepts substitute security as described in this paragraph for the ten percent retention, the contractor is entitled to receive all

interest or income earned by this security as it accrues and all such security in lieu of retention shall be returned to the contractor by the agent within sixty days after final completion and acceptance of all material, equipment and work covered by the contract if the contractor has furnished the agent satisfactory receipts for all labor and material billed and waivers of liens from any and all persons holding claims against the work. In no event shall the agent accept a time certificate of deposit of a bank or shares of a savings and loan institution in lieu of the retention specified unless accompanied by a signed and acknowledged waiver of the bank or savings and loan association of any right or power to setoff against either the agent or the contractor in relationship to the certificates or shares assigned.

6. In any instance where the agent has accepted substitute security as provided in paragraph 5, any subcontractor undertaking to perform any part of this public work is entitled to provide substitute security to the contractor on terms and conditions similar to those described in paragraph, and this security is in lieu of any retention under the subcontract.

7. Notwithstanding paragraphs 1 through 6, retention is not required for job-order-contracting construction services contracts, except that the agent may elect to require retention for a job-order-contracting construction services contract. If the agent elects to require retention, paragraphs 1 through 6 apply to the job-order-contracting construction services contract, except that:

(a) Retention shall be five percent of each payment instead of ten percent reducing to five percent.

(b) Retention applicable to each job order shall be released within sixty days after final completion of the job order and acceptance of the work under the job order.

(c) No retention on the job order may be released until that time.

(d) The retention percentage shall not be increased.

C. No contract for construction services may materially alter the rights of any contractor, subcontractor or material supplier to receive prompt and timely payment required to be included in the contract under subsection B.

D. The contract shall be signed by the agent and the contractor.

E. A contract for the procurement of construction services shall include a provision that provides for negotiations between the agent and the contractor for the recovery of damages related to expenses incurred by the contractor for a delay for which the agent is responsible, that is unreasonable under the circumstances and that was not within the contemplation of the parties to the contract. This section shall not be construed to void any provision in the contract that requires notice of delays, provides for arbitration or other procedures for settlement or provides for liquidated damages.

F. The contractor shall pay to the contractor's subcontractors or material suppliers and each subcontractor shall pay to the subcontractor's subcontractor or material supplier, within seven days of receipt of each progress payment, unless otherwise agreed in writing by the parties, the respective amounts allowed the contractor or subcontractor on account of the work performed by the subcontractors, to the extent of each subcontractor's interest, except that no contract for construction services may materially alter the rights of any contractor, subcontractor or material supplier to receive prompt and timely payment as provided under this section the payments to subcontractors or material suppliers shall be based on payments received pursuant to this section. Any diversion by the contractor or subcontractor of payments received for work performed on a contract, or failure to reasonably account for the application or use of those payments, constitutes grounds for disciplinary action by the registrar of contractors. The subcontractor or material supplier shall notify the registrar of contractors and the agent in writing of any payment less than the amount or percentage approved for the class or item of work as set forth in this section.

G. A subcontractor may notify the agent in writing requesting that the subcontractor be notified by the agent in writing within five days from payment of each progress payment made to the contractor. The subcontractor's request remains in effect for the duration of the subcontractor's work on the project.

H. Nothing in this chapter prevents the contractor or subcontractor, at the time of application and certification to the owner or contractor, from withholding the application and certification to the owner or contractor for payment to the subcontractor or material supplier for unsatisfactory job progress, defective construction work or materials not remedied, disputed work or materials, third party claims filed or reasonable evidence that a claim will be filed, failure of a subcontractor to make timely payments for labor, equipment and materials, damage to the contractor or another subcontractor, reasonable evidence that the subcontract cannot be completed for the unpaid balance of the subcontract sum or a reasonable amount for retention that does not exceed the actual percentage retained by the owner.

I. If any payment to a contractor is delayed after the date due interest shall be paid at the rate of one percent per month or fraction of a month on the unpaid balance as may be due.

J. If any periodic or final payment to a subcontractor is delayed by more than seven days after receipt of the periodic or final payment by the contractor or subcontractor, the contractor or subcontractor shall pay the subcontractor or material supplier interest, beginning on the eighth day, at the rate of one percent per month or fraction of a month on the unpaid balance as may be due.

K. Notwithstanding anything to the contrary in this section, this section applies only to amounts payable in a construction services contract for construction and does not apply to amounts payable in a construction services contract for design services, preconstruction services, finance services, maintenance services, operations services and other related services.

A.R.S. § 34-610. Accounting standards; statutory applicability.

A. Any construction by a county pursuant to this chapter shall comply with the uniform accounting system prescribed for counties by the auditor general pursuant to section 41-1279.21. Any construction by a city or a town pursuant to this chapter shall comply with generally accepted accounting principles.

B. Any building, structure, addition or alteration may be constructed without complying with this chapter if the construction, including construction of buildings or structures on public or private property, is required as a condition of development of private property and is authorized by section 9-463.01 or 11-806.01. For the purposes of this subsection, building does not include police, fire, school, library or other public buildings.

C. Sections 34-102, 34-103 and 34-104, section 34-201, subsections a through h, J and K and sections 34-202, 34-203, 34-221, 34-222, 34-223 and 34-224 do not apply to procurement by an agent of construction-manager-at-risk construction services, ***design-build construction services*** and job-order-contracting construction services.

D. Section 34-201, subsections I and L and sections 34-225 and 34-226 apply to procurement by an agent of construction-manager-at-risk construction services, ***design-build construction services*** and job-order-contracting construction services.

A.R.S. § 28-6713. Bids for construction, reconstruction, equipment or supplies; procedure; bond; exceptions.

A. Except as provided in subsection F of this section, in a county with a population of one hundred fifty thousand persons or more as determined by the most recent United States decennial census or the most recent special census as provided in section 28-6532, bids for all items of construction or reconstruction involving an expenditure equal to or greater than the amount determined pursuant to subsection B of this section, all purchases or other acquisition of equipment involving an expenditure of more than five thousand dollars and all purchases of supplies and materials involving an expenditure of two thousand five hundred dollars or more shall be called for by advertising in a newspaper of general circulation in the county for two consecutive publications if it is a weekly newspaper, or for two publications of at least six but not more than ten days apart if it is a daily newspaper. The advertisement shall state specifically the character of the work to be done and the kind and quality of materials or supplies to be furnished.

B. Bids shall be called pursuant to subsection A of this section for all items of construction or reconstruction involving an expenditure of:

1. In fiscal year 1985-1986, thirty-five thousand dollars.

2. In fiscal year 1986-1987 and each fiscal year thereafter, the amount provided in paragraph 1 of this subsection adjusted by the annual percentage change in the GDP price deflator as defined in section 41-563.

C. If the board of supervisors receives a satisfactory bid, it shall contract with the lowest responsible bidder after the contractor or supplier gives any bond required by title 34, chapter 2, article 2, except that in counties with a population of more than one million persons according to the most recent United States decennial census, in determining the lowest responsible bidder under this section, the board may consider, for no more than five projects, the time of completion proposed by the bidder, the value over time of completed services and facilities and the value over time of interrupted services, if the board determines that this procedure will serve the public interest by providing a substantial fiscal benefit or that the use of the traditional awarding of contracts is not practicable for meeting desired construction standards or delivery schedules and if the formula for considering the time of completion is specifically stated in the bidding information. The board may reject any or all bids and readvertise.

* * *

G. Subsections A, B and C of this section do not apply to procurement of construction-manager-at-risk, ***design-build*** and job-order-contracting construction services pursuant to title 34, chapter 6.

A.R.S. § 15-213. Procurement practices of school districts; definitions.

A. The state board of education shall adopt rules prescribing procurement practices for all school districts in this state as follows:

* * *

J. No more than one hundred twenty days after the effective date of this amendment to this section, the state board of education and the auditor general shall adopt rules authorizing school districts to procure construction services by construction-manager-at-risk, ***design-build*** and job-order-contracting methods of project delivery. The rules adopted shall require each school district that uses construction-manager-at-risk, ***design-build*** or job-order-contracting to procure construction services to submit, on or before January 15 of each year, a report to the state board of education, the governor, the president of the senate, the speaker of the house of representatives, the secretary of state and the director of the department of library, archives and public records on the total benefits associated with the use of such procurement methods. The report shall include a statement of costs incurred by the school district in securing these services, any overall administrative benefits realized through the use of the procurement methods, the number of projects begun in the preceding calendar year using that procurement method, the cost and description of each project, any anticipated cost savings resulting from the use of that procurement method and any actual cost savings realized through the use of that procurement method and any other information that the state board of education deems necessary and appropriate. The report is a requirement for participation in construction-manager-at-risk, ***design-build*** or job-order-contracting procurement agreements.

K. Until the rules prescribed in subsection J of this section are adopted, school districts may procure construction services by the construction-manager-at-risk, ***design-build*** and job-order-contracting methods of project delivery as provided in title 41, chapter 23, except adopted, school districts may procure construction services by the construction-manager-at-risk, ***design-build*** and job-order-contracting methods of project delivery as provided in title 41, chapter 23, except that the rules adopted by the director of the department of administration.

A.R.S. § 28-7361. Definitions.

In this article, unless the context otherwise requires:

1. ***"Design-build" means the process of entering into and managing a contract between the department and another party in which the other party agrees to both design and build a highway, a structure, a facility or other items specified in the contract.***

2. ***"Design-builder"*** means any individual, partnership, joint venture, corporation or other legal entity that is appropriately licensed in this state and that furnishes the necessary design services, in addition to construction of the work, whether by itself or through subcontracts, including subcontracts for architectural and engineering services.

* * *

A.R.S. § 28-7363. Design-build method of project delivery; report.

A. Notwithstanding any other law, the department may use the design-build method of project delivery on a project if the department makes a determination in writing that it is appropriate and in the best interests of the department to use ***the design-build method of project delivery*** for that project, except that:

1. The department shall not enter into a contract to operate any structure, facility or other item pursuant to this article.

2. The department shall not Enter into contracts for more than two ***design-build*** projects in each fiscal year.

3. Each ***design-build project*** shall be a specific, single project with a minimum construction cost of forty million dollars. The department shall not artificially combine or aggregate projects in order to circumvent the minimum construction cost requirement. For the purposes of this paragraph, "specific, single project" means a project that is constructed at a single location, at a common location or for a common purpose.

4. The department shall not commence any ***design-build*** project after June 30, 2007. For the purposes of this paragraph, a project is commenced on the date the department and ***design-builder*** execute the contract for the project. If the department and ***design-builder*** execute a ***design-build*** contract on or before June 30, 2007, services and construction under the contract may be rendered in whole or in part after June 30, 2007.

B. the estimated cost of the project shall not include the cost to procure any right-of-way or other cost of condemnation. The cost to procure any right-of-way or other cost of condemnation remains at all times the responsibility of the department. The department shall obtain all necessary rights-of-way.

C. The department is responsible for preparation and acquisition of all environmental documents, including the scope of any remediation and required clearances.

D. If construction of a ***design-build*** project involves railroad facilities, the railroad shall approve the use of the ***design-build*** delivery method before the department awards the ***design-build*** contract.

E. To ensure fair, uniform, clear and effective procedures that will deliver a quality project on time and within budget, the director, in conjunction with the appropriate and affected professionals and contractors, may adopt procedures for procuring a project using the ***design-build*** method of project delivery.

F. The provisions of sections 28-6923 and 28-6924 relating to bid, performance and payment bonds and to change orders, progress payments, contract retentions, definitions and authority to award contracts apply to department ***design-build*** projects for transportation facilities pursuant to this article.

G. On or before January 15 of each year, the director of the department of transportation shall transmit to the governor, the president of the senate, the speaker of the house of representatives, the secretary of state and the director of the department of library, archives and public records a report on the total benefits associated with the use of ***design-build*** in procuring construction services. The report shall include a statement of costs incurred in securing the services, any overall administrative benefits realized through the use of ***design-build***, the number of projects begun in the preceding calendar year using ***design-build*** and the cost and description of each project, any anticipated cost savings resulting from the use of that procurement method and any actual cost savings realized through the use of that procurement method.

Note: AZ ST § 28-7363

Sec. 23. Delayed repeal; ***design-build pilot program election***.

A. The following are repealed from and after December 31, 2005:

1. Laws 1998, chapter 278, section 5, as amended by Laws 1999, chapter 207, section 1.

2. Laws 1998, chapter 278, section 6, as amended by Laws 1999, chapter 101, section 3.

3. Laws 1998, chapter 278, section 7.

4. Laws 1998, chapter 278, section 8, as amended by Laws 1999, chapter 207, section 3.

5. Laws 1998, chapter 278, section 9, as amended by Laws 1999, chapter 207, section 4.

6. Laws 1998, chapter 278, sections 10 and 11.

B. Notwithstanding this act, for ***design-build*** projects authorized by Laws 1998, chapter 278, as amended, the department of transportation, the department of administration and counties and cities described in Laws 1998, chapter 278, section 6, as amended by Laws 1999, chapter 101, section 3, may elect to proceed with ***design-build projects*** authorized by Laws 1998, chapter 278, as amended, pursuant to Laws 1998, chapter 278, as amended, or pursuant to this act.

A.R.S. § 28–7364. Design-build criteria.

The department shall use the following criteria as the minimum basis for determining when to use ***the design-build method of project delivery***:

1. The extent to which it can adequately define the project requirements.
2. The time constraints for delivery of the project.
3. The capability and experience of potential teams with the ***design-build method*** of project delivery.
4. The suitability of the project for use of ***the design-build method of project delivery*** in the areas of time, schedule, costs and quality.
5. The capability of the department to manage the project, including the employment of experienced personnel or outside consultants.
6. The capability of the department to oversee the project with persons who are familiar with the ***design-build*** method of project delivery.
7. Other criteria the department deems relevant.

A.R.S. § 28–7365. Design-build; two-phase solicitation.

A. If the department determines that ***the design-build method of project delivery*** is appropriate, the department shall establish a two-phase procedure for awarding ***the design-build contract***.

B. During phase one, and before solicitation, the director shall appoint a selection team of at least three persons. At least one-half of the selection team shall be architects or engineers who are registered pursuant to section 32-121. The selection team members may be either department employees or outside consultants. The selection team shall also include a licensed contractor who is not involved in the project. Any architect or engineer who is serving on the selection team and who is not a department employee shall not be otherwise involved in the project. The department shall prepare documents for a request for qualifications.

C. The request for qualifications shall include all of the following:

1. The minimum qualifications of the ***design-builder***.
2. A scope of work statement and schedule.
3. Documents defining the project requirements.
4. The form of contract to be awarded.

5. The selection criteria for compiling a short list and the number of firms to be included on the short list. at least three but not more than five firms shall be included on the short list.

6. A description of the phase two requirements and subsequent management needed to bring the project to completion.

7. The maximum time allowable for design and construction.

8. The department's estimated cost of design and construction.

D. The selection team shall evaluate the ***design-build qualifications*** of responding firms and shall compile a short list of firms in accordance with technical and qualifications-based criteria. The number of firms on the short list shall be the number of firms specified in the request for qualifications, except that, if a smaller number of firms responds to the solicitation or if one or more of the firms on the short list drop out so that only two firms remain on the short list, the selection team may proceed with the selection process with the remaining firms if at least two firms remain or the selection team may readvertise as the selection team deems necessary.

E. During phase two, the department shall issue a request for proposals to the ***design-builders*** on the short list. The request shall include:

1. The scope of work, including programmatic, performance and technical requirements, conceptual design, specifications and functional and operational elements for the delivery of the completed project, which shall all be prepared by an architect or engineer, as appropriate, who is registered pursuant to section 32-121.

2. A description of the qualifications required of the ***design-builder*** and the selection criteria, including the weight or relative order, or both, of each criterion.

3. Copies of the contract documents that the successful proposer will be expected to sign.

4. The maximum time allowable for design and construction.

5. The department's estimated cost of design and construction.

6. The requirement that a proposal be segmented into two parts, a technical proposal and a price proposal. Each proposal shall be in a separately sealed, clearly identified package and shall include the date and time of the submittal deadline. The technical proposal shall include a schedule, schematic design plans and specifications, technical reports, calculations, permit requirements, applicable development fees and other data requested in the request for proposals. The price proposal shall contain all design, construction, engineering, inspection and construction costs of the proposed project.

7. The date, time and location of the public opening of the sealed price proposals.

8. Other information relevant to the project.

F. After reviewing the proposals, the department shall proceed as follows:

1. The selection team shall score the technical proposals using the selection criteria in the request for proposals. The technical review team shall then submit a technical proposal score for each ***design-builder*** to the department. The technical review team shall reject any proposal it deems to be nonresponsive.

2. The department shall announce the technical proposal score for ***each design-builder***, shall publicly open the sealed price proposals and shall divide each ***design-builder's price*** by the score that the selection team has given to it to obtain an adjusted score. The ***design-builder*** selected shall be that responsive and responsible ***design-builder*** whose adjusted score is the lowest.

3. If a time factor is included with the selection criteria in the request for proposals package, the department may also adjust the bids using a value of the time factor established by the department. the value of the time factor shall be a value per day. The adjustment shall be based on the total time value. The total time value is the ***design-builder's*** proposed number of days to complete the project multiplied by the factor. The time adjusted price is the total time value plus the bid amount. This adjustment shall be used for selection purposes only and shall not affect the department's liquidated damages schedule or incentive and disincentive program. An adjusted score shall then be obtained by dividing each ***design-builder's*** time adjusted price by the score given by the technical review team. The department shall select the responsive and responsible ***design-builder*** whose adjusted score is the lowest.

4. Unless all proposals are rejected, the board shall award the contract to the responsive and responsible ***design-builder*** with the lowest adjusted score. The board reserves the right to reject all proposals.

5. ***The department shall award a stipulated fee equal to two-tenths of one percent of the department's estimated cost of design and construction to each short list responsible proposer who provides a responsive, but unsuccessful proposal.*** If the department does not award a contract, all responsive proposers shall receive the stipulated fee. If the department cancels the contract before reviewing the technical proposals, ***the department shall award each design-builder on the selected short list a stipulated fee equal to two-tenths of one percent of the department's estimated cost of design and construction.*** The department shall pay the stipulated fee to each proposer within ninety days after the award of the contract or the decision not to award a contract. In consideration for paying the stipulated fee, the department may use any ideas or information contained in the proposals in connection with any contract awarded for the project, or in connection with a subsequent procurement, without any obligation to pay any additional compensation to the unsuccessful proposers. Notwithstanding the other provisions of this paragraph, an unsuccessful short list proposer may elect to waive the stipulated fee. If an unsuccessful short list proposer elects to waive the stipulated fee, the department may not use ideas and information contained in the proposer's proposal.

A.R.S. § 41–2503. Definitions.

In this chapter, unless the context otherwise requires:

1. "Architect services" means those professional architect services that are within the scope of architectural practice as provided in title 32, chapter 1.

2. "Business" means any corporation, partnership, individual, sole proprietorship, joint stock company, joint venture or other private legal entity.

3. "Change order" means a written order which is signed by a procurement officer and which directs the contractor to make changes that the changes clause of the contract authorizes the procurement officer to order.

4. "Construction":

(a) Means the process of building, altering, repairing, improving or demolishing any public structure or building, or other public improvements of any kind to any public real property.

(b) Does not include the routine operation, routine repair or routine maintenance of existing facilities, structures, buildings or real property.

5. "Construction-manager-at-risk" means a project delivery method in which:

(a) There is a separate contract for design services and a separate contract for construction services.

(b) The contract for construction services may be entered into at the same time as the contract for design services or at a later time.

(c) Design and construction of the project may be in sequential phases or concurrent phases.

(d) Finance services, maintenance services, operations services, preconstruction services and other related services may be included.

6. "construction services" means either of the following for construction-manager-at-risk, ***design-build*** and job-order-contracting project delivery methods:

(a) Construction, excluding services, through the construction-manager-at-risk or job-order-contracting project delivery methods.

(b) ***A combination of construction and,*** as elected by the purchasing agency, one or more related services, such as finance services, maintenance services, operations services, ***design services*** and preconstruction services, as those services are authorized in the definitions of construction-manager-at-risk, ***design-build*** or job-order-contracting in this section.

7. "Contract" means all types of state agreements, regardless of what they may be called, for the procurement of materials, services or construction or the disposal of materials.

8. "Contract modification" means any written alteration in the terms and conditions of any contract accomplished by mutual action of the parties to the contract.

9. "Contractor" means any person who has a contract with a state governmental unit.

10. "Data" means documented information, regardless of form or characteristic.

11. "Department" means the department of administration.

12. "Design-bid-build" means a project delivery method in which:

(a) There is a sequential award of two separate contracts.

(b) The first contract is for design services.

(c) The second contract is for construction.

(d) Design and construction of the project are in sequential phases.

(e) Finance services, maintenance services and operations services are not included.

13. "Design-build" means a project delivery method in which:

(a) There is a single contract for design services and construction services.

(b) Design and construction of the project may be in sequential phases or concurrent phases.

(c) Finance services, maintenance services, operations services, design services, preconstruction services and other related services may be included.

14. "Design requirements":

(a) Means at a minimum the purchasing agency's written description of the project or service to be procured, including:

(i) The required features, functions, characteristics, qualities and properties.

(ii) The anticipated schedule, including start, duration and completion.

(iii) The estimated budgets applicable to the specific procurement for design and construction and, if applicable, for operation and maintenance.

(b) May include:

(i) Drawings and other documents illustrating the scale and relationship of the features, functions and characteristics of the project, which shall all be prepared by an architect or engineer, as appropriate, who is registered pursuant to section 32-121.

(ii) Additional design information or documents that the purchasing agency elects to include.

15. "Design services" means architect services, engineer services or landscape architect services.

16. "Designee" means a duly authorized representative of the director.

17. "Director" means the director of the department of administration.

18. "Employee" means an individual drawing a salary from a state governmental unit, whether elected or not, and any noncompensated individual performing personal services for any state governmental unit.

19. "Engineer services" means those professional engineer services that are within the scope of engineering practice as provided in title 32, chapter 1.

20. "Finance services" means financing for a construction services project.

21. "General services administration contract" means contracts awarded by the United States government general services administration.

22. "Grant" means the furnishing of financial or other assistance, including state funds or federal grant funds, by any state governmental unit to any person for the purpose of supporting or stimulating educational, cultural, social or economic quality of life, if there is no substantial involvement between the state governmental unit and the recipient during performance. Grant does not include an agreement whose primary purpose is to procure a specific end product in the form of materials, services or construction. A contract resulting from such an agreement is not a grant but a procurement contract.

23. "Job-order-contracting" means a project delivery method in which:

(a) The contract is a requirements contract for indefinite quantities of construction.

(b) The construction to be performed is specified in job orders issued during the contract.

(c) Finance services, maintenance services, operations services, preconstruction services, design services and other related services may be included.

24. "Landscape architect services" means those professional landscape architect services that are within the scope of landscape architectural practice as provided in title 32, chapter 1.

25. "Maintenance services" means routine maintenance, repair and replacement of existing facilities, structures, buildings or real property.

26. "Materials":

(a) Means all property, including equipment, supplies, printing, insurance and leases of property.

(b) Does not include land, a permanent interest in land or real property or leasing space.

27. "Operations services" means routine operation of existing facilities, structures, buildings or real property.

28. "Owner" means a state purchasing agency or state governmental unit.

29. "Person" means any corporation, business, individual, union, committee, club, other organization or group of individuals.

30. "Preconstruction services" means advice during the design phase.

31. "Procurement":

(a) Means buying, purchasing, renting, leasing or otherwise acquiring any materials, services, construction or construction services. Procurement also

(b) Includes all functions that pertain to the obtaining of any material, services or construction, including description of requirements, selection and solicitation of sources, preparation and award of contract, and all phases of contract administration.

32. "Procurement officer":

(a) Means any person duly authorized to enter into and administer contracts and make written determinations with respect to the contracts.

(b) Includes an authorized representative acting within the limits of the authorized representative's authority.

33. "Purchasing agency" means any state governmental unit which is authorized by this chapter or rules adopted pursuant to this chapter, or by way of delegation from the director, to enter into contracts.

34. "Services":

(a) Means the furnishing of labor, time or effort by a contractor or subcontractor which does not involve the delivery of a specific end product other than required reports and performance.

(b) Does not include employment agreements or collective bargaining agreements.

35. "State governmental unit" means any department, commission, council, board, bureau, committee, institution, agency, government corporation or other establishment or official of the executive branch or corporation commission of this state.

36. "Subcontractor" means a person who contracts to perform work or render service to a contractor or to another subcontractor as a part of a contract with a state governmental unit.

37. "Using agency" means any state governmental unit which utilizes any materials, services or construction procured under this chapter.

A.R.S. § 41-2573. Bid Security.

A. As a guarantee that the contractor will enter into a contract bid security is required for all construction procured pursuant to section 41-2533 and all construction services procured pursuant to section 41-2578, subsection F if the purchasing agency estimates that the budget for construction, excluding the cost of any finance services, maintenance services, operations services and any other related services included in the contract, will be more than the amount established by section 41-2535, subsection . Bid security shall be a certified check, cashier's check or surety bond for ten percent of the amount of the bid.

B. Bid security shall be submitted in the following amounts:

1. For design-bid-build construction services, ten percent of the contractor's bid.

2. For ***design-build construction services*** awarded by competitive sealed proposals pursuant to section 41-2578, subsection F, ten percent of the purchasing agency's construction budget for the project as stated in the request for proposals, excluding finance services, maintenance services, operations services or any other related services included in the contract.

3. For job-order-contracting construction services awarded by competitive

sealed proposals pursuant to section 41-2578, subsection F, the amount prescribed by the purchasing agency in the request for proposals, but not more than ten percent of the purchasing agency's estimated budget for construction during the first year under the contract, excluding any finance services, maintenance services, operations services or any other related services included in the contract.

C. Nothing in this section prevents a state governmental unit from requiring such bid security in relation to any construction contract. The surety bond shall be executed and furnished as required by title 34, chapter 2, and the conditions and provisions of the surety bond regarding the surety's obligations shall follow the form required by section 34-606.

D. If the invitation for bids or request for proposals requires security, noncompliance requires that the bid be rejected unless, pursuant to rules, it is determined that the bid fails to comply in a nonsubstantial manner with the security requirements.

E. After the bids or proposals are opened, they are irrevocable for the period specified in the invitation for bids or request for proposals, except as provided in section 41-2533, subsection F and section 41-2578 subsection F. If a bidder is permitted to withdraw its bid before award, no action may be had against the bidder or the bid security.

A.R.S. § 41-2578. Procurement of specified professional and construction services; definition.

* * *

B. This state shall provide notice, in accordance with rules, of each procurement of professional services or construction services specified in this section and shall award contracts for these services on the basis of demonstrated competence and qualifications for the type of professional services or construction services pursuant to procedures prescribed in this section.

* * *

E. The procurement officer shall enter into negotiations for a contract with the highest qualified person or firm for the professional services or construction services. the negotiations shall include consideration of compensation and other contract terms that the officer determines to be fair and reasonable to this state. In making this decision, the procurement officer shall take into account the estimated value, the scope, the complexity and the nature of the professional services or construction services to be rendered. If the procurement officer is unable to negotiate a satisfactory contract with the person or firm considered to be the most qualified, at a price and on othercontract terms the procurement officer determines to be fair and reasonable to this state, the procurement officer shall formally terminate negotiations with that person or firm . The procurement officer may undertake negotiations with the next most qualified person or firm in sequence until an agreement is reached or a determination is made to reject all persons or firms on the short list. If a contract for construction services is entered into pursuant to this subsection, construction shall not commence until the purchasing agency and contractor agree in writing on fixed price or a guaranteed maximum price for the construction to be commenced.

F. As an alternative to subsection E of this section, the procurement officer may award ***design-build construction services*** or job-order-contracting construction services as follows:

1. The procurement officer shall use the selection committee appointed for the contract pursuant to subsection c of this section.

2. The procurement officer shall issue a request for proposals to the persons or firms on the short list developed pursuant to subsection c of this section.

3. For ***design-build construction services*** and job-order-contracting construction services, the request for proposals shall include:

(a) The purchasing agency's project schedule and project budget or life cycle budget for a procurement that includes maintenance services or operations services.

(b) A statement that the contract will be awarded to the person or firm whose proposal receives the highest number of points under a scoring method.

(c) A description of the scoring method, including a list of the factors in the scoring method and the number of points allocated to each factor. The factors in the scoring method shall include:

(i) For ***design-build construction services*** only, demonstrated compliance with the design requirements.

(ii) Offeror qualifications.

(iii) Offeror financial capacity.

(iv) Compliance with the purchasing agency's project schedule.

(v) For ***design-build construction services*** only, if the request for proposals specifies that the purchasing agency will spend its project budget and not more than its project budget and is seeking the best proposal for the project budget, compliance of the offeror's price or life cycle price for procurements that include maintenance services, operations services or finance services with the purchasing agency's budget as prescribed in the request for proposals.

(vi) For ***design-build construction services*** if the request for proposals does not contain the specifications prescribed in item (v) and for job-order-contracting construction services, the price or life cycle price for procurements that include maintenance services, operations services or finance services.

(vii) A offeror quality management plan.

(viii) Other evaluation factors as determined by the purchasing agency, if any.

(d) For ***design-build construction services*** only, The design requirements.

(e) A requirement that each offeror submit separately a technical proposal and a price proposal and that the offeror's entire proposal be responsive to the requirements in the request for proposals for ***design-build construction services***, The price in the price proposal shall be a fixed price or a guaranteed maximum price.

(f) A statement that in applying the scoring method the selection committee will separately evaluate the technical proposal and the price proposal and will evaluate and score the technical proposal before opening the price proposal.

(g) If the purchasing agency conducts discussions pursuant to paragraph 5 of this subsection, a statement that discussions will be held and a requirement that each offeror submit a preliminary technical proposal before the discussions are held.

4. If the purchasing agency determines to conduct discussions pursuant to paragraph 5 of this subsection, each offeror shall submit a preliminary technical proposal to the purchasing agency before those discussions are held.

5. If determined by the purchasing agency and included by the purchasing

agency in the request for proposals, the selection committee shall conduct discussions with all persons or firms that submit preliminary technical proposals. Discussions shall be for the purpose of clarification to assure full understanding of, and responsiveness to, the solicitation requirements. Offerors shall be accorded fair treatment with respect to any opportunity for discussion and for clarification by the owner. Revision of preliminary technical proposals shall be permitted after submission of preliminary technical proposals and before award for the purpose of obtaining best and final proposals. In conducting any discussions, information derived from proposals submitted by competing offerors shall not be disclosed to other competing offerors.

6. After completion of any discussions pursuant to paragraph 5 of this subsection or if no discussions are held, each offeror shall submit separately its final technical proposal and its price proposal.

7. Before opening any price proposal, the selection committee shall open the final technical proposals, evaluate the final technical proposals and score the final technical proposals using the scoring method in the request for proposals. No other factors or criteria may be used in the evaluation and scoring.

8. After completion of the evaluation and scoring of all final technical proposals, the selection committee shall open the price proposals, evaluate the price proposals, score the price proposals and complete the scoring of the entire proposals using the scoring method in the request for proposals. No other factors or criteria may be used in the evaluation and scoring.

9. The procurement officer shall award the contract to the responsive and responsible offeror whose proposal receives the highest score under the method of scoring in the request for proposals. No other factors or criteria may be used in the evaluation.

10. The contract file shall contain the basis on which the award is made.

11. For ***design-build construction services*** only, the procurement officer shall award a stipulated fee equal to a percentage, as prescribed in the request for proposals, of the purchasing agency's project budget, as prescribed in the request for proposals, but not less than two-tenths of one percent of the project budget to each short list offeror who provides a responsive, but unsuccessful, proposal. If the procurement officer does not award a contract, all responsive short list offerors shall receive the stipulated fee. the procurement officer shall pay the stipulated fee to each offeror within ninety days after the award of the initial contract or the decision not to award a contract. In consideration for paying the stipulated fee, the procurement officer may use any ideas or information contained in the proposals in connection with any contract awarded for the project, or in connection with a subsequent procurement, without any obligation to pay any additional compensation to the unsuccessful offerors. Notwithstanding the other provisions of this paragraph, an unsuccessful short list offeror may elect to waive the stipulated fee. If an unsuccessful short list offeror elects to waive the stipulated fee, the purchasing agency may not use ideas and information contained in the offeror's proposal, except that this restriction does not prevent the purchasing agency from using any idea or information if the idea or information is also included in a proposal of an offeror that accepts the stipulated fee.

G. Until award and execution of a contract by a purchasing agency, only the name of each person or firm on the short list developed pursuant to subsection c of this section shall be available to the public. All other information received by the purchasing agency in response to the request for qualifications or contained in the proposals shall be confidential in order to avoid disclosure of the contents that may be prejudicial to competing offerors during the selection process. The proposals shall be open to public inspection after the contract is awarded and the purchasing agency has executed the contract. To the extent that the offeror designates and the purchasing agency concurs, trade secrets and other proprietary data contained in a proposal remain confidential.

H. A purchasing agency may cancel a request for qualifications or a request for proposals or reject in whole or in part any or all proposals as specified in the solicitation if it is in the best interest of the purchasing agency. The purchasing agency shall make the reasons for cancellation or rejection part of the contract file.

I. Notwithstanding any other law:

1. The contractor for construction-manager-at-risk, ***design-build*** or job-order-contracting construction services is not required to be registered to perform design services pursuant to title 32, chapter 1 if the person or firm actually performing the design services on behalf of the contractor is appropriately registered.

2. The contractor for construction-manager-at-risk, ***design-build*** or job-order-contracting construction services is not required to be licensed to perform construction pursuant to title 32, chapter 10 if the firm actually performing the construction on behalf of the contractor is appropriately licensed.

3. Each project under a ***design-build construction services*** contract or a construction-manager-at-risk construction services contract shall be a specific, single project. For the purposes of this paragraph, "specific, single project" means a project that is constructed at a single location, at a common location or for a common purpose.

K. Notwithstanding anything to the contrary in this section or this title, a purchasing agency shall not:

1. Enter into a contract as contractor to provide construction-manager-at-risk construction services, ***design-build construction services*** or job-order-contracting construction services.

2. Contract with itself, with another purchasing agency, with this state or with any other governmental unit of this state or the federal government for the purchasing agency to provide construction-manager-at-risk construction services, ***design-build construction services*** or job-order-contracting construction services.

A.R.S. § 41-2579. Project delivery methods for design and construction services; report.

A. A purchasing agency may procure design services, construction and construction services, as applicable, under any of the following project delivery methods:

1. Design-bid-build.
2. Construction-manager-at-risk.
3. Design-build.
4. Job-order-contracting.

B. For the design-bid-build project delivery method, the director shall procure:

1. Design services pursuant to section 41-2578.
2. Construction by competitive sealed bidding, except as otherwise provided in section 41-2532.

C. The director shall procure construction services under the construction-manager-at-risk, ***design-build*** and job-order-contracting project delivery methods pursuant to section 41-2578.

D. The director shall procure design services relating to a construction-manager-at-risk construction services project pursuant to section 41-2578.

E. For job-order-contracting construction services projects, if the director does not include design services in the job-order-contracting construction services contract, the director shall procure any design services relating to job-order-contracting construction services projects under the contract pursuant to section 41-2578.

F. On or before January 15 of each year, any state entity that uses construction-manager-at-risk, ***design-build*** or job-order-contracting to procure construction services in a calendar year shall transmit to the governor, the president of the senate, the speaker of the house of representatives, the secretary of state and the director of the department of library, archives and public records a report on the total benefits associated with the use of construction-manager-at-risk, ***design-build*** or job-order-contracting to procure construction services. The report shall include a statement of costs incurred in securing the services, any overall administrative benefits realized through the use of the procurement methods, the number of projects begun in the preceding calendar year using the procurement methods and the cost and description of each project, any anticipated cost savings resulting from the use of the procurement methods and any actual cost savings realized through the use of the procurement method.

A.R.S. § 48-4204. County Stadium Districts. Constructing and operating a stadium and other structures; regulating alcoholic beverages.

A. From the taxes and surcharges levied pursuant to article 2 of this chapter for use with respect to major league baseball spring training, the district may acquire land and construct, finance, furnish, maintain, improve, operate, market and promote the use of existing or proposed major league baseball spring training facilities or stadiums and other structures, utilities, roads, parking areas or buildings necessary for full use of the training facilities or stadiums for sports and other purposes and do all things necessary or convenient to accomplish those purposes. The board shall require that any project undertaken by the district include financial participation from the county or municipality in which the project is located, from a private party or from any combination of these entities which equals or exceeds one-half of the amount to be expended or distributed by the district. Capital improvement funds expended at any time after June 1, 1991 by a county, municipality or private party for a purpose authorized by this section may be deemed financial participation with respect to any project the district may undertake.

B. From the taxes and charges levied or identified pursuant to section 48-4237 for use with respect to multipurpose facilities and from other monies lawfully available to the district, the district may acquire land and construct, finance, furnish, maintain, improve, operate, market and promote the use of multipurpose facilities and other structures, utilities, roads, parking areas or buildings necessary for full use of the multipurpose facilities and do all things necessary or convenient to accomplish those purposes. Public funds identified in section 48-4237, including funds distributed pursuant to section 42-5031, may only be used for the components for a multipurpose facility which are owned by the district or which are publicly owned.

C. Title 34 applies to the district, except that regardless of the funding source for design and construction of facilities and structures the district may establish alternative systems and procedures, ***including the use of the design-build method of construction or the use of qualifications-based selection*** of contractors with experience in stadium design or construction to expedite the design and construction of any of its facilities or structures or any facilities or structures leased to it or used by it pursuant to an intergovernmental agreement. For purposes of this subsection:

1. ***"design-build" means*** a process of entering into and managing a contract between the district and another party in which the other party agrees ***to both design and build a structure***, facility or other items specified in the contract.

2. "qualifications-based selection" means a process of entering into and managing a contract between the district and another party in which the other party is selected by the district on the basis of the party's qualifications and experience in designing or constructing facilities structures or other items similar to those the district is authorized to construct or lease. The other party may be selected by direct selection of by public competition.

D. For purposes of financing, designing, constructing or operating facilities or structures, the district is not the agent of any municipality participating in the funding of such facilities or structures.

III. Prompt Payment Acts.

A.R.S. § 41-2577. Progress payments.

A. Progress payments may be made by this state to the contractor on the basis of a duly certified and approved estimate of the work performed during a preceding period of time as set by rule, except that a percentage of all estimates shall be retained as provided in section 41-2576. The progress payments shall be paid on or before fourteen days after the estimate of the work is certified and approved * * *

B. The contractor shall pay to the contractor's subcontractors or material suppliers and each subcontractor shall pay to the subcontractor's subcontractor or material supplier, within seven days of receipt of each progress payment, unless otherwise agreed in writing by the parties, the respective amounts allowed the contractor or subcontractor on account of the work performed by the contractor's or subcontractor's subcontractors, to the extent of each such subcontractor's interest therein, except that no contract for construction may materially alter the rights of any contractor, subcontractor or material supplier to receive prompt and timely payment as provided under this section. These payments to subcontractors or material suppliers shall be based on payments received pursuant to this section. Any diversion by the contractor or subcontractor of payments received for work performed on a contract, or failure to reasonably account for the application or use of such payments, constitutes grounds for disciplinary action by the registrar of contractors. The subcontractor or material supplier shall notify the registrar of contractors and the purchasing agency in writing of any payment less than the amount or percentage approved for the class or item of work as set forth in this section.

* * *

F. If any periodic or final payment to a subcontractor is delayed by more than seven days after receipt of periodic or final payment by the contractor or subcontractor, the contractor or subcontractor shall pay his the contractor's or subcontractor's subcontractor or material supplier interest, beginning on the eighth day, at the rate of one percent per month or a fraction of a month on such unpaid balance as may be due.

G. Notwithstanding anything to the contrary in this section, this section applies only to amounts payable in a construction services contract for construction and does not apply to amounts payable in a construction services contract for design services, preconstruction services, finance services, maintenance services, operations services or any other related services included in the contract.

IV. Payment and Performance Bonds.

A.R.S. § 34-608. Construction-manager-at-risk, design-build and job-order-contracting construction services surety bond required; suit on bond; limitations.

A. Except if specifically exempted by statute, before an agent executes any contract with any person or firm for construction-manager-at-risk construction services, ***design-build construction services*** or job-order-contracting construction services, the person or firm shall furnish to the agent entering into the contract the following bonds, except that the bonds shall be furnished only on and at the same time as execution of a contract or an amendment to a contract that commits the contractor to provide construction for a fixed price, a guaranteed maximum price or any other fixed amount within a designated time frame:

1. ***A performance bond*** in an amount equal to the full contract amount conditioned on the faithful performance of the contract in accordance with plans, specifications and conditions of the contract, except that, for job-order-contracting construction services, the performance bond shall cover the full amount of construction under the job-order-contracting construction services contract, ***shall not include any design services,*** preconstruction services, finance services, maintenance services, operations services or other related services included in the contract, shall initially be based on the agent's estimate of the amount of construction that will be done under the contract and, for multiyear contracts, may be a single bond for the full term of the contract or a separate bond for each year of the contract, as determined by the agent, and except that, for construction-manager-at-risk construction services and ***design-build construction services***, the amount of the performance bond shall be the price of construction and shall not include the cost of any design services, preconstruction services, finance services, maintenance services, operations services or any other related services included in the contract. ***The performance bond shall cover performance of construction included in the contract and shall not cover performance of any design services***, preconstruction services, finance services, maintenance services, operations services or any other related services included in the contract. the performance bond shall be solely for the protection of the public body awarding the contract.

2. ***A payment bond*** in an amount equal to the full contract amount solely for the protection of claimants supplying labor or materials to the contractor or the contractor's subcontractors in the prosecution of the construction ***and not for the protection of persons providing any design services***, preconstruction services, finance services, maintenance services, operations services or other related services provided for in the contract, except that:

(a) For job-order-contracting construction services, the payment bond shall cover the full amount of construction under the job-order-contracting construction services contract, shall not include any design services, preconstruction services, finance services, maintenance services, operations services or other related services included in the contract, shall initially be based on the agent's estimate of the amount of construction that will be done under the contract and, for multiyear contracts, may be a single bond for the full term of the contract or a separate bond for each year of the contract, as determined by the agent.

(b) For construction-manager-at-risk construction services and ***design-build construction services***, the amount of the payment bond shall be the price of construction and shall not include the cost of any design services, preconstruction services, finance services, maintenance services, operations services or any other related services included in the contract.

B. Each bond shall include a provision allowing the prevailing party in a suit on the bond to recover as a part of the judgment any reasonable attorney fees as may be fixed by the court.

C. Notwithstanding any other statute, each bond shall be executed solely by a surety company or companies holding a certificate of authority to transact surety business in this state issued by the director of the department of insurance pursuant to title 20, chapter 2, article 1. The bonds shall not be executed by an individual surety or sureties, even if the requirements of section 7-101 are satisfied. The bonds shall be payable to the public body concerned.

D. the bonds shall be filed in the office of the department, board, commission, institution, agency or other contracting body awarding the contract.

E. It is illegal for a request for qualifications or a request for proposals pursuant to section 34-603, or any person acting or purporting to act on behalf of the contracting body, to require that bonds be furnished by a particular surety company, or through a particular agent or broker.

F. The conditions and provisions in the payment bond regarding the surety's obligations shall follow the following form:

Now, therefore, the condition of this obligation is that if the principal promptly pays all monies due to all persons supplying labor or materials to the principal or the principal's subcontractors in the prosecution of the construction provided for in the contract, this obligation is void. Otherwise it remains in full force and effect. Provided, however, that this bond is executed pursuant to title 34, chapter 6, Arizona Revised Statutes, and all liabilities on this bond shall be determined in accordance with the provisions, conditions and limitations of title 34, chapter 6, Arizona Revised Statutes, to the same extent as if they were copied at length in this agreement. The prevailing party in a suit on this bond shall recover as a part of the judgment reasonable attorney fees that may be fixed by the court.

G. The conditions and provisions in the performance bond regarding the surety's obligations shall follow the following form:

Now, therefore, the condition of this obligation is that if the principal faithfully performs and fulfills all of the undertakings, covenants, terms, conditions and agreements of the contract during the original term of the contract and any extension of the contract, with or without notice to the surety, and during the life of any guaranty required under the contract, and also performs and fulfills all of the undertakings, covenants, terms, conditions and agreements of all duly authorized modifications of the contract that may hereafter be made, notice of which modifications to the surety being hereby waived, the above obligation is void. Otherwise it remains in full force and effect. Provided, however, that this bond is executed pursuant to title 34, chapter 6, Arizona Revised Statutes, and all liabilities on this bond shall be determined in accordance with title 34, chapter 6, Arizona Revised Statutes, to the extent as if it were copied at length in this agreement. The prevailing party in a suit on this bond shall recover as part of the judgment reasonable attorney fees that may be fixed by the court. ***The performance under this bond is limited to the construction to be performed under the contract and does not include any design services***, preconstruction services, finance services, maintenance services, operations services or any other related services included in the contract.

H. If the prime contract or specifications require any persons supplying labor or materials in the prosecution of the work to furnish payment or performance bonds, these bonds shall be executed solely by a surety company or companies holding a certificate of authority to transact surety business in this state issued by the director of the department of insurance pursuant to title 20, chapter 2, article 1. Notwithstanding any other statute, the bonds shall not be executed by an individual surety or sureties, even if the requirements of section 7-101 are satisfied.

I. All bonds given by a contractor and surety pursuant to this section, regardless of their actual form, are deemed by law to be in the form required and set forth in this section.

J. For ***design-build*** or job-order-contracting construction services, if a person or firm that is not licensed to perform construction pursuant to title 32, chapter 10 has a construction services contract with the agent and has contracted for the construction portion of the construction services contract with a firm that is licensed to perform construction pursuant to title 32, chapter 10:

1. The person or firm holding the contract shall provide the payment bond and the performance bond.

2. The firm licensed to perform construction is not required to provide the payment bond or the performance bond.

A.R.S. § 34-609. Payment bonds for construction-manager-at-risk, design-build and job-order-contracting construction services.

A. Every claimant who has furnished labor or material in the prosecution of the construction provided for in a contract for construction-manager-at-risk construction services, ***design-build construction services*** and job-order-contracting construction services in respect of which a payment bond is furnished under section 34-608, and who has not been paid in full before the expiration of ninety days after the day on which the last of the labor was done or performed by the claimant or material was furnished or supplied by the claimant for which the claim is made, shall have the right to sue on the payment bond for the amount, or the balance, unpaid at the time of institution of the suit and to prosecute the action to final judgment for the sums justly due the claimant, and have execution thereon, provided, however, that any claimant having a direct contractual relationship with a subcontractor of the contractor furnishing the payment bond but no contractual relationship express or implied with the contractor shall have a right of action on the payment bond on giving the contractor only a written preliminary twenty day notice, as provided for in section 33-992.01, subsection C, paragraphs 1, 2, 3 and 4 and subsections E and H, and on giving written notice to the contractor within ninety days from the date on which the claimant performed the last of the labor or furnished or supplied the last of the material for which the claim is made, stating with substantial accuracy the amount claimed and the name of the party to whom the material was furnished or supplied or for whom the labor was done or performed. The notice shall be served by registered or certified mail, postage prepaid, in an envelope addressed to the contractor at any place the contractor maintains an office or conducts business, or at the contractor's residence.

B. Every suit instituted under this section shall be brought in the name of the claimant, but no suit may be commenced after the expiration of one year from the date on which the last of the labor was performed or materials were supplied by the person bringing this suit.

C. The contracting body and the agent in charge of its office shall furnish to anyone making written application therefor and who states that it has supplied labor or materials for work, and payment therefor has not been made, or that it is being sued on any bond, or that it is the surety on the bond, a certified copy of the bond and the contract for which it was given, which copy is prima facie evidence of the contents, execution and delivery of the original. Applicants shall pay for these certified copies such reasonable fees as the contracting body or the agent in charge of its office fixes to cover the actual cost of preparation of the copies.

A.R.S. § 41-2574. Contract performance and payment bonds.

A. The following bonds or security is required and is binding on the parties to the contract if the value of a construction award exceeds the amount established by section 41-2535:

1. A performance bond that is executed and furnished as required under title 34, chapter 2 article 2 or chapter 6, as applicable, in an amount equal to one hundred percent of the price specified in the contract conditioned on the faithful performance of the contract in accordance with the plans, specifications and conditions of the contract, except that, for job-order-contracting construction services, The performance bond shall cover the full amount of construction under the job-order-contracting construction services contract, shall not include any design services, preconstruction services, finance services, maintenance services, operations services or other related services included in the contract, shall initially be based on the purchasing agency's estimate of the amount of construction that will be done under the contract and, for multiyear contracts, may be a single bond for the full term of the contract or a separate bond for each year of the contract, as determined by the purchasing agency, and except that for construction-manager-at-risk construction services and ***design-build construction services***, the amount of the performance bond shall be the price of construction services, excluding the cost of any design services, preconstruction services, finance services, maintenance services, operations services and other related services included in the contract. The performance bond shall cover performance of construction included in the contract and shall not cover performance of any design services, preconstruction services, finance services, maintenance services, operations

services or other related services included in the contract. This bond is solely for the protection of this state. The conditions and provisions of the performance bond regarding the surety's obligations shall follow the form required under section 34-222, subsection G or section 34-608, subsection G, as applicable.

2. A payment bond that is executed and furnished as required by title 34, chapter 2, article 2 or chapter 6, as applicable, in an amount equal to one hundred percent of the price specified in the contract for the protection of all persons supplying labor or material to the contractor or its subcontractors for the performance of the construction provided for in the contract, except that, for job-order-contracting construction services, the payment bond shall cover the full amount of construction under the job-order-contracting construction services contract, shall not include any design services, preconstruction services, finance services, maintenance services, operations services or other related services included in the contract, shall initially be based on the purchasing agency's estimate of the amount of construction that will be done under the contract and, for multiyear contracts, may be a single bond for the full term of the contract or a separate bond for each year of the contract, as determined by the purchasing agency, and except that, for construction-manager-at-risk construction services and ***design-build construction services***, the amount of the payment bond shall be the price of construction services, excluding the cost of any design services, preconstruction services, finance services, maintenance services, operations services or any other related services included in the contract. The conditions and provisions of the payment bond regarding the surety's obligations shall follow the form required under section 34-222, subsection F or section 34-608, subsection F, as applicable.

B. For design-bid-build construction, the bonds prescribed in subsection A of this section shall be provided on and at the same time as execution of the contract. For construction-manager-at-risk, ***design-build*** and job-order-contracting construction services, the bonds prescribed in subsection A of this section shall be provided only on and at the same time as execution of a contract or an amendment to a contract that commits the contractor to provide construction for a fixed price, guaranteed maximum price or other fixed amount within a designated time frame. For ***design-build*** or job-order-contracting construction services, if a person or firm that is not licensed to perform construction pursuant to title 32, chapter 10 has a construction services contract with the purchasing agency and has contracted for the construction portion of the construction services contract with a firm that is licensed to perform construction pursuant to title 32, chapter 10:

1. The person or firm holding the contract shall provide the payment bond and the performance bond.

2. The firm licensed to perform construction is not required to provide the payment bond or the performance bond.

C. If the prime contract or specifications require any persons supplying labor or materials in the prosecution of the work to furnish payment or performance bonds, these bonds shall be executed solely by a surety company or companies holding a certificate of authority to transact surety business in this state issued by the director of the department of insurance pursuant to title 20, chapter 2, article 1. Notwithstanding the provisions of any other statute, the bonds shall not be executed by an individual surety or sureties, even if the requirements of section 7-101 are satisfied.

ARKANSAS

I. Attorney General Opinions.

The Arkansas Attorney General has issued an opinion that design/build violates state competitive bidding laws for projects over $10,000. The opinion also questions whether design/build would violate the state architect/engineer selection law. Ark. Att'y Gen. Op. No. 93-051 (1993).

II. Public Procurement Laws.

Ark. Stat. Ann. § 19-11-229. Competitive sealed bidding.

(a) Definition. "Competitive sealed bidding" means a method of procurement which requires:

(1) Issuance of an invitation for bids with a purchase description and all contractual terms and conditions applicable to the procurement;

(2) Public, contemporaneous opening of bids at a predesignated time and place;

(3) Unconditional acceptance of a bid without alteration or correction, except as authorized in §§ 19-11-204 and 19-11-228 – 19-11-240;

(4) Award to the responsive and responsible bidder who has submitted the lowest bid that meets the requirements and criteria set forth in the invitation for bids; and

(5) Public notice.

(b) Conditions for use. Contracts exceeding an estimated purchase price of ten thousand dollars ($10,000) shall be awarded by competitive sealed bidding unless a determination is made in writing by the agency purchasing official or the State Purchasing Director that this method is not practicable and advantageous, and specifically states the reasons therefor. The director may provide by regulation that it is not practicable to procure specified types of commodities or services by competitive sealed bidding. Factors to be considered in determining whether competitive sealed bidding is not practicable shall include whether:

(1) Purchase descriptions are suitable for award on the basis of the lowest evaluated bid price; and (2) The available sources, the time and place of performance, and other relevant circumstances are appropriate for the use of competitive sealed bidding.

(c) Two-Step Sealed Bidding. Where it is considered impractical to initially prepare a purchase description to support an award based on price, an invitation for bids may be issued requesting the submission of unpriced technical proposals to be followed by an invitation for bids limited to those bidders whose technical proposals meet the requirements set forth in the first invitation.

(d) Public Notice. Notice inviting bids shall be given not less than five (5) calendar days nor more than thirty (30) calendar days preceding the date for the opening of bids by publishing such notice at least one (1) time in at least one (1) newspaper having general circulation in the state, but in all instances adequate notice shall be given. The notice shall include a general description of the commodities or services to be procured and shall state where invitations for bid may be obtained. The notice shall also state the date, time, and place of bid opening.

(e) Bid Opening. Bids shall be opened publicly in the presence of one (1) or more witnesses at the time and place designated in the invitation for bids. Each bid, together with the name of the bidder, shall be recorded and open to public inspection.

(f) Bid Evaluation. Bids shall be evaluated based on the requirements set forth in the invitation for bids. These requirements may include criteria to determine acceptability such as inspection, testing, quality, workmanship, delivery, and suitability for a particular purpose and criteria affecting price such as life-cycle or total ownership costs. The invitation for bids shall set forth the evaluation criteria to be used. No criteria may be used in bid evaluation that were not set forth in the invitation for bids.

(g) Correction or Withdrawal of Bids. Correction of patent or provable errors in bids which do not prejudice other bidders, or withdrawal of bids, may be allowed only to the extent permitted under regulations promulgated by the State Purchasing Director and upon written approval of the Attorney General or a designee of such officer. No award shall be made on the basis of a corrected bid if the corrected bid exceeds the next lowest bid of a responsible bidder.

(h) Award. The contract shall be awarded with reasonable promptness by written notice to the lowest responsive and responsible bidder whose bid meets the requirements and criteria set forth in the invitation for bids. All other bidders requesting to be notified of the award decision shall be promptly notified of the decision.

(i) Cancellation of Invitation for Bids. An invitation for bid may be cancelled, or any or all bids may be rejected in writing by the State Purchasing Director or the agency purchasing official.

APPENDIX E

CALIFORNIA

I. Public Procurement Laws.

AB 470, a design/build bill passed by the state Assembly by a 73-4 margin in 1999 was vetoed by Governor Gray Davis. The bill sought to authorize school districts to use design/build contracts for schools and related facilities. The bill was opposed by the California Seismic Safety Commission, who felt that design/build puts schools at greater risk during an earthquake. Governor Davis stated in his October 10, 1999 veto letter, "I am supportive of the ***design-build*** process in concept. However, I cannot support this bill because it prohibits the withholding of retention proceeds by school districts using ***design-build*** contracts necessary to protect taxpayers if the contractor does not fulfill his responsibilities under the contract." The section in question by the Governor stated that a school district electing to use design/build may not withhold retention. A nearly identical bill, AB 958, is pending in the Assembly and deals with transit issues. The primary difference in the two bills is that AB 958 does not contain the "no retainage" clause.

Cal. Gov. Code § 5956.6. Agreements with private entities; provisions

(a) For purposes of facilitating projects, the agreements specified in Section 5956.4 may include provisions for the lease of rights-of-way in, and airspace over, property owned by a governmental agency, for the granting of necessary easements, and for the issuance of permits or other authorizations to enable the private entity to construct infrastructure facilities supplemental to existing government-owned facilities. Infrastructure constructed by a private entity pursuant to this chapter shall, at all times, be owned by a governmental agency, unless the governmental agency, in its discretion, elects to provide for ownership of the facility by the private entity during the term of the agreement. The agreement shall provide for the lease of those facilities to, or ownership by, the private entity for up to 35 years. In consideration therefor, the agreement shall provide for complete reversion of the privately constructed facility to the governmental agency at the expiration of the lease at no charge to the governmental agency. Subsequent to the expiration of the lease or ownership period, the governmental agency may continue to charge fees for use of the infrastructure facility. If, after the expiration of the lease or ownership period, the governmental agency continues to lease airspace rights to the private entity, it shall do so at fair market value.

(b) The agreement between the governmental agency and the private entity shall include, but need not be limited to, provisions to ensure the following:

(1) Compliance with the California Environmental Quality Act (Division 13 (commencing with Section 21000) of the Public Resources Code). Neither the act of selecting a proposed project or a private entity, nor the execution of an agreement with a private entity, shall require prior compliance with the act. However, appropriate compliance with the act shall thereafter occur before project development commences.

(2) Security for the construction of the facility to ensure its completion, and contractual provisions that are necessary to protect the revenue streams of the project.

(3) Adequate financial resources of ***the private entity to design, build, and operate the facility***, after the date of the agreement.

* * *

Cal. Gov. Code § 8169.5. Acquiring real property and constructing facilities in Capitol Area; Financing; Bids.

(a) In furtherance of the Capitol Area Plan, the objectives of Resolution Chapter 131 of the Statutes of 1991, and the legislative findings and declarations contained in Chapter 193 of the Statutes of 1996, relative to the findings by the Urban Land Institute, the director may purchase, exchange, or otherwise acquire real property and construct facilities, including any improvements, betterments, and related facilities, within the jurisdiction of the Capitol Area Plan in the City of Sacramento pursuant to this section. The total authorized scope of the project shall consist of up to approximately 1,470,200 gross square feet of office space and approximately 742,625 gross square feet of parking structures for use by the State Department of Education, the State Department of Health Services, and the Department of General Services as anchor tenants on blocks 171, 172, 173, 174, and 225, along with related additional parking on block 224, within the Capitol area. The acquisition and construction authorized pursuant to this section may not cause the displacement of any state or legislative employee parking spaces in the blocks specified in this subdivision unless the Department of General Services makes available existing state-owned parking spaces, acquires parking spaces, or constructs replacement parking that results in the affected employees' parking spaces being located at a reasonable distance from their place of employment.

(b) Subject to paragraphs (2) and (3) of subdivision (c), the department may contract for the lease, lease-purchase, lease with an option to purchase, acquisition, design, ***design-build,*** construction, construction management, and other services related to the design and construction of the office and parking facilities authorized to be acquired pursuant to subdivision (a).

(c) (1) The State Public Works Board may issue revenue bonds, negotiable notes, or negotiable bond anticipation notes pursuant to Chapter 5 (commencing with Section 15830) of Part 10b of Division 3 to finance all costs associated with acquisition, design, and construction of office and parking facilities for the purposes of this section. The State Public Works Board and the department may borrow funds for project costs from the Pooled Money Investment Account pursuant to Sections 16312 and 16313. In the event the bonds authorized by the project are not sold, the State Department of Education, the State Department of Health Services, and the Department of General Services, as determined by the Department of Finance, shall commit a sufficient amount of their support appropriations to repay any loans made for the project from the Pooled Money Investment Account. It is the intent of the Legislature that this commitment shall be included in future Budget Acts until all outstanding loans from the Pooled Money Investment Account are repaid either through the proceeds from the sale of bonds or from an appropriation.

(2) (A) If the department proposes to acquire the facilities ***on a design-build basis***, prior to the department entering into an agreement pursuant to subdivision (b) to design and build the facilities on blocks 171, 172, 173, 174, and 225, as specified in subdivision (a), the department shall submit to the Legislature a copy of all documents that shall be the basis upon which bids will be solicited and awarded to design and build the facilities. The documents shall include the following:

(i) The request for qualifications.

(ii) Site development guidelines.

(iii) Architectural and all system design requirements for the facilities.

(iv) Notwithstanding any other provision of law, the recommended specific criteria and process by which the contractor shall be selected.

(v) The performance criteria and standards for the architecture and all components and systems of the facilities.

(B) The information in the documents shall be provided in at least as much detail as was prepared for the San Francisco Civic Center Complex project and shall cover the quality of materials, equipment, and workmanship to be used in the facilities. These documents shall also include a detailed and specific space program for the facilities that identifies the specific spatial needs of the state agencies.

(C) If the department proposes to contract for construction separate from design, the department shall, prior to commencing work on working drawings for the facilities on blocks 171, 172, 173, 174, and 225, submit to the Legislature a copy of the preliminary plans for the facilities and a detailed and specific space program for the facilities that identifies the specific spatial needs of the state agencies.

(D) Regardless of how the department proposes to acquire the facilities, the department also shall submit all of the following information, which may be included in the bid documents:

(i) A final estimated cost for design, construction, and other costs.

(ii) How the department would manage the contracts entered into for this project to ensure compliance with contract requirements and to ensure that the state receives the highest level of quality workmanship and materials for the funds spent on the project.

(3) The department shall submit to the Legislature the information required to be submitted pursuant to paragraphs (2) and (6) on or before December 1, 1998. Except for those contracts and agreements necessary to prepare the information required by paragraphs (2) and (6), the department shall not solicit bids to enter into ***any agreement to design and build*** or otherwise acquire the facilities or commence work on working drawings on blocks 171, 172, 173, 174, or 225 sooner than the later of April 1, 1999, or 120 days after the department submits to the Legislature the information required to be submitted pursuant to paragraphs (2) and (6). The Legislative Analyst shall evaluate the information submitted to the Legislature and shall prepare a report to the Joint Committee on Rules within 60 days of receiving the documents submitted to the Legislature. It is the intent of the Legislature that the Joint Committee on Rules meet prior to the date the department is authorized to ***solicit bids to design and build*** or otherwise acquire the facilities or commence work on working drawings for the purposes of discussing the report from the Legislative Analyst and adopting a report with any recommendations to the department on changes to the site design criteria, performance criteria, and specifications and specific criteria for determining the winning bidder. If the Joint Committee on Rules adopts a report prior to the date the department is authorized to ***solicit bids to design and build*** or otherwise acquire the facilities or commence work on working drawings, the department may solicit the bids or commence the work when the report is adopted by the Joint Committee on Rules. The Senate Committee on Rules and the Speaker of the Assembly may designate members of their respective houses to monitor the progress of the preparation of the documents to be submitted pursuant to paragraph (2). The department shall prepare periodic progress reports and meet with the designated members or their representatives, as necessary, while preparing the documents.

(4) The amount of revenue bonds, negotiable notes, or negotiable bond anticipation notes to be sold may equal, but shall not exceed, the cost of planning, preliminary plans, working drawings, construction, construction management and supervision, other costs relating to the design and construction of the facilities, and any additional sums necessary to pay interim and permanent financing costs. The additional amount may include interest and a reasonable required reserve fund.

(5) Authorized costs of the facilities for preliminary plans, working drawings, construction, and other costs shall not exceed three hundred ninety-two million dollars ($ 392,000,000). Notwithstanding Section 13332.11, the State Public Works Board may authorize the augmentation of the amount authorized under this paragraph by up to 10 percent of the amount authorized.

(6) The net present value of the cost to acquire and operate the facilities authorized by subdivision (a) may not exceed the net present value of the cost to lease and operate an equivalent amount of comparable office space over the same time period. The department shall perform this analysis and shall obtain interest rates, discount rates, and Consumer Price Index figures from the Treasurer and submit its analysis with the documents submitted pursuant to paragraph (2) of subdivision (c). For purposes of this analysis, the department shall compare the cost of acquiring and operating the proposed facilities with the avoided cost of leasing and operating an equivalent amount of comparable office space that will no longer need to be leased because either (A) agencies will no longer occupy currently leased facilities when they occupy the proposed facilities, or (B) agencies will no longer occupy currently leased facilities when they occupy state-owned space being vacated by state agencies occupying the proposed facilities. The analysis shall also include the cost of any unique improvement associated with the moving of an agency into any state-owned space that would be vacated by agencies moving into the proposed facilities. However, these costs shall not include the cost of renovating or modernizing vacated state-owned space that is necessary to accommodate state agencies in general purpose office space. This paragraph shall not be construed as authorizing any renovation of state-owned space.

(d) The director may execute and deliver a contract with the State Public Works Board for the lease of the facilities described in this section that are financed with the proceeds of the board's bonds, notes, or bond anticipation notes issued in accordance with this section.

Cal. Pub. Con. Code § 1104. Responsibility for completeness and accuracy of plans and specifications.

No local public entity, charter city, or charter county shall require a bidder to assume responsibility for the completeness and accuracy of architectural or engineering plans and specifications on public works projects, ***except on clearly designated design build projects.*** Nothing in this section shall be construed to prohibit a local public entity, charter city, or charter county from requiring a bidder to review architectural or engineering plans and specifications prior to submission of a bid, and report any errors and omissions noted by the contractor to the architect or owner. The review by the contractor shall be confined to the contractor's capacity as a contractor, and not as a licensed design professional.

Cal. Gov. Code § 13332.19. (Operative term contingent)Approval as prerequisite to expenditure of funds for design-build project; Reversion of appropriated funds; Augmentations; Certification of compliance with legislative approval; Excess costs; Definitions.

(a) Except as otherwise specified in paragraphs (1) and (2), no funds appropriated for ***a design-build capital outlay project*** may be expended until the Department of Finance and the State Public Works Board have approved performance criteria or performance criteria and concept drawings for the project to be financed from the appropriation for capital outlay.

This subdivision shall not apply to either of the following:

(1) Amounts for acquisition of land.

(2) Amounts appropriated specifically for preliminary surveys, studies, and planning.

(b) Any appropriated amounts for the ***design-build phase of a design-build project***, where funds have been expended ***on the design-build phase*** by any state agency prior to the approval of the performance criteria or the performance criteria and concept drawings by the State Public Works Board, and all amounts not approved by the State Public Works Board under this section shall be reverted to the fund from which the appropriation was made. If the Director of Finance or his or her authorized representative requests ***review of the design-build bid package***, no ***design-build*** project for which an appropriation is made shall be put out to bid until the bid package has been approved by the Department of Finance. No substantial change shall be made from the performance criteria or performance criteria and concept drawings as approved by the State Public Works Board and the Department of Finance without written approval by the Department of Finance. Any proposed bid alternates shall be approved by the Department of Finance.

(c) The State Public Works Board shall defer all augmentations in excess of 20 percent of the amount appropriated ***for each design-build capital*** outlay project until the Legislature makes additional funds available for the specific project.

(d) Any augmentation of the ***design-build phase of a design-build*** project shall be consistent with the intent of subdivision (e) of Section 13332.11.

(e) Prior to State Public Works Board action ***on any design-build capital outlay*** appropriation, the Department of Finance shall certify, in writing, to the Chairperson of the Joint Legislative Budget Committee, the chairpersons of the respective fiscal committees, and the legislative members of the board that the requested action is in accordance with the legislatively approved scope and cost. If, pursuant to other provisions of this section, the Department of Finance approves changes to the approved scope or cost, or both, the department shall report the changes and associated cost implications. The reports also shall include all proposed or potential augmentations in excess of 10 percent of the amount appropriated for the ***costs of the design-build phase*** of the capital outlay project.

(f) The State Public Works Board shall defer action with respect to approval of performance criteria or performance criteria and concept drawings, when it is determined that ***the estimated cost of the total design-build project*** approved by the Legislature is in excess of 20 percent of the amount appropriated.

(g) For the purposes of this section, the following definitions shall apply:

(1) "Concept drawings" means any schematic drawings or architectural renderings that are prepared, in addition to performance criteria, in such detail as the

Director of General Services determines necessary to sufficiently describe the state' s needs.

(2) ***"Design-build bid package"*** means the performance criteria, any concept drawings deemed necessary by the Department of General Services, the form of contract, and all other documents and information that serve as the basis on which bids or proposals will be solicited to the ***design-build*** entities.

(3) ***"Design-build phase"*** means the period following the award of a contract to a ***design-build*** team in which the ***design-build*** entity completes the design and construction activities necessary to fully complete the project in compliance with the terms of the contract.

(4) ***"Performance criteria"*** means the information that fully describes the scope of the proposed project and includes, but is not limited to, the size, type, and design character of the buildings and site, the performance specifications that describe the required quality of materials, equipment, and workmanship, and any other information deemed necessary to sufficiently describe the state's needs.

Cal. Gov. Code § 14016. Agreement with City of San Bernardino for purchase, lease-purchase, lease with option to purchase, or exchange of property.

(a) The Director of General Services, as agent for the Department of Transportation, may enter into an agreement to purchase, lease-purchase, or lease with an option to purchase or exchange real property in the City of San Bernardino for the purpose of providing office and parking facilities, and any other improvements, betterments, and facilities related thereto, in order to allow consolidation of the offices of the Department of Transportation in the Riverside/San Bernardino region.

(b) The Department of Transportation shall take necessary actions to ensure that its annual budget includes all payments which may be necessary to satisfy the obligation for the office and parking facilities authorized pursuant to subdivision (a).

(c) Following the procurement and occupancy of the building and facilities described in subdivision (a), the Department of Transportation may lease or the office building at 247 West 3rd Street in the City of San Bernardino. The net proceeds of that lease or sale shall be applied toward any obligations undertaken pursuant to subdivision (a).

(d) Notwithstanding subdivision (a), the office and parking facilities obtained by the Department of General Services pursuant to subdivision (a) shall be and remain under the jurisdiction and control of, and shall be operated and maintained by, the Department of Transportation. Construction or acquisition of any office and parking facilities utilizing the financing methods authorized by subdivision (a) shall be done by the Department of General Services as agent of the Department of Transportation. The Department of General Services, in consultation with the Department of Transportation, may contract for design, ***design/build,*** and construction, construction management, and other services related to the design and construction of the office and parking facilities obtained pursuant to subdivision (a).

Cal. Gov. Code § 14661. (Operative term contingent) Procedure for procurement of state office and other facilities when legislature has authorized design-build process.

(a) Notwithstanding any provision of the Public Contract Code or any other provision of law, when the Legislature authorizes ***the use of the design-build construction***

procurement process for a specific project, the Director of General Services may contract and procure state office facilities and other buildings, structures, and related facilities pursuant to this section.

(b) For purposes of this section, ***"design-build"*** means a procurement process in which both the design and construction of a project are procured from a single entity.

(c) For purposes of this section, ***"design-build entity"*** means a partnership, corporation, or other legal entity that is able to provide appropriately licensed contracting, architectural, and engineering services as needed.

(d) Prior to contracting for the procurement of state office facilities and other state buildings and structures, the director shall:

(1) Prepare a program setting forth the scope of the project that may include, but is not limited to, the size, type, and desired design character of the buildings and site, performance specifications covering the quality of materials, equipment, and workmanship, or any other information deemed necessary to describe adequately the state's needs. The performance specifications shall be prepared by a design professional duly licensed and registered in the State of California.

(2) (A) Establish a competitive prequalification and ***selection process for design-build entities***, including any subcontractors listed at the time of bid, that clearly specifies the prequalification criteria, as well as recommend the manner in which the winning entity will be selected.

(B) Prequalification shall be limited to consideration of all of the following criteria:

(i) Possession of all required licenses, registration, and credentials in good standing that are required to design and construct the project.

(ii) Submission of evidence that establishes that ***the design-build entity*** members have completed, or demonstrated the capability to complete, projects of similar size, scope, or complexity, and that proposed key personnel have sufficient experience and training to competently manage and complete the design and construction of the project.

(iii) Submission of a proposed project management plan that establishes that ***the design-build entity*** has the experience, competence, and capacity needed to effectively complete the project.

(iv) Submission of evidence that establishes that ***the design-build entity*** has the capacity to obtain all required payment and performance bonding, liability insurance, and errors and omissions insurance, as well as a financial statement that assures the department that the ***design-build*** entity has the capacity to complete the project.

(v) Provision of a declaration certifying that applying members of ***the design-build entity*** have not had a surety company finish work on any project within the last five years.

(vi) Provision of information and a declaration providing detail concerning all of the following:

(I) Any construction or design claim or litigation totaling more than five hundred thousand dollars ($500,000) or 5 percent of the annual value of work performed, whichever is less, settled against any ***member of the design-build entity*** over the last five years.

(II) Serious violations of the Occupational Safety and Health Act, as provided in Part 1 (commencing with Section 6300) of Division 5 of the Labor Code, settled against any member of ***the design-build entity.***

(III) Violations of federal or state law, including, but not limited to, those laws governing the payment of wages, benefits, or personal income tax withholding, or of Federal Insurance Contributions Act (FICA) withholding requirements, state disability insurance withholding, or unemployment insurance payment requirements, settled against ***any member of the design-build entity*** over the last five years. For the purposes of this subclause, only ***violations by a design-build member*** as an employer shall be deemed applicable, unless it is shown that the ***design-build*** entity member, in his or her capacity as an employer, had knowledge of his or her subcontractor's violations or failed to comply with the conditions set forth in subdivision (b) of Section 1775 of the Labor Code.

(IV) Information required by Section 10162 of the Public Contract Code.

(V) Violations of the Contractors' State License Law (Chapter 9 (commencing with Section 7000) of Division 3 of the Business and Professions Code), excluding alleged violations or complaints.

(VI) Any conviction of ***any member of the design-build entity*** of submitting a false or fraudulent claim to a public agency over the last five years.

(VII) Provision of a declaration that ***the design-build entity*** will comply with all other provisions of law applicable to the project, including, but not limited to, the requirements of Chapter 1 (commencing with Section 1720) of Part 7 of Division 2 of the Labor Code.

(c) The director, when ***requested by the design-build entity***, shall hold in confidence any information required by clauses (I) to (vi), inclusive.

(D) Any declaration required under subparagraph (B) shall state that reasonable diligence has been used in its preparation and that it is true and complete to the best of the signer's knowledge. A person who certifies as true any material matter that he or she knows to be false is guilty of a misdemeanor and shall be punished by not more than one year in a county jail, by a fine of not more than five thousand dollars ($ 5,000), or by both the fine and imprisonment.

(3) (A) Recommend, as he or she deems in the best interests of the state, to the Legislature one of the following methods as the process to be used for the selection of the winning entity:

(i) ***A design-build competition*** based upon performance, price, and other criteria set forth by the department in the solicitation for proposals. The department shall establish technical criteria and methodology, including price, to evaluate proposals and shall describe the criteria and methodology in the request for ***design-build*** proposals. ***Award shall be made to the design-build entity*** whose proposal is judged as providing the best value in meeting the interest of the department and meeting the objectives of the project. A project with an approved budget of ten million dollars ($10,000,000) or more may be awarded pursuant to this clause.

(ii) ***A design-build competition*** based upon performance and other criteria set forth by the department in the solicitation for proposals. Criteria used in this evaluation of proposals may include, but need not be limited to, items such as proposed design approach, life-cycle costs, project features, and functions. However, any criteria and methods used to evaluate proposals shall be limited to those contained in the ***request for design-build proposals. Award shall be made to the design-build entity*** whose proposal is judged as providing the best value, for the lowest price, meeting the interests of the department and meeting the objectives of the project. A project with an approved budget of ten million dollars ($10,000,000) or more may be awarded pursuant to this clause.

(iii) ***A design-build competition*** based upon program requirements and a detailed scope of work, including any concept drawings and specifications set forth by the department in the solicitation for proposals. Award shall be made on the basis of the lowest responsible bid. A project with an approved budget of two hundred fifty thousand dollars ($250,000) or more may be awarded pursuant to this clause.

(B) Notwithstanding any other provision of law, no project valued under two hundred fifty thousand dollars ($250,000) shall be awarded pursuant to subparagraph (A).

(C) The legislation providing final authorization to construct a specific project ***using the design-build construction procurement process*** shall specify one of the methods described in clauses (I) to (iii), inclusive, of subparagraph (A) for the selection of the winning entity.

(e) The Legislature recognizes that ***the design-build entity*** is charged with performing both design and construction. Because a ***design-build contract*** may be awarded prior to the completion of the design, it is often impracticable for the ***design-build entity*** to list all subcontractors at the time of the award. As a result, the subcontractor listing requirements contained in Chapter 4 (commencing with Section 4100) of Part 1 of Division 2 of the Public Contract Code can create a conflict with the implementation of the ***design-build process*** by requiring all subcontractors to be listed at a time when a sufficient set of plans may not be available. It is the intent of the Legislature to establish a clear process for the selection and award of subcontracts entered into pursuant to this section in a manner that retains protection for subcontractors while enabling ***design-build projects*** to be administered in an efficient fashion. Therefore, all of the following requirements shall apply to subcontractors, licensed pursuant to Chapter 9 (commencing with Section 7000) of Division 3 of the Business and Professions Code, that are employed on ***design-build*** projects undertaken pursuant to this section:

(1) The department, ***in each design-build request for proposal***, may identify types of subcontractors, by subcontractor license classification, that will be listed by the ***design-build entity*** at the time of the bid. In selecting the subcontractors that will be listed by the ***design-build entity***, the department shall limit the identification to only those license classifications deemed essential for proper completion of the project. In no event, however, may the department specify more than five licensed subcontractor classifications. In addition, at its discretion, the ***design-build entity*** may list an additional two subcontractors, identified by subcontractor license classification, that will perform design or construction work, or both, on the project. In no event shall the ***design-build*** entity list at the time of bid a total amount of subcontractors that will perform design or construction work, or both, in a total of more than seven subcontractor license classifications on a project. All subcontractors that are listed at the time of bid shall be afforded all of the protection contained in Chapter 4 (commencing with Section 4100) of Part 1 of Division 2 of the Public Contract Code. All subcontracts that were not listed by the ***design-build*** entity at the time of bid shall be awarded in accordance with paragraph (2).

(2) All subcontracts that were not to be ***performed by the design-build entity*** in accordance with paragraph (1) shall be competitively bid and awarded by the

design-build entity, in accordance with the ***design-build*** process set forth by the department in the ***design-build*** package. The ***design-build*** entity shall do all of the following:

(A) Provide public notice of the availability of work to be subcontracted in accordance with Section 10140 of the Public Contract Code.

(B) Provide a fixed date and time on which the subcontracted work will be awarded in accordance with Section 10141 of the Public Contract Code.

(C) As authorized by the department, establish reasonable prequalification criteria and standards, limited in scope to those detailed in paragraph (2) of subdivision (d).

(D) Provide that the subcontracted work shall be awarded to the lowest responsible bidder.

(F) For purposes of this section, "best interests of the state" shall mean a ***design-build*** process that is projected by the director to reduce the project delivery schedule and total cost of a project while maintaining a high level of quality workmanship and materials, when compared to the traditional design-bid-build process.

(G) This section shall not be construed and is not intended to extend or limit the authority specified in Section 19130.

(H) ***Any design-build entity that is selected*** to design and build a project pursuant to this section shall possess or obtain sufficient bonding consistent with applicable provisions of the Public Contract Code. Nothing in this section shall prohibit a general or engineering contractor from being designated the lead entity on a ***design-build*** entity for the purposes of purchasing necessary bonding to cover the activities of the ***design-build*** entity.

(I) Any payment or performance bond written for the purposes of this section shall use a bond form developed by the Department of General Services. In developing the bond form, the department shall consult with the surety industry to achieve a bond form that is consistent with surety industry standards, while protecting the interests of the state.

Cal. Gov. Code § 14669.20. Sacramento County facilities; Bonds.

(a) Subject to paragraphs (2) and (3) of subdivision (b), the director may purchase, exchange, or otherwise acquire real property and construct facilities, including any improvements, betterments, and related facilities, within the County of Sacramento or the City of West Sacramento pursuant to this section, for use by the Department of Corrections and other state agencies. The department may enter into a lease-purchase, or a lease with an option to purchase the facilities, or may contract for the acquisition, design, ***design-build,*** construction, construction management, and other services related to the design and construction of the office and parking facilities authorized to be acquired pursuant to this section. The total authorized scope of the project shall not exceed 750,000 gross square feet of office facilities and the total authorized costs of the facilities, including land acquisition, preliminary plans, working drawings, construction, and other costs shall not exceed one hundred sixty million dollars ($160,000,000), excluding any additional sums necessary to pay interim and permanent financing costs.

Notwithstanding Section 13332.11, the State Public Works Board may authorize augmentation of the costs authorized in the annual Budget Act pursuant to this paragraph by up to 10 percent of the amount authorized. The department shall acquire the facility on a site that permits future expansion of the facility authorized by this section, if evaluations of future Department of Corrections' headquarters workload indicates that future expansion of the facility may be warranted.

(b) (1) The State Public Works Board may issue revenue bonds, negotiable notes, or negotiable bond anticipation notes pursuant to Chapter 5 (commencing with Section 15830) of Part 10b of Division 3 to finance all costs associated with acquisition, design, and construction of office and parking facilities for the purposes of this section. The State Public Works Board and the department may borrow funds for project costs from the Pooled Money Investment Account pursuant to Sections 16312 and 16313. In the event the bonds authorized by the project are not sold, the Department of Corrections shall commit a sufficient amount of its support appropriation to repay any loans made for the project from the Pooled Money Investment Account. It is the intent of the Legislature that this commitment shall be included in future Budget Acts until all outstanding loans from the Pooled Money Investment Account are repaid either through the proceeds from the sale of bonds or from an appropriation.

(2) (A) If the department proposes to ***acquire the facilities on a design-build basis***, prior to the department entering into an agreement pursuant to subdivision (a) to design and build the facilities, the department shall submit to the Legislature a copy of all documents that shall be the basis upon which bids will be solicited and awarded to design and build the facilities. The documents shall include the following:

(i) The request for qualifications.

(ii) Site development guidelines.

(iii) Architectural and all system design requirements for the facilities.

(iv) Notwithstanding any other provision of law, the recommended specific criteria and process by which the contractor shall be selected.

(v) The performance criteria and standards for the architecture and all components and systems of the facilities.

(B) The information in the documents shall be provided in at least as much detail as was prepared for the San Francisco Civic Center Complex project and shall cover the quality of materials, equipment, and workmanship to be used in the facilities. These documents shall also include a detailed and specific space program for the facilities that identifies the specific spatial needs of the state agencies.

(C) If the department proposes to contract for construction separate from design, the department shall, prior to commencing work on working drawings for the facilities authorized in this section, submit to the Legislature a copy of the preliminary plans for the facilities and a detailed and specific space program for the facilities that identifies the specific spatial needs of the state agencies.

(D) Regardless of how the department proposes to acquire the facilities, the department also shall submit all of the following information, which may be included in the bid documents:

(i) A final estimated cost for design, construction, and other costs.

(ii) How the department would manage the contracts entered into for this project to ensure compliance with contract requirements and to ensure that the state receives the highest level of quality workmanship and materials for the funds spent on the project.

(E) If the department proposes to acquire the facility pursuant to a lease-purchase agreement or a lease with an option to purchase, prior to executing an agreement to lease the facilities, the department shall submit to the Legislature a report that includes the site for the proposed facilities, the size of the facilities, the site characteristics, including, but not limited to, its proximity to public transit, the proposed lease terms, and any proposed terms concerning the exercise of any option to purchase the facilities.

(3) The department shall submit to the Legislature the information required to be submitted pursuant to paragraphs (2) and (5), notwithstanding Section 7550.5. Except for those contracts and agreements necessary to prepare the information required by paragraphs (2) and (5), the department shall not solicit bids to enter into any agreement to design and build or otherwise acquire the facilities, commence work on working drawings, or enter into an agreement to lease the facilities authorized in this section sooner than 120 days after the department submits to the Legislature the information required to be submitted pursuant to paragraphs (2) and (5). The Legislative Analyst shall evaluate the information submitted to the Legislature and shall prepare a report to the Joint Legislative Budget Committee within 60 days of receiving the documents submitted to the Legislature. It is the intent of the Legislature that the Joint Legislative Budget Committee meet prior to the date the department is authorized to solicit bids to ***design and build*** or otherwise acquire the facilities, commence work on working drawings, or enter into an agreement to lease the facilities for the purposes of discussing the report from the Legislative Analyst and adopting a report with any recommendations to the department on changes to the site design criteria, performance criteria, and specifications and specific criteria for determining the winning bidder. If the Joint Legislative Budget Committee adopts a report prior to the date the department is authorized to ***solicit bids to design and build*** or otherwise acquire the facilities, commence work on working drawings, or enter into an agreement to lease the facilities, the department may solicit the bids, commence the work, or enter into an agreement to lease the facilities when the report is adopted by the Joint Legislative Budget Committee. The department shall submit periodic progress reports to the Joint Legislative Budget Committee while preparing the documents, including a draft of any request for proposals.

(4) The amount of revenue bonds, negotiable notes, or negotiable bond anticipation notes to be sold may equal, but shall not exceed, the cost of land acquisition, planning, preliminary plans, working drawings, construction, construction management and supervision, other costs relating to the design and construction of the facilities, and any additional sums necessary to pay interim and permanent financing costs. The additional amount may include interest and a reasonable required reserve fund.

(5) The net present value of the cost to acquire and operate the facilities authorized in this section may not exceed the net present value of the cost to lease and operate an equivalent amount of comparable office space over the same time period. The department shall perform this analysis and shall obtain interest rates, discount rates, and Consumer Price Index figures from the Treasurer and submit its analysis with the documents submitted pursuant to paragraph (2). For purposes of this analysis, the department shall compare the cost of acquiring and operating the proposed facilities with the avoided cost of leasing and operating an equivalent amount of comparable office space that will no longer need to be leased because the agencies will no longer occupy currently leased facilities when they occupy the proposed facilities.

(C) The director may execute and deliver a contract with the State Public Works Board for the lease of the facilities described in this section that are financed with the proceeds of the board's bonds, notes, or bond anticipation notes issued in accordance with this section.

Cal. Gov. Code § 14669.5. Additional amendment to joint powers agreement with Community Redevelopment Agency of City of Los Angeles; Authority to borrow.

(a) Notwithstanding any other provisions of law, the Director of General Services may enter into an additional amendment to the existing joint powers agreement with the Community Redevelopment Agency of the City of Los Angeles dated June 30, 1982, in connection with the financing, planning, acquisition, equipping, furnishing, and construction or renovation of an additional office building or buildings and parking facilities in the City of Los Angeles in near proximity to the Ronald Reagan State Building and in connection therewith may enter into a lease-purchase agreement or agreements, an agreement for the appointment of a bond trustee, and an agreement or agreements for the Department of General Services to act as agent for acquisition, planning, and construction or renovation matters, each of which agreements shall be with the joint powers authority created under the joint powers agreement. The amendment to the existing joint powers agreement shall provide that any authorization under the joint powers agreement for the joint powers authority to acquire property by means of condemnation proceedings shall not include the power to condemn the property on which was located, as of January 1, 1991, that certain homeless women's housing and day care center commonly known as the Downtown Women's Center, which property is legally described as set forth below. The lease-purchase agreement or agreements may provide for space to be used for private commercial purposes. The director shall notify the chairperson of the Joint Legislative Budget Committee or his or her designee, and the chairpersons of the committees in each house which consider appropriations, of his or her intention to execute the lease-purchase agreement or agreements at least 20 days prior to its execution.

The property referred to above is legally described as follows:

[legal description deleted here]

(b) In as much as it is in the best interest of the people of the State of California to consolidate state offices in the City of Los Angeles as described in subdivision (a), at the earliest opportunity, ***a "design-build" concept may be utilized*** in meeting the objective of this section.

(c) Notwithstanding any other provision of law, the joint powers authority described in subdivision (a) shall have the authority to borrow from the Pooled Money Investment Account as provided in Sections 16312 and 16313.

Cal. Gov. Code § 14669.8. Additional amendment to joint powers agreement with San Francisco Redevelopment Agency; Authority to borrow.

(a) Notwithstanding any other provision of law, the Director of General Services may enter into an amendment to the existing joint powers agreement with the San Francisco Redevelopment Agency in connection with the redevelopment of the 350 McAllister/455 Golden Gate block in the City and County of San Francisco. The redevelopment shall include, but not be limited to, demolition of existing structures, renovation, financing, planning, acquisition, construction and equipping, and furnishing of new state office buildings and parking facilities, and any betterments, improvements, and facilities related thereto, in the San Francisco Civic Center Area. In connection therewith, the director may enter into a lease-purchase agreement, an agreement for the appointment of a bond trustee, any other documents and agreements in connection with the financing by sale of bonds or otherwise of the development, and an agreement for the department to act as agent for acquisition, planning, and construction matters, each of which agreements shall be with the joint powers authority created under the joint powers agreement. In connection with the development any agreements authorized by this section or any work or expenses related thereto, the joint powers authority may use any funds lawfully available to it for those purposes, and the department is empowered to use and expend those funds in accordance with the terms of any agreement between the department and the joint powers authority for the carrying out of the works on the development. The Treasurer shall be the agent for sale, as defined in Chapter 9 (commencing with Section 5700) of Division 6 of Title 1, for any financing authorized by this section.

(b) Inasmuch as it is in the best interest of the people of the State of California to consolidate state offices in the San Francisco Civic Center Area as described in subdivision (a), at the earliest opportunity, ***a "design-build" concept may be utilized*** in meeting the objective of this section.

(c) Notwithstanding any other provision of law, the joint powers authority described in subdivision (a) shall have the authority to borrow from the Pooled Money Investment Account as provided in Sections 16312 and 16313.

Cal. Gov. Code § 14669.12. Amendment of joint powers agreement with Oakland Redevelopment Agency for development of state-owned office space in Oakland; Lease-purchase agreement; Design-build concept.

(a) Notwithstanding any other provision of law, the Director of General Services may enter into an amendment to the existing joint powers agreement with the Oakland Redevelopment Agency in connection with the development of approximately 700,000 gross usable square feet of new state-owned office space in the City of Oakland. This development shall include, but not be limited to, the financing, planning, acquisition, construction, and equipping and furnishing of new state office buildings and parking facilities, and any betterments, improvements, and facilities related thereto, in the City of Oakland. The state office building and parking facility in the City of Oakland developed pursuant to this subdivision shall be named and shall be known as the Elihu M. Harris Building.

(b) In connection with the development described in subdivision (a), the Director of General Services may enter into a lease-purchase agreement, an agreement for the appointment of a bond trustee, and other documents and agreements in connection with the financing, by sale of bonds or otherwise, of this development, and an agreement for the Department of General Services to act as agent for acquisition, planning, and construction matters. Each agreement shall be with the joint powers authority created under the joint powers agreement.

(c) In connection with the development or any agreements provided for by this section, or any work or expenses related thereto, the joint powers authority may use any funds lawfully available to it for the purposes of this section, and the Director of General Services is authorized to use and expend these funds in accordance with the terms of any agreement between the director and the joint powers authority for the carrying out of these works on the development.

(d) Inasmuch as it is in the best interest of the people of the State of California to consolidate state offices in the City of Oakland, as well as other San Francisco Bay area locations, at the earliest possible opportunity, ***a "design-build" concept may be utilized*** in meeting the objective of this section.

(e) Notwithstanding any other provision of law, the joint powers authority described in this section may borrow from the Pooled Money Investment Account as provided for in Sections 16312 and 16313.

University of California Competitive Bidding.

Cal. Pub. Cont. Code § 10503. Preparation of estimates; Plans and specifications; Documents for solicitation.

Before entering into any contract for a project, the Regents of the University of California shall cause to be prepared estimates and either:

(a) Complete plans and specifications setting forth such directions as will enable a competent mechanic or other builder to carry them out.

(b) Documents for the ***solicitation of bids on a design-and-build basis***, including: (1) a program setting forth the scope of the project, the size, type, and desired design character of the buildings and the site, (2) a set of performance specifications covering the quality of materials, equipment, and workmanship, (3) a maximum acceptance cost, and (4) a method and grading system for evaluating contractor proposals on the basis of a preliminary design, outline specifications, a price, and the financial condition and relevant experience of the contractor and the contractor's architect.

* * *

(e) Documents for the solicitation of bids under such other contracting mode as the regents determine to be in the best interest of the university, provided that such proposals be compared on a uniform basis and that award be made as determined by the published selection standards.

Local Agency Public Construction Act

Cal. Pub. Contract Code § 20133. (Operative until January 1, 2001) Alternative bidding procedure for Alameda, Sacramento, Santa Clara, Solano, and Tulare Counties.

(a) This section provides an alternative and optional procedure on bidding on building construction projects applicable only in the Counties of Alameda, Sacramento, Santa Clara, Solano, and Tulare, upon the approval of the appropriate board of supervisors.

(b) (1) If a board of supervisors elects to proceed under this section, it shall, before entering into any contract requiring advertising for bids for a project, cause to be prepared estimates, and prepare documents, for the ***solicitation of bids on a design and build basis***.

(2) For the purposes of this section, ***"design and build" means a method of procuring design and construction from a single source.*** The selection of the single source occurs before the development of complete plans and specifications.

(c) The request for submittals shall include all of the following:

(1) A clear precise description of the services to be provided and work to be performed.

(2) A description of the format that submittals shall follow and the elements they shall contain, including the qualifications and relevant experience of the design professional and the contractor, and the criteria that shall be used in evaluating the submittal, including the bid price.

(3) The date on which the submittals are due, and the timetable that will be used in reviewing and evaluating the submittals.

(d) In addition to the information required in paragraph (2) of subdivision (c), bidders shall submit their proposals with the construction bid price and all cost information in a separate sealed envelope.

(e) All submittals received prior to the closing time stated in the request for submittal shall be reviewed to determine those that meet the format requirements and the standards specified in the request for submittal.

(f) The contract shall be awarded to the lowest responsible bidder meeting the standards of the request for submittal.

(g) For the purposes of this section, selections of design professionals shall meet the standards of Chapter 10 (commencing with Section 4525) of Division 5 of Title 1 of the Government Code.

(h) Each county utilizing the procedures set forth in this section shall file, on or before September 1, 1998, and again on or before September 1, 2000, with the Committees on Local Government of the Senate and the Assembly, a report containing the following information:

(1) A description of each project ***procured through the design and build process*** authorized by this section, including but not limited to, all of the following:

(A) The type of facility.

(B) The gross square footage of the facility.

(C) The company or contractor who was awarded the project.

(D) The estimated and actual length of time to complete the project.

(E) The estimated and actual project cost.

(F) A description of the relative merits of projects authorized pursuant to this section and similar projects procured pursuant to existing requirements of this code.

(G) A description of any written protest concerning any aspect of the solicitation, bid, proposal, or award of projects pursuant to this section, including the resolution of the project.

(2) Other pertinent information that each county believes is instructive in evaluating whether the method of procurement should be continued or expanded, or both.

(i) This section shall be applicable only to any project for which the costs specified in any contract awarded pursuant to this section do not exceed fifty million dollars ($50,000,000).

(j) Contracts awarded pursuant to this section shall be valid until the project is completed, within the period specified in the contract entered into by the county prior to January 1, 2001.

(k) This section shall be applicable only to a project that is under the supervision of a licensed general building contractor within the meaning of Section 7057 of the Business and Professions Code.

(l) This section shall remain in effect only until January 1, 2001, and as of that date is repealed, unless a later enacted statute, that is enacted on or before January 1, 2001, deletes or extends that date.

Cal. Pub. Cont. Code § 10708. Agreement with contractor for design and construction of project (California State University).

When, in the opinion of the trustees, the best interests of the California State University dictate, the trustees may enter into ***an agreement with a contractor to provide***

all or significant portions of the design services and construction of a project under this chapter. ***The contractor shall design the project*** pursuant to the scope of services set forth in the request for proposals, ***build the project***, and present the completed project to the trustees for their approval and acceptance. Work under this section shall be carried out by a contractor chosen by a competitive bidding process that employs selection criteria in addition to cost. Any design work performed pursuant to this section shall be prepared and signed by an architect certificated pursuant to Chapter 3 (commencing with Section 5500) of Division 3 of the Business and Professions Code.

Cal. Pub. Cont. Code § 20175. Local Agency Public Construction Act. (Operative until January 1, 2001)Optional procedures.

(a) This section provides an alternative and optional procedure on bidding on building construction projects applicable only in the City of West Sacramento and the City of Davis, upon the approval of the city council of the respective city.

(b) (1) If the city council elects to proceed under this section, it shall, before entering into any contract requiring advertising for bids for a project, cause to be prepared estimates, and prepare documents, for ***the solicitation of bids on a design-and-build basis.***

(2) For the purposes of this section, ***"design and build" means a method of procuring design and construction from a single source.*** The selection of the single source occurs before the development of complete plans and specifications.

(c) The request for submittals shall include all of the following:

(1) A clear and precise description of the services to be provided and work to be performed.

(2) A description of the format that submittals shall follow and the elements they shall contain, including the qualifications and relevant experience of the design professional and the contractor, and the criteria that shall be used in evaluating the submittal, including the bid price.

(3) The date on which the submittals are due, and the timetable that will be used in reviewing and evaluating the submittals.

(d) In addition to the information required in paragraph (2) of subdivision (c), bidders shall submit their proposals with the construction bid price and all cost information in a separate sealed envelope.

(e) All submittals received prior to the closing time stated in the request for submittal shall be reviewed to determine those that meet the format requirements and the standards specified in the request for submittal.

(f) The contract shall be awarded to the lowest responsible bidder meeting the standards of the request for submittal.

(g) For the purposes of this section, selections of design professionals shall meet the standards of Chapter 10 (commencing with Section 4525) of Division 5 of Title 1 of the Government Code.

(h) The City of West Sacramento or the City of Davis, if it chooses to utilize this section, shall file, on or before September 1, 1998, and again on or before September 1, 2000, with the Committees on Local Government of the Senate and the Assembly, a report containing the following information:

(1) A description of each project ***procured through the design-and-build process*** authorized by this section including, but not limited to, all of the following:

(A) The type of facility.

(B) The gross square footage of the facility.

(C) The company or contractor who was awarded the project.

(D) The estimated and actual length of time to complete the project.

(E) The estimated and actual project cost.

(F) A description of the relative merits of projects authorized pursuant to this section and similar projects procured pursuant to existing requirements of this code.

(G) A description of any written protest concerning any aspect of the solicitation, bid, proposal, or award of projects pursuant to this section, including the resolution of the project.

(2) Other pertinent information that the city believes is instructive in evaluating whether the method of procurement should be continued or expanded, or both.

(i) This section shall be applicable only to any project for which the costs specified in any contract awarded pursuant to this section do not exceed fifty million dollars ($50,000,000).

(j) Contracts awarded pursuant to this section shall be valid until the project is completed, within the period specified in the contract entered into by the city prior to January 1, 2001.

(k) This section shall be applicable only to a project that is under the supervision of a licensed general building contractor within the meaning of Section 7057 of the Business and Professions Code.

(l) Nothing in this section shall limit any existing authority, whether explicit or implied, for cities to utilize ***design-build*** in contracting for public improvements.

(m) This section shall remain in effect only until January 1, 2001, and as of that date is repealed, unless a later enacted statute, that is enacted on or before January 1, 2001, deletes or extends that date.

Cal. Pub. Cont. Code § 20221.1. District's election to design and construct extension of facilities to airport [San Francisco Bay Area Rapid Transit (BART) District]. Notwithstanding any other provision of law, the governing board of the San Francisco Bay Area Rapid Transit District may, at its discretion, ***elect to design and construct*** all or any part of the extension of the district's transit facilities to the San Francisco International Airport. If the board elects to design and construct the project, it shall comply with all of the following procurement procedures:

(a) A request for qualifications shall be prepared and submitted to an adequate number of sources, as determined by the district, in its discretion, to permit reasonable competition consistent with the nature and requirements of that portion of the extension ***the district has elected to design and construct*** in accordance with this section. In addition, notice of the request for qualifications shall be published at least once in a newspaper of general circulation, which publication shall be at least 10 days before responses to the request for qualifications are received. The request for qualifications shall identify the minimum standards that the district has determined, in its discretion, shall be met or exceeded by ***a contractor to successfully design and construct the project*** described in this section. Those standards may include, but are not limited to, technical experience and capability, financial condition and capacity, organization, and personnel.

(b) All respondents determined by the district to meet or exceed the minimum standards set forth in the request for qualifications shall be considered prequalified. The district may, in its discretion, repeat the prequalification process described in subdivision (a). If the district has prequalified at least three respondents, it may invite those prequalified to submit sealed bids for the project based upon plans and specifications prepared by the district. Only those who have been prequalified are eligible for award of a contract. Award shall be made to the prequalified bidder submitting the lowest responsive bid.

(c) The board may reject any and all bids and request new bids.

Cal. Pub. Util. Code §§ 132410. Powers and duties [Pasadena Metro Blue Line Construction Authority.]

(a) The authority has all of the powers necessary for planning, acquiring, leasing, developing, jointly developing, owning, controlling, using, jointly using, disposing of, designing, procuring, and building the project, including, but not limited to, all of the following:

(1) Acceptance of grants, fees, and allocations from the state, local agencies, and private entities.

(2) Acquiring, through purchase or through eminent domain proceedings, any property necessary for, incidental to, or convenient for, the exercise of the powers of the authority.

(3) Incurring indebtedness, secured by pledges of revenue available for project completion.

(4) Contracting with public and private entities for the planning, design, and construction of the project. These contracts may be assigned separately or may be combined to include any or all tasks necessary for completion of the project.

(5) Entering into cooperative or joint development agreements with local governments or private entities. These agreements may be entered into for the purpose of sharing costs, selling or leasing land, air, or development rights, providing for the transferring of passengers, making pooling arrangements, or for any other purpose that is necessary for, incidental to, or convenient for the full exercise of the powers granted the authority. For purposes of this paragraph, "joint development" includes, but is not limited to, an agreement with any person, firm, corporation, association, or organization for the operation of facilities or development of projects adjacent to, or physically or functionally related to, the project.

(6) Relocation of utilities, as necessary for completion of the project . . .

Cal. Pub. Util. Code § 185036. California High-Speed Rail Service. Additional powers. Upon approval by the Legislature, by the enactment of a statute, or approval by the voters of a financial plan providing the necessary funding for the construction of a high-speed network, the authority may do any of the following:

(a) Enter into contracts with private or public entities for the design, construction and operation of high-speed trains. The contracts may be separated into individual tasks or segments or may include all tasks and segments, ***including a design-build or design-build-operate contract.***

(b) Acquire rights-of-way through purchase or eminent domain.

(c) Issue debt, secured by pledges of state funds, federal grants, or project revenues. The pledge of state funds shall be limited to those funds expressly authorized by statute or voter-approved initiatives.

(d) Enter into cooperative or joint development agreements with local governments or private entities.

(e) Set fares and schedules.

(f) Relocate highways and utilities.

Cal. Pub. Cont. Code § 20301.5.

(a) Notwithstanding Section 20301, ***the district may let a design-and-build contract*** for any project for a transit center or station, transit park-and-ride lot, bus and light rail maintenance facility, or administrative office building, or any combination of those, upon approval by the board of directors.

(b) (1) If the board of directors elects to proceed under subdivision (a), before entering into any contract that requires advertising for bids for a project, the board shall cause to be prepared estimates, and shall prepare documents, for the ***solicitation of bids on a design-and-build basis.***

(2) For the purposes of this section, ***"design and build" means a method of procuring design and construction from a single source.*** The selection of the single source shall occur before the development of complete plans and specifications.

(c) The request for submission of bids shall include all of the following:

(1) A clear, precise description of the services to be provided and work to be performed.

(2) A description of the format that submittals shall follow and the elements they shall contain, including the qualifications and relevant experience of the design professional and the contractor, and the criteria that shall be used in evaluating the submittal, including the bid price.

(3) A requirement that bidders submit their proposals with the construction bid price and all cost information in a separate sealed envelope.

(4) The date on which the submittals are due, and the timetable that will be used in reviewing and evaluating the submittals.

(d) All submittals received prior to the closing time stated in the request for submittal shall be reviewed to determine those that meet the format requirements and the standards specified in the request for submittal.

(e) The contract shall be awarded to the lowest responsible bidder meeting the standards of the request for submittal.

(f) For the purposes of this section, selections of design professionals shall meet the standards set forth in Chapter 10 (commencing with Section 4525) of Division 5 of Title 1 of the Government Code.

(g) This section shall apply only to a project that is under the supervision of a licensed general building contractor, as defined in Section 7057 of the Business and Professions Code.

Cal. Pub. Res. Code § 5080.50 [Leland Stanford Mansion State Historical Park Construction] Contracts for preservation and restoration.

Notwithstanding any provision of law to the contrary, the Director of Parks and Recreation may contract for construction for the preservation and restoration of the Leland Stanford Mansion State Historical Park and related facilities ***using the design-build process*** set forth in Section 14661 of the Government Code.

Cal. Pub. Res. Code § 5080.51 [Leland Stanford Mansion State Historical Park Construction] Selection of the design-build entity; authorized methods.

The director is authorized to use the method for selection of ***the design-build entity*** set forth in clause (I) of subparagraph (A) of paragraph (3) of subdivision (d) of Section 14661 of the Government Code, notwithstanding that the approved project budget may not be ten million dollars ($10,000,000) or more.

Cal. Pub. Res. Code § 5080.53 [Leland Stanford Mansion State Historical Park Construction] Donation of services; acceptance.

The director may accept donations of services, including architectural, engineering, construction management, or other professional services, construction labor and materials, or any other donations, in addition to donations of funds to be integrated into the project on terms and conditions that, in the discretion of the department, will contribute to the successful prosecution and completion of the restoration and preservation project. The director may, in addition to any other criteria authorized by Section 14661 of the Government Code, include the donation of goods and services as a ***basis for selection for the design-build entity.***

Cal. Pub. Res. Code § 5080.54 [Leland Stanford Mansion State Historical Park Construction]Advertising for design-build proposals; approval of scope and cost of project.

In addition to the approvals required by Section 13332.19 of the Government Code, prior to advertising ***for design-build proposals***, the director shall submit to the State Public Works Board the program required by paragraph (1) of subdivision (d) of Section 14661 of the Government Code and Section 5080.52 of this code and shall obtain from the board approval of the scope and cost for the project.

Cal. Pub. Res. Code § 5080.55 [Leland Stanford Mansion State Historical Park Construction]Reports on program submitted to Joint Legislative Budget Committee.

Notwithstanding Section 7550.5 of the Government Code, the director shall submit a report containing the program required by paragraph (1) of subdivision (d) of Section 14661 of the Government Code and Section 5080.52 of this code to the Joint Legislative Budget Committee for review and comment not less than 30 days prior to advertising for the submittal of ***design-build proposals*** for the project. In addition to any other requirements for the program contained in paragraph (1) of subdivision (d) of Section 14661 of the Government Code and Section 5080.52 of this code, the report shall contain the criteria, including the relative weight to be assigned to the criteria, to be used for the evaluation of ***the design-build proposals*** and selection of ***the design-build entity***.

COLORADO

C.R.S. § 19-2-411.5. Department of Human Services. Juvenile facility—contract for design, construction, and operation.

(1) The department of human services is hereby authorized to contract with a single private entity for ***the design, construction, and operation*** of a five-hundred-bed residential juvenile facility to house juveniles who are in the custody of the department of human services. The juvenile facility shall follow an academic correctional model, providing educational, vocational, and positive developmental programming. The juvenile facility shall be constructed in a campus-style design and located on the parcel of real property formerly known as the Lowry bombing range. The state shall retain ownership of the juvenile facility constructed and operated pursuant to this section. Nothing in this section shall be construed as requiring that the parcel of real property formerly known as the Lowry bombing range be used exclusively for the juvenile facility constructed pursuant to this section.

(2) On or before January 31, 1998, the department shall issue a request for ***proposals for the design, construction, and operation of the juvenile facility***. The executive director of the department of human services shall establish the time and manner for submitting proposals and shall select a contractor on or before June 1, 1998. In choosing a contractor, the executive director of the department of human services shall ensure that the contractor and the contract meet the following requirements:

(a) The executive director of the department of human services shall select the lowest responsible bid by the contractor most qualified to operate the juvenile facility on an academic correctional model, subject to available appropriations. Prior to final selection, the executive director shall confirm that the contractor has the qualifications, experience, and management personnel necessary to carry out the terms of the contract.

(b) The contractor shall agree to indemnify the state and the department of human services, including their officials and agents, against any and all liability including but not limited to any civil rights claims. The department of human services shall require proof of satisfactory insurance, the amount of which shall be determined by the department of human services following consultation with the division of insurance in the department of regulatory agencies.

(c) The proposed juvenile facility and the management plan for juveniles housed at the facility shall meet the requirements of applicable court orders and state law.

(d) The contractor shall be responsible for a range of dental, medical, and psychological services and diet, education, and work programs at least equal to those services and programs provided by the department of human services at comparable state juvenile facilities. The work and education programs shall be designed to reduce recidivism.

(e) The department of human services shall monitor the juvenile facility, and the contractor shall bear the costs of monitoring.

(3) Upon completion of the construction of the juvenile facility contracted for pursuant to this section, the contract for operation of the facility shall be subject to annual renewal. The initial contract for design, construction, and operation of the juvenile facility and subsequent contracts for operation of the juvenile facility entered into pursuant to this section shall specify the responsibilities to be retained by

the department of human services with regard to juveniles housed at the juvenile facility and the responsibilities to be exercised by the contractor.

(4) The contractor shall require applicants for employment at the juvenile facility to submit a set of fingerprints to the Colorado bureau of investigation for a criminal background check, and the Colorado bureau of investigation may accept such fingerprints. For the purpose of conducting background checks, to the extent authorized by federal law, the Colorado bureau of investigation may exchange with the department any state, multistate, and federal criminal history records of individuals who apply for employment at the juvenile facility.

(5) Beginning twelve months after the juvenile facility constructed pursuant to this section begins operations, and annually thereafter, the division of youth corrections shall calculate the recidivism rate for juveniles who complete the program offered by the juvenile facility. In calculating the recidivism rate, the division shall include any juvenile who commits a criminal offense, either as a juvenile or as an adult, within three years after leaving the facility. The division shall report the recidivism rate to the general assembly.

Department of Transportation.

C.R.S. § 43-1-1401. Legislative declaration.

(1) The general assembly hereby finds and declares that:

(a) The increased population growth and economic activity within the state has resulted in the significant and growing demand for increased construction and reconstruction of highways and other transportation projects within the state to facilitate the movement of people, goods, and information;

(b) As a result of the increased federal and state funding provided to the department of transportation in recent years for transportation projects, together with the increasing number, size, and complexity of planned transportation projects, the department will benefit from the use of a faster, more efficient, and more cost-effective contractor selection and procurement process to design and construct transportation projects;

(c) A ***design-build selection*** and procurement process will provide the department of transportation with: A savings of time, cost, and administrative burden; improved quality expectations with respect to the schedule and budget of transportation projects, as well as completion of such projects; and a reduction in the risks associated with transportation projects, including reduced duplication of expenses and improved coordination of efforts to meet the transportation needs of Colorado.

(2) The general assembly intends that this part 14 authorize the department of transportation to enter design-build contracts and to use an adjusted score ***design-build selection*** and procurement process for particular transportation projects regardless of the minimum or maximum cost of such projects, based on the individual needs and merits of such projects, and subject to approval by the transportation commission. The general assembly also intends that the department's use of an adjusted score ***design-build contract*** process shall not prohibit use of the low bid process currently used by the department pursuant to part 1 of article 92 of title 24 and part 14 of article 30 of title 24, C.R.S.

C.R.S. § 43-1-1402. Definitions.

As used in this part 14:

(1) "Adjusted score ***design-build contract*** process" means a process to award contracts based on the lowest adjusted score of proposals submitted to the department.

(2) "Best value" means the overall maximum value of a proposal to the department after considering all of the evaluation factors described in the specifications for the transportation project or the request for proposals, including but not limited to the time needed for performance of the contract, innovative design approaches, the scope and quality of the work, work management, aesthetics, project control, and the total cost of the transportation project.

(3) ***"Design-build contract"*** means the procurement of both the design and the construction of a transportation project in a single contract with ***a single design-build*** firm or a combination of such firms that are capable of providing the necessary design and construction services.

(4) ***"Design-build firm"*** means any company, firm, partnership, corporation, association, joint venture, or other entity permitted by law to practice engineering, architecture, or construction contracting in the state of Colorado.

(5) "Transportation project" means any project that the department is authorized by law to undertake including but not limited to a highway, tollway, bridge, mass transit, intelligent transportation system, traffic management, traveler information services, or any other project for transportation purposes.

C.R.S. § 43-1-1403. Authority to use a design-build contract process.
Notwithstanding any other provision of law to the contrary, the department may select a ***design-build firm*** and award a ***design-build contract*** for a transportation project as provided in this part 14. The department may include a warranty provision in ***any design-build contract*** that requires the ***design-build firm*** to perform maintenance services on the completed transportation project.

C.R.S. § 43-1-1404. Criteria.

(1) The department may use a ***design-build contract*** for a transportation project if the design work for such project must be performed before a potential bidder can develop a price or cost proposal for such project and if the chief engineer of the engineering, design, and construction division determines that using a ***design-build contract*** is appropriate. The chief engineer shall consider the following factors in making a determination pursuant to this subsection (1):

(a) The extent to which the transportation project requirements are adequately defined;

(b) The time constraints for completing the transportation project;

(c) The capability and experience of potential ***design-build firms***;

(d) The suitability of the transportation project to a ***design-build contract***; and

(e) The capability of the department to manage the ***design-build contract***.

(2) The department may use a ***design-build contract*** regardless of the estimated minimum or maximum cost of a transportation project.

C.R.S. § 43-1-1405. Public notice procedures.
At least forty-five days prior to the anticipated date of selecting ***a design-build firm*** for a transportation project, the department shall publish a public notice at least twice in one or more daily newspapers of general circulation in the state. The public notice shall set forth a general description of the transportation project.

C.R.S. § 43-1-1406. General procedures.

(1) The department shall describe in the specifications for the transportation project the particular ***design-build contract*** and selection procedures to be used in awarding such contract, including but are not limited to the following:

(a) A scope of work statement that defines the transportation project and provides prospective ***design-build firms*** with sufficient information regarding the department's requirements for the transportation project;

(b) If the department uses an adjusted score ***design-build contract*** process to select ***a design-build firm***, a scope of work statement that is flexible and that identifies the end result that the department wants to achieve. The department may determine the adjustment factors and methods it will use to adjust scores and shall state such factors and methods in the specifications for the transportation project. The department may also provide a general concept of the transportation project to potential ***design-build firms***. Adjusted score ***design-build procedures*** shall consist of the following two phases:

(I) In the first phase, the department shall issue a request for qualifications within the time specified in section 43-1-1405 to solicit proposals that include information on the ***design-build firm's*** qualifications and its technical approach to the proposed transportation project. The department shall include appropriate evaluation factors in the request for qualifications, including the factors set forth in section 24-30-1403 (2), C.R.S. The department shall not include cost-related or price-related factors in the request for qualifications. In accordance with the time requirements specified in the department's rules, the department shall develop a short list of the highest qualified ***design-build firms*** from the proposals submitted in response to the request for qualifications.

(II) In the second phase, the department shall issue a request for proposals to the ***design-build firms*** included on the short list developed pursuant to subparagraph (I) of this paragraph (b) in accordance with the time requirements specified in the department's rules. The request for proposals shall include:

(A) A request to separately submit a sealed technical proposal and a sealed cost proposal for the transportation project;

(B) The required content of the technical proposal to be submitted by the ***design-build firm***, including design concepts for the transportation project, the proposed solutions to the requirements addressed in the department's scope of work statement, or both;

(C) Any other evaluation factors the department considers appropriate, including the estimated cost of the transportation project; and

(D) Any formula the department determines is appropriate to adjust the total score of a design-build firm's proposal.

(2) Except as provided in this subsection (2), the department shall allow the preference to Colorado residents provided in section 8-19-101, C.R.S., in awarding an adjusted score ***design-build contract*** pursuant to this part 14. In evaluating and selecting a proposal for a ***design-build contract*** under this part 14, the department shall assign greater value to a proposal in proportion to the extent such proposal commits to using Colorado residents to perform work on the transportation project. If, however, the department determines that compliance with this subsection (2) may cause the denial of federal moneys that would otherwise be available for

the transportation project or if such compliance would otherwise be inconsistent with the requirements of federal law, the department shall suspend the preference granted under this subsection (2) only to the extent necessary to prevent denial of federal moneys or to eliminate the inconsistency with federal law.

(3) The department may use any basis for awarding a design-build contract pursuant to this part 14 that it deems appropriate so long as the basis for awarding such contract is adequately described in the specifications for the transportation project or the request for proposals. Such basis may include awarding a contract to the design-build firm whose proposal provides the best value to the department.

(4) The department may cancel any request for qualifications, request for proposals, or other solicitation issued pursuant to this part 14 or may reject any or all proposals in whole or in part when the department determines that such cancellation or rejection is in the best interest of the department.

(5) If the department awards a design-build contract pursuant to this part 14, the department shall execute a design-build contract with the successful ***design-build firm*** and shall give notice to said firm to commence work on the transportation project.

C.R.S. § 43-1-1407. Stipulated fee.

At its discretion, the department may award a stipulated fee to ***the design-build firms*** that submit responsive proposals but that are not awarded the ***design-build contract*** for a transportation project. The department shall not be required to award such stipulated fee, but if it elects to award such fee for a transportation project, the department shall identify the availability and the amount of such fee in its request for proposals.

C.R.S. § 43-1-1408. Commission approval required.

The department shall obtain approval from the transportation commission prior to using an adjusted score ***design-build contract*** process for any transportation project.

C.R.S. § 43-1-1409. Rule-making authority.

The department may adopt rules in accordance with sections 43-1-110 and 24-4-103, C.R.S., to establish requirements for the procurement of ***design-build contracts*** that it determines necessary or appropriate, including but not limited to rules implementing the ***design-build*** selection and contract procedures, subcontracting, and the warranty provisions of this part 14.

C.R.S. § 43-4-605. Powers of the authority -inclusion or exclusion of property—of rural transportation system alignment—fund created. [Rural Transportation Authorities.]

(1) In addition to any other powers granted to the authority pursuant to this part 6, the authority has the following powers:

* * *

(f) To finance, construct, operate, or maintain rural transportation systems within or without the boundaries of the authority; except that the authority shall not construct rural transportation systems in any territory located outside the boundaries of the authority and within the boundaries of a municipality as the boundaries of the municipality exist on the date the authority is created without the consent of the governing body of such municipality or outside the boundaries of the authority and within the unincorporated boundaries of a county as the unincorporated boundaries of the county exist on the date the authority is created without the consent of the governing body of such county;

* * *

(5) Any rural transportation system constructed by an authority under this part 6 that is funded, in whole or in part, from the highway users tax fund and that may be reasonably expected to exceed one hundred fifty thousand dollars in the aggregate for any fiscal year shall be subject to the construction bidding provisions in part 7 of article 1 of title 29, C.R.S. If the state is involved in the construction of such rural transportation system, the construction bidding provisions in article 92 of title 24, C.R.S., shall apply. ***Nothing herein shall be construed to affect the ability of such entities to enter into design-build contracts under applicable state laws***

* * *

Colorado Department of Transportation Regulations.

(The Colorado Department of Transportation issued regulations effective December 30, 1999, entitled "Rules to Establish Requirements for Procurement by the Colorado Department of Transportation (CDOT) of Design-build Contracts for Transportation Projects. The Rules are in 22 sections).

CONNECTICUT

Conn. Gen. Stat. § 4b-24. Commissioner of Public Works' duties re state realty; disclosure of principals; audit of contractors. [Department of Public Works]

In acting as the determining authority in fulfilling the needs of the various departments and agencies of state government, except the Legislative Department, and choosing the method of acquisition which shall be pursued in the open competitive market, the commissioner shall:

* * *

(4) The commissioner may designate projects to be accomplished on a total cost basis for (A) new facilities to provide for the substantial space needs of a requesting agency, or (B) the installation of mechanical or electrical equipment systems in existing state facilities. If the commissioner designates a project as a designated total cost basis project, ***the commissioner may enter into a single contract with a private developer which includes such project elements as site acquisition, if applicable, architectural design and construction.*** All contracts for such designated projects shall be based on competitive proposals received by the commissioner, who shall give notice of such project, and specifications therefor, by advertising, at least once, in a newspaper having a substantial circulation in the area in which such project is to be located. The commissioner shall determine all other requirements and conditions for such proposals and awards and shall have sole responsibility for all other aspects of such contracts. If applicable, such contracts shall state clearly the responsibilities of the developer to deliver a completed and acceptable facility on a date certain, the maximum cost of the project and, as a separate item, the cost of site acquisition. No such contract may be entered into by the commissioner without the prior approval of the State Properties Review Board and unless funding has been authorized pursuant to the general statutes or a public or special act

* * *

Connecticut Substitute Senate Bill No. 1127 (1999).

[Related to a convention center, hotel, sportsplex and parking project, within the city of Hartford, including multipurpose sports and entertainment facility with a minimum of 40,000 seats and with capacity for expansion to a minimum of 50,000 seats].

* * *

(7) "Design professional" means each duly licensed architect, engineer or other design professional experienced in the design of major multipurpose sports and entertainment facilities and related improvements retained by the Secretary from time to time to prepare plans and specifications and perform related professional services in connection with the overall project and related development activities.

* * *

(24) "Prime contractor" means the general contractor, construction manager or other construction professional with primary responsibility for construction activi-

ties with respect to the convention center, the sportsplex or the parking facilities, as the case may be.

* * *

Sec. 31. Except as otherwise limited by sections 26 to 46, inclusive, of this act, the secretary may:

(1) Acquire, by condemnation, purchase, lease, lease-purchase or otherwise, the real property for the sportsplex site and the parking facilities site;

(2) Select, engage and compensate surveyors, appraisers, engineers, architects and other providers of professional, management, financial or technical services and undertake or arrange for engineering, architectural, environmental, legal, fiscal and economic investigations and studies, surveys, designs, plans, working drawings, specifications, procedures and other activities incidental thereto;

(3) Lease or sublease as lessor or lessee or sublessor or sublessee any real property in connection with the sportsplex project, the parking project or the convention center facilities;

(4) Enter into agreements, pursuant to which the state may be obligated, among other things, to (A) acquire or provide the sportsplex site and the parking facilities site, (B) provide bond proceeds or other state moneys with which to pay project costs, and (C) plan, undertake, perform or otherwise provide for the site preparation, the implementation of the infrastructure the coordination of public and private facilities for sportsplex and convention center parking;

(5) Plan, design, develop, construct, finish, furnish, equip, replace, alter, restore, reconstruct, improve or enlarge and enhance the sportsplex and the parking facilities and engage in other activities incidental thereto, and, subject to section 32 of this act, enter into such construction, construction management, ***design-build*** or other types of contracts or arrangements with respect to the foregoing, including such requirements with respect to guaranteed maximum price, adherence to the project schedule, assumption of force majeure and completion risk, surety and performance bonding, insurance, letters of credit and financial guarantees and other assurances of performance and completion as the secretary may deem prudent, expedient and in the best interests of the state;

* * *

(c) Notwithstanding any provision of the general statutes applicable to a state facility or public works project to the contrary, the secretary may select the prime contractor on a negotiated basis, provided that the secretary shall invite not less than three qualified general contractors or construction managers with current construction experience with major projects of comparable scope to submit proposals to act as prime contractor, which invitation may include minimum requirements established by the Secretary pursuant to subdivision (5) of section 31 of this act.

(d) Notwithstanding any provision of the general statutes applicable to a state facility or public works project to the contrary, except as otherwise provided in subsection (c) of this section, all construction contracts entered into by the Secretary with respect to the sportsplex and the parking facilities shall be entered into on an open-bid or negotiated basis, except that in the discretion of the secretary, where the estimated expense of any such construction contract is more than one hundred thousand dollars, the secretary may determine that selection of a contractor for a particular construction contract through the prequalification and public letting process provided for in subsection (e) of this section is warranted, is expected to be cost effective and is in the public interest of the state.

(e) (1) With respect to any construction contract that is to be publicly let pursuant to subsection (d) of this section, the secretary shall identify a list of potentially responsible qualified bidders for the particular contract, to which list the secretary may add additional names. Thereafter, the secretary shall give notice to those on the list of the work required and of the invitation to prequalify. The invitation to prequalify shall contain such information as the secretary shall deem appropriate and a statement of the time and place where the responses shall be received. Upon receipt of such responses, the secretary shall select each bidder which has shown itself able to post surety bonds required by such contract and has demonstrated that it possesses the financial, managerial and technical ability and the integrity necessary and without conflict of interest for faithful and efficient performance of the work provided for therein. The secretary shall evaluate whether a bidder is responsible and qualified based on its experience with projects similar to that for which the bid is to be submitted, and based on objective written criteria which shall be approved by the secretary and included in the request for prequalification with respect to such contract. In lieu of such prequalification process, the secretary may deem to be prequalified those bidders listed on any list of qualified bidders maintained by the Department of Public Works or the Department of Transportation if the secretary determines that such list is appropriate in light of the scope and type of contract to be let.

(2) The secretary shall thereafter give notice to those so prequalified of the time and place where the public letting shall occur and shall include in such notice such information of the work required as appropriate. At the time and place as set forth in the notice, the secretary shall publicly open and read the bids. The secretary shall not award any construction contract after public letting except to the responsible qualified bidder submitting the lowest bid and in compliance with the bid requirements. The secretary may, however, waive any informality in a bid, and may either reject all bids and again advertise for bids, or, enter into such construction contract on a negotiated basis.

(3) The invitation to bid and the construction contract awarded shall contain such other terms and conditions, and such provisions for penalties, as the secretary may deem appropriate.

DELAWARE

I. Public Procurement Laws.

Department of Transportation.
Del. Code Ann. § 2003. Demonstration projects.

(a) Number. — The Secretary shall entertain and solicit proposals from, and may negotiate and enter into ***agreements with, private entities, or consortia thereof, for up to 4 demonstration projects using in whole or in part private sources of financing involving (i) all or a portion of the study, planning, design, construction, operation and maintenance of transportation systems***, or (ii) the repair and/or expansion, operation and maintenance of existing transportation systems, or any combination of the foregoing, provided that the Secretary shall not enter into any agreements regarding transportation systems in the area of New Castle County known as Churchmans Crossing (formerly known as Metroform) until such time as road improvements in such area which are: (i) included in the Capital Improvements Program as of the date of the enactment of this chapter or (ii) under construction as of the date of the enactment of this chapter, have been completed.

(b) Eligibility. — The Secretary may entertain and solicit proposals from any source whatsoever; provided, however, that the Secretary shall only enter into agreements regarding a transportation system (i) for which construction funding has been authorized by the Delaware General Assembly in the applicable Bond and Capital Improvements Act (except that no agreement may be entered into which contains (A) direct or indirect expenditures or loans on the part of the State in excess of the total sum appropriated by the Delaware General Assembly for said transportation system or (B) credit enhancements which pledge the full faith and credit of the State); and/or (ii) for which the General Assembly has provided specific or categorical funding authorization for purposes of implementing this chapter; and (iii) which is consistent with § 8419(2)(a) of Title 29, applicable provisions of the Department's long range transportation plan, any applicable recommendations developed by the Cabinet Committee on State Planning pursuant to Chapter 91 of Title 29, and applicable provisions of the Federal Clean Air Act [42 U.S.C. § 1857 et seq. and 42 U.S.C. § 7551].

(c) Proposals. — (1) The Secretary shall solicit proposals through a request for proposals accompanied by material explaining the Public-Private Initiatives Program enacted hereunder and describing the selection process and criteria.

(2) Potential projects may be identified and proposed by any potential contracting party. However, the Secretary may identify in the request for proposals specific systems, corridors, or routes for improvement. Any provisions of the Code to the contrary notwithstanding, proposals may provide for the ***design-build mode of infrastructure development***.

(3) Proprietary information contained in proposals not selected for demonstration projects and records of negotiations in progress shall be exempt from public disclosure.

(d) Fees authorized. — To offset a portion of the costs of initiating this program and reviewing proposals received for demonstration projects under this chapter, the Department is authorized to assess a non-refundable Proposal Review Fee for each proposal not to exceed $50,000.00.

(e) Selection and approval. — (1) The demonstration projects shall be selected by a demonstration project committee, chaired by the Secretary, consisting of the Secretary, the Director of Financial Management and Budget, the Chief Engineer of the Department of Transportation, and up to 4 other persons to be appointed by the Secretary. The demonstration projects shall be selected without regard to the provisions of Chapter 69 of Title 29. ***Each proposal shall be weighed on its own merits and ranked according to the selection criteria stipulated in the request for proposals***, provided that upon receipt of all proposals the demonstration project committee may group similar types of project proposals together for purposes of evaluation and selection, and provided further that the proposals selected by such committee from any such group of proposals must be those with the highest ranking within that group, and provided further that such committee may elect not to select any proposals from an established group of proposals, and provided further that as to similar proposals or proposals that are mutually exclusive so that the undertaking of one would preclude the need, desirability, or ability of undertaking the other, only the proposal with the highest ranking among such proposals shall be selected, and, subject to approval as set forth above, proceed to negotiations. Each of the agreements shall be negotiated individually as a stand-alone project.

* * *

Del. S.B. 242 (1999). Department of Transportation.

Section 111. ***Design-Build Contracts***. Notwithstanding other provisions of state law to the contrary, the Department of Transportation is hereby authorized to utilize a ***design-build contract*** mechanism for not more than four transportation construction projects as an experiment in using this procurement method for transportation infrastructure improvements. Except as required to implement the award and administration of a ***design-build contract***, the provisions of Chapter 69 of Title 29, Delaware Code shall apply to such contracts. The Co-chairs of the Joint Legislative Committee on Capital Improvement Program shall approve the four aforementioned projects.

FLORIDA

I. Public Procurement Laws.

Fla. Stat. § 235.211 Educational facilities contracting and construction techniques.

(1) Boards may employ procedures to contract for construction of new facilities, or major additions to existing facilities, that will include, but not be limited to:

(a) Competitive bids.

(b) ***Design-build*** pursuant to § 287.055.

(c) Selecting a construction management entity, pursuant to the process provided by § 287.055

* * *

(3) Contractors, ***design-build firms***, contract management entities, program management entities, or any other person under contract to construct facilities or major additions to facilities may use any construction techniques allowed by contract and not prohibited by law, including, but not limited to, those techniques known as fast-track construction scheduling, use of components, and systems building process.

(4) Except as otherwise provided in this section and § 481.229, the services of a registered architect must be used for the development of plans for the erection, enlargement, or alteration of any educational facility. The services of a registered architect are not required for a minor renovation project for which the construction cost is less than $50,000 or for the placement or hookup of relocatable educational facilities that conform with standards adopted under § 235.26(2) and (3). However, boards must provide compliance with building code requirements and ensure that these structures are adequately anchored for wind resistance as required by law. Boards are encouraged to consider the reuse of existing construction documents or design criteria packages where such reuse is feasible and practical. Notwithstanding § 287.055, a board may purchase the architectural services for the design of educational or ancillary facilities under an existing contract agreement for professional services held by a school board in the State of Florida, provided that the purchase is to the economic advantage of the purchasing board, the services conform to the standards prescribed by rules of the Commissioner of Education, and such reuse is not without notice to, and permission from, the architect of record whose plans or design criteria are being reused. The department shall review these plans for compliance with the state requirements for educational facilities. Rules adopted under this section must establish uniform prequalification, selection, bidding, and negotiation procedures applicable to construction management contracts and ***the design-build process***. This section does not supersede any small, woman-owned or minority-owned business enterprise preference program adopted by a board. Except as otherwise provided in this section, the negotiation procedures applicable to construction management contracts and ***the design-build process*** must conform to the requirements of § 287.055. A board may not modify any rules regarding construction management contracts or ***the design-build process***.

Fla. Stat. § 235.217. SMART (Soundly Made, Accountable, Reasonable, and Thrifty) Schools Clearinghouse.

* * *

(3) The clearinghouse shall:

(a) Recommend to the Governor, the Legislature, and the State Board of Education frugal construction standards that ensure appropriate industry standards and optimal life cycles and that may include, but are not limited to, standards for optimal size of core facility space, ***design-build,*** performance contracting, energy efficiency, and life-cycle systems costing.

Fla. Stat. § 287.055. Acquisition of professional architectural, engineering, landscape architectural, or surveying and mapping services; definitions; procedures; contingent fees prohibited; penalties.

(1) Short Title.—This section shall be known as the "Consultants' Competitive Negotiation Act."

(2) Definitions.—For purposes of this section:

(a) "Professional services" means those services within the scope of the practice of architecture, professional engineering, landscape architecture, or registered surveying and mapping, as defined by the laws of the state, or those performed by any architect, professional engineer, landscape architect, or registered surveyor and mapper in connection with his or her professional employment or practice.

(b) "Agency" means the state, a state agency, a municipality, a political subdivision, a school district, or a school board. The term "agency" does not extend to a nongovernmental developer that contributes public facilities to a political subdivision under § 380.06 or §§ 163.3220-163.3243.

(c) "Firm" means any individual, firm, partnership, corporation, association, or other legal entity permitted by law to practice architecture, engineering, or surveying and mapping in the state.

(d) "Compensation" means the total amount paid by the agency for professional services.

(e) "Agency official" means any elected or appointed officeholder, employee, consultant, person in the category of other personal service or any other person receiving compensation from the state, a state agency, municipality, or political subdivision, a school district or a school board.

(f) "Project" means that fixed capital outlay study or planning activity described in the public notice of the state or a state agency under paragraph (3)(a). A project may include:

1. A grouping of minor construction, rehabilitation, or renovation activities.

2. A grouping of substantially similar construction, rehabilitation, or renovation activities.

(g) A "continuing contract" is a contract for professional services entered into in accordance with all the procedures of this act between an agency and a firm whereby the firm provides professional services to the agency for projects in which construction costs do not exceed $500,000, for study activity when the fee for such professional service does not exceed $25,000, or for work of a specified nature as outlined in the contract required by the agency, with no time limitation except that the contract must provide a termination clause.

(h) ***A "design-build firm"*** means a partnership, corporation, or other legal entity that:

1. Is certified under § 489.119 to engage in contracting through a certified or registered general contractor or a certified or registered building contractor as the qualifying agent; or

2. Is certified under § 471.023 to practice or to offer to practice engineering; certified under § 481.219 to practice or to offer to practice architecture; or certified under § 481.319 to practice or to offer to practice landscape architecture.

(i) ***A "design-build contract"*** means a single contract with a ***design-build firm*** for the design and construction of a public construction project.

(j) A "design criteria package" means concise, performance-oriented drawings or specifications of the public construction project. The purpose of the design criteria package is to furnish sufficient information to permit ***design-build firms*** to prepare a bid or a response to an agency's request for proposal, or to permit an agency to enter into a negotiated ***design-build contract***. The design criteria package must specify performance-based criteria for the public construction project, including the legal description of the site, survey information concerning the site, interior space requirements, material quality standards, schematic layouts and conceptual design criteria of the project, cost or budget estimates, design and construction schedules, site development requirements, provisions for utilities, stormwater retention and disposal, and parking requirements applicable to the project.

(k) A "design criteria professional" means a firm who holds a current certificate of registration under chapter 481 to practice architecture or landscape architecture or a firm who holds a current certificate as a registered engineer under chapter 471 to practice engineering and who is employed by or under contract to the agency for the providing of professional architect services, landscape architect services, or engineering services in connection with the preparation of the design criteria package.

(3) Public Announcement and Qualification Procedures.—

(a) Each agency shall publicly announce, in a uniform and consistent manner, each occasion when professional services must be purchased for a project the basic construction cost of which is estimated by the agency to exceed the threshold amount provided in § 287.017 for CATEGORY FIVE or for a planning or study activity when the fee for professional services exceeds the threshold amount provided in § 287.017 for CATEGORY TWO, except in cases of valid public emergencies certified by the agency head. The public notice must include a general description of the project and must indicate how interested consultants may apply for consideration.

(b) Each agency shall encourage firms engaged in the lawful practice of their professions that desire to provide professional services to the agency to submit annually statements of qualifications and performance data.

(c) Any firm or individual desiring to provide professional services to the agency must first be certified by the agency as qualified pursuant to law and the regulations of the agency. The agency must find that the firm or individual to be employed is fully qualified to render the required service. Among the factors to be considered in making this finding are the capabilities, adequacy of personnel, past record, and experience of the firm or individual.

(d) Each agency shall evaluate professional services, including capabilities, adequacy of personnel, past record, experience, whether the firm is a certified minority business enterprise as defined by the Florida Small and Minority Business Assistance Act of 1985, and other factors determined by the agency to be applicable to its particular requirements. When securing professional services, an agency must endeavor to meet the minority business enterprise procurement goals under n2 § 287.0945.

(e) The public must not be excluded from the proceedings under this section.

(4) Competitive Selection.—

(a) For each proposed project, the agency shall evaluate current statements of qualifications and performance data on file with the agency, together with those that may be submitted by other firms regarding the proposed project, and shall conduct discussions with, and may require public presentations by, no fewer than three firms regarding their qualifications, approach to the project, and ability to furnish the required services.

(b) The agency shall select in order of preference no fewer than three firms deemed to be the most highly qualified to perform the required services. In determining whether a firm is qualified, the agency shall consider such factors as the ability of professional personnel; whether a firm is a certified minority business enterprise; past performance; willingness to meet time and budget requirements; location; recent, current, and projected workloads of the firms; and the volume of work previously awarded to each firm by the agency, with the object of effecting an equitable distribution of contracts among qualified firms, provided such distribution does not violate the principle of selection of the most highly qualified firms. The agency may request, accept, and consider proposals for the compensation to be paid under the contract only during competitive negotiations under subsection (5).

(c) This subsection does not apply to a professional service contract for a project the basic construction cost of which is estimated by the agency to be not in excess of the threshold amount provided in § 287.017 for CATEGORY FIVE or for a planning or study activity when the fee for professional services is not in excess of the threshold amount provided in § 287.017 for CATEGORY TWO.

(d) Nothing in this act shall be construed to prohibit a continuing contract between a firm and an agency.

(5) Competitive Negotiation.—

(a) The agency shall negotiate a contract with the most qualified firm for professional services at compensation which the agency determines is fair, competitive, and reasonable. In making such determination, the agency shall conduct a detailed analysis of the cost of the professional services required in addition to considering their scope and complexity. For any lump-sum or cost-plus-a-fixed-fee professional service contract over the threshold amount provided in § 287.017 for CATEGORY FOUR, the agency shall require the firm receiving the award to execute a truth-in-negotiation certificate stating that wage rates and other factual unit costs supporting the compensation are accurate, complete, and current at the time of contracting. Any professional service contract under which such a certificate is required must contain a provision that the original contract price and any additions thereto will be adjusted to exclude any significant sums by which the agency determines the contract price was increased due to inaccurate, incomplete, or noncurrent wage rates and other factual unit costs. All such contract adjustments must be made within 1 year following the end of the contract.

(b) Should the agency be unable to negotiate a satisfactory contract with the firm considered to be the most qualified at a price the agency determines to be fair, competitive, and reasonable, negotiations with that firm must be formally terminated. The agency shall then undertake negotiations with the second most qualified firm. Failing accord with the second most qualified firm, the agency must terminate negotiations. The agency shall then undertake negotiations with the third most qualified firm.

(c) Should the agency be unable to negotiate a satisfactory contract with any of the selected firms, the agency shall select additional firms in the order of their competence and qualification and continue negotiations in accordance with this subsection until an agreement is reached.

* * *

(10) Applicability to ***Design-Build Contracts***.—

(a) Except as provided in this subsection, this section is not applicable to the procurement of ***design-build contracts*** by any agency, and the agency must award ***design-build contracts*** in accordance with the procurement laws, rules, and ordinances applicable to the agency.

(b) The design criteria package must be prepared and sealed by a design criteria professional employed by or retained by the agency. If the agency elects to enter into a professional services contract for the preparation of the design criteria package, then the design criteria professional must be selected and contracted with under the requirements of subsections (4) and (5). A design criteria professional who has been selected to prepare the design criteria package is not eligible to render services under ***a design-build contract*** executed pursuant to the design criteria package.

(c) Except as otherwise provided in § 240.209(3) or § 337.11(7), the Department of Management Services shall adopt rules for the award of ***design-build contracts*** to be followed by state agencies. Each other agency must adopt rules or ordinances for the award of ***design-build contracts***. Municipalities, political subdivisions, school districts, and school boards shall award ***design-build contracts*** by the use of a competitive proposal selection process as described in this subsection, or by the use of a qualifications-based selection process pursuant to subsections (3), (4), and (5) for entering into a contract whereby the selected firm will subsequently establish a guaranteed maximum price and guaranteed completion date. If the procuring agency elects the option of qualifications-based selection, during the selection of the ***design-build firm*** the procuring agency shall employ or retain a licensed design professional appropriate to the project to serve as the agency's representative. Procedures for the use of a competitive proposal selection process must include as a minimum the following:

1. The preparation of a design criteria package for the design and construction of the public construction project.

2. The qualification and selection of no fewer than three ***design-build firms*** as the most qualified, based on the qualifications, availability, and past work of the firms, including the partners or members thereof.

3. The criteria, procedures, and standards for the evaluation of ***design-build contract*** proposals or bids, based on price, technical, and design aspects of the public construction project, weighted for the project.

4. The solicitation of competitive proposals, pursuant to a design criteria package, from those ***qualified design-build firms*** and the evaluation of the responses or bids submitted by those firms based on the evaluation criteria and procedures established prior to the solicitation of competitive proposals.

5. For consultation with the employed or retained design criteria professional concerning the evaluation of the responses or bids submitted by the ***design-build firms***, the supervision or approval by the agency of the detailed working drawings of the project; and for evaluation of the compliance of the project construction with the design criteria package by the design criteria professional.

6. In the case of public emergencies, for the agency head to declare an emergency and authorize negotiations with the best qualified ***design-build firm*** available at that time.

Fla. Stat. § 337.11. Contracting authority of department; bids; emergency repairs, supplemental agreements, and change orders; combined design and construction contracts; progress payments; records; requirements of vehicle registration.

(1) The department shall have authority to enter into contracts for the construction and maintenance of all roads designated as part of the State Highway System or the State Park Road System or of any roads placed under its supervision by law . . .

* * *

(7) (a) If the head of the department determines that it is in the best interests of the public, the department may combine the design and construction phases of a building, a major bridge, or a rail corridor project into a single contract. Such contract is referred to as ***a design-build contract***.

(b) The department shall adopt by rule procedures for administering ***design-build contracts***. Such procedures shall include, but not be limited to:

1. Prequalification requirements.
2. Public announcement procedures.
3. Scope of service requirements.
4. Letters of interest requirements.
5. Short-listing criteria and procedures.
6. Bid proposal requirements.
7. Technical review committee.
8. Selection and award processes.

(c) The department must receive at least three letters of interest in order to proceed with a request for proposals. The department shall request proposals from no fewer than three of ***the design-build firms*** submitting letters of interest. ***If a design-build firm*** withdraws from consideration after the department requests proposals, the department may continue if at least two proposals are received.

II. Public Bidding Laws.

Fla. Stat. § 255.20 Local bids and contracts for public construction works; specification of state-produced lumber.

(1) A county, municipality, special district as defined in chapter 189, or other political subdivision of the state seeking to construct or improve a public building, struc-

ture, or other public construction works must competitively award to an appropriately licensed contractor each project that is estimated in accordance with generally accepted cost-accounting principles to have total construction project costs of more than $200,000. For electrical work, local government must competitively award to an appropriately licensed contractor each project that is estimated in accordance with generally accepted cost-accounting principles to have a cost of more than $50,000. As used in this section, the term "competitively award" means to award contracts based on the submission of sealed bids, proposals submitted in response to a request for proposal, proposals submitted in response to a request for qualifications, or proposals submitted for competitive negotiation. ***This subsection expressly allows contracts for construction management services, design/build contracts, continuation contracts based on unit prices, and any other contract arrangement with a private sector contractor permitted by any applicable municipal or county ordinance, by district resolution, or by state law.*** For purposes of this section, construction costs include the cost of all labor, except inmate labor, and include the cost of equipment and materials to be used in the construction of the project. Subject to the provisions of subsection (3), the county, municipality, special district, or other political subdivision may establish, by municipal or county ordinance or special district resolution, procedures for conducting the bidding process . . .

Fla. Stat. § 235.211 Educational facilities contracting and construction techniques.

* * *

(4) Except as otherwise provided in this section and § 481.229, the services of a registered architect must be used for the development of plans for the erection, enlargement, or alteration of any educational facility. The services of a registered architect are not required for a minor renovation project for which the construction cost is less than $ 50,000 or for the placement or hookup of relocatable educational facilities that conform with standards adopted under § 235.26(2) and (3). However, boards must provide compliance with building code requirements and ensure that these structures are adequately anchored for wind resistance as required by law. Boards are encouraged to consider the reuse of existing construction documents or design criteria packages where such reuse is feasible and practical. Notwithstanding § 287.055, a board may purchase the architectural services for the design of educational or ancillary facilities under an existing contract agreement for professional services held by a school board in the State of Florida, provided that the purchase is to the economic advantage of the purchasing board, the services conform to the standards prescribed by rules of the Commissioner of Education, and such reuse is not without notice to, and permission from, the architect of record whose plans or design criteria are being reused. Plans shall be reviewed for compliance with the state requirements for educational facilities. Rules adopted under this section must establish uniform prequalification, selection, bidding, and negotiation procedures applicable to construction management contracts and ***the design-build process***. This section does not supersede any small, woman-owned or minority-owned business enterprise preference program adopted by a board. Except as otherwise provided in this section, the negotiation procedures applicable to construction management contracts and ***the design-build process*** must conform to

the requirements of § 287.055. A board may not modify any rules regarding construction management contracts or ***the design-build process.***

Fla. Stat. § 337.14. Application for qualification; certificate of qualification; restrictions; request for hearing.

(1) Any person desiring to bid for the performance of any construction contract in excess of $250,000 which the department proposes to let must first be certified by the department as qualified pursuant to this section and rules of the department. The rules of the department shall address the qualification of persons to bid on construction contracts in excess of $250,000 and shall include requirements with respect to the equipment, past record, experience, financial resources, and organizational personnel of the applicant necessary to perform the specific class of work for which the person seeks certification.

* * *

(7) No "contractor" as defined in § 337.165(1)(d) or his or her "affiliate" as defined in § 337.165(1)(a) qualified with the department under this section may also qualify under § 287.055 or § 337.105 to provide construction, engineering, and inspection services to the department. This limitation shall not apply to ***any design-build*** prequalification under § 337.11(7).

III. Architect and Engineer Licensing Laws.

Fla. Stat. § 471.003. Qualifications for practice, exemptions.

(1) No person other than a duly registered engineer shall practice engineering or use the name or title of "registered engineer" or any other title, designation, words, letters, abbreviations, or device tending to indicate that such person holds an active registration as an engineer in this state.

(2) The following persons are not required to register under the provisions of §§ 471.001-471.037 as a registered engineer:

* * *

(i) Any general contractor, certified or registered pursuant to the provisions of chapter 489, when negotiating or performing services ***under a design-build contract*** as long as the engineering services offered or rendered in connection with the contract are offered and rendered by an engineer licensed or registered in accordance with this chapter.

Fla. Stat. § 481.229 (3). Exceptions; exemptions from licensure.

(3) Notwithstanding the provisions of this part, a general contractor who is certified or registered pursuant to the provisions of chapter 489 is not required to be licensed as an architect when negotiating or performing services ***under a design-build contract*** as long as the architectural services offered or rendered in connection with the contract are offered and rendered by an architect licensed in accordance with this chapter.

Fla. Stat. § 481.329 (3). Exceptions; exemptions from licensure.

(3) None of the provisions of this part shall apply to any general contractor certified or registered pursuant to the provisions of chapter 489 when negotiating or

performing services under ***a design-build contract***, as long as the landscape architectural services offered or rendered in connection with the contract are offered and rendered by a landscape architect licensed in accordance with this part, or by an architect licensed in accordance with part I or by an engineer licensed in accordance with chapter 471.

IV. Contractor Licensing Laws.

Fla. Stat. § 489.103. Exemptions.
This part does not apply to:

* * *

(11) A registered architect or engineer acting within the scope of his or her practice or any person exempted by the law regulating architects and engineers, including persons doing design work as specified in § 481.229(1)(b); provided, however, that an architect or engineer shall not act as a contractor unless properly licensed under this chapter.

* * *

(16) An architect or landscape architect licensed pursuant to chapter 481 or an engineer licensed pursuant to chapter 471 who offers or renders ***design-build services*** which may require the services of a contractor certified or registered pursuant to the provisions of this chapter, as long as the contractor services to be performed under the terms of the ***design-build contract*** are offered and rendered by a certified or registered general contractor in accordance with this chapter.

V Mechanic Lien Laws.

Fla. Stat. § 713.01. Definitions.
As used in this part, the term:
(1) "Architect" means a person or firm that is authorized to practice architecture pursuant to chapter 481 or a general contractor who provides architectural services under ***a design-build contract*** authorized by § 481.229(3).

* * *

(7) "Contractor" means a person other than a materialman or laborer who enters into a contract with the owner of real property for improving it, or who takes over from a contractor as so defined the entire remaining work under such contract. The term "contractor" includes an architect, landscape architect, or engineer who improves real property pursuant to ***a design-build contract*** authorized by § 489.103(16).

* * *

(9) "Engineer" means a person or firm that is authorized to practice engineering pursuant to chapter 471 or a general contractor who provides engineering services ***under a design-build contract*** authorized by § 471.003(2)(l).

GEORGIA

I. Public Procurement Laws.

H. B. No. 1079 (AS PASSED HOUSE AND SENATE)(2000).
Title 36 of the Official Code of Georgia Annotated, relating to local government, is amended by inserting at the end thereof a new chapter to read as follows:
CHAPTER 91 ARTICLE 1.

Ga. Code Ann. § 36-91-1.
As used in this chapter, the term:

* * *

(3) 'Competitive sealed bidding' means a method of soliciting public works construction contracts whereby the award is based upon the lowest responsive, responsible bid in conformance with the provisions of subsection (b) of Code Section 36-91-21.
(4) 'Competitive sealed proposals' means a method of soliciting public works contracts whereby the award is based upon criteria identified in a request for proposals in conformance with the provisions of subsection (c) of Code Section 36-91-21.

* * *

(7) 'Governmental entity' means a county, municipal corporation, consolidated government, authority, board of education, or other public board, body, or commission but shall not include any authority, board, department, or commission of the state, or a public transportation agency as defined by Chapter 9 of Title 32.

* * *

Ga. Code Ann. § 36-91-20.
(a) All public works construction contracts subject to this chapter entered into by a governmental entity with private persons or entities shall be in writing and on file and available for public inspection at a place designated by such governmental entity. Municipalities and consolidated governments shall execute and enter into contracts in the manner provided in applicable local legislation or by ordinance.

* * *

(c) Governmental entities are authorized to utilize any construction delivery method, provided that all public works construction contracts subject to the requirements of this chapter that place the bidder or offeror at risk for construction and require labor and building materials in the execution of the contract shall be awarded on the basis of competitive sealed bidding or competitive sealed proposals. Governmental entities shall have the authority to reject any and all bids or proposals and to waive technicalities and informalities.

* * *

(f) Any governmental entity may, in its discretion, adopt a process for mandatory prequalification of prospective bidders or offerors; provided, however, that:
(1) Criteria for prequalification must be reasonably related to the project or the quality of work;

(2) Criteria for prequalification must be available to any prospective bidder or offeror requesting such information;

(3) Any prequalification process must include a method of notifying prospective bidders or offerors of the criteria for prequalification; and

(4) Any prequalification process must include a procedure for a disqualified bidder to respond to his or her disqualification to a representative of the governmental entity; provided, however, that such procedure shall not be construed to require the governmental entity to provide a formal appeals procedure.

Ga. Code Ann. § 36-91-21.

* * *

(b) Any competitive sealed bidding process shall comply with the following requirements:

(1) The governmental entity shall publicly advertise an invitation for bids;

(2) Bidders shall submit sealed bids based on the criteria set forth in such invitation;

(3) The governmental entity shall open the bids publicly and evaluate such bids without discussions with the bidders; and

(4) The contract shall be awarded to the lowest responsible and responsive bidder whose bid meets the requirements and criteria set forth in the invitation for bids; provided, however, that if the bid from the lowest responsible and responsive bidder exceeds the funds budgeted for the public works construction contract, the governmental entity may negotiate with such apparent low bidder to obtain a contract price within the budgeted amount. Such negotiations may include changes in the scope of work and other bid requirements.

(c) (1) In making any competitive sealed proposal, a governmental entity shall:

(A) Publicly advertise a request for proposals, which request shall include conceptual program information

in the request for proposals describing the requested services in a level of detail appropriate to the project delivery method selected for the project, as well as the relative importance of the evaluation factors;

(B) Open all proposals received at the time and place designated in the request for proposals so as to avoid disclosure of contents to competing offerors during the process of negotiations; and

(C) Make an award to the responsible and responsive offeror whose proposal is determined in writing to be the most advantageous to the governmental entity, taking into consideration the evaluation factors set forth in the request for proposals. The evaluation factors shall be the basis on which the award decision is made. The contract file shall indicate the basis on which the award is made.

* * *

HAWAII

Haw. Rev. Stat. § 206X-7. Agreement with developer. [Honolulu Convention Center Authority]

(a) The authority shall enter into ***an agreement with a developer for the development and construction of the convention center*** and all other improvements. The agreement shall include provision for access to, or ownership of, adequate parking facilities in favor of the authority within the convention center district to accommodate the activities of the convention center facility. The development and construction of the improvements within the convention center district shall conform with the convention center development plan approved by the authority . . .

IDAHO

I. Public Procurement Laws.

Idaho Code § 67-2309. Written plans and specifications for work to be made by officials — Availability.

All officers of the state of Idaho, the separate counties, cities, towns, villages or school districts within the state of Idaho, all boards or trustees thereof or other persons required by the statutes of the state of Idaho to advertise for bids on contracts for the construction, repair or improvement of public works, public buildings, public places or other work, shall make written plans and specifications of such work to be performed or materials furnished, and such plans and specifications shall be available for all interested and prospective bidders therefor, providing that such bidders may be required to make a reasonable deposit upon obtaining a copy of such plans and specifications; all plans and specifications for said contracts or materials shall state, among other things pertinent to the work to be performed or materials furnished, the number, size, kind and quality of materials and service required for such contract, and such plans and specifications shall not specify or provide the use of any articles of a specific brand or mark, or any patented apparatus or appliances when other materials are available for such purpose and when such requirements would prevent competitive bidding on the part of dealers or contractors in other articles or materials of equivalent value, utility or merit. ***The design-build method of construction may be employed by public officials*** in contracts for the construction, repair, or improvement of public works, public buildings, public places or other work. For purposes of this section, ***a design-build contract is a contract between a public entity*** and a nongovernmental party in which the nongovernmental party contracting with the public entity agrees ***to both design and build*** a structure, roadway or other item specified in the contract. In any action which shall arise under this section, the court may assess a civil penalty not to exceed five hundred dollars ($500) to be paid by the public entity.

Idaho Code § 67-5711A. Design-build contracting authorized.

Notwithstanding any other provisions of law to the contrary, the director of the department of administration, or his designee, is authorized and empowered, subject to the approval of the permanent building fund council, to employ ***the use of the design-build method of construction*** in the letting of any and all contracts for the construction, alteration, equipping, furnishing and repair of any and all buildings, improvements, or other public works of the state of Idaho. For the purposes of this section, ***a design-build contract*** is a contract between the state of Idaho and a nongovernmental party in which the nongovernmental party contracting with the state of Idaho agrees ***to both design and build*** the structure, roadway, or other items specified in the contract.

ILLINOIS

I. Licensing Board Rules.

Illinois Board Rule. 69 Ill. Adm. Code 1150.85

Section 1150.85 Acts Constituting the Practice of Architecture Pursuant to Section 5 of the Act

a) ***Design/Build***

1) The ***design/build*** project delivery process is a method whereby an entity signs a single contract to provide a combination of architectural and construction services.

2) The ***design/build entity*** will not be required to register as a professional design firm pursuant to Section 21 of the Act only if the services in the ***design/build project*** delivery process are provided by the entity in accordance with the following:

A) An architect licensed or a professional design firm registered in Illinois independently contracts with the entity and participates substantially in all material aspects of the offering and providing of architectural services relating to any bid process, contract negotiations, design, consultation, development, preparation and coordination of construction documents, and verification of adherence to construction documents and completion.

B) At the time of offering services, a written disclosure shall be given to the client by the entity identifying the architect who will be engaged by and is contractually responsible to the entity offering ***design/build*** project services.

C) The entity agrees that the architect will have direct supervision of the architectural work and the architect's services will not be terminated on the project without the immediate replacement by another architect mutually agreed to by the client and the entity.

D) A ***design/build entity*** shall not offer to provide or provide architectural services, unless the ***design/build*** entity is an Illinois licensed architect or professional design firm. Offering to provide architectural services shall include, but shall not necessarily be limited to, any tender of architectural services either independently or in combination with construction services by any sign, card, advertisement or other device which might indicate to the public that the entity is entitled to provide architectural services.

II. Public Bidding Laws.

70 ILCS 3615/4.06(b)(1). Illinois Regional Transportation Act. Public bidding.

* * *

(b) (1) In connection with ***two-phase design/build selection procedures*** authorized in this Section, a Service Board may authorize, by the affirmative vote of two-thirds of the then members of the Service Board, the use of competitive selection and the prequalification of responsible bidders consistent with applicable federal regulations and this subsection (b).

(2) ***Two-phase design/build*** selection procedures shall consist of the following:

(i) A Service Board shall develop, through licensed architects or licensed engineers, a scope of work statement for inclusion in the solicitation for phase-one proposals that defines the project and provides prospective offerors with sufficient information regarding the Service Board's requirements. The statement shall include criteria and preliminary design, and general budget parameters and general schedule or delivery requirements to enable the offerors to submit proposals which meet the Service Board's needs. When ***the two-phase design/build selection procedure*** is used and the Service Board contracts for development of the scope of work statement, the Service Board shall contract for architectural or engineering services as defined by and in accordance with the Architectural, Engineering, and Land Surveying Qualifications Based Selection Act [30 ILCS 535/1 et seq.] and all applicable licensing statutes.

(ii) The evaluation factors to be used in evaluating phase-one proposals must be stated in the solicitation and must include specialized experience and technical competence, capability to perform, past performance of the offeror's team (including the architect-engineer and construction members of the team) and other appropriate technical and qualifications factors. Each solicitation must establish the relative importance assigned to the evaluation factors and the subfactors that must be considered in the evaluation of phase-one proposals on the basis of the evaluation factors set forth in the solicitation. Each ***design/build*** team must include a licensed design professional independent from the Service Board's licensed architect or engineer and a licensed design professional must be named in the phase-one proposals submitted to the Service Board.

(iii) On the basis of the phase-one proposal the Service Board shall select as the most highly qualified the number of offerors specified in the solicitation and request the selected offerors to submit phase-two competitive proposals and cost or price information. Each solicitation must establish the relative importance assigned to the evaluation factors and the subfactors that must be considered in the evaluation of phase-two proposals on the basis of the evaluation factors set forth in the solicitation. A Service Board ***may negotiate with the selected design/build team*** after award but prior to contract execution for the purpose of securing better terms than originally proposed, provided the salient features of the ***design/build solicitation*** are not diminished. Each phase-two solicitation evaluates separately (A) the technical submission for the proposal, including design concepts or proposed solutions to requirements addressed within the scope of work, and (B) the evaluation factors and subfactors, including cost or price, that must be considered in the evaluations of proposals.

(iv) ***A design/build solicitation*** issued under the procedures in this subsection (b) shall state the maximum number of offerors that are to be selected to submit competitive phase-two proposals. The maximum number specified in the solicitation shall not exceed 5 unless the Service Board with respect to an individual solicitation determines that a specified number greater than 5 is in the best interest of the Service Board and is consistent with the purposes and objectives of ***the two-phase design/build selection process.***

(v) All designs submitted as part of the two-phase selection process and not selected shall be proprietary to the preparers.

III. Public Procurement Law (Architects, Engineers, Land Surveyors).

30 ILCS 535/75. Design/build project — Contracting.

Nothing in this Act shall be deemed to prohibit a State agency from contracting for ***a design/build project.***

INDIANA

None.

IOWA

None.

KANSAS

I. Public Procurement Laws.

K.S.A. § 75-3741(b) (1977). Univ. of Kansas Medical Center.

In 1977, Kansas enacted a special law permitting use of design/build for a parking garage at the University of Kansas Medical Center. K.S.A. § 75-3741(b) (1977). The new section exempted the project from competitive bidding laws and stated that,

> "Notwithstanding the provisions of subsection (a) to the contrary for the purposes of constructing the multi-level parking facility at the University of Kansas medical center as auth orized by 1977 Senate Bill No. 430, such project shall be prepared for competitive bids pursuant to K.S.A. 1976 Supp. 75-3739 and K.S.A. 75-3740, 75-3741, and 75-3742, without compliance with this act or with the act of which this act is amendatory."

II. Attorney General Opinions.

In 1978, the Kansas Attorney General gave an opinion which said that since K.S.A. § 75-3741(b) (1977) required that the contract be let on the basis of plans and specifications which are approved by the director of architectural services, and since all plans and specifications for the proposed facility would be reviewed and approved by the director before construction begins, this sequential review process met the requirements of Kansas competitive bidding laws. Ks. Att'y. Gen. Op. No. 78-62 (February 9, 1978).

KENTUCKY

I. Public Procurement Laws.

Ky. Rev. Stat. Ann. § 45A.045(11). Authority of cabinet — Authority to promulgate administrative regulations. [Finance and Administration Cabinet]

(1) The Finance and Administration Cabinet shall serve as the central procurement and contracting agency of the Commonwealth . . .

(c) The Finance and Administration Cabinet shall attempt in every practicable way to ensure that state agencies are fulfilling their business needs through the application of the best value criteria . . .

(11) For capital construction projects, subject to the provisions of this code, the bidding may be on whichever of the following methods, in the judgment of the secretary of the Finance and Administration Cabinet, offers the lowest real cost to the taxpayer:

(a) A total design-bid basis;

(b) A package system commonly referred to as "turnkey";

(c) Phase bidding commonly referred to as "fast track"; or

(d) Construction management.

Bids shall be reviewed by the engineering staff to assure quality and value, and compliance with bid procedures. All specifications shall be written to promote competition.

Ky. Rev. Stat. Ann. § 164 A.575. Purchasing procedures—Inventories—Sales of surplus property—Bidding procedures—Acquisition and disposition of real property and capital construction projects for University of Kentucky Community College System. [Higher Educational Facilities]

(1) The governing boards of each institution may elect to purchase interest in real property, contractual services, rentals of all types, supplies, materials, equipment, printing, and services, except that competitive bids may not be required for:

(a) Contractual services where no competition exists;

(b) Food, clothing, equipment, supplies, or other materials to be used in laboratory and experimental studies;

(c) Instructional materials available from only one (1) source;

(d) Where rates are fixed by law or ordinance;

(e) Library books;

(f) Commercial items that are purchased for resale;

(g) Professional, technical, scientific, or artistic services, but contracts shall be submitted in accordance with KRS 45A.690 to 45A.725;

(h) All other commodities, equipment, and services which, in the reasonable discretion of the board, are available from only one (1) source; and

(i) Interests in real property . . .

(9) For capital construction projects, ***the bidding may be on a total design-bid basis, a package system commonly referred to as "turnkey,"*** or construction management, whichever in the judgment of the board offers the lowest real cost to the taxpayer. Bids shall be reviewed by the institution's engineering staff to assure quality and value, and compliance with bid procedures. All specifications shall be written to promote competition.

* * *

II. Architect & Engineer Licensing Law.

Ky. Rev. Stat. Ann. § 322.010. Engineers. Definitions for chapter.
As used in this chapter, unless the context requires otherwise:

* * *

(4) "Engineering" means any professional service or creative work, the adequate performance of which requires engineering education, training, and experience as an engineer.

(a) "Engineering" shall include:

1. Consultation, investigation, evaluation, planning, certification, and design of engineering works and systems;

a. Engineering design and engineering work ***associated with design/build projects***;

* * *

III. Attorney General Opinions.

The issue was whether a county may procure, utilizing the "design/build" construction procurement method. The opinions concluded that if there are sufficiently definite specifications describing the scope of construction sought, so that meaningfully competitive bids could be submitted for substantially the same facility by different bidders, for a sum certain which includes design costs, and bids are invited for design and construction of such project in accordance with competitive bidding requirements, procurement of both design and construction, in combination, might lawfully be obtained as a single procurement. The opinion warns that it should not be taken as generally approving use of the "design/build" approach in obtaining construction of public projects at the county level. 1992-1993 Ky. Op. Atty. Gen. 2-269, Ky. OAG 92-143; 1992 WL 541069 (Ky.A.G.)

Kentucky Society of Professional Engineers and that the Kentucky Department of Transportation developed and adopted the "turnkey" procedure for bidding highway construction projects where project specifications are sent out to prequalified contractors who also provide planning and design services by engineers. The opinion concludes that the adoption of the "turnkey" concept violated state law. Nothing authorized the state DOT to employ the "turnkey" technique in road building. The opinion concluded that state law requires in effect that the engineer's plans and specifications be completed before the Bureau of Highways may advertise for bids. 1980-1981 Ky. Op. Atty. Gen. 2-869, Ky. OAG 81-349; 1981 WL 142018 (Ky.A.G.)

APPENDIX E

LOUISIANA

I. Public Procurement Laws.

La. Rev. Stat. Ann. § 30:2307(A)(19). Resource Recovery and Development Authority. Powers, duties, and functions.

A. The authority shall have the power to:

* * *

(3) ***Enter into contracts*** or cooperative agreements with political subdivisions, municipal or parish governments, state and federal agencies, private or quasi-public organizations, investor-owned electric public utilities, electric cooperatives or industries to provide waste management services and ***to plan, design, construct, manage, operate and maintain*** solid waste disposal or resource recovery processing and allied facilities on their behalf where necessary to fulfill the purposes of this Chapter.

* * *

(5) ***Design and construct improvements*** to upgrade solid waste properties it acquires and operates, or acquires and closes, on behalf of local or regional authorities. Design, construct, acquire, alter, improve, expand or extend waste management facilities or projects including provision for the inspection and supervision thereof and the engineering, architectural, legal, fiscal and economic investigations and studies, surveys, designs, plans, working drawings, specifications, procedures and any other actions incidental thereto, ***or contract therefor with individuals or firms or consortiums of such individuals or firms.***

(6) ***Contract for services in the performance of architectural and engineering design, the supervision of design and construction***, system management and facility management, for such professional or technical services as may require either prequalification of a contractor or the submission ***by any person, firm or consortium or association of persons*** or firms of a proposal or similar written communication of the authority, whenever such services are in the discretion of the authority, deemed necessary, desirable or convenient in carrying out the purposes of the authority.

* * *

(19) Develop and utilize a competitive negotiation procedure to secure ***full-service arrangements for the design, construction, and operation of resource recovery facilities*** and systems as deemed necessary, desirable, or convenient by the authority. Use of the competitive negotiation procedure for this purpose shall exempt the authority from the provisions of R.S. 39:1551-39:1736, R.S. 39:1481-39:1526, R.S. 38:2310-38:2316, and R.S. 38:2181-38:2225.

* * *

La. Rev. Stat. Ann. § 48.250.2. Department of Transportation. Design-build contracts; administration.

A. Notwithstanding any law to the contrary or the requirements of this Part, if the secretary determines in his discretion that it is in the best interest of the taxpayers, the

Department of Transportation and Development may formulate, develop, and implement a pilot program to study the feasibility of ***combining the design and construction phases of a transportation facility***, including but not limited to highways, interchanges, bridges, or buildings ***into a single contract***.

B. (1) The department shall adopt and promulgate rules in accordance with the Administrative Procedure Act for administering the ***pilot program for design-build contracts***. Such procedures shall include but not be limited to:

(a) Prequalification requirements of competitors ***for design-build projects***.

(b) Public announcement procedures for solicitation of interested ***design-build competitors***.

(c) Scope of service requirements to be met by ***the successful designer-builder***.

(d) Requirements of letters of interest by competitors for ***the design-build contract.***

(e) Criteria and procedures for choosing a short list from the interested competitors for requesting the submission of technical proposals.

(f) Requirements for bid proposals by competitors for ***design-build contracts.***

(g) Composition and appointment by the secretary to the technical review committee grading and judging the technical proposals for ranking and recommendation to the chief engineer.

(h) Selection and process of award by the chief engineer and execution of ***the design-build contract*** by the secretary for a stipulated sum certain.

(2) The department shall report to the Senate and House Committees on Transportation, Highways and Public Works annually on the status of any and all projects undertaken under this pilot program.

C. This pilot program shall be limited in duration to allow only one project , at a cost not to exceed five million dollars for such project, to be performed ***under the design-build method.***

D. There shall be no challenge by any legal process to the choice of ***the successful designer-builder*** other than for fraud, bias for pecuniary or personal reasons not related to the interest of the taxpayers, or arbitrary and capricious selection by the chief engineer. Once the designer-builder has been chosen and a contract or a stipulated sum certain executed, the price of ***the design-build contract shall not be increased*** other than for inflation as prescribed in the contract and for site or other conditions of which the designer-builder had no knowledge and should not have had knowledge as a reasonable possibility existing at the site or concerning the design and construction.

II. Attorney General Opinions.

The issue was whether the Housing Authority of Lafourche Parish could accept a proposal from a private developer of a low income housing project to develop a turnkey project. Determined that the Housing Authority was exempt and could accept a turnkey proposal. La. Atty. Gen. Op. No. 83-799; 1983 WL 177178 (La.A.G.)

The issue was whether a contract which requires the rendition of professional services as a relatively small part of a contract dealing with the supplying of materials and the performance of non-professional services, does the presence of the small professional services component make the Public Bid Law inapplicable? In a situation where a contract has a very substantial materials component, a substantial non-professional services component, and a relatively small professional services component, it is our opinion that the Public Bid Law is applicable. Held that to conclude otherwise would be to invite widespread abuse of the Public Bid Law. 1988-89 La. Op. Atty. Gen. 74, La. Atty. Gen. Op. No. 88-392; 1989 WL 454231 (La.A.G.)

MAINE

I. Public Procurement Laws.

House Proposal No. 431. Arts and Technology Center.
(This was special project legislation, passed on an emergency basis due to a challenge to the legal authority of the Maine School Administrative District No. 49 to proceed with a design/build project).

Whereas, Private and Special Law 1995, chapter 32 and Private and Special Law 1997, chapter 59 authorized Maine School Administrative District No. 49 to enter into a ***design-build*** contract or other alternative construction arrangements to construct an arts and technology center in the Town of Fairfield; and

Whereas, Maine School Administrative District No. 49 conducted a ***design-build*** competition and requested a number of ***design-build firms*** to submit proposals for the design and construction of the arts and technology center; and

Whereas, the building committee of Maine School Administrative District No. 49 has selected a proposal ***for the design and construction*** of the arts and technology center that combines new construction and reconstruction of existing school facilities; and

Whereas, a question has arisen whether the concept design of the project is within the scope of the original authorization of Private and Special Law 1995, chapter 32 and Private and Special Law 1997, chapter 59; and

Whereas, Maine School Administrative District No.49 cannot proceed with the design and construction of the project until its authority has been clarified; and

Whereas, in the judgment of the Legislature, these facts create an emergency within the meaning of the Constitution of Maine and require the following legislation as immediately necessary for the preservation of the public peace, health and safety; now, therefor,

Be it enacted by the People of the State of Maine as follows:

Sec. 1. Authority to construct arts and technology center. Subject to approval of the voters of Maine School Administrative District No. 49 at a referendum conducted on or after January 19, 1999, the Board of School Directors of Maine School Administrative District No. 49 is authorized to construct and equip a locally funded arts and technology center employing a design for the project that includes a technology suite, an auditorium using the structure of the existing junior high school gymnasium, a new junior high school gymnasium and other additions and improvements to the Lawrence Junior-Senior High School complex, with the total cost of the project not to exceed $ 5,500,000, including $ 4,700,000 from the Town of Fairfield's Bert Williamson Fund previously paid to the district by the municipal officers of the Town of Fairfield, with the balance of the total project costs to be derived from conditional and unconditional gifts obtained by the district through fund raising, available funds of the district from sources other than taxation and interest earnings on the Williamson Fund and other funds described in this section. The additional operating costs of the project during its first 2 years must be paid from conditional and unconditional gifts obtained by the district through fund raising, available funds of the district from sources other than taxation and local appropriations without state participation approved by the voters of the district pursuant to the Maine Revised Statutes, Title 20-A, section 15614, subsection 3. The district is authorized to enter into ***a design-build contract*** or to employ other alternative construction arrangements for the project. Notwithstanding the provisions of Title 20-A, sections 15903, 15905-A, 15908 and 15910, or other applicable laws, this project is not subject to approval by the Commissioner of Education or the Department of Administrative and Financial Services, Bureau of General Services. This Act may not be construed as waiving any building construction standards related to health or safety that apply to school construction projects, and the specifications for the project must be approved by the State Fire Marshal and the Department of Human Services.

MARYLAND

I. Public Procurement Laws.

Md. Code Ann. art. 3 § 3-602. State Finance and Procurement Code. Capital Projects.

* * *

(g) Alternative construction methods.—Total project funding may utilize alternative construction methods, such as:

(1) ***design/build*** which involves a single solicitation to design and build the facility; or

(2) "fast track" in which design and construction are implemented concurrently.

Md. Ann. Code art. 29, § 3-102. Washington Suburban Sewer District. Competitive bidding and construction.

(a) Definitions. —

(1) In this section the following words have the meanings indicated.

(2) ***"Design/build contract"*** means a contract that provides for both architectural and engineering design services and construction services as a part of a single contract.

(3) "Facilities construction contract" means a contract that provides services for the construction of:

(i) Water or wastewater treatment plants;

(ii) Water or wastewater pumping stations and related force mains within the pumping station site limits;

(iii) Water storage facilities; and

(iv) Wastewater storage facilities or buildings.

(4) "Pipeline contract" means a contract that provides services for the construction of a pipeline.

(b) Applicability of section.—This section only ***applies to design/build contracts*** and construction contracts.

(c) Design and construction of system. —

(1) Whenever the WSSC decides, after an opportunity for a hearing has been given, to proceed with the design or construction of a water supply or sanitary sewer system, the WSSC shall comply with the requirements of this section.

(2) The WSSC shall advertise, by notice in newspapers and technical press as the WSSC considers proper, for bids or proposals for the design and construction of the system, in parts or as a whole, as the WSSC considers advisable.

(3) ***Design/build contracts*** shall be used only for facilities construction contracts with costs exceeding $2 million and may not be used for pipeline contracts.

(4) The WSSC:

(i) May readvertise the work or any part of the work if the WSSC considers that the prices quoted are unreasonable; or

(ii) Subject to paragraph (5) of this subsection, may do any part or all of the work by day labor.

(5) Notwithstanding paragraph (6) of this subsection, at any time the WSSC may expend $15,000 or less by day labor for construction work without advertising or receiving bids or proposals.

(6) The WSSC shall award contracts by:

(i) Competitive sealed bids in accordance with subsection (d) of this section; or

(ii) Competitive sealed proposals in accordance with subsection (e) of this section.

(7) The WSSC may reject any bid or proposal.

(8) The WSSC:

(i) May adopt regulations to establish a prequalification process for bidders or offerors; and

(ii) Shall adopt regulations to govern discussions held under subsection (e) (3) of this section.

(9) (i) Except as prohibited under subparagraph (ii) of this paragraph, the WSSC may enter into any type of contract under this section that promotes the best interest of the WSSC.

(ii) The WSSC:

1. May not enter a cost-plus-percentage-of-cost contract; and

2. Shall prohibit a cost-plus-percentage-of-cost subcontract under a WSSC contract.

(d) Invitations to bid. —

(1) (i) In this subsection the following words have the meanings indicated.

(ii) "Evaluated bid price" means the price of a bid after adjustment in accordance with objective measurable criteria.

(iii) "Objective measurable criteria" means standards that enable the WSSC to compare the economy, effectiveness, or value of the subject of the bids.

(2) If a contract is awarded based on competitive sealed bids, the WSSC shall seek bids by issuing an invitation to bid.

(3) Subject to paragraphs (4) through (6) of this subsection, an invitation to bid shall:

(i) Include the specifications of the contract; and

(ii) State whether the contract will be awarded based on the lowest bid price or the lowest evaluated bid price.

(4) If a contract will be awarded on an evaluated bid price, the invitation to bid shall include the objective measurable criteria by which the lowest evaluated bid price will be determined.

(5) The WSSC shall award contracts based on competitive sealed bids to the responsible bidder who submits the lowest bid price or lowest evaluated bid price, as appropriate.

(6) If the WSSC determines that an initial preparation of specifications for price bids is impractical, the invitation for bids may:

(i) Include a request for unpriced technical offers or samples; and

(ii) Direct the bidder to submit a:

1. Sealed price bid with the unpriced technical offer or sample; or

2. Price bid after the WSSC evaluates the technical offer or sample and finds that the offer or sample is acceptable under the criteria set forth in the invitation to bid.

(7) If an invitation to bid includes a request for unpriced technical offers or samples, the WSSC shall:

(i) Consider the price bid of a bidder whose technical offer or sample is acceptable;

(ii) Return unopened the price bid of a bidder whose technical offer or sample is unacceptable; and

(iii) Award the contract to the responsible bidder whose technical offer or sample is acceptable and who submits the lowest bid or lowest evaluated bid, as specified in the invitation to bid.

(e) Requests for proposals. —

(1) If a contract is awarded based on competitive sealed proposals, the WSSC shall seek proposals by issuing a request for proposals.

(2) A request for proposals shall include:

(i) A statement describing the scope of the contract;

(ii) The factors, including price, that will be used in evaluating proposals; and

(iii) The relative importance of each factor.

(3) After receipt of proposals, but before the contract is awarded, the WSSC may conduct discussions with an offeror to:

(i) Obtain the best price for the WSSC; and

(ii) Ensure full understanding of the requirements of the WSSC, as set forth in the request for proposals and in the proposal.

(4) The WSSC:

(i) Shall treat all responsible offerors fairly and equally; and

(ii) May allow each responsible offeror to revise the offeror's initial proposal and submit a best and final offer.

(5) The WSSC shall award a contract based on competitive sealed proposals to the responsible offeror who submits the proposal or best and final offer that the WSSC determines is the most advantageous to the WSSC, considering the evaluation factors set out in the request for proposals.

* * *

APPENDIX E

MASSACHUSETTS

I. Public Procurement Laws.

Mass. Gen. Laws ch. 29 § 7E. Capital Facilities Budgets; Duties of Governor and Deputy Commissioner of Capital Planning and Operation.

* * *

The governor in his capital facility budget and the commissioner of capital asset management and maintenance, in his recommendation to the governor of a capital facility budget, shall include in such requests for each building project contained therein, for which the using agency is a state agency, a recommendation as to the need for and where appropriate, a request for, a study and program as a prerequisite to contracting for, performance of, or allotment or expenditure of funds for any design or construction-related activities. If a study or program is not recommended the governor and deputy commissioner shall include the reasons therefor. They shall also include a recommendation as to the mode of procurement of such facility, including but not limited to, sequential, construction management, ***turnkey, design/build procurement***, and the phasing of such procurement, including but not limited to approval of design and construction stages as separate or combined phases, which will most efficiently, economically and best serve the interests of the commonwealth. When an alternative mode of procurement is recommended, the governor and deputy commissioner shall also recommend the method by which design and construction services shall be procured for such project, provided that such method shall be compatible with the policies and procedures for the selection of designers in sections thirty-eight A 1/2 to thirty-eight O, inclusive, of chapter seven and with the policies and procedures for the selection of contractors in sections forty-four A to forty-four H, inclusive of chapter one hundred and forty-nine, to the extent feasible. If the governor or the deputy commissioner should recommend a mode of procurement other than the sequential mode or a phasing of procurement other than approval of design and construction as a combined phase, each shall state in detail the reasons therefor.

* * *

Mass. Gen. Laws ch. 149 § 44A(1). Environmental Technology Systems. Award of Contracts for Construction, etc. of Public Buildings; Definitions; Posting Bond.

* * *

"Public Agency" means a department, agency, board, commission, authority, or other instrumentality of the commonwealth or political subdivision of the commonwealth, or two or more subdivisions thereof but not including the Massachusetts Bay Transportation Authority;

* * *

"Proprietary environmental technology systems", systems, in the town of Nantucket, including solid waste related equipment, supporting structures, and buildings, designed, manufactured, and produced under exclusive individual right to sell such product, pertaining to solid waste related environmental protection or remediation. Such systems shall include, but not be limited to, sequential, turnkey, construction management, ***design/build procurement***, and the phasing of such procurement, including approval of design and construction stages as separate or combined phases. (2) Every contract for the construction, reconstruction, installation, demolition, maintenance or repair of any building by a public agency estimated to cost more than twenty-five thousand dollars except for a pumping station to be constructed as an integral part of a sewer construction or water construction project bid under the provisions of section thirty-nine M of chapter thirty shall be awarded to the lowest responsible and eligible general bidder on the basis of competitive bids in accordance with the procedure set forth in the provisions of section forty-four A to forty-four H, inclusive. When the general court has approved the use of an alternative mode of procurement of construction for a project pursuant to section seven E of chapter twenty-nine, the awarding authority responsible for procuring construction services for such project shall follow the policies and procedures of this section and of section forty-four B to forty-four H, inclusive, to the extent compatible with the mode of construction procurement selected. The term "pumping station" as used in this section shall mean a building or other structure which houses solely pumps and appurtenant electrical and plumbing fixtures.

* * *

Mass. Gen. Laws ch. 7, § 42B. Director of Project Management; Powers and Duties.

The duties and responsibilities of the director shall include, but not be limited to, the following: review and comment on all long range capital facilities development plans and capital budget requests for building projects by any state agency or building authority for purpose of assisting in the development of schedules, cost estimates and projections; review of said plans and requests for technical feasibility; where appropriate, recommendation that a study or program be conducted; and recommendation to the commissioner of methods which might be used for the design and construction of new facilities or major additions to existing facilities.

The methods recommended shall include the latest developments in construction as well as standard methods, for the purpose of insuring quality, timeliness and economy of construction, such techniques to include but not be limited to construction management, fast-tracked or phased construction, ***turnkey procurement and design and build procurement.*** The director shall also recommend to the commissioner the method for procuring design and construction services when an alternative construction method is recommended; such recommendation shall be in writing and contain the reasons for not complying with the standard selection and bidding laws provided that the legislature shall approve the method for procuring design and/or construction services for such project and provided that such procurement method shall comply with the policies and procedures of sections thirty B through thirty P, inclusive, of chapter seven and sections forty-four A through L, inclusive, of chapter one hundred and forty-nine, to the extent feasible.

The director shall develop guidelines regarding the types of projects that would most benefit from use of alternative construction methods and shall periodically evaluate their effectiveness.

* * *

Mass. S.B. No. 1945 (1999). Highway Department.

* * *

Section 6. (1) As used in this act, the following words shall have the following meanings unless the context clearly requires otherwise:
"Department", the department of highways.
"Developer", the private entity ***selected to design, build, maintain, and operate*** state highway route 3 north which private entity may be comprised of, but not limited to, a civil engineering firm, a highway construction firm, a financial services firm, a law firm and a consulting firm.

* * *

"Route 3 north" and "facility", that portion of state highway route 3, approximately 21 miles in length, extending from and including its interchange with interstate highway route 95 in the town of Burlington north to the commonwealth's border with the state of New Hampshire including, but not limited to, its interchanges with other highways and roads, bridges and other facilities.
"Secretary", the secretary of transportation and construction or his designee.

* * *

Each proposal shall be evaluated by criteria determined by the secretary and set forth in the request for proposals including, but not limited to, the proposed cost of the project and the financial benefit to the commonwealth, the reputation, industry experience and financial capacity of the developer, the proposed design of the facility, the time schedule proposed for completion of the project, local citizen and government concerns, environmental concerns relative to the project, benefits to the traveling public, the developer's ability to ensure labor harmony during the length of the project and such other criteria as the secretary deems appropriate. The secretary may request oral presentations by such developers as he deems necessary for the purpose of understanding, clarifying and improving the terms contained in any such proposals. An oral presentation shall include a written component, including minutes of the meeting in which the presentation took place, which shall be made public after the execution of the development agreement. The secretary shall, within 60 days of the receipt of proposals from the qualified developers, select the developer which he determines best meets the selection criteria for the benefit of the commonwealth. Within five days of the selection of the developer, the secretary shall notify the joint committee on transportation of his choice. Within 15 days of that notification, the secretary and the selected developer shall make a presentation of the details of the development agreement to the joint committee on transportation. Any of the deadlines contained in this section may be extended up to 60 days, in 15-day incrementa, upon the provision of written notice by the secretary to the joint committee on transportation.
If the secretary selects a developer which did not submit the proposal offering the

lowest overall cost, the secretary shall explain the reason for the selection in writing to the joint committee on transportation not later than five days prior to the execution of the development agreement.

(4) The secretary may enter into a binding development agreement with the selected developer, which development agreement shall:

(a) ***provide for the design and construction*** of the project;

(b) specify a design and construction schedule with project milestones and an enforceable project completion date, subject to delays beyond the control of the developer;

(c) specify the cost of the project, as an aggregate total and separated into cost for each identified project segment, with maintenance costs annualized and separated from the cost of construction;

(d) provide for a lease of the facility to the developer or a lease of the facility back to the department for a term not to exceed 30 years upon the completion and final acceptance of the project; but the developer shall retain the primary responsibility for all reconstruction, capital maintenance and operational maintenance work, if any, to be performed during the lease period. If the developer proposes that a third party perform any such work on its behalf, then the third party shall be approved in advance by the department;

* * *

(j) require that the developer secure and maintain bonding and liability insurance coverage in amounts appropriate to protect the project's viability in accordance with subsection 14;

* * *

(6) The developer shall be ***required to prepare plans and specifications*** for the project subject to review by an independent value engineering expert selected by the secretary.

* * *

(9) An agreement with a developer shall require ***the developer's prime contractor to obtain a labor and material payment bond***, in accordance with section 29 of chapter 149 of the General Laws, covering all construction, reconstruction, or maintenance, including capital maintenance, work of the project and shall require the payment of prevailing wages for labor performed on the project in accordance with sections 26 to 27H, inclusive, of said chapter 149.

(15) This section shall be specific to the state highway route 3 north project. Within three months of completion of construction of the project, the secretary shall file a report with the house and senate committees on ways and means, the joint committee on transportation and the clerks of the house and senate. Said report shall detail the actual costs incurred by the developer in the design and construction stages of the project and the aggregate cost expected to be incurred by the department, broken down into annual amounts for each year covered under the development agreement, including any anticipated costs for annual maintenance. If actual yearly costs differ from those contained in the report, the secretary shall submit a written explanation for the difference in his annual budget request to the house and senate committees on ways and means. ***The report***

shall also include an analysis comparing the aggregate cost with the expenses the department would have incurred under a traditional design-bid-build approach.

The secretary shall include in the report a recommendation of whether it would be beneficial for the commonwealth to utilize the construction approach authorized in this section in conjunction with other large construction projects. If the recommendation is positive, the report shall include a comprehensive list of criteria which other projects should meet in order to access this method, and a detailed list of lessons learned on this project which can be incorporated into future projects. No further projects may be undertaken utilizing the method authorized in this section unless and until the secretary has filed the report required by this section and has received legislative authorization to expand this approach to other projects.

* * *

MICHIGAN

I. Public Procurement Law.

Michigan S.B. No. 372 (1999). State Transportation Department.
[For the state transportation department and certain transportation purposes for the fiscal year ending September 30, 2000]

* * *

Sec. 602. The legislature encourages the department to work with the road construction industry to develop performance and road construction warranties for construction contracts. The development of warranties shall include warranties on materials, workmanship, performance criteria, and ***design/build*** projects. The department will report by September 30, 2000, to the house and senate appropriations subcommittees on transportation and to the house and senate fiscal agencies on the status of efforts to develop performance and road construction warranties.

* * *

APPENDIX E

MINNESOTA

I. Public Procurement Laws.

Minn. Stat. § 160.262. Recreational vehicle lanes.

* * *

Subd. 4. ***Design-build bridges*** for nonmotorized vehicles. For streets and highways, the commissioner shall allow for the acceptance of performance-specification bids, made by the lowest responsible bidder, ***for constructing design-build bridges*** for bicycle paths, bicycle trails, and pedestrian facilities that are: (1) designed and used primarily for nonmotorized transportation, but may allow for motorized wheelchairs, golf carts, necessary maintenance vehicles and, when otherwise permitted by law, rule, or ordinance, snowmobiles; and

(2) located apart from any road or highway or protected by barriers, provided that a design-built bridge may cross over and above a road or highway.

Minn. Stat. § 161.32 Contracting for work on trunk highway.

* * *

Subd. 1b. Lowest responsible bidder. Trunk highway construction contracts, ***including design-build contracts, must be awarded to the lowest responsible bidder***, taking into consideration conformity with the specifications, the purpose for which the contract or purchase is intended, the status and capability of the vendor, and other considerations imposed in the call for bids. The commissioner may decide which is the lowest responsible bidder for all contracts and may use the principles of life-cycle costing, where appropriate, in determining the lowest overall bid. Any or all bids may be rejected. In a case where competitive bids are required and where all bids are rejected, new bids, if solicited, must be called for as in the first instance, unless otherwise provided by law.

* * *

Minn. Stat. § 471.371. Contracts for construction of treatment works. [Municipal Waste Water Treatment Plants]

Subdivision 1. Repealed, 1991 c 212 s 4

Subd. 2. ***Authorization of design and construct contracts.*** Notwithstanding the provisions of any law or charter to the contrary, ***any municipality*** authorized by law to enter into a contract for the design and/or construction of water or wastewater treatment facilities ***may advertise for sealed bids for the design and construction thereof under a single contract.*** Prior to such advertisement the municipality shall prepare or cause to be prepared documents which shall serve as a basis for the comparison of bids and any contract to be entered into. These documents shall be prepared by a professional engineer in sufficient detail, including hydraulic flow and organic loading calculations, design capacity, effluent limits, design life, and the treatment alternatives for the wastewater treatment facility, for the bidder to describe the probable cost, scope of work, equipment and materials of construction; and the documents shall include performance standards for the construction and performance standards for the operation of the facility which must be met for specified conditions and time periods, prior to final acceptance of the facility by the municipality. The documents shall require the bidder to furnish estimates of the annual operation and maintenance costs of the facility, conceptual plans and specifications and any other information deemed relevant for contract award.

In awarding the contract, the municipality shall take into consideration the performance guarantee, completion date, construction cost, capacity of the facility, design life estimated annual operation and maintenance cost, and other relevant factors.

The provisions of any law which require the Minnesota pollution control agency to approve all plans and specifications on a municipal or regional water or wastewater treatment facility prior to calling for construction bids shall not be applicable to contracts authorized by this section. However, after bids have been received and evaluated by the governing body, the best bid determined, and the contract awarded, a municipality shall not, by the terms of the awarded contract, allow construction to commence until all legal requirements are met and the plans and specifications for construction of a wastewater treatment facility have been approved by the Minnesota pollution control agency or, in the case of a water treatment facility, the plans and specifications for construction have been approved by the Minnesota department of health. Upon award of the contract the municipality shall require the successful bidder to furnish detailed plans and specifications and shall provide for termination of the contract and may provide for penalties if such plans and specifications are insufficient to permit the municipality to satisfy the requirements of any federal or state permit.

Subd. 3. Limitations. The provisions of subdivision 2 shall not in any way limit the application and effect of laws governing the practice of architecture, professional engineering, or land surveying in this state, including sections 326.02 to 326.15, and 541.051.

Subd. 4. Definitions. As used in this section, "municipality" has the meaning given to it in section 471.345; ***"contract" includes not only construction work but also all necessary design services***, including process and mechanical equipment, provisions for the start-up of the new facility, performance guarantee, and the other necessary and related items to make an operable plant; and "facility" or "facilities" shall, in addition to the treatment facility, include collection and distribution systems.

Subd. 5. Contract security and insurance. Each awarded contract:

(1) shall require a payment and performance bond for the construction portion of the contract;

(2) shall require the successful bidder to guarantee the performance of the facility to the level required by a permit for the operation of the facility, for 12 months after the date operation begins; and

(3) may allow construction progress payments by the municipality to the successful bidder.

Subd. 6. Repealed, 1991 c 212 s 4

Minn. Stat. § 473.556. Powers of commission. [Metropolitan Sports Facilities Commission]

Subdivision 1. General. The commission shall have all powers necessary or convenient to discharge the duties imposed by law, including but not limited to those specified in this section.

* * *

Subd. 5. Facility operation. The commission may equip, improve, operate, manage, maintain, and control the metrodome, met center, basketball and hockey arena and sports facilities constructed, remodeled, or acquired under the provisions of sections 473.551 to 473.599.

* * *

Subd. 7. Contracts. The commission may contract for materials, supplies, and equipment in accordance with section 471.345, except that ***the commission may employ persons, firms, or corporations to perform one or more or all of the functions of architect, engineer, construction manager, or contractor for both design and construction,*** with respect to all or any part of a project to build or remodel sports facilities. Contractors shall be selected through the process of public bidding, provided that it shall be permissible for the commission to narrow the listing of eligible bidders to those which the commission determines to possess sufficient expertise to perform the intended functions. Any construction manager or contractor shall certify, before the contracts are finally signed, a construction price and completion date to the commission and shall post a bond in an amount at least equal to 100 percent of the certified price, to cover any costs which may be incurred over and above the certified price, including but not limited to costs incurred by the commission or loss of revenues resulting from incomplete construction on the completion date. The commission shall secure surety bonds as required in section 574.26, securing payment of just claims in connection with all public work undertaken by it. Persons entitled to the protection of the bonds may enforce them as provided in sections 574.28 to 574.32, and shall not be entitled to a lien on any property of the commission under the provisions of sections 514.01 to 514.16.

* * *

Minn. Stat. § 473.3993. Light rail transit facility plans; definitions.[Metropolitan Transit (light rail systems)]

Subdivision 1. Application. The definitions in this section apply to section 473.3994.

Subd. 2. Preliminary design plan. "Preliminary design plan" means a light rail transit plan that identifies: (1) preliminary plans for the physical design of facilities, including location, length, and termini of routes; general dimension, elevation, alignment, and character of routes and crossings; whether the track is elevated, on the surface, or below ground; approximate station locations; and related park and ride, parking, and other transportation facilities; and a plan for handicapped access; and

(2) preliminary plans for intermodal coordination with bus operations and routes; ridership; capital costs; operating costs and revenues, and sources of funds for operating subsidies; funding for final design, construction, and operation; and an implementation method.

The preliminary design plan includes the preliminary or draft environmental impact statement for the light rail transit facilities proposed.

Subd. 2a. Preliminary engineering plan. "Preliminary engineering plan" means a light rail transit plan that includes the items in the preliminary design plan for the facilities proposed for construction, but with greater detail and specificity to satisfy final environmental impact statement requirements.

Subd. 3. Final design plan. "Final design plan" means a light rail transit plan that includes the items in the preliminary design plan and the preliminary engineering plan for the facilities proposed but with greater detail and specificity needed for construction. The final design plan must include, at a minimum:

(1) final plans for the physical design of facilities, including the right-of-way definition; environmental impacts and mitigation measures; intermodal coordination with bus operations and routes; and civil engineering plans for vehicles, track, stations, parking, and access, including handicapped access; and (2) final plans for civil engineering for electrification, communication, and other similar facilities; operational rules, procedures, and strategies; capital costs; ridership; operating costs and revenues, and sources of funds for operating subsidies; financing for construction and operation; an implementation method; and other similar matters.

The final design plan must be stated with sufficient particularity and detail to allow the proposer to begin the acquisition and construction of operable facilities. ***If a design-build implementation method is proposed, instead of civil engineering plans the final design plan must state detailed design criteria and performance standards for the facilities.***

The commissioner of transportation may use a design-build method of project development and construction for light rail transit. Notwithstanding any law to the contrary, the commissioner may award a design-build contract on the basis of requests for proposals or requests for qualifications without bids. "Design-build method of project development and construction" means a project delivery system in which a single contractor is responsible for both the design and construction of the project and bids the design and construction together. * * *

II. Appropriations Bills.

Minnesota S. B. No. 2221 (1999).

* * *

Sec. 19. Administration.3,554,000.

$3,386,000 is to complete design documents and site preparation for the new facility for the bureau of criminal apprehension in St. Paul for which site acquisition and preliminary design money were appropriated in Laws 1998, chapter 04, section 13, subdivision 11. The commissioner may use ***a design-build method*** of project development and construction for this project. The commissioner may award ***a design-build contract*** on the basis of requests for proposals or requests for qualifications without bids. This is a one-time appropriation.

APPENDIX E

MISSISSIPPI

I. Public Procurement Laws.

Miss. Code Ann. § 49-17-205. Wastewater Projects. Application of competitive bid provisions.

If the district finds and records on its minutes that the acquisition or ***construction of any collection facilities or treatment facilities***, or any interest therein, or any portion thereof, or any property or any interest therein or any portion thereof, which is authorized by sections 49-17-161 through 49-17-209 is available or can be acquired or contracted for, ***from or with only a single source, person, firm or corporation,*** then such acquisition or contract may be made or entered into without meeting the requirements of any law relating to acquisitions, purchases or contracts by competitive bids. If, after advertising for competitive bids as to other proposed purchases, acquisitions or contracts, only one (1) bid is received, the district may reject the bid and negotiate privately any purchase, contract or acquisition for a consideration not exceeding that proposed in the bid.

Miss. Code Ann.§ 31-7-13(m)(ix). Bidding requirements.

All agencies and governing authorities shall purchase their commodities and printing; contract for fire insurance, automobile insurance, casualty insurance (other than workers' compensation) and liability insurance; contract for garbage collection or disposal; contract for solid waste collection or disposal; contract for sewage collection or disposal; and contract for public construction as herein provided.

* * *

(ix) Waste disposal facility construction contracts. Construction of incinerators and other facilities for disposal of solid wastes in which products either generated therein, such as steam, or recovered therefrom, such as materials for recycling, are to be sold or otherwise disposed of; provided, however, in constructing such facilities a governing authority or agency shall publicly issue requests for proposals, advertised for in the same manner as provided herein for seeking bids for public construction projects, ***concerning the design, construction, ownership, operation and/or maintenance of such facilities,*** wherein such requests for proposals when issued shall contain terms and conditions relating to price, financial responsibility, technology, environmental compatibility, legal responsibilities and such other matters as are determined by the governing authority or agency to be appropriate for inclusion; and after responses to the request for proposals have been duly received, ***the governing authority or agency may select the most qualified proposal or proposals on the basis of price, technology and other relevant factors and from such proposals, but not limited to the terms thereof, negotiate and enter contracts with one or more of the persons or firms submitting proposals.***

Miss. Code Ann. § 49-17-345. Wastewater Projects. Applicability of competitive bid provisions.

If the authority finds and records on its minutes that the acquisition or ***construction of any collection facilities or treatment facilities,*** or any interest therein, or any portion thereof, or any property or any interest therein or any portion thereof, which is authorized by sections 49-17-301 through 49-17-353 is available or can be acquired or contracted for, ***from or with only a single source***, person, firm or corporation, then such acquisition or contract may be made or entered into without meeting the requirements of any law relating to acquisitions, purchases or contracts by competitive bids. If, after advertising for competitive bids as to other proposed purchases, acquisitions or contracts, only one (1) bid is received, the authority may reject the bid and negotiate privately any purchase, contract or acquisition for a consideration not exceeding that proposed in the bid.

II. Attorney General Opinions.

The issue was whether a governmental entity, the Warren County Port Commission in this instance, could employ the design-construct or turnkey concept. In this case, the design-construct or turnkey concept would entail a request for proposals which would include a request for design services, plans, specifications, bidding documents, and construction services. The proposals would then be evaluated on, among other things, the total cost of the proposal. Mississippi Code Annotated section 31-7-13 states that the plans and specifications for a proposed public construction project are to be filed with the clerk of the board of the governing authority, thereby enabling those interested in bidding on the project to do so with full knowledge of precisely what has been proposed for construction. Determined that there was no authority for a public entity to use the design-construct or turnkey concept in accepting bids for the proposed construction project. It would be difficult to make accurate comparisons of the proposals when there is no common basis, i.e. specific plan, by which to compare. 1994 WL 175613 (Miss.A.G.); 1994 WL 175613 (Miss.A.G.)

MISSOURI

I. Public Procurement Laws.

R.S.Mo. § 221.500. Detention Facility.

1. The commissioner of administration is authorized to enter into a contract with a developer ***for the design and construction of a minimum security correctional facility*** in any city located within at least four counties.

2. The commissioner shall select the five ***most qualified design and construction teams*** and ask each of the five to submit a schematic design proposal for the facility along with a detailed construction bid. The commissioner shall select the proposer who provides the lowest and best proposal based on the preestablished criteria.

MONTANA

None.

NEBRASKA

I. Architect & Engineer Licensing Regulations.

Board Rule 7.2. Design-Build.
An organization may offer architectural or engineering services combined with construction services under the following conditions:
7.2.1. An architect and/or engineer licensed to practice in Nebraska participates substantially in all material aspects of the ***design-build*** offering. To participate substantially in all material aspects means having a proprietary, or exclusive, role in the architectural and/or engineering services required for the offering;
7.2.2. There is written disclosure at time of offering that the architect or engineer licensed to practice in Nebraska is engaged by and contractually responsible to such organization. To be contractually responsible means to have a written agreement with, or be a corporate part of, the organization;
7.2.3. Such organization agrees that the architect or engineer licensed to practice in Nebraska shall have direct supervision of the work and that such architect's or engineer's services will not be terminated without the consent of the entity engaging the organization; and
7.2.4. The rendering of architectural and/or engineering services by such licensed architect and/or engineer will conform to the E&A Act and E&A Rules.

II. Architect & Engineer Licensing Laws.

Neb. Rev. Stat. § 81-3449. Practice of architecture; exempted activities.
The provisions of the Engineers and Architects Regulation Act regulating the practice of architecture do not apply to the following activities:

* * *

(8) The offering by an organization of a combination of services involved in the practice of ***architecture and construction services*** if:

(a) An architect or person otherwise permitted under subdivision (10) of this section to offer architectural services participates substantially in all material aspects of the offering;

(b) There is written disclosure at the time of the offering that an architect is engaged by and contractually responsible to such organization;

(c) Such organization agrees that the architect will have direct supervision of the work and that such architect's services will not be terminated without the consent of the person engaging the organization; and

(d) The rendering of architectural services by such architect will conform to the act and the rules and regulations;

* * *

NEVADA

I. State Public Works Board.

Nev. Rev. Stat. § 341.171. Single contract for design and construction of capital improvement: Legislative approval; regulations [Effective October 1, 2003.]

1. ***The board may***, with the approval of the interim finance committee when the legislature is not in regular or special session, or with the approval of the legislature by concurrent resolution when the legislature is in regular or special session, ***let to a contractor licensed under chapter 624 of NRS a single contract for both the design and construction of a project of capital improvement.*** The board shall for the purpose prepare a comprehensive sketch plan and narrative of the scope of the work involved in a project.

2. The board shall adopt regulations establishing procedures for:

(a) The determination of the qualifications of contractors to bid for contracts for the design and construction of such projects. The board shall consult with the American Institute of Architects and the Associated General Contractors, or the successor of either if the named organization ceases to exist, before adopting procedures under this paragraph.

(b) The board's approval of designs and architects employed in a project.

(c) The bidding and awarding of contracts for the design and construction of projects based on a final cost of the project which the contractor guarantees will not be exceeded.

(d) The scheduling and controlling of projects.

Nev. Rev. Stat. § 373.117. Regional Transportation Commissions.
Authority of commission and certain counties and cities to establish or operate public transit system; special procedures for procurement and requirements for development of fixed guideway system.

1. A regional transportation commission, a county whose population is less than 100,000 or an incorporated city within such a county may establish or operate a public transit system consisting of:

(a) Regular routes and fixed schedules to serve the public;

(b) Nonemergency medical transportation of persons to facilitate their use of a center as defined in NRS 435.170, if the transportation is available upon request and without regard to regular routes or fixed schedules;

(c) Nonmedical transportation of disabled persons without regard to regular routes or fixed schedules; or

(d) In a county whose population is less than 100,000 or an incorporated city within such a county, nonmedical transportation of persons if the transportation is available by reservation 1 day in advance of the transportation and without regard to regular routes or fixed schedules.

* * *

4. Notwithstanding the provisions of chapter 332 of NRS or NRS 625.530, ***a regional transportation commission may utilize a turnkey procurement process to select a person to design, build, operate and maintain, or any combination thereof, a fixed guideway system***, including, without limitation, any minimum operable segment thereof. The commission shall determine whether to utilize

turnkey procurement for a fixed guideway project before the completion of the preliminary engineering phase of the project. In making that determination, the commission shall evaluate whether turnkey procurement is the most cost effective method of constructing the project on schedule and in satisfaction of its transportation objectives.

5. Notwithstanding the provisions of chapter 332 of NRS, a regional transportation commission may ***utilize a competitive negotiation procurement process*** to procure rolling stock for a fixed guideway project. The award of a contract under such a process must be made to the person whose proposal is determined to be the most advantageous to the commission, based on price and other factors specified in the procurement documents.

6. If a commission develops a fixed guideway project, the department of transportation is hereby designated to serve as the oversight agency to ensure compliance with the federalsafety regulations for rail fixed guideway systems set forth in 49 C.F.R. Part 659.

7. As used in this section:

(a)"Fully regulated carrier" means a common carrier or contract carrier of passengers or household goods who is required to obtain from the transportation services authority a certificate of public convenience and necessity or a contract carrier's permit and whose rates, routes and services are subject to regulation by the transportation services authority.

(b)"Minimum operable segment" means the shortest portion of a fixed guideway system that is technically capable of providing viable public transportation between two end points.

(c)"Public transit system" means a system employing motor buses, rails or any other means of conveyance, by whatever type of power, operated for public use in the conveyance of persons.

(d)***"Turnkey procurement" means a competitive procurement process*** by which a person is selected by a regional transportation commission, based on evaluation criteria established by the commission, ***to design, build, operate and maintain, or any combination thereof***, a fixed guideway system, or a portion thereof, in accordance with performance criteria and technical specifications established by the commission.

II. Public Works Projects.

Nev. Rev. Stat. Ann. § 338.010 Definitions.

[Note: Effective only until October 1, 2003.]

As used in this chapter:

1. "Day labor" means all cases where public bodies, their officers, agents or employees, hire, supervise and pay the wages thereof directly to a workman or workmen employed by them on public works by the day and not under a contract in writing.

2. ***"Design-build contract" means a contract between a public body and a design-build team in which the design-build team agrees to design and construct a public work.***

3. ***"Design-build team" means an entity that consists of:***

(a) At least one person who is licensed as a general engineering contractor or a general building contractor pursuant to chapter 624 of NRS; and

(b) For a public work that consists of:

(1) A building and its site, at least one person who holds a certificate of registration to practice architecture pursuant to chapter 623 of NRS.

(2) Anything other than a building and its site, at least one person who holds a certificate of registration to practice architecture pursuant to chapter 623 of NRS or is licensed as a professional engineer pursuant to chapter 625 of NRS.

4. "Design professional" means a person with a professional license or certificate issued pursuant to chapter 623, 623A or 625 of NRS.

5. "Eligible bidder" means a person who is:

(a) Found to be a responsible and responsive contractor by a local government which requests bids for a public work in accordance with paragraph (b) of subsection 1 of NRS 338.1373; or

(b) Determined by a public body which awarded a contract for a public work pursuant to NRS 338.1375 to 338.1389, inclusive, to be qualified to bid on that contract pursuant to NRS 338.1379 or was exempt from meeting such qualifications pursuant to NRS 338.1383.

6. "Local government" means every political subdivision or other entity which has the right to levy or receive money from ad valorem or other taxes or any mandatory assessments, and includes, without limitation, counties, cities, towns, boards, school districts and other districts organized pursuant to chapters 244A, 309, 318, 379, 474, 541, 543 and 555 of NRS, NRS 450.550 to 450.750, inclusive, and any agency or department of a county or city which prepares a budget separate from that of the parent political subdivision.

7. "Offense" means failing to:

(a) Pay the prevailing wage required pursuant to this chapter;

(b) Pay the contributions for unemployment compensation required pursuant to chapter 612 of NRS; or

(c) Provide and secure compensation for employees required pursuant to chapters 616A to 617, inclusive, of NRS.

8. "Prime contractor" means a person who:

(a) Contracts to construct an entire project;

(b) Coordinates all work performed on the entire project;

(c) Uses his own work force to perform all or a part of the construction, repair or reconstruction of the project; and

(d) Contracts for the services of any subcontractor or independent contractor or is responsible for payment to any contracted subcontractors or independent contractors.

9. "Public body" means the state, county, city, town, school district or any public agency of this state or its political subdivisions sponsoring or financing a public work.

10. "Public work" means any project for the new construction, repair or reconstruction of:

(a) A project financed in whole or in part from public money for:

(1) Public buildings;

(2) Jails and prisons;

(3) Public roads;

(4) Public highways;

(5) Public streets and alleys;

(6) Public utilities which are financed in whole or in part by public money;

(7) Publicly owned water mains and sewers;

(8) Public parks and playgrounds;

(9) Public convention facilities which are financed at least in part with public funds; and

(10) Any other publicly owned works and property whose cost as a whole exceeds $20,000. Each separate unit that is a part of a project is included in the cost of the project to determine whether a project meets that threshold.

(b) A building for the University and Community College System of Nevada of which 25 percent or more of the costs of the building as a whole are paid from money appropriated by this state or from federal money.

11. "Specialty contractor" means a contractor whose operations as such are the performance of construction work requiring special skill and whose principal contracting business involves the use of specialized building trades or crafts.

12. "Stand-alone underground utility project" means an underground utility project that is not integrated into a larger project, including, without limitation:

(a) An underground sewer line or an underground pipeline for the conveyance of water, including facilities appurtenant thereto; and

(b) A project for the construction or installation of a storm drain, including facilities appurtenant thereto, that is not located at the site of a public work for the design and construction of which a public body is authorized to contract with a ***design-build*** team pursuant to subsection 2 of NRS 338.1711.

13. "Wages" means:

(a) The basic hourly rate of pay; and

(b) The amount of pension, health and welfare, vacation and holiday pay, the cost of apprenticeship training or other similar programs or other bona fide fringe benefits which are a benefit to the workman.

14. "Workman" means a skilled mechanic, skilled workman, semiskilled mechanic, semiskilled workman or unskilled workman. The term does not include a design professional.

Nev. Rev. Stat. Ann. § 338.1385. Advertising for bids required for certain projects; soliciting bids by mail required for certain projects; plans and specifications concerning project must be available for public inspection; rejection of bids; exemptions.

[Note: This Section is effective until October 1, 2003.]

1. Except as otherwise provided in subsection 7 and NRS 338.1906 and 338.1907, this state, or a local government that awards a contract for the construction, alteration or repair of a public work in accordance with paragraph (a) of subsection 1 of NRS 338.1373, or a public officer, public employee or other person responsible for awarding a contract for the construction, alteration or repair of a public work who represents the state or the local government, shall not:

(a) Commence such a project for which the estimated cost exceeds $100,000 unless it advertises in a newspaper of general circulation in this state for bids for the project; or

(b) Divide such a project into separate portions to avoid the requirements of paragraph (a).

2. Except as otherwise provided in subsection 7, a public body that maintains a list of properly licensed contractors who are interested in receiving offers to bid on public works projects for which the estimated cost is more than $25,000 but less

than $100,000 shall solicit bids from not more than three of the contractors on the list for a contract of that value for the construction, alteration or repair of a public work. The public body shall select contractors from the list in such a manner as to afford each contractor an equal opportunity to bid on a public works project. A properly licensed contractor must submit a written request annually to the public body to remain on the list. Offers for bids which are made pursuant to this subsection must be sent by certified mail.

3. Each advertisement for bids must include a provision that sets forth:

(a) The requirement that a contractor must be qualified pursuant to NRS 338.1379 to bid on the contract or must be exempt from meeting such qualifications pursuant to NRS 338.1383; and

(b) The period during which an application to qualify as a bidder on the contract must be submitted. 4. Approved plans and specifications for the bids must be on file at a place and time stated in the advertisement for the inspection of all persons desiring to bid thereon and for other interested persons. Contracts for the project must be awarded on the basis of bids received.

5. Any bids received in response to an advertisement for bids may be rejected if the person responsible for awarding the contract determines that:

(a) The bidder is not a qualified bidder pursuant to NRS 338.1379, unless the bidder is exempt from meeting such qualifications pursuant to NRS 338.1383;

(b) The bidder is not responsive;

(c) The quality of the services, materials, equipment or labor offered does not conform to the approved plan or specifications; or

(d) The public interest would be served by such a rejection.

6. Before the state or a local government may commence a project subject to the provisions of this section, based upon a determination that the public interest would be served by rejecting any bids received in response to an advertisement for bids, it shall prepare and make available for public inspection a written statement containing:

(a) A list of all persons, including supervisors, whom the state or the local government intends to assign to the project, together with their classifications and an estimate of the direct and indirect costs of their labor;

(b) A list of all equipment that the state or the local government intends to use on the project, together with an estimate of the number of hours each item of equipment will be used and the hourly cost to use each item of equipment;

(c) An estimate of the cost of administrative support for the persons assigned to the project;

(d) An estimate of the total cost of the project; and

(e) An estimate of the amount of money the state or the local government expects to save by rejecting the bids and performing the project itself.

7. This section does not apply to:

(a) Any utility subject to the provisions of chapter 318 or 710 of NRS;

(b) Any work of construction, reconstruction, improvement and maintenance of highways subject to NRS 408.323 or 408.327;

(c) Normal maintenance of the property of a school district; or

(d) The Las Vegas Valley water district created pursuant to chapter 167, Statutes of Nevada 1947, the Moapa Valley water district created pursuant to chapter 477, Statutes of Nevada 1983 or the Virgin Valley water district created pursuant to chapter 100, Statutes of Nevada 1993; or

(e) The design and construction of a public work for which a public body contracts with a design-build team pursuant to NRS 338.1711 to 338.1727, inclusive.

Nev. Rev. Stat. Ann. § 338.143. Advertising for bids required for certain projects of local governments; soliciting bids by mail required for certain projects of local governments; plans and specifications concerning project must be available for public inspection; rejection of bids; exemptions.

[Note: This Section effective until October 1, 2003.]

1. Except as otherwise provided in subsection 6 and NRS 338.1907, a local government that awards a contract for the construction, alteration or repair of a public work in accordance with paragraph (b) of subsection 1 of NRS 338.1373, or a public officer, public employee or other person responsible for awarding a contract for the construction, alteration or repair of a public work who represents that local government, shall not:

(a) Commence such a project for which the estimated cost exceeds $100,000 unless it advertises in a newspaper of general circulation in this state for bids for the project; or

(b) Divide such a project into separate portions to avoid the requirements of paragraph (a).

2. Except as otherwise provided in subsection 6, a local government that maintains a list of properly licensed contractors who are interested in receiving offers to bid on public works projects for which the estimated cost is more than $25,000 but less than $100,000 shall solicit bids from not more than three of the contractors on the list for a contract of that value for the construction, alteration or repair of a public work. The local government shall select contractors from the list in such a manner as to afford each contractor an equal opportunity to bid on a public works project. A properly licensed contractor must submit a written request annually to the local government to remain on the list. Offers for bids which are made pursuant to this subsection must be sent by certified mail.

3. Approved plans and specifications for the bids must be on file at a place and time stated in the advertisement for the inspection of all persons desiring to bid thereon and for other interested persons. Contracts for the project must be awarded on the basis of bids received.

4. Any bids received in response to an advertisement for bids may be rejected if the person responsible for awarding the contract determines that:

(a) The bidder is not responsive or responsible;

(b) The quality of the services, materials, equipment or labor offered does not conform to the approved plan or specifications; or

(c) The public interest would be served by such a rejection.

5. Before a local government may commence a project subject to the provisions of this section, based upon a determination that the public interest would be served by rejecting any bids received in response to an advertisement for bids, it shall prepare and make available for public inspection a written statement containing:

(a) A list of all persons, including supervisors, whom the local government intends to assign to the project, together with their classifications and an estimate of the direct and indirect costs of their labor;

(b) A list of all equipment that the local government intends to use on the project, together with an estimate of the number of hours each item of equipment will be used and the hourly cost to use each item of equipment;

(c) An estimate of the cost of administrative support for the persons assigned to the project;

(d) An estimate of the total cost of the project; and

(e) An estimate of the amount of money the local government expects to save by rejecting the bids and performing the project itself.

6. This section does not apply to:

(a) Any utility subject to the provisions of chapter 318 or 710 of NRS;

(b) Any work of construction, reconstruction, improvement and maintenance of highways subject to NRS 408.323 or 408.327;

(c) Normal maintenance of the property of a school district;

(d) The Las Vegas Valley water district created pursuant to chapter 167, Statutes of Nevada 1947, the Moapa Valley water district created pursuant to chapter 477, Statutes of Nevada 1983 or the Virgin Valley water district created pursuant to chapter 100, Statutes of Nevada 1993; or

(e) The design and construction of a public work for which a public body contracts with a design-build team pursuant to NRS 338.1711 to 338.1727, inclusive.

Nev. Rev. Stat. Ann. § 338.155. Provisions relating to payment of certain design professionals.

[Note: This Section effective until October 1, 2003.]

If a public body enters into a contract with a design professional who is not a member of a design-build team for the provision of services in connection with a public work, the contract:

1. Must set forth:

(a) The specific period within which the public body must pay the design professional.

(b) The specific period and manner in which the public body may dispute a payment or portion thereof that the design professional alleges is due.

(c) The terms of any penalty that will be imposed upon the public body if the public body fails to pay the design professional within the specific period set forth in the contract pursuant to paragraph (a). (d) That the prevailing party in an action to enforce the contract is entitled to reasonable attorney's fees and costs.

2. May set forth the terms of any discount that the public body will receive if the public body pays the design professional within the specific period set forth in the contract pursuant to paragraph (a) of subsection 1.

Nev. Rev. Stat. Ann. § 338.1711. Public body must contract with prime contractor for certain projects; public body may contract with design-build team, prime or specialty contractor or nonprofit organization for certain projects.

[Note: This Section effective until October 1, 2003.]

1. Except as otherwise provided in this section, a public body shall contract with a prime contractor for the construction of a public work for which the estimated cost exceeds $100,000.

2. ***A public body may contract with a design-build team for the design and construction of a public work that is a discrete project if the public body determines that:***

(a) The public work is:

(1) A plant or facility for the treatment and pumping of water or the treatment and disposal of wastewater or sewage, the estimated cost of which exceeds $100,000,000; or

(2) Any other type of public work, except a stand-alone underground utility project, the estimated cost of which exceeds $30,000,000; and

(b) Contracting with a design-build team will enable the public body to:

(1) Design and construct the public work at a cost that is significantly lower than the cost that the public body would incur to design and construct the public work using a different method;

(2) Design and construct the public work in a shorter time than would be required to design and construct the public work using a different method, if exigent circumstances require that the public work be designed and constructed within a short time; or

(3) Ensure that the design and construction of the public work is properly coordinated, if the public work is unique, highly technical and complex in nature.

3. In a county whose population is 400,000 or more, a public body that is responsible for financing public works may, for its own public works and those financed by a different public body, including, without limitation, an airport if the airport is owned and operated as a department of the public body, ***contract with a design-build team once in each fiscal year for the design and construction of a public work*** if the public body determines that:

(a) The estimated cost of the public work is at least $5,000,000 but less than $30,000,000; and

(b) Contracting with a design-build team will enable the public body to:

(1) Design and construct the public work at a cost that is significantly lower than the cost that the public body would incur to design and construct the public work using a different method;

(2) Design and construct the public work in a shorter time than would be required to design and construct the public work using a different method, if exigent circumstances require that the public work be designed and constructed within a short time; or

(3) Ensure that the design and construction of the public work is properly coordinated, if the public work is unique, highly technical and complex in nature.

4. Notwithstanding the provisions of subsections 1, 2 and 3, a public body may contract with:

(a) A nonprofit organization for the design and construction of a project to restore, enhance or develop wetlands.

(b) A prime contractor, specialty contractor or ***design-build*** team with respect to a public work if the public body determines that the public work is:

(1) Not part of a larger public work; and

(2) Limited in scope to:

(I) Removal of asbestos;

(II) Replacement of equipment or systems for heating, ventilation and air-conditioning;

(III) Replacement of a roof;

(IV) Landscaping; or

(V) Restoration, enhancement or development of wetlands.

Nev. Rev. Stat. Ann. § 338.1713. Public meeting; notice.

[Note: This Section effective until October 1, 2003.]

1. A public body ***shall not contract with a design-build team*** with respect to a public work unless the governing body of the public body makes the determinations, at

a public hearing, that are required pursuant to subsection 2, 3 or 4 of NRS 338.1711, as applicable.

2. A public body that is required to hold a public hearing pursuant to this section shall publish notice of the hearing at least once each week for 3 consecutive weeks in:

(a) A newspaper of general circulation published in the county in which the public work is proposed to be constructed or, if there is no suchnewspaper, in a newspaper of general circulation in the county published in this state; and

(b) A newspaper of general circulation in this state.

Nev. Rev. Stat. Ann. § 338.1715. Procedures for selecting prime contractor, specialty contractor or design-build team.

[Note: This Section effective until October 1, 2003.]

1. A public body that is required to contract with a prime contractor pursuant to subsection 1 of NRS 338.1711 or elects to contract with a specialty contractor pursuant to subsection 4 of NRS 338.1711 shall select the prime contractor or specialty contractor, as appropriate, in accordance with the procedures for bidding that are set forth in:

(a) The provisions of NRS 338.1375 to 338.1389, inclusive; or

(b) NRS 338.143, 338.145 and 338.147, if the public body is a local government that elects to award a contract for a public work in accordance with paragraph (b) of subsection 1 of NRS 338.1373. ***2. A public body that contracts with a design-build team pursuant to NRS 338.1711 and 338.1713 shall select the design-build team in accordance with NRS 338.1721 to 338.1727, inclusive.***

Nev. Rev. Stat. Ann. § 338.1717. Employment of architect or engineer as consultant.

[Note: This Section effective until October 1, 2003.]

A public body may employ a registered architect or licensed professional engineer as a consultant to assist the public body in overseeing the construction of a public work. An architect or engineer so employed shall not:

1. Construct the public work; or

2. Assume overall responsibility for ensuring that the construction of the public work is completed in a satisfactory manner.

Nev. Rev. Stat. Ann. § 338.1721. Qualifications of design-build team.

[Note: This Section effective until October 1, 2003.]

To qualify to participate in a project for the design and construction of a public work, a design-build team must:

1. Obtain a performance bond and payment bond as required pursuant to NRS 339.025;

2. Obtain insurance covering general liability and liability for errors and omissions;

3. Not have been found liable for breach of contract with respect to a previous project, other than a breach for legitimate cause;

4. Not have been disqualified from being awarded a contract pursuant to NRS 338.017, 338.1387, 338.145 or 408.333; and

5. Ensure that the members of the design-build team possess the licenses and certificates required to carry out the functions of their respective professions within this state.

Nev. Rev. Stat. Ann. § 338.1723. Advertisement for preliminary proposals; publication. [Note: This Section effective until October 1, 2003.]

1. A public body shall advertise for preliminary proposals for the design and construction of a public ***work by a design-build team*** at least twice each week for 3 consecutive weeks in:

(a) A newspaper of general circulation published in the county in which the public work is proposed to be constructed or, if there is no such newspaper, in a newspaper of general circulation in the county published in this state; and

(b) A newspaper of general circulation in this state.

2. A request for preliminary proposals published pursuant to subsection 1 must include, without limitation:

(a) A description of the public work to be designed and constructed;

(b) ***Separate estimates of the costs of designing and constructing the public work***;

(c) The dates on which it is anticipated that the separate phases of the design and construction of the public work will begin and end;

(d) A statement setting forth the place and time in which ***a design-build team*** desiring to submit a proposal for the public work may obtain the information necessary to submit a proposal, including, without limitation, the extent to which designs must be completed for both preliminary and final proposals and any other requirements for the design and construction of the public work that the public body determines to be necessary;

(e) A list of the requirements set forth in NRS 338.1721;

(f) ***A list of the factors that the public body will use to evaluate design-build teams who submit a proposal*** for the public work, including, without limitation:

(1) The relative weight to be assigned to each factor; and

(2) A disclosure of whether the factors that are not related to cost are, when considered as a group, more or less important in the process of evaluation than the factor of cost;

(g) Notice that ***a design-build team*** desiring to submit a proposal for the public work must include with its proposal the information used by the public body to determine finalists among the ***design-build*** teams submitting proposals pursuant to subsection 2 of NRS 338.1725 and a description of that information;

(h) A statement that ***a design-build team*** whose prime contractor holds a certificate of eligibility to receive a preference in bidding on public works issued pursuant to NRS 338.1389 or 338.147 should submit a copy of the certificate of eligibility with its proposal;

(i) A statement as to whether ***a design-build team*** that is selected as a finalist pursuant to NRS 338.1725 but is not awarded the design-build contract pursuant to NRS 338.1727 will be partially reimbursed for the cost of preparing a final proposal and, if so, an estimate of the amount of the partial reimbursement; and

(j) The date by which preliminary proposals must be submitted to the public body, which must not be less than 30 days or more than 60 days after the date on which the request for preliminary proposals is first published in a newspaper pursuant to subsection 1.

Nev. Rev. Stat. Ann. § 338.1727. Request for final proposals; submission of final proposals; selection or rejection of final proposals; awarding of contract; duties of design-build team.

[Note: This Section effective until October 1, 2003.]

1. After selecting the finalists pursuant to NRS 338.1725, the public body shall provide to each finalist a request for final proposals for the public work. The request for final proposals must:

(a) Set forth the factors that the public body will use to ***select a design-build team*** to design and construct the public work, including the relative weight to be assigned to each factor; and

(b) Set forth the date by which final proposals must be submitted to the public body.

2. A final proposal submitted by ***a design-build team*** pursuant to this section must be prepared thoroughly, be responsive to the criteria that the public body will use to ***select a design-build team*** to design and construct the public work described in subsection 1 and comply with the provisions of NRS 338.141. If the cost of construction is a factor in the ***selection of a design-build team***, a ***design-build*** team whose prime contractor has submitted with its proposal a certificate of eligibility to receive a preference in bidding on public works issued pursuant to NRS 338.1389 or 338.147 shall be deemed to have submitted a better proposal than a competing ***design-build*** team whose prime contractor has not submitted such a certificate of eligibility if the amount proposed by the ***design-build*** team is not more than 5 percent higher than the amount proposed by the competing ***design-build*** team.

3. At least 30 days after receiving the final proposals for the public work, the public body shall:

(a) Select the most cost-effective and responsive final proposal, using the criteria set forth pursuant to subsection 1; or

(b) Reject all the final proposals.

4. If a public body selects a final proposal pursuant to paragraph (a) of subsection 3, the public body shall, at its next regularly scheduled meeting:

(a) Review and ratify the selection.

(b) ***Award the design-build contract to the design-build team*** whose proposal is selected.

(c) Partially reimburse the unsuccessful finalists if partial reimbursement was provided for in the request for preliminary proposals pursuant to paragraph (i) of subsection 2 of NRS 338.1723. The amount of reimbursement must not exceed, for each unsuccessful finalist, three percent of the total amount to be paid to the ***design-build*** team as set forth in the ***design-build*** contract.

(d) Make available to the public a summary setting forth the factors used by the public body to select the successful ***design-build*** team and the ***ranking of the design-build teams*** who submitted final proposals. The public body shall not release to a third party, or otherwise make public, financial or proprietary information submitted by a ***design-build*** team.

5. A contract awarded pursuant to this section must specify:

(a) An amount that is the maximum amount that the public body will pay for the performance of all the work required by the contract, excluding any amount related to costs that may be incurred as a result of unexpected conditions or occurrences as authorized by the contract;

(b) An amount that is the maximum amount that the public body will pay for the performance of the professional services required by the contract; and

(c) A date by which performance of the work required by the contract must be completed.

6. ***A design-build team*** to whom a contract is awarded pursuant to this section shall:

(a) Assume overall responsibility for ensuring that the design and construction of the public work is completed in a satisfactory manner; and

(b) Use the work force of the prime contractor on ***the design-build team*** to construct at least 15 percent of the public work.

Nev. Rev. Stat. Ann. § 338.410. "Contractor" defined.

[Note: This Section effective until October 1, 2003.]

"Contractor" means:

1. A person who:

(a) Is licensed pursuant to the provisions of chapter 624 of NRS or performs such work that he is not required to be licensed pursuant to chapter 624 of NRS; and

(b) Contracts with a public body to provide labor, materials or services for a public work.

2. A ***design-build*** team that contracts with a public body to design and construct a public work pursuant to NRS 338.1711 to 338.1727, inclusive.

Nev. Rev. Stat. Ann. § 373.117. Authority of commission and certain counties and cities to establish or operate public transit system; special procedures for procurement and requirements for development of fixed guideway system.

1. A regional transportation commission, a county whose population is less than 100,000 or an incorporated city within such a county may establish or operate a public transit system consisting of:

(a) Regular routes and fixed schedules to serve the public;

(b) Nonemergency medical transportation of persons to facilitate their use of a center as defined in NRS 435.170, if the transportation is available upon request and without regard to regular routes or fixed schedules;

(c) Nonmedical transportation of disabled persons without regard to regular routes or fixed schedules; or

(d) In a county whose population is less than 100,000 or an incorporated city within such a county, nonmedical transportation of persons if the transportation is available by reservation 1 day in advance of the transportation and without regard to regular routes or fixed schedules.

2. A regional transportation commission may lease vehicles to or from or enter into other contracts with a private operator for the provision of such a system.

3. In a county whose population is less than 400,000, such a system may also provide service which includes:

(a) Minor deviations from the regular routes and fixed schedules required by paragraph (a) of subsection 1 on a recurring basis to serve the public transportation needs of passengers. The deviations must not exceed one-half mile from the regular routes.

(b) The transporting of persons other than those specified in paragraph (b), (c) or (d) of subsection 1 upon request without regard to regular routes or fixed schedules, if the service is provided by a common motor carrier which has a certificate of public convenience and necessity issued by the transportation services authority pursuant to NRS 706.386 to 706.411, inclusive, and the service is subject to the rules and regulations adopted by the transportation services authority for a fully regulated carrier.

4. Notwithstanding the provisions of chapter 332 of NRS or NRS 625.530, ***a regional transportation commission may utilize a turnkey procurement process to select a person to design, build, operate and maintain, or any combination thereof, a fixed guideway system***, including, without limitation, any minimum operable segment thereof. The commission shall determine ***whether to utilize turnkey procurement*** for a fixed guideway project before the completion of the preliminary engineering phase of the project. In making that determination, the commission shall ***evaluate whether turnkey procurement is the most cost effective method*** of constructing the project on schedule and in satisfaction of its transportation objectives.

5. Notwithstanding the provisions of chapter 332 of NRS, a regional transportation commission may utilize a competitive negotiation procurement process to procure rolling stock for a fixed guideway project. The award of a contract under such a process must be made to the person whose proposal is determined to be the most advantageous to the commission, based on price and other factors specified in the procurement documents.

6. If a commission develops a fixed guideway project, the department of transportation is hereby designated to serve as the oversight agency to ensure compliance with the federal safety regulations for rail fixed guideway systems set forth in 49 C.F.R. Part 659.

7. As used in this section:

(a) "Fully regulated carrier" means a common carrier or contract carrier of passengers or household goods who is required to obtain from the transportation services authority a certificate of public convenience and necessity or a contract carrier's permit and whose rates, routes and services are subject to regulation by the transportation services authority.

(b) "Minimum operable segment" means the shortest portion of a fixed guideway system that is technically capable of providing viable public transportation between two end points.

(c) "Public transit system" means a system employing motor buses, rails or any other means of conveyance, by whatever type of power, operated for public use in the conveyance of persons.

(d) "Turnkey procurement" means a competitive procurement process by which a person is selected by a regional transportation commission, based on evaluation criteria established by the commission, to design, build, operate and maintain, or any combination thereof, a fixed guideway system, or a portion thereof, in accordance with performance criteria and technical specifications established by the commission.

Nev. Rev. Stat. Ann. § 408.3876. "Design-build contract" defined.
[This Section effective until October 1, 2003.]

"Design-build contract" means a contract between the department and a design-build team in which the design-build team agrees to design and construct a project.

Nev. Rev. Stat. Ann. § 408.3877. "Design-build team" defined.
[This Section effective until October 1, 2003.]

"Design-build team" means an entity that consists of:

1. At least one person who is licensed as a general engineering contractor or a general building contractor pursuant to chapter 624 of NRS; and

2. At least one person who holds a certificate of registration to practice architecture pursuant to chapter 623 of NRS or is licensed as a professional engineer pursuant to chapter 625 of NRS.

Nev. Rev. Stat. Ann. § 408.3878. "Prime contractor" defined.
[This Section effective until October 1, 2003.]

"Prime contractor" means a person who:

1. Contracts to construct an entire project;

2. Coordinates all work performed on the entire project;

3. Uses his own workforce to perform all or a part of the construction of the project; and

4. Contracts for the services of any subcontractor or independent contractor or is responsible for payment to any contracted subcontractors or independent contractors.

Nev. Rev. Stat. Ann. § 408.3879. "Project" defined.
[This Section effective until October 1, 2003.]

"Project" means a project for the construction, reconstruction or improvement of a highway.

Nev. Rev. Stat. Ann. § 408.388. Projects for which department may contract with design-build team.
[This Section effective until October 1, 2003.]

1. The department may contract ***with a design-build team*** for the design and construction of a project if the department determines that:

(a) Except as otherwise provided in subsection 2, the estimated cost of the project exceeds $30,000,000; and

(b) Contracting with ***a design-build team*** will enable the department to:

(1) Design and construct the project at a cost that is significantly lower than the cost that the department would incur to design and construct the project using a different method;

(2) Design and construct the project in a shorter time than would be required to complete the project using a different method, if exigent circumstances require that the project be designed and constructed within a short time; or

(3) Ensure that the design and construction of the project is properly coordinated, if the project is unique, highly technical and complex in nature.

2. Notwithstanding the provisions of subsection 1, the department may, once in each fiscal year, ***contract with a design-build team*** for the design and construction of a project the estimated cost of which is at least $5,000,000 but less than $30,000,000 if the department makes thedeterminations otherwise required pursuant to paragraph (b) of subsection 1.

Nev. Rev. Stat. Ann. § 408.3881. Public meeting; notice.
[Expires by limitation on October 1, 2003.]

1. The department shall not contract ***with a design-build team*** with respect to a project unless the board makes the determinations, at a public meeting, that are required pursuant to NRS 408.388.

2. If the department is required to hold a public meeting pursuant to this section, the department shall publish notice of the meeting at least once each week for 3 consecutive weeks in:

(a) A newspaper of general circulation published in each county in which the project is proposed to be constructed or, if there is no such newspaper, in a newspaper of general circulation in each county published in this state; and

(b) A newspaper of general circulation in this state.

Nev. Rev. Stat. Ann. § 408.3882. Procedure for selecting design-build team.
[This Section effective until October 1, 2003.]

If the department desires to contract with ***a design-build team*** pursuant to NRS 408.388 and 408.3881, the department must select ***the design-build team*** in accordance with NRS 408.3883 to 408.3886, inclusive.

Nev. Rev. Stat. Ann. § 408.3883. Advertisement for preliminary proposals; publication.
[Expires by limitation on October 1, 2003.]

1. The department shall advertise for preliminary proposals for the design and construction of a project by ***a design-build team*** at least twice each week for 3 consecutive weeks in:

(a) A newspaper of general circulation published in each county in which the project is proposed to be constructed or, if there is no such newspaper, in a newspaper of general circulation in each county published in this state; and

(b) A newspaper of general circulation in this state.

2. A request for preliminary proposals published pursuant to subsection 1 must include, without limitation:

(a) A description of the proposed project;

(b) Separate estimates of the costs of designing and constructing the project;

(c) The dates on which it is anticipated that the separate phases of the design and construction of the project will begin and end;

(d) A statement setting forth the place and time in which ***a design-build team*** desiring to submit a proposal for the project may obtain the information necessary to submit a proposal, including, without limitation, the extent to which designs must be completed for both preliminary and final proposals and any other ***requirements for the design and construction*** of the project that the department determines to be necessary;

(e) A list of the requirements set forth in NRS 408.3884;

(f) A list of the factors that the department will use ***to evaluate design-build teams*** who submit a proposal for the project, including, without limitation:

(1) The relative weight to be assigned to each factor; and

(2) A disclosure of whether the factors that are not related to cost are, when considered as a group, more or less important in the process of evaluation than the factor of cost;

(g) Notice that ***a design-build team*** desiring to submit a proposal for the project must include with its proposal the information used by the department to determine finalists among ***the design-build teams submitting proposals*** pursuant to subsection 2 of NRS 408.3885 and a description of that information;

(h) A statement that ***a design-build team*** whose prime contractor holds a certificate of eligibility to receive a preference in bidding on public works issued pursuant to NRS 338.1389 or 338.147 should submit a copy of the certificate of eligibility with its proposal;

(i) A statement as to whether a bidding ***design-build team*** that is selected as a finalist pursuant to NRS 408.3885 but is not awarded ***the design-build contract*** pursuant to NRS 408.3886 will be partially reimbursed for the cost of preparing a final proposal and, if so, an estimate of the amount of the partial reimbursement; and

(j) The date by which preliminary proposals must be submitted to the department, which must not be less than 30 days or more than 60 days after the date on which the request for preliminary proposals is first published in a newspaper pursuant to subsection 1.

Nev. Rev. Stat. Ann. § 408.3884. Qualifications of design-build team.

[This Section effective until October 1, 2003.]

To qualify to participate in the design and construction of a project for the department, ***a design-build team must:***

1. Obtain a performance bond and payment bond as the department may require;
2. Obtain insurance covering general liability and liability for errors and omissions;
3. Not have been found liable for breach of contract with respect to a previous project, other than a breach for legitimate cause;
4. Not have been disqualified from being awarded a contract pursuant to NRS 338.017, 338.1387, 338.145 or 408.333; and
5. Ensure that the members of ***the design-build team*** possess the licenses and certificates required to carry out the functions of their respective professions within this state.

Nev. Rev. Stat. Ann. § 408.3885. Selection of finalists based on preliminary proposals.

[This Section effective until October 1, 2003.]

1. At least 30 days after the date by which preliminary proposals must be submitted to the department, the department shall select at least three but not more than five finalists from among ***the design-build teams*** that submitted preliminary proposals. If the department does not receive at least three preliminary proposals from ***design-build teams*** that the department determines to be qualified pursuant to this section and NRS 408.3884, the department may not contract with ***a design-build team*** for the design and construction of the project.
2. The department shall select finalists pursuant to subsection 1 by:

(a) Verifying that ***each design-build team*** which submitted a preliminary proposal satisfies the requirements of NRS 408.3884; and

(b) Conducting an evaluation of the qualifications of ***each design-build team*** that submitted a preliminary proposal, including, without limitation, an evaluation of:

(1) The professional qualifications and experience of the members of ***the design-build team***;

(2) The performance history of the members of the ***design-build*** team concerning other recent, similar projects completed by those members, if any;

(3) The safety programs established and the safety records accumulated by the members of the ***design-build*** team; and

(4) The proposed plan of ***the design-build team*** to manage the design and construction of the project that sets forth in detail the ability of ***the design-build team*** to design and construct the project.

Nev. Rev. Stat. Ann. § 408.3886. Request for final proposals; submission of final proposals; selection or rejection of final proposals; awarding contract; duties of design-build team.

[This Section effective until October 1, 2003.]

1. After selecting the finalists pursuant to NRS 408.3885, the department shall provide to each finalist a request for final proposals for the project. The request for final proposals must:

(a) Set forth the factors that the department will use to ***select a design-build team*** to design and construct the project, including the relative weight to be assigned to each factor; and

(b) Set forth the date by which final proposals must be submitted to the department.

2. A final proposal submitted by ***a design-build team*** pursuant to this section must be prepared thoroughly, be responsive to the criteria that the department will use to ***select a design-build team*** to design and construct the project described in subsection 1 and comply with the provisions of NRS 338.141. If the cost of construction is a factor in the ***selection of a design-build team, a design-build team*** whose prime contractor has submitted with its proposal a certificate of eligibility to receive a preference in bidding on public works issued pursuant to NRS 338.1389 or 338.147 shall be deemed to have submitted a better proposal than a competing ***design-build team*** whose prime contractor has not submitted such a certificate of eligibility if the amount ***proposed by the design-build team*** is not more than 5 percent higher than the amount proposed by the ***competing design-build team***.

3. At least 30 days after receiving the final proposals for the project, the department shall:

(a) Select the most cost-effective and responsive final proposal, using the criteria set forth pursuant to subsection 1; or

(b) Reject all the final proposals.

4. If the department selects a final proposal pursuant to paragraph (a) of subsection 3, the department shall hold a public meeting to:

(a) Review and ratify the selection.

(b) ***Award the design-build contract to the design-build team*** whose proposal is selected.

(c) Partially reimburse the unsuccessful finalists if partial reimbursement was provided for in the request for preliminary proposals pursuant to paragraph (i) of subsection 2 of NRS 408.3883. The amount of reimbursement must not exceed, for each unsuccessful finalist, three percent of the total amount to be paid to the design-build team as set forth in the design-build contract.

(d) Make available to the public a summary setting forth the factors used by the department to select the successful ***design-build*** team and the ranking of ***the design-build teams*** who submitted final proposals. The department shall not release to a third party, or otherwise make public, financial or proprietary information ***submitted by a design-build team.***

5. A contract awarded pursuant to this section must specify:

(a) An amount that is the maximum amount that the department will pay for the performance of all the work required by the contract, excluding any amount related to costs that may be incurred as a result of unexpected conditions or occurrences as authorized by the contract;

(b) An amount that is the maximum amount that the department will pay for the performance of the professional services required by the contract; and

(c) A date by which performance of the work required by the contract must be completed.

6. ***A design-build team*** to whom a contract is awarded pursuant to this section shall:

(a) Assume overall responsibility for ensuring that the design and construction of the project is completed in a satisfactory manner; and

(b) Use the work force of the prime contractor on the ***design-build*** team to construct at least 15 percent of the project.

Nev. Rev. Stat. Ann. § 408.3887. Employment of architect or engineer as consultant during construction.

[This Section effective until October 1, 2003.]

The department may employ a registered architect or licensed professional engineer as a consultant to assist the department in overseeing the construction of a project. An architect or engineer so employed shall not:

1. Construct, reconstruct or improve the highway; or

2. Assume overall responsibility for ensuring that the construction of the project is completed in a satisfactory manner.

II. Licensing Board Regulations.

Nev. Admin. Code § 623.495. Determination of practice as principal.

The following provisions apply to all applicants for registration:

1. A stockholder, director or officer of a corporation who is authorized to engage in the practice of architecture is not considered to be practicing as a principal solely because of his position in the corporation.

2. No employee of a corporation is considered to be practicing as a principal unless he is a director or officer of the corporation.

3. In any firm composed of general partners as well as limited partners or associates, or both, the general partners are considered by the board to be practicing as principals. Persons who share in the profits of a firm but not in the losses, or who are not liable for the acts of the firm, will not be considered to be practicing as principals.

4. ***Architects who are employed by firms engaged in the design and construction of buildings***, whether by the same or affiliated organizations, ***are not considered to be practicing as principals. Practical training received in firms engaged in design and construction is not considered to be training in the office of a registered architect who is practicing as a principal.***

5. A registered architect who:

(a) Occupies a position of management in an organization which employs architects and candidates for the architect's examination; and

(b) Performs architectural services in connection with buildings owned or occupied by the organization, will not be considered by the board to be practicing as a principal but the architects and candidates employed by the organization will receive credit for training received under subsections 9 and 10 of NAC 623.415 and subsection 8 of NAC 623.445.

NEW HAMPSHIRE

N.H. Rev. Stat. Ann. § 228:4(I)(f). Major Capital Projects.

I. Each state capital budget project whose estimated cost is more than $25,000 shall be built under contracts awarded to the lowest responsible bidder through competitive bidding. The following are excluded from this competitive bidding requirement:

(a) Normal highway and bridge maintenance and improvements, unless they involve federal funds. This includes state aided town road and bridge projects.

(b) Projects executed under RSA 481 with approval of the governor and council.

(c) Projects for the department of fish and game and the department of resources and economic development whose estimated total cost is not more than $250,000. Such projects may be done on a force account basis, by contracts awarded through competitive bidding, by short term rental of construction equipment, or by any combination of these methods. These departments are authorized to rent construction equipment for periods not exceeding 6 months at rates the departments deem competitive through the use of quotes or bids.

(d) [Repealed.]

(e) In an emergency, projects may be done on a force account basis upon the recommendation of the commissioner, with the approval of the governor and council.

(f) Projects may be built through lease purchase arrangements based on a request for proposal provided that selection and award is based on an objective standard and that there are measurable criteria for evaluation. ***Projects may be built under the design build concept*** based on a request for proposal provided that selection and award is based on an objective standard and that there are measurable criteria for evaluation only if such projects are expressly designated as design build and authorized as such by the capital budget while the general court is in session or by the fiscal committee when the general court is out of session.

* * *

NEW JERSEY

None.

NEW MEXICO

I. Public Procurement Laws.

New Mex. Stat. § 13-1-111. Competitive sealed proposals; conditions for use.
Except as provided in Subsection G of Section 13-1-119.1 NMSA 1978, when a state agency or a local public body is procuring professional services or a ***design and build project delivery system,*** or when the state purchasing agent, a central purchasing office or a designee of either officer makes a written determination that the use of competitive sealed bidding for items of tangible personal property or services is either not practicable or not advantageous to the state agency or a local public body, a procurement shall be effected by competitive sealed proposals. Competitive qualifications-based proposals shall be used for procurement of professional services of architects, engineers, landscape architects, construction managers and surveyors who submit proposals pursuant to Sections 13-1-120 through 13-1-124 NMSA 1978.

New Mex. Stat. § 13-1-119.1
Public works project delivery system; design and build projects authorized.

A. Except for road and highway construction or reconstruction projects, a ***design and build project delivery system*** may be authorized when the state purchasing agent or a central purchasing office makes a determination in writing that it is appropriate and in the best interest of the state or local public body to use the system on a specific project with a maximum allowable construction cost of more than ten million dollars ($10,000,000). The determination shall be issued only after the state purchasing or central purchasing office has taken into consideration the following criteria, which shall be used as the minimum basis in determining when to use the ***design and build process:***

(1) the extent to which the project requirements have been or can be adequately defined;

(2) time constraints for delivery of the project;

(3) the capability and experience of potential teams with ***the design and build process;***

(4) the suitability of the project for use of ***the design and build process*** as concerns time, schedule, costs and quality; and

(5) the capability of the using agency to manage the project, including experienced personnel or outside consultants, and to oversee the project with persons who are familiar with ***the design and build process.***

B. When a determination has been made by the state purchasing agent or a central purchasing office that it is appropriate to use ***a design and build project delivery system, the design and build team*** shall include, as needed, a New Mexico registered engineer or architect and a contractor properly licensed in New Mexico for the type of work required.

C. Except as provided in Subsections F and G of this section, for each proposed state or local ***public works design and build project,*** a two-phase procedure for awarding ***design and build contracts*** shall be adopted and shall include at a minimum the following:

(1) during phase one, and prior to solicitation, documents shall be prepared for a request for qualifications by a registered engineer or architect, either in-house or

selected in accordance with Sections 13-1-120 through 13-1-124 NMSA 1978, and shall include minimum qualifications, a scope of work statement and schedule, documents defining the project requirements, the composition of the selection committee and a description of the phase-two requirements and subsequent management needed to bring the project to completion. ***Design and build qualifications*** of responding firms shall be evaluated and a maximum of five firms shall be short listed in accordance with technical and qualifications-based criteria; and

(2) during phase two, the short-listed firms shall be invited to submit detailed specific technical concepts or solutions, costs and scheduling. Unsuccessful firms may be paid a stipend to cover proposal expenses. After evaluation of these submissions, selection shall be made and the contract awarded to the highest ranked firm.

D. Except as provided in Subsections F and G of this section, to ensure fair, uniform, clear and effective procedures that will strive for the delivery of a quality project on time and within budget, the secretary, in conjunction with the appropriate and affected professional associations and contractors, shall promulgate regulations applicable to all using agencies, which shall be followed by all using agencies when procuring ***a design and build project delivery system.***

E. A state agency shall make the decision on ***a design and build project delivery system*** for a state public works project, and a local public body shall make that decision for a local public works project. A state agency shall not make the decision on ***a design and build project delivery system*** for a local public works project.

F. The requirements of Subsections C and D of this section and the minimum construction cost requirement of Subsection A of this section do not apply to a ***design and build project delivery system*** and the services procured for the project if:

(1) the maximum allowable construction cost of the project is two hundred thousand dollars ($200,000) or less; and

(2) the only requirement for architects, engineers, landscape architects or surveyors is limited to either site improvements or adaption for a pre-engineered building or system.

G. The procurement of ***a design and build project delivery system*** qualifying for exemptions pursuant to Subsection F of this section, including the services of any architect, engineer, landscape architect, construction manager or surveyor needed for the project, shall be accomplished by competitive sealed bids pursuant to Sections 13-1-102 through 13-1-110 NMSA 1978.

NEW YORK

I. Public Procurement Laws.

N.Y. Gen Mun. Law § 120-w. Contracts and agreements for solid waste management, collection and disposal.

1. Definitions. As used in this section:

(a) "Municipality" means a town, city, county, or village or any designated public agency thereof, or a garbage or sanitary district established under the Nassau county civil divisions act; or any two or more of the foregoing which are acting jointly in connection with a solid waste management facility, or a public authority.

(b) "Solid waste management-resource recovery facility" or "facility" or "project" means any facility, plant, works, system, building, structure, improvement machinery, equipment, fixture or other real or personal property which is to be used, occupied or employed beyond the initial solid waste collection process for the storage, processing, or disposal of solid waste or the recovery by any means of any material or energy product or resource therefrom including but not limited to recycling centers, transfer stations, baling facilities, rail haul or barge haul facilities, processing systems, resource recovery facilities or other facilities for reducing solid waste volume, sanitary landfills, plants and facilities for compacting, composting or pyrolization of solid wastes, incinerators, and other solid waste disposal, reduction or conversion facilities. For the purpose of this section, solid waste management-resource recovery facilities include solid waste recovery and management projects as defined in subdivision two of section 51-0903 of the environmental conservation law.

(c) "Resource recovery" means the separation, extraction and recovery of useable materials or energy from solid waste through source separation, recycling centers or other programs, projects or facilities.

(d) "Person" means a municipality or other governmental body, public corporation or authority, private corporation, partnership or individual.

(e) "Source separation" means the segregation of recyclable materials from the solid waste stream at the point of generation for separate collection, sale or other disposition.

2. Notwithstanding the provisions of any other law, general, special or local relating to the length, duration and terms of contracts which a municipality may enter into, ***any municipality may enter into a contract with any person, upon such terms and conditions as may be agreed upon, for the design, construction, operation, financing, ownership or maintenance of a solid waste management-resource recovery facility,*** for the processing or disposal of solid waste or for a system of collection and disposal of municipal solid waste through resource recovery which may include source separation, for a period not to exceed twenty-five years, and, except in a city having a population of one million or more, for collection and disposal of municipal solid wastes by means other than resource recovery for a period not to exceed five years as provided for in this section.

3. The share of the cost to be paid by municipalities acting jointly shall be determined in any manner which may be agreed upon, and such share shall be included in the annual budget of the participating municipality as an expense and levied against the taxable real property in the municipality or municipalities.

4. Construction, financing and operation of solid waste management-resource recovery facilities.

(a) Notwithstanding the provisions of any other law, general, special or local relating to the length, duration and terms of contracts a municipality may enter into, any county, city, town or village, or any combination thereof, is hereby authorized and empowered from time to time to enter into contracts, leases or rental agreements, with, or grant licenses, permits, concessions or any other authorizations, to any person, upon such terms and conditions for such consideration and for such term or duration, not to exceed twenty-five years, as may be agreed upon by any county, city, town or village or any combination thereof, and such person, whereby, for any purpose or purposes hereinafter referred to, such person is granted the right to construct, maintain, use, occupy or carry on activities in the whole or any part of a solid waste management-resource recovery facility on a site authorized by any county, city, town or village or any combination thereof.

(b) Prior to or after the expiration or termination of the term or duration of any contract, lease, rental agreement, license, permit, concession, or other authorization, entered into or granted pursuant to the provisions of this section, any county, city, town or village or any combination thereof, in accordance with the requirements and conditions of this section, may from time to time enter into amended, supplemental, new, additional or further contracts, leases or rental agreements with, and grant new, additional, supplemental, or further licenses, permits, concessions, or other authorizations to the same or any other person for any purpose or purposes referred to herein.

(c) The person, entering into any contract, lease, rental agreement license, permit, concession, or other authorization referred to herein, with any county, city, town or village or any combination thereof, may be granted the rights hereinbefore referred to for any purpose or purposes as shall provide for the benefit of the people of any county, city, town or village or any combination thereof, and the improvement of their health and welfare, by the utilization of solid waste management-resource recovery systems, technology, processes and procedures, and for any purpose which aids in the financing of the construction and operation of a solid waste management-resource recovery facility for the benefit of any county, city, town or village or any combination thereof.

(d) The products generated by any solid waste management-resource recovery facility may be sold, utilized or otherwise disposed of, pursuant to contract, lease, rental agreement, license, permit, concession, or other authorization, between any municipality and any person, upon such terms and conditions for such consideration, and for such term or duration, not to exceed twenty-five years, as may be agreed upon by any municipality and such person.

(e) It is the intent of the legislature that overall cost should in all cases be a major criterion in the selection of contractors for award of contracts pursuant to this section and that, wherever practical, such contracts which include construction work should be procured through competitive bidding procedures as prescribed by sections one hundred one and one hundred three of this chapter. It is further the intent of the legislature to acknowledge the highly complex and innovative nature of resource recovery technology for processing mixed solid waste, the relative newness of the variety of resource recovery systems now available, ***the desirability of a single point of responsibility for the development of facilities*** and the economic and ***technical utility of contracts for resource recovery projects which include in their scope various combinations of design, construction, operation, management and/or maintenance responsibilities over prolonged periods of time*** and that in some instances it may be beneficial to the municipality to ***award a contract on the basis of factors other than cost alone, including but not limited to facility design,*** system reliability, energy efficiency, compatibility with source separation and other recycling systems and environmental protection. Accordingly, and notwithstanding the provisions of any general, special or local law or charter, a contract entered into between a municipality and any person pursuant to this section may be awarded pursuant to public bidding in compliance with sections one hundred one and one hundred three of this chapter or pursuant to the following provisions for the award of a contract based on evaluation of proposals submitted in response to a request for proposals prepared by or for the municipality:

(1) The municipality shall require that each proposal to be submitted shall include:

(i) information relating to the experience of the proposer on the basis of which said proposer purports to be qualified to carry out all work required by a proposed contract; the ability of the proposer to secure adequate financing; and proposals for project staffing, implementation of work tasks, and the carrying out of all responsibilities required by a proposed contract;

(ii) a proposal clearly identifying and specifying all elements of cost which would become charges to the municipality, in whatever form, in return for the fulfillment by the proposer of all tasks and responsibilities established by the request for the proposal for the full lifetime of a proposed contract, including, as appropriate, but not limited to the cost of planning, design, construction, operation, management and/or maintenance of any facility; and clearly identifying and specifying all elements of revenue which would accrue to the municipality from the sale of any material or energy produced or from any other source; provided, that the municipality may prescribe the form and content of such proposal and that, in any event, the proposer must submit sufficiently detailed information to permit a fair and equitable evaluation by the municipality of such proposal; and provided, further, that the municipality may set maximum allowable cost limits in any form in the request for proposals; and

(iii) such other information as the municipality may determine to have a material bearing on its ability to evaluate any proposal in accordance with subparagraph five of this paragraph;

(2) To provide for public information and review:

(i) prior to the issuance of a request for proposals pursuant to this paragraph, the municipality shall publish notice of such issuance in the official newspaper of the municipality, if any, in at least one newspaper of general circulation, in the state register and in the environmental notice bulletin. Concurrent with the publication of such notice a draft request for proposals shall be filed with the clerk or chief executive officer of the municipality;

(ii) after allowing a sixty day comment period and an additional ten days to review such comments, the municipality may publish the final request for proposals and concurrent with such publication shall publish notice of such issuance in the manner specified in item (i) of this subparagraph. Concurrent with the publication of the final request for proposals a set of comments filed in relation to the draft request for proposals and findings related to the substantive elements of such comments shall be filed along with the request for proposals with the clerk or chief executive officer of the municipality and in the municipal public library or libraries in proximity to the proposed site or sites;

(3) Proposals received in response to such request for proposals shall be evaluated by the municipality as to net cost or, if a net revenue is projected, net revenue, and, in a manner consistent with provisions set forth in the request for proposals, may be evaluated on the basis of additional factors, including but not limited to the technical evaluation of facility design, system reliability, energy balance and efficiency, environmental protection, overall employment and recovery of materials. The evaluation of such proposals and the determination of whether a proposer is "responsible" may include, but shall not be limited to, consideration, in a manner consistent with provisions set forth in the request for proposals, of the record of the proposer in complying with existing labor standards and recognizing state and federally approved apprentice training programs, and the willingness of the proposer to provide for meaningful participation of minority group persons and business enterprises in the conduct of the work;

(4) ***The municipality*** may make a contract award to any responsible proposer selected pursuant to subparagraph three of this paragraph based on a determination by the municipality that the selected proposal is most responsive to the request for proposals and ***may negotiate with any proposer;*** provided, however, that if an award is made to any proposer whose total proposal does not provide either the lowest net cost or, if a net revenue is projected, the greatest net revenue of any proposal received, the municipality shall adopt a resolution after public hearing which includes particularized findings relevant to factors evaluated pursuant to subparagraph three indicating that the municipality's requirements are met by such award and that such action is in the public interest.

(f) Whenever a municipality enters into a contract pursuant to this section for the construction of a solid waste management-resource recovery project, the provisions of section two hundred twenty of the labor law shall be applicable to such construction work.

* * *

7. Every contract entered into between a municipality and a project developer pursuant to the provisions of subparagraph four of paragraph (e) of subdivision four of this section, for construction of a solid waste management-resource recovery building by the project developer, shall contain provisions that such building shall be constructed through construction contracts awarded through public competitive bidding in accordance with paragraphs (a) through (g) of this subdivision; that ***the project developer shall furnish a bond guaranteeing prompt payment of moneys that are due to all persons furnishing labor and materials pursuant to the requirements of such construction contracts,*** and that a copy of such payment bond shall be kept by the municipality and shall be open to public inspection; provided, however, that the requirements of this subdivision shall not apply when the cost of such construction is less than five thousand dollars.

(a) The project developer shall advertise for bids for such construction contracts in a daily newspaper having general circulation in the county in which such public solid waste management-resource recovery building is to be located. Such advertisement shall contain a statement of the time when and place where all bids received pursuant to such notice will be publicly opened and read. An employee of the municipality shall be designated to open the bids at the time and place specified in the notice. All bids received shall be publicly opened and read at the time and place so specified. At least five days shall elapse between the publication of such advertisement and date on which the bids are opened.

(b) When the entire cost of constructing such building shall exceed fifty thousand dollars, the project developer shall prepare separate specifications for the following subdivisions of such work, so as to permit separate and independent bidding upon each subdivision: (i) plumbing and gas fittings; (ii) steam heating, hot water heating, ventilating and air conditioning apparatus; and (iii) electric wiring and standard illuminating fixtures.

(c) After public competitive bidding the project developer shall award one or more separate contracts for each of the above subdivisions of such work, whenever separate specifications are required pursuant to paragraph (b) of this subdivision, and one or more contracts for the remainder of such work. The project developer may award such contracts at different times. Contracts awarded pursuant to this paragraph shall be awarded by the project developer to the lowest responsible and responsive bidder and shall be contracts of the project developer and not of the municipality which shall have no obligation or liabilities, whatsoever, thereunder. The project developer shall have the responsibility for the supervision, coordination, and termination of such contracts, unless otherwise specified in contractual terms between the project developer and the municipality.

(d) In determining whether a prospective contractor is responsible and responsive, the project developer may require that prospective contractors:

(i) have adequate financial resources or the ability to obtain such resources;

(ii) be able to comply with the required or proposed delivery or performance schedule;

(iii) have a satisfactory record of performance;

(iv) have the necessary organization, experience, operational controls, and technical skills, or the ability to obtain them;

(v) have the necessary production, construction and technical equipment and facilities, or the ability to obtain them;

(vi) be eligible to receive an award under applicable laws and regulations and be otherwise qualified.

(e) The project developer may reject any bid of a bidder which the project developer determines to be nonresponsible or nonresponsive to the advertisement for bids.

(f) The project developer may, in its discretion, reject all bids, and may revise bid specifications and may readvertise for bids as provided herein.

(g) Only as used in this subdivision:

(i) "project developer" means any private corporation, partnership, or individual or combination thereof which has submitted a proposal in response to a request for proposals issued pursuant to subparagraph two of paragraph (e) of subdivision four of this section;

(ii) "construction" shall include reconstruction, rehabilitation or improvement;

(iii) "solid waste management-resource recovery building" means a building of a solid waste management-resource recovery facility. Such building shall not include the system to be used for the purposes of receiving, processing, handling or storing solid waste, the products and by-products derived therefrom, or materials used in such processing or handling and any equipment or property involving proprietary or trade secrets.

N.Y. Gen Mun. Law § 101. Separate specifications for certain public work.

1. Every officer, board or agency of a political subdivision or of any district therein, charged with the duty of preparing specifications or awarding or entering into contracts for the erection, construction, reconstruction or alteration of buildings, when the entire cost of such work shall exceed fifty thousand dollars, ***shall prepare separate specifications*** for the following three subdivisions of the work to be performed:

a. Plumbing and gas fitting;

b. Steam heating, hot water heating, ventilating and air conditioning apparatus; and

c. Electric wiring and standard illuminating fixtures.

2. ***Such specifications shall be drawn so as to permit separate and independent bidding upon each of the above three subdivisions of work.*** All contracts awarded by any political subdivision or by an officer, board or agency thereof, or of any district therein, for the erection, construction, reconstruction or alteration of buildings, or any part thereof, shall award the three subdivisions of the above specified work separately in the manner provided by section one hundred three of this chapter. Nothing in this section shall be construed to prevent any political subdivision from performing any such branches of work by or through their regular employees, or in the case of public institutions, by the inmates thereof.

3. In the county of Erie, specifications for public work pertaining to the construction of a county stadium may provide for assignment of responsibility for supervision and coordination of any of the contracts for such work to a single responsible and reliable person, firm or corporation.

4. (Added, L 1985) In the county of Albany, specifications for work pertaining to the construction of a civic center may provide for assignment of responsibility for supervision and coordination of any or all of the contracts for such work to a single responsible and reliable person, firm or corporation.

New York State University Education Fund.
N.Y. Educ. Law § 373. General powers and duties of fund

The fund shall have the following powers in addition to those specifically conferred elsewhere in this article:

* * *

8. To design, construct, acquire, reconstruct, rehabilitate and improve academic buildings, dormitories and other facilities for the state university in accordance with sections three hundred seventy-five and three hundred seventy-six of this chapter;

9. In connection with such design, construction, acquisition, reconstruction, rehabilitation and improvement, (i) to install or cause to be installed water, sewer, gas, electrical, telephone, heating, air conditioning and other utility services, including appropriate connections, (ii) to construct or cause to be constructed walkways, roadways and other improvements on adjacent real property made available by the state university, and (iii) to landscape adjacent real property or cause the same to be landscaped;

10. To maintain, repair and keep up the real property held by it and the facilities constructed, acquired, reconstructed, rehabilitated or improved pursuant to this article until responsibility therefor is transferred to the state university in accordance with section three hundred seventy-nine of this chapter;

11. Subject to the terms of any lease or agreement with third parties, with respect to approved university-related economic development facilities or approved university related economic development projects to develop, manage, operate, and/or lease, solely or in cooperation with one or more entities or individuals, or to surrender to the appropriate state official, for other public use or for sale, lease or other disposition in accordance with law, real property made available to the fund for its corporate purposes by the state university whenever the state university trustees determine that such real property is unnecessary for present or foreseeable future needs of the state university;

* * *

15. To engage the services of construction, engineering, architectural, legal and financial consultants, surveyors and appraisers, on a contract basis or as employees, for professional service and technical assistance and advice; and
16. To do any and all things necessary or convenient to carry out its corporate purposes and exercise the powers given and granted to it in this article.
17. [As amended by L 1999, ch 405] ***To construct, design/build (using a competitive selection process with due consideration of cost),*** develop, manage, lease and/or operate real property and facilities made available to the fund for its corporate purposes by the state university related to academic incubator facilities and a pharmaceutical research, development , which may also include a manufacturing facility located at the state university[of] New York college of technology at Farmingdale.
17. [As amended by L 1999, ch 624] ***To construct, design/build, develop, manage and/or operate real property and facilities*** made available to the fund for its corporate purposes by the state university related to academic incubator and approved university-related economic development facilities;

* * *

New York State University Construction Fund
N.Y. Educ. Law § 376. Letting of construction contracts

1. After May first, nineteen hundred sixty-two, the fund as agent for the state university shall construct, acquire, reconstruct, rehabilitate and improve, or cause to be constructed, acquired, reconstructed, rehabilitated and improved, all academic buildings, dormitories and other facilities to be constructed, acquired, reconstructed, rehabilitated or improved at state-operated institutions or statutory or contract colleges under the jurisdiction of the state university, provided that legislation or appropriations authorizing the same (i) have been requested by the state university trustees, (ii) have been recommended by the governor in a budget bill relating to a state fiscal year commencing on or after April first, nineteen hundred sixty-two which specifies the facilities to be constructed, acquired, reconstructed, rehabilitated or improved and the total estimated cost for each such facility, and (iii) have been approved by the legislature for such state fiscal year. With respect to state fiscal years commencing on or after April first, nineteen hundred sixty-three, the budget bill referred to in the preceding sentence shall include in addition to the items stated the date when it is desired that the construction, acquisition, reconstruction, rehabilitation or improvement of each facility referred to therein be completed, and all work for such purposes shall be

performed in such manner as to assure completion, so far as practicable, by the dates specified.

* * *

5. No contract for the construction, acquisition, reconstruction, rehabilitation or improvement of academic buildings, dormitories and other facilities shall be awarded by any letting agency unless the state university trustees shall have approved the architectural concept of the facility to be constructed, acquired, reconstructed, rehabilitated or improved and unless the fund shall have approved the proposed terms of such contract, including the detailed plans and specifications for such facility.

6. Each contract for the construction, acquisition, reconstruction, rehabilitation or improvement of academic buildings, dormitories and other facilities shall include a provision that the architect who designed the facility, or an architect or engineer, or, for projects authorized by section three hundred seventy-two-a of this article a construction manager or other consultant, retained specifically for the purpose of supervision, shall supervise the work to be performed through to completion and shall see to it that the materials furnished and the work performed are in accordance with the drawings, plans, specifications and contract therefor.

* * *

11. (Added, L 1999) ***Notwithstanding any other law, rule or regulation, upon the approval of the trustees, the fund may solicit proposals and award contracts for design/build projects to an entity or combination of entities for approved university related economic development facilities.*** Such approval shall include a justification stating the financial schedule and/or other benefits to the state university and the public over traditional competitive bidding procedures. The fund shall ***award such design/build contracts*** in accordance with the fund's guidelines for the award of procurement contracts based on a clear and detailed scope of work. Such guidelines shall be subject to review and approval by the state comptroller and shall establish procedures and criteria for public advertisement, qualification of proposers and ***award for design/build contracts.*** The fund shall issue a report on or before June first, two thousand to the governor, the temporary president of the senate, the speaker of the assembly, the director of the budget and the chairs of the senate finance committee and the assembly ways and means committee. This subdivision shall expire and be deemed repealed June one, two thousand three.

NORTH CAROLINA

I. Architect & Engineer Licensing Laws.

N.C. Gen. Stat. § 83A-13. Exemptions.

(a) Nothing in this Chapter shall be construed to prevent the practice of general contracting under the provisions of Article 1 of Chapter 87, or the practice by any person who is qualified under law as a "registered professional engineer" of such architectural work as is incidental to engineering projects or utilities, or the practice of any other profession under the applicable licensure provisions of the General Statutes.

(b) ***Nothing in this Chapter shall be construed to prevent a duly licensed general contractor, professional engineer or architect, acting individually or in combination thereof, from participating in a "Design/Build" undertaking including the preparation of plans and/or specifications and entering individual or collective agreements with the owner*** in order to meet the owner's requirements for pre-determined costs and unified control in the design and construction of a project, and for the method of compensation for the design and construction services rendered; provided, however, that nothing herein shall be construed so as to allow the performance of any such services or any division thereof by one who is not duly licensed to perform such service or services in accordance with applicable licensure provisions of the General Statutes; provided further, that full disclosure is made in writing to the owner as to the duties and responsibilities of each of the participating parties in such agreements; and, provided further, nothing in this Chapter shall prevent the administration by any of the said licensees of construction contracts and related services or combination of services in connection with the construction of buildings.

II. State Procurement Laws.

N.C. Gen. Stat. § 136-28.1. Letting of contracts to bidders after advertisement; exceptions.

(a) All contracts over eight hundred thousand dollars ($800,000) that the Department of Transportation may let for construction or repair necessary to carry out the provisions of this Chapter shall be let to a responsible bidder after public advertising under rules and regulations to be made and published by the Department of Transportation. The right to reject any and all bids shall be reserved to the Board of Transportation. Contracts for construction or repair for federal aid projects entered into pursuant to this section shall not contain the standardized contract clauses prescribed by 23 U.S.C. § 112(e) and 23 C.F.R. § 635.131(a) for differing site conditions, suspensions of work ordered by the engineer or significant changes in the character of the work. The Department of Transportation shall use only the contract provisions provided in the North Carolina Department of Transportation, Standard Specifications for Roads and Structures, January 1, 1984, except as each may be changed or provided for by rule adopted by the Board of Transportation in accordance with the Administrative Procedure Act.

* * *

(j) Notwithstanding any other provision of law, the Board of Transportation ***may award up to three contracts annually for construction of transportation projects***

on a design-build basis. These contracts may be awarded after a determination by the Department of Transportation that delivery of the projects must be expedited and that it is not in the public interest to comply with normal design and construction contracting procedures. Prior to the ***award of a design-build contract,*** the Secretary of Transportation shall report to the Joint Legislative Transportation Oversight Committee and to the Joint Legislative Commission on Governmental Operations on the nature and scope of the project and the reasons an ***award on a design-build basis*** will best serve the public interest.

N.C. Gen. Stat. § 143-128. Requirements for certain building contracts.

(a) Preparation of specifications.—Every officer, board, department, commission or commissions charged with responsibility of preparation of specifications or awarding or entering into contracts for the erection, construction, alteration or repair of any buildings for the State, or for any county, municipality, or other public body, ***must have prepared separate specifications for each of the following subdivisions*** or branches of work to be performed:

(1) Heating, ventilating, air conditioning and accessories (separately or combined into one conductive system) and/or refrigeration for cold storage (where the cold storage cooling load is 15 tons or more of refrigeration), and all work kindred thereto.

(2) Plumbing and gas fittings and accessories, and all work kindred thereto.

(3) Electrical wiring and installations, and all work kindred thereto.

(4) General work relating to the erection, construction, alteration, or repair of any building above referred to, which work is not included in the above-listed three subdivisions or branches.

All such specifications must be so drawn as to permit separate and independent bidding upon each of the subdivisions or branches of work enumerated above. The above enumeration of subdivisions or branches of work shall not be construed to prevent any officer, board, department, commission or commissions from preparing additional separate specifications for any other category of work.

* * *

(d) Single-prime and alternative contracts.—The State, a county, municipality, or other public body may accept bids under the single-prime contract system or a contracting method approved by the State Building Commission under G.S. 143-135.26. If the State, county, municipality, or other public body accepts bids under the single-prime contract system, it must also seek bids for the project under the separate prime contract system, except as otherwise authorized under G.S. 143-135.26, and award the contract to the lowest responsible bidder or bidders for the total project, taking into consideration quality, performance and the time specified in the bids for the performance of the contract. When bids are accepted under the single-prime contract system all bidders must identify on their bid the contractors they have selected for the subdivisions or branches of work for:

(1) Heating, ventilating, and air conditioning;

(2) Plumbing;

(3) Electrical; and

(4) General.

No contractor whose bid is accepted shall substitute any person as subcontractor in the place of the subcontractor listed in the original bid, except with the approval of the awarding authority for good cause shown by the contractor. The terms, conditions, and requirements of each contract between the contractor and a subcontractor performing work under a subdivision or branch of work listed in this subsection shall be substantially the same as the terms, conditions, and requirements of the contract between the contractor and the State, county, municipality, or other public body.

* * *

III. Appropriations Bills.

1991 North Carolina Laws Ch. 689 (H.B. 83).

* * *

Sec. 239. (a) General Purposes. The appropriations hereby made by the 1991 General Assembly for capital improvements from the proceeds of the $200,000,000 State of North Carolina Prison and Youth Services Facilities Bonds authorized by Chapter 935 of the 1989 Session Laws (the "bond act") and approved by the qualified voters of the State who voted thereon on November 6, 1990, as said bonds may be issued from time to time (the "bonds"), are for the purposes of financing the cost of $112,500,000 of State prison facilities and youth services facilities, including, without limitation, the cost of constructing capital facilities, renovating or reconstructing existing facilities, acquiring equipment related thereto, purchasing land, paying costs of issuance of bonds and notes and paying contractual services necessary for the partial implementation of the purposes of the bond act, all as defined in and authorized by the bond act and as more particularly described in this section. The particular projects within the purposes under the bond act to be financed by the $87,500,000 balance of the $200,000,000 bond authorization may, as authorized by the bond act, be determined by legislative action of the General Assembly during the 1991 Session or any subsequent session.

(b) Appropriation Procedures. The appropriations hereby made by the 1991 General Assembly for the purposes under the bond act shall be disbursed for the particular projects authorized by this section.

* * *

(f) Administration. With respect to facilities authorized for the Department of Correction, the Office of State Budget and Management ***may contract for*** and supervise all aspects of administration, technical assistance, design, construction or demolition of ***prison facilities*** in order to implement the providing of prison facilities under the provisions of this act without being subject to the requirements of the following statutes and rules implementing those statutes: G.S. 143-135.26(1), 143-128, 143-129, 143-131, 143-132, 143-134, 143-135.26, 143-64.10 through 143-64.13, 113A-1 through 113A-10, 113A-50 through 113A-66, 133-1.1(b), 133-1.1(g), and 143-408.1; provided, however, of the funds allocated under the provisions of this act for the construction of prison facilities, the Office of State Budget and Management shall have a verifiable ten percent (10%) goal for participation by minority and women-owned businesses. All contracts for the design, construction, or demolition of prison facilities shall include a penalty for failure to complete the work by a specified date. The proposals for prison facilities authorized in this section shall be invited by advertisement in newspapers having general circulation in the State. The form of advertisement shall be prepared in the form of Section 301 of the State Construction Manual of the Department of Administration, and shall be published in one issue of the newspaper. A minimum of at least seven full days shall lapse between the date of publication and the date of the opening of bids. Initiation of the advertisement shall be by the Office of State Budget and Management. The Office of State Budget and Management shall consider alternative delivery systems that could expedite the delivery of prison facilities. ***Such delivery systems as design-build, using modular or conventional building systems, shall be considered. However, in order for such alternatives to be used, the Department of Correction must approve the proposed design for operational programming and cost of operations and maintenance.***

* * *

N.C. Session Laws 1997-443, § 32.11. Department of Transportation; Design-Build.

Provides that the Department of Transportation ***may enter into a design-build-warrant contract*** to develop, with Federal Highway Administration participation, a Congestion Avoidance and Reduction for Autos and Trucks (CARAT) system of traffic management in the Charlotte-Mecklenburg urban areas. Notwithstanding any other provision of law, contractors, their employees, and Department of Transportation employees involved in this project only do not have to be licensed by occupational licensing boards, and for the purpose of entering into contracts, the Department of Transportation is exempted from the provisions of G.S. 136-28.1, 143-52, 143-53, 143-58, 143-128, and 143-129; these exemptions are limited and available only to the extent necessary to comply with federal rules, regulations, and policies for completion of this project. The Department shall report quarterly to the Joint Legislative Transportation Oversight Committee on the project. Session Laws from 1993 and 1995 contained similar provisions.

NORTH DAKOTA

None.

OHIO

I. Architectural and Engineering Licensing Laws.

Ohio Rev. Code Ann. § 4703.182. Contractor may provide architectural services through arrangement with registered architect; unauthorized practice prohibited.

(A) Notwithstanding any other provision of this chapter, ***any person who performs labor or who provides services pursuant to a construction contract may offer or provide architectural services*** in connection with the person's own labor or services if all of the following apply:

(1) The architectural services are provided and performed by an architect registered under this chapter or by a firm holding a certificate of authorization granted pursuant to this chapter through either direct employment or independent contract with the person contracting with the owner and offering the services.

(2) The architectural services are offered by the person who intends to perform the actual construction of an improvement to real property ***as part of a design-build contract.***

(3) The contractor, the architect registered under this chapter, or the firm holding a certificate of authorization granted pursuant to this chapter meets the requirements of section 153.70 of the Revised Code ***when design-build services are provided to a public authority*** as defined in section 153.65 of the Revised Code.

(B) Except as otherwise provided in this section, no person shall offer or engage in the practice of architecture without a certificate of authorization for any construction contract that is to be constructed by another person, or hold himself or herself out to the public as engaging in the practice of architecture or as providing architectural services without a certificate of authorization.

(C) As used in this section:

(1) "Construction contract" means any oral or written agreement that involves any activity in connection with the erection, alteration, repair, replacement, renovation, installation, or demolition of any building, structure, highway, or bridge.

(2) "Design-build" means a project delivery system for improvements to real property by which a person is solely responsible contractually as a contractor to an owner for both the design and construction of the improvement, which design and construction may include a performance-based specification established by the owner rather than a specific design as an improvement goal. As used in this division, "contractor" has the same meaning as in section 4113.61 of the Revised Code.

Ohio Rev. Code Ann. § 4733.161. Contractor may provide engineering services through arrangement with registered engineer; unauthorized practice prohibited.

(A) Notwithstanding any other provision of this chapter, ***any person who performs labor or who provides services pursuant to a construction contract may offer or provide engineering services*** in connection with the person's own labor or services if all of the following apply:

(1) The engineering services are provided and performed by an engineer registered under this chapter or by a firm holding a certificate of authorization granted pursuant to this chapter through either direct employment or independent contract with the person contracting with the owner and offering the services.

(2) The engineering services are offered by the person who intends to perform the actual construction of an improvement to real property ***as part of a design-build contract.***

(3) The contractor, the engineer registered under this chapter, or the firm holding a certificate of authorization granted pursuant to this chapter meets the requirements of section 153.70 of the Revised Code ***when design-build services are provided to a public authority*** as defined in section 153.65 of the Revised Code.

(B) Except as otherwise provided in this section, no person shall offer or engage in the practice of engineering without a certificate of authorization for any construction contract that is to be constructed by another person, or hold himself or herself out to the public as engaging in the practice of engineering or as providing engineering services without a certificate of authorization.

(C) As used in this section, "construction contract" and ***"design-build" have the same meaning as in section 4703.182 of the Revised Code.***

II. Public Procurement Laws.

Ohio Rev. Code Ann. § 5517.011. Pilot program for combining design and construction elements of project into single contract.

(A) (1) Notwithstanding section 5517.01 of the Revised Code, the director of transportation may establish ***a pilot program*** to expedite the sale and construction of no more than six special projects by ***combining the design and construction elements of a highway or bridge project into a single contract.*** Except in regard to those requirements relating to providing plans, the director shall award contracts under this section in accordance with section 5525.01 of the Revised Code.

(2) On or before December 31, 2000, the director shall prepare and submit to the general assembly a report evaluating the experience of the department of transportation with each project under this division and contract under division (B) of this section, including whether the department realized any cost or time savings. Regarding those projects and contracts, the report shall include a discussion of the number and cost of change orders, the quality of work performed, the number of bids received, the impact on minority and female contract participation, and other issues the director considers appropriate. The director also may make recommendations regarding the continuation of the program, including the need for any changes.

(3) After completion of the sixth project, no projects shall be commenced under this division unless the general assembly either approves additional projects to further study the effectiveness of the procedures or makes the program permanent.

(B) In addition to the six projects under division (A) of this section, during the period beginning July 1, 1999, and ending June 30, 2001, ***the director may expand the pilot program to more contracts combining the design and construction elements of highway or bridge projects.*** The total dollar value of contracts made under this division shall not exceed two hundred fifty million dollars. The director may seek either bids or technical proposals for contracts under this division.

(1) When the director determines to ***award a single contract for a design-build project*** under this division through the receipt of bids, except for those requirements relating to providing plans, the director shall award contracts in accordance with Chapter 5525. of the Revised Code. When the director determines to award ***a single contract for a design-build project*** under this division through

the receipt of technical proposals, the director shall advertise and ***select the design-build team*** using a value-based selection process combining technical qualifications and competitive bidding elements.

(2) If the director elects to utilize the competitive bid option for ***design-build projects,*** the director shall prepare and distribute a scope of work document upon which the bidders shall base their bids.

(3)(a) If the director elects to utilize a value-based ***selection process for design-build projects*** through the receipt of technical proposals, the director shall restrict usage of this method annually to no more than sixty million dollars and no more than three projects, whose per-project estimate must exceed ten million dollars. The director shall prepare conceptual documents for review by interested parties, accept letters of interest, and select the three most qualified ***design-build teams*** to submit a technical proposal. The criteria for selecting the three finalists shall include the qualifications and experience of the ***design-build team,*** including the proposed personnel to be utilized, equipment usage, and general proposed project approach. The schedule of activities and financial resources of ***the design-build team*** also shall be factors in the selection process. In addition, the director shall take into consideration ***the design-build team's*** affirmative action policies and record with regard to employees and subcontracts.

(b) After the director selects the three finalists, the finalists shall prepare both a technical proposal and a price proposal. The technical proposal shall state the finalist's qualifications and experience, including prior performance by ***the design-build team*** on similar projects, the identity of the members of each team, and a detailed project approach and schedule, including innovative design and construction techniques. The finalists shall submit the price proposal separately as requested by the director.

The director first shall review the submitted technical proposals and ascribe a numerical score to each proposal. The technical numerical scores shall be equated to a percentage adjustment to be applied to the finalists' price proposals, using a predetermined schedule of adjustment made known to the finalists at the time of advertising. ***In no case shall the technical proposal rating exceed twenty-five percent of the value-based technical and price selection criteria.*** The director shall reserve the right to consider a technical proposal as being nonresponsive, thereby eliminating that finalist from further consideration. Upon completion of the rating of technical proposals, the director shall apply to the price proposals the percentage adjustments predetermined from the numerical scores assigned to the technical proposals. Unless all proposals are rejected, the director shall select the finalist with the lowest adjusted price. The adjusted price shall be used for selection only. The contract shall be based on the price proposal as submitted. The department shall compensate each responsive finalist not selected in an amount generally equal to one-fourth of one percent of the unadjusted price proposal amount submitted by the selected finalist or by an amount the director establishes at the time of advertising.

III. Attorney General Opinions.

The issue was whether the various competitive bidding requirements that appear in R.C. Chapter 153 (public improvements) apply to the construction of an addition to an existing joint township district hospital that is performed by a design/build firm.

Concluded that a joint township district hospital board is required to comply with the provisions of R.C. 153.12-.14 when awarding a contract for the construction of an addition to an existing joint township district hospital but that a joint township district hospital board is not required to comply with the provisions of R.C. 153.50-.571 when awarding a contract for the construction of an addition to an existing joint township district hospital. The attorney general stated that the term "design-build" is commonly understood as referring to an enterprise that, in addition to undertaking the construction of particular buildings or structures, renders architectural and planning services in conjunction therewith. 1987 Ohio Op. Atty. Gen. No. 87-079; Cite as: 1987 WL 289438 (Ohio A.G.)

OKLAHOMA

I. Public Procurement Laws.

Senate Bill No. 1250 (1999). 61 Okla. Stat. § 202.

BE IT ENACTED BY THE PEOPLE OF THE STATE OF OKLAHOMA:

[*1] SECTION 1. AMENDATORY 61 O.S. 1991, Section 202, is amended to read as follows: Section 202. As used in the Public Building Construction and Planning Act:

1. "Administrator" means the State Construction Administrator of the Department of Central Services;

2. "Construction" means the process of planning, acquiring, designing, building, equipping, altering, repairing, improving, or demolishing any structure or appurtenance thereto including facilities, utilities, or other improvements to any real property but not including highways, bridges, airports, railroads, tunnels, sewers not related to a structure or appurtenance thereto, or dams;

* * *

5. "Design-Build" means a project delivery method whereby this State acquires both design and construction services in the same contract from a single legal entity referred to as the Design-Builder, without the bid component of the traditional design-bid-build process;

6. "Department" means the Department of Central Services;

7. "Design Consultant" means any individual, firm, corporation, association, partnership, copartnership, or any other legal entity possessing the required qualifications to provide licensed architectural, registered engineering, or registered land surveying services for any construction project;

8. "Division" means the Construction and Properties Division of the Department of Central Services;

9. "Energy performance index or indices" (EPI) means a number describing the energy requirements at the building boundary of a structure, per square foot of floor space or per cubic foot of occupied volume, as appropriate under defined internal and external ambient conditions over an entire seasonal cycle. As experience develops on the energy performance achieved with state construction, the indices (EPI) will serve as a measure of structure performance with respect to energy consumption;

10. "Life cycle costs" means the cost of owning, operating, and maintaining the structure over the life of the structure. This may be expressed as an annual cost for each year of the facility's use;

11. "Procurement" means buying, purchasing, renting, leasing, or otherwise acquiring or disposing of supplies, services, or construction;

12. "Shared savings financing" means the financing of energy conservation measures and maintenance services through a private firm which may own any purchased equipment for the duration of a contract. Such contract shall specify that the private firm will be recompensed either out of a negotiated portion of the savings resulting from the conservation measures and maintenance services provided by the private firm or, in the case of a cogeneration project, through the payment of a rate for energy lower than would otherwise have been paid for the same energy from current sources;

13. "State agency" means an agency, board, commission, counsel, court, office, officer, bureau, institution, unit, division, body or house of the executive or judicial branches of government of this State, whether elected or appointed, excluding only political subdivisions.

[*2] SECTION 2. NEW LAW A new section of law to be codified in the Oklahoma Statutes as Section 202.1 of Title 61, unless there is created a duplication in numbering, reads as follows:

A. The design-build and at-risk construction management project delivery methods shall not be used without the written approval of the Director of the Department of Central Services or an act of the Legislature specifying design-build or at-risk construction management for a particular project. In all instances where the ***design-build project*** or at-risk construction management delivery method is authorized, construction administration shall be performed by the State Construction Administrator, the Administrator's designee or designees, or otherwise by contract or contract provision approved by the Director of the Department of Central Services for construction administration by another party.

B. The provisions of subsection A of this section shall not apply to projects by contract pursuant to an interagency agreement under Section 581 of Title 74 of the Oklahoma Statutes or to projects a state agency performs solely with the staff of the agency.

C. The Department of Central Services shall, pursuant to the Administrative Procedures Act and with regard for such recommendations or other assistance as it may receive from the Advisory Task Force on Public Construction and Renovation Laws, promulgate rules to effect procedures and processes necessary to the fulfillment of its responsibilities under subsection A of this section.

[*3] SECTION 3. AMENDATORY Section 2, Chapter 69, O.S.L. 1999, is amended to read as follows:

Section 2. A. Members appointed to the Advisory Task Force on Public Construction and Renovation Laws shall serve at the pleasure of the respective appointing authority in terms expiring with the termination of the task force, which shall occur on June 30, 2002. Vacancies shall be filled by the respective appointing authority.

B. The President Pro Tempore of the Senate shall designate a member of the task force to serve as Chair, and the Speaker of the House of Representatives shall designate a member of the task force to serve as Vice Chair. The Chair and Vice Chair shall serve at the pleasure of the respective appointing authority.

C. Eight members shall constitute a quorum. The task force shall conduct its business in compliance with the Oklahoma Open Meeting Act and with the Oklahoma Open Records Act. Upon termination of the task force, staff shall transfer all task force records to the Legislative Service Bureau for disposition as required by law.

D. Staff services for the task force shall be provided by the Senate and the House of Representatives. The Department of Central Services shall provide additional staff assistance upon request of the Chair.

E. Members of the task force shall receive no compensation for serving on the task force but shall receive travel reimbursement as follows:

1. Legislative members shall be reimbursed for their necessary travel expenses incurred in the performance of their duties in accordance with Section 456 of Title 74 of the Oklahoma Statutes;

2. Nonlegislative members who are State officers or employees shall be reimbursed by their respective agencies for their necessary travel expenses incurred in the performance of their duties in accordance with the State Travel Reimbursement Act; and

3. Members who are not otherwise officers or employees of the state shall be reimbursed by the Department of Central Services in accordance with the State Travel Reimbursement Act.

[*4] SECTION 4. AMENDATORY Section 3, Chapter 69, O.S.L. 1999, is amended to read as follows:

Section 3. A. The purpose of the Advisory Task Force on Public Construction and Renovation Laws shall be to make specific recommendations to the Legislature and to the Department of Central Services for giving greater clarity and coherence to laws relating to the construction and renovation of public buildings. ***However, the purview of the task force shall not extend to roads and highways*** except insofar as the task force may wish to recommend that certain provisions in Title 61 of the Oklahoma Statutes relating to roads and highways be deleted or be moved to Title 69 or elsewhere. B. The scope of task force responsibility shall include, but not be limited to, assisting the Department of Central Services with the development of rules relating to the public building construction and planning act, to be promulgated pursuant to the administrative procedures act, and to examining and possibly making recommendations concerning:

1. Bonds required on public works contracts;
2. Purchasing preferences;
3. Consultant contracts;
4. The Oklahoma Lighting Energy Conservation Act;
5. Provisions of Senate Bill No. 1070 of 1996;
6. The Public Competitive Bidding Act of 1974;
7. The Public Building Construction and Planning Act; and
8. Miscellaneous provisions found in Title 61 of the Oklahoma Statutes.

C. In performing its duties pursuant to this section, ***the task force shall invite comment regarding,*** and give consideration to, a variety of approaches, strategies, or methods for the efficient accomplishment of public construction projects, such as ***"design/build"*** and construction management. The task force shall also consider the usefulness of lease-purchase procedures to the foregoing. The chair may appoint committees of task force members, and, if the chair so desires, other willing persons, to prepare recommendations for consideration by the task force. Other persons so appointed shall not receive compensation or travel reimbursement.

D. The task force shall, from time to time report its specific recommendations, in the form of proposed statutory revisions and rule provisions, together with such explanation and comment as the task force may deem appropriate, to the President Pro Tempore of the Senate and the Speaker of the House of Representatives , or their designees, and to the Director of the Department of Central Services.

E. For further consideration and discussion of its recommendations, the task force shall continue to make itself available to the Legislature and the Department of Central Services until June 30, 2002.

[*5] SECTION 5. Sections 3 and 4 of this act shall not be codified in the Oklahoma Statutes. [*6] SECTION 6. It being immediately necessary for the preservation of the public peace, health and safety, an emergency is hereby declared to exist, by reason whereof this act shall take effect and be in full force from and after its passage and approval.

HISTORY: Approved by the Governor June 6, 2000.

OREGON

I. Public Procurement Laws.

Ore. Stat. § 279.015. Competitive bidding; exceptions; exemptions.

(1) Subject to the policies and provisions of ORS 279.005 and 279.007, all public contracts shall be based upon competitive bids or proposals except:

* * *

(6) In granting exemptions pursuant to subsection (2)(a) and (b) of this section, the director or board shall:

(a) Where appropriate, direct the ***use of alternate contracting and purchasing practices*** that take account of market realities and modern or innovative contracting and purchasing methods, which are also consistent with the public policy of encouraging competition.

(b) Require and approve or disapprove written findings by the public contracting agency that support the awarding of a particular public contract or a class of public contracts, without the competitive requirements of subsection (1) of this section. The findings must show that the exemption of a contract or class of contracts complies with the requirements of subsection (2)(a) and (b) of this section.

(7) No written agreement under ORS chapter 190 is necessary under subsection (1)(g) of this section if the arrangement is between or among units of local government.

Department of Transportation. Toll Roads and Bridges.
Ore. Rev. Stat. § 383.001. Findings.

The Legislative Assembly finds that:

(1) The development, improvement, expansion and maintenance of an efficient, safe and well-maintained system of roads, highways and other transportation facilities is essential to the economic well-being and high quality of life of the people of this state.

(2) Public sources of revenues, including federal funding, to provide an efficient transportation system have not kept pace with the state's growing population and growing transportation needs, and all available alternative sources of funding should be utilized to supplement available public sources of revenues.

(3) Because public funding sources are not providing the state with sufficient funds to meet all of its transportation needs, ***private funding should be encouraged as an additional source of funding*** for transportation projects and facilities.

(4) Various alternatives for utilizing the funds of private entities in the acquisition, design, construction, reconstruction, operation and maintenance of transportation facilities exist, ***including arrangements whereby private entities obtain exclusive agreements to design, build, own, lease or operate with private funds all or a portion of transportation projects and facilities*** in exchange for the right to receive certain revenues generated from the operation and utilization of such transportation projects and facilities.

(5) Another important alternative for the funding of transportation facilities is the use of federal funds pursuant to ***23 U.S.C. 129(a),*** as amended by section 112 of the Intermodal Surface Transportation Efficiency Act of 1991, which established a program authorizing federal participation in construction of publicly or privately owned toll highways, bridges and tunnels.

(6) The federal legislation allows for a mix of federal funding and private funding of transportation facilities, allowing the states to leverage available federal funds as a means for attracting private capital.

(7) Legislation for the utilization of private funding of transportation facilities should be flexible enough to permit the Department of Transportation to obtain the advantages of any available alternative under which the acquisition, design, construction, reconstruction, operation, maintenance and repair of transportation facilities can be financed in whole or in part or in combination by any available sources of private or public funding.

(8) The funding of transportation facilities through the imposition of tolls on those who use such facilities is a fair and impartial means of assessing the costs of improvements against those who most benefit from such improvements, and is consistent with public policy.

(9) Joint endeavors of public and private entities do the following:

(a) Take advantage of private sector efficiencies in designing, constructing and operating transportation projects.

(b) Allow for the rapid formation of capital necessary for funding transportation projects.

(c) Require continued compliance with environmental requirements and applicable state and federal laws that all publicly financed projects must address.

Ore. Rev. Stat. § 383.005. Agreements for tollway projects; operation of projects.

(1) For purposes of the acquisition, design, construction, reconstruction, operation or maintenance and repair of tollway projects, ***the Department of Transportation may enter into any combination of contracts, agreements and other arrangements with any one or more private entities*** or units of government, or any combination thereof, including but not limited to the following:

(a) ***Design-build contracts*** with private entities pursuant to which a portion or all aspects of the design, construction and installation of all or any portion of a tollway project are accomplished by the private entity;

(b) Lease agreements, lease-purchase agreements and installment sale arrangements for the lease, sale or purchase of real and personal property for tollway projects by the state from private entities or units of government or by private entities or units of government from the state;

(c) Licenses, franchises or other agreements for the periodic or long-term operation or maintenance of a tollway project;

(d) Financing agreements for a tollway project pursuant to which the department makes any loan, grant, guaranty or other financing arrangement with a private entity or unit of government; and

(e) Agreements for purchase or acquisition of fee ownership, easements, rights of way or any other interests in land upon which a tollway project is to be built.

(2) The department may operate tollway projects and impose and collect tolls on any tollway project the department operates. Any private entity or unit of government that operates a tollway Project pursuant to an agreement with the department may impose and collect tolls on the tollway project, subject to the limits established by the department pursuant to ORS 383.021.

Ore. Rev. Stat. § 279.029.

(1) After the bids are opened as required by ORS 279.027, and after a determination is made that a contract is to be awarded, the public contracting agency shall award the contract to the lowest responsible bidder.

(2) In determining the lowest responsible bidder, a public contracting agency shall, for the purpose of awarding the contract, add a percent increase on the bid of a nonresident bidder equal to the percent, if any, of the preference given to that bidder in the state in which the bidder resides.

(3) The Oregon Department of Administrative Services on or before January 1 of each year shall publish a list of states that give preference to in-state bidders with the percent increase applied in each such state. The public contracting agency may rely on the names of states and percentages so published in determining the lowest responsible bidder without incurring any liability to any bidder.

(4) The successful bidder shall:

(a) Promptly execute a formal contract.

(b) If the contract is for a public improvement, execute and deliver to the public contracting agency a good and sufficient bond, to be approved by the public contracting agency, in a sum equal to the contract price for the faithful performance of the contract. In lieu of a surety bond, the public contracting agency may permit the successful bidder to submit a cashier's check or certified check in an amount equal to 100 percent of the contract price. ***If the public improvement contract is with a single person to provide both design and construction of a public improvement, the obligation of the surety bond, or the obligation of the bidder on the cashier's check or certified check, for the faithful performance of the contract required by the paragraph, shall be also for the preparation and completion of the design and related services covered under the contract.*** Notwithstanding when a cause of action, claim or demand accrues or arises, the surety or the bidder on the cashier's check or certified check shall not be liable after final completion of the contract, or longer if defined in the contract, for damages of any nature, economic or otherwise and including corrective work, attributable to ***the design aspect of a design-build project***, or for the costs of design revisions needed to implement corrective work.

(5) In cases of emergency, or where the interest or property of the public contracting agency probably would suffer material injury by delay or other cause, the requirement of furnishing a good and sufficient bond for the faithful performance of any public contract may be excused, if a declaration of such emergency is made and concurred in by all members of the governing board of the public contracting agency.

(6) As used in this section:

(a) 'Lowest responsible bidder' means the lowest bidder who has substantially complied with all prescribed public bidding procedures and requirements and who has not been disqualified by the public contracting agency under ORS 279.037.

(b) 'Resident bidder' means a bidder that has paid unemployment taxes or income taxes in this state during the 12 calendar months immediately preceding submission of the bid, has a business address in this state and has stated in the bid whether the bidder is a 'resident bidder' pursuant to this subsection.

(c) 'Nonresident bidder' means a bidder who is not a ' resident bidder' as defined by paragraph (b) of this subsection.

Ore. Admin. Rules 125-065-0100. Contract Provisions.
The Department of Administrative Services shall develop and maintain a standard contract form which shall be used for all architect, engineer, and related service contracts, which can be obtained from the Facilities Division. In using the standard contract form, state agencies shall abide by the following contract provisions:

(1) Except as otherwise required by law, no consultant contract for architectural, engineering, or related services shall be awarded which contains fee provisions or fee schedules that are based on or limited to:

(a) Cost-plus-a-percentage-of-cost; or

(b) A percentage of construction or project costs.

(2) Except in cases of emergency as defined in ORS 279.011(4), ***no building materials, supplies or equipment for any building,*** structure or facility constructed by or for the Department ***shall be sold by*** or purchased from any person or ***firm employed as a consultant by the Division*** to provide architectural, engineering, or related services for such building, structure or facility, ***unless the consultant is providing:***

(a) Construction manager/general contractor services; or

(b) ***Design-build services***, or where that portion of the contract relating to the acquisition of building materials, supplies or equipment was awarded pursuant to applicable law governing the award of such contracts.

Ore. Admin. Rules 137-035-0080. Prohibited Fee Provisions, Purchases.

(1) Except as otherwise required by law, no consultant contract shall be awarded which contains fee provisions or fee schedules that are based on or limited to:

(a) Cost-plus-a-percentage-of-cost; or

(b) A percentage of construction or project costs.

(2) Except in:

(a) Cases of emergency as defined in ORS 279 .011(4); or

(b) The particular instances noted below, no building materials, supplies or equipment for any building, structure or facility constructed by or for a public contracting agency shall be sold by or purchased from any person or firm employed as a consultant by the public contracting agency to provide consultant services for such building, structure or facility. The prohibition stated in this OAR 137-035-0080(1) shall not apply:

(A) Where a consultant is providing architectural or engineering services under a contract with a public contracting agency to provide:

(i) Construction manager/general contractor services; or

(ii) ***Design build services***; or

(B) Where that portion of the consultant contract relating to the acquisition of building materials, supplies or equipment was awarded pursuant to applicable law governing the award of such contracts.

Ore. Admin. Rules. 333-675-0000 Submission of Project Plans and Specifications for Review.

(1) Any party proposing to make certain alterations or additions to an existing health care or residential care facility, or to construct new facilities shall, before commencing such alteration, addition or new construction, submit plans and specifications to the Construction Plans Review Program, Health Division, Suite 550, 800 N.E. Oregon, #21, Portland, OR 97232, for preliminary inspection and

approval or recommendations with respect to compliance with rules authorized by ORS 441.055, 443.420 and for compliance to National Fire Protection Association standards when the facility is also to be Medicare or Medicaid certified.

(2) Project plans and specifications must be submitted for review to the division when the project conforms to one or more of the following criteria:

(a) When a new structure or addition to an existing structure is proposed;

(b) When alterations to a building wing or service area, or a mechanical or electrical system serving it, exceed either 25 percent of equivalent replacement cost, $50,000 for a hospital project, or $25,000 for a nursing home or residential care project;

(c) When a clinically-related health or ancillary service, or dietary or laundry service is to be initiated or relocated within the facility; or when significant changes in the use of rooms within such areas will occur;

(d) When a project involves the correction of licensure or fire and life safety code deficiencies issued by the Oregon Health Division, Senior and Disabled Services Division, or Office of State Fire Marshal, and the Health Division requests submission of plans prior to making such corrections;

(e) When an existing building is to be converted for first time use as a licensed facility or changed in its usage from one licensure category to another having differing physical requirements.

(3) Waivers of review: The division may waive review of construction plans, and all or part of the review fee, despite criteria of section (2) of this rule, when significant division rules do not exist for the project type planned; the facilities will be temporary or mobile; or plans have previously been approved for an identical or very similar facility. For projects similar or identical to prior approved projects, the review fee may be reduced up to 50 percent of the normal fee.

(4) Schematic plans may, at the discretion of the project sponsor, be submitted for review by the division prior to the submission of finalized construction documents. When conversion of building use is proposed, an initial onsite visit may be scheduled in lieu of a schematic plans submission. Schematic plans may accompany a project sponsor's certificate of need application when such review is also required. Payment of a portion of the plans review fee is required in the amount of 1/3 of the full fee described in OAR 333-675-0010 and is to accompany the submission of such plans. A completed copy of Form PR-1 shall also accompany the submission.

(5) A project design conference may be requested by the owner or project architect. Design conferences are intended to assist the sponsor in interpretation of applicable requirements, provide design consultation, and resolve individual problems in meeting applicable standards. Submission of a written program for the project, if available, is also recommended.

(6) Finalized construction drawings and specifications must normally be submitted for review and approval prior to the initiation of related construction. Such submission must be accompanied by payment of the review fee outlined in OAR 333-675-0010. It is suggested that project sponsors contact the division regarding the number of plans required to be submitted, which may vary. A completed copy of Form PR-1 shall also accompany the submission. When the project involves fast track design and construction methods, ***design-build contracts***, or other alternatives which do not allow for submission of full contract documents at the same time, the

division may allow for such irregularities; but it is the responsibility of the project sponsor to negotiate approval of such submission methods prior to the start of construction.

(7) Plans of corrections: Project sponsors shall submit written response to deficiencies identified in construction document reviews, indicating method(s) being used for their correction. If sufficient documentation of corrections is received by the division, subsequent onsite inspections of the project may be waived by the Health Division. Licensure and building code agencies, however, may have final inspection requirements and should be contacted prior to project occupancy.

(8) Time period for reviews: The division will normally issue plan review comments to project sponsors within 15 working days of receipt of plans and the appropriate review fee. When unusual conditions do not allow for review to be completed within this time period, the division will inform the project sponsor of the approximate date such review will be completed.

PENNSYLVANIA

I. Public Procurement Laws.

The Department of General Services of the Commonwealth.

62 Pa. Stat. § 103. Definitions.

Subject to additional definitions contained in subsequent provisions of this part which are applicable to specific provisions of this part, the following words and phrases when used in this part shall have the meanings given to them in this section unless the context clearly indicates otherwise:

"Change order." A written order signed by the contracting officer directing the contractor to make changes which the changes clause of the contract authorizes the contracting officer to order. The change order may be either with the consent of the contractor or a unilateral order by the contracting officer.

"Commonwealth agency." An executive agency, an independent agency or a State-affiliated entity.

"Construction." The process of building, altering, repairing, improving or demolishing any public structure or building or other public improvements of any kind to any public real property. The term does not include the routine operation or maintenance of existing structures, buildings or real property.

"Contract." A type of written agreement, regardless of what it may be called, for the procurement or disposal of supplies, services or construction.

"Contract modification." A written alteration in specifications, delivery point, rate of delivery, period of performance, price, quantity or other provisions of any contract accomplished by mutual action of the parties to the contract.

"Contracting officer." A person authorized to enter into and administer contracts and make written determinations with respect to contracts.

"Department." The Department of General Services of the Commonwealth.

"Design/build contract." A construction contract in which the contractor is responsible for both the design and construction of any public structure or building or other public improvements of any kind to any public real property.

* * *

62 Pa. Stat. § 322. Specific construction powers, duties and procedures.

The following procedure shall apply to construction to be completed by the department which costs more than the amount established by the department under section 514 (relating to small procurements) for construction procurement unless the work is to be done by Commonwealth agency employees or by inmates or patients of a Commonwealth agency institution:

(1) The Commonwealth agency or State-related institution shall notify the department to have plans and specifications for the project.

(2) Promptly after the notice in such cases or promptly after any appropriation made to it becomes available, the department shall, if necessary, select an architect and/or an engineer in accordance with the selection procedures of section 905 (relating to procurement of design professional services) to design the work and prepare the specifications therefor. ***The department may, as an alternative, enter into a design/build contract*** in accordance with section 511 (relating to methods of source selection). ***Such design/build contracts shall be subject to the requirements of this act and the provisions of the act of August 15, 1961 (P.L. 987, No. 442), known as***

the Pennsylvania Prevailing Wage act. Design/build contracts shall also be subject to the act of May 1, 1913 (P.L. 155, No. 104), entitled "An act regulating the letting of certain contracts for the erection, construction, and alteration of public buildings," to the extent provided in paragraph (6).

(3) The department shall enter into a contract with the architect or engineer which shall provide all of the following:

(i) A date for the completion of the plans and specifications.

(ii) That the plans and specifications must meet with the approval of the Commonwealth agency or State-related institution for which the building is being erected, altered or enlarged and, in the case of an administrative board or commission of the Commonwealth agency with which the board or commission is respectively connected, to the extent of the type and general character of the building, design of the floor layouts, medical equipment or other equipment of a nature peculiar to the building for which the plans and specifications are being prepared.

(iii) That the plans, drawings and specifications must be approved by the department.

(iv) That the plans and specifications must be approved by the Department of Labor and Industry, the Department of Health and the Department of Environmental Protection to the extent to which those Commonwealth agencies, respectively, have jurisdiction to require the submission to them for approval of certain features of the building.

The architect or engineer in preparing plans and specifications shall consult with the department and the department shall insist upon the prompt completion of the plans and specifications within the time prescribed in the architect's or engineer's contracts unless it shall specifically agree in writing to an extension thereof.

(4) The enforcement of all contracts provided for by this section shall be under the control and supervision of the department. The department shall have the authority to engage the services of a construction management firm to coordinate the work of the total project. All questions or disputes arising between the department and any contractor with respect to any matter pertaining to a contract entered into with the department or any part thereto or any breach of contract arising thereunder shall be submitted to final and binding arbitration as provided by the terms of the contract, which finding shall be final and not subject to further appeal, or, if not so provided, shall be referred to the Board of Claims as set forth in the act of May 20, 1937 (P.L. 728, No. 193), referred to as the Board of Claims Act, whose decision and award shall be final and binding and conclusive upon all parties thereto except that either party shall have the right to appeal from the decision and award as provided by law.

(5) The department shall examine all bills on account of the contracts entered into under the provisions of this section, and, if they are correct, the department shall certify that the materials have been furnished or that the work or labor has been performed in a workmanlike manner and in accordance with the contract, approve the bills and issue its requisition therefor or forward its certificate to the proper Commonwealth agency or State-related institution as the case may be.

Progress payments and final payments shall not estop the department from pursuing its lawful remedies for defects in workmanship or materials or both and other damages.

(6) For construction contracts where the total construction costs are less than $25,000, the department shall not be required to comply with the act of May 1, 1913

(P.L. 155, No. 104), entitled "An act regulating the letting of certain contracts for the erection, construction, and alteration of public buildings," and the department may award such contracts in accordance with section 511. All projects equal to or exceeding $25,000 shall be subject to the act of May 1, 1913 (P.L. 155, No. 104), entitled "An act regulating the letting of certain contracts for the erection, construction, and alteration of public buildings." ***Whenever the department enters into a single contract for a project***, in the absence of good and sufficient reasons the contractor shall pay each subcontractor, within 15 days of receipt of payment from the department, an amount equal to the percentage of completion allowed to the contractor on the account of the subcontractor's work. The contractor shall also require the subcontractor to make similar payments to his subcontractors.

(7) The department shall have the right to engage the services of any architect or consulting or supervising engineer or engineers whom it may deem necessary for the proper designing of or inspection or supervision of projects constructed, altered or enlarged by the department under this section in accordance with the selection procedures of section 905.

(8) Changes in scope in the plans or specifications, or both, may be made after their approval only with the consent of the Governor and all of the Commonwealth agencies and State-related institutions whose approval of the original plans or specifications, or both, was necessary under this section.

(9) If the appropriation is to a Commonwealth agency, other than the department or State-related institution, the department shall award and enter into the contract as agent for the Commonwealth agency or State-related institution to which the appropriation was made.

II. Architecture & Engineer Licensing Laws.

63 Pa. Stat. § 34.3. Definitions.

The following words and phrases when used in this act shall have, unless the context clearly indicates otherwise, the meanings given to them in this section:

"ACCREDITED SCHOOL." Any educational institution with an academic program in architecture accredited by a national architectural accrediting body or approved by the board.

"ARCHITECT." An individual licensed under this act to engage in the practice of architecture within the Commonwealth of Pennsylvania.

"BOARD." The Architects Licensure Board established by this act.

"CERTIFICATE." The certificate of licensure issued by the board indicating that the individual named therein is an architect.

"Design-build." A project delivery method whereby a design-build entity signs a single contract to provide a combination of architectural and construction services to a client.

"Design-build entity." An entity which provides by single contract to a client a combination of architectural and construction services.

"PRACTICE OF ARCHITECTURE." The rendering or offering to render certain services, hereinafter described, in connection with the design and construction of a structure or group of structures which have as their principal purpose human habitation or use, and the utilization of space within and surrounding such structures. The services referred to in the previous sentence include planning, providing preliminary studies, designs, drawings, specifications, and other design documents, construction management and administration of construction contracts. The foregoing shall not be deemed to include the practice of engineering as such, for which separate registration is required under the provisions of the act of May 23, 1945 (P.L. 913, No. 367), known as the "Professional Engineers Registration Law," excepting only engineering work incidental to the practice of architecture.

"PROFESSIONAL DEGREE." A first professional degree in architecture conferred by an accredited school.

63 Pa. Stat. § 34.13. Firm practice.

(a) An individual architect or a group of architects in Pennsylvania may practice architecture in one of the following forms of architectural firms:

(1) sole proprietorship;
(2) partnership;
(3) professional association;
(4) professional corporation;
(5) business corporation;
(6) limited liability company; or
(7) limited liability partnership.

(b) A partnership may engage in the practice of architecture in Pennsylvania provided it complies with the provisions of 15 Pa.C.S. Ch. 83 (relating to general partnerships) and that it also satisfies the following requirements:

(1) At least two-thirds of the partners are licensed under the laws of any state to practice architecture, engineering or landscape architecture.

(2) At least one-third of the partners are licensed under the laws of any state to practice architecture.

(c) A professional association may engage in the practice of architecture in Pennsylvania provided it complies with the provisions of 15 Pa.C.S. Ch. 93 (relating to professional associations) and must also satisfy the following requirements:

(1) At least two-thirds of the members of the board of governors are licensed under the laws of any state to practice architecture, engineering or landscape architecture.

(2) At least one-third of the members of the board of governors are licensed under the laws of any state to practice architecture.

(d) A professional corporation shall comply with the provisions of 15 Pa.C.S. Ch. 29 (relating to professional corporations).

(e) A business corporation may engage in the practice of architecture in Pennsylvania, provided that it complies with the provisions of 15 Pa.C.S. Pt. II Subpt. B (relating to business corporations) and that it also satisfies the following requirements:

(1) At least two-thirds of the directors are licensed under the laws of any state to practice architecture, engineering or landscape architecture.

(2) At least one-third of the directors are licensed under the laws of any state to practice architecture.

(3) At least two-thirds of all classes of voting stock issued and outstanding at any one time shall be owned by an individual or individuals licensed under the laws of any state to practice architecture, engineering or landscape architecture.

(4) At least one-third of all classes of voting stock issued and outstanding at any one time shall be owned by an individual or individuals licensed under the laws of any state to practice architecture.

(f) A limited liability company may engage in the practice of architecture in this Commonwealth provided that it complies with the provisions of 15 Pa.C.S. Ch. 89 (relating to limited liability companies), and that it also satisfies the following requirements:

(1) At least two-thirds of the members if managed by members or at least two-thirds of the managers if managed by managers are licensed under the laws of any state to practice architecture, engineering or landscape architecture.

(2) At least one-third of the members if managed by members or at least one-third of the managers if managed by managers are licensed under the laws of any state to practice architecture.

(3) At least two-thirds of all classes of voting membership at any one time shall be owned by an individual or individuals licensed under the laws of any state to practice architecture, engineering or landscape architecture.

(4) At least one-third of all classes of voting membership at any one time shall be owned by an individual or individuals licensed under the laws of any state to practice architecture.

(g) A partnership which has elected to be a limited liability partnership and to be governed by the provisions of 15 Pa.C.S. Ch. 82 (relating to registered limited liability partnerships) may engage in the practice of architecture in this Commonwealth provided it complies with the provisions of 15 Pa.C.S. Ch. 82 and that it also satisfies the following requirements:

(1) At least two-thirds of the partners are licensed under the laws of any state to practice architecture, engineering or landscape architecture.

(2) At least one-third of the partners are licensed under the laws of any state to practice architecture.

(h) A partnership, professional association, corporation, limited liability company or limited liability partnership engaged in the practice of architecture having fewer than three partners, governors, shareholders or directors, members or managers under this section shall have at least one partner, governor, shareholder or director, member or manager who is an individual duly certified to practice architecture in the Commonwealth pursuant to this act.

(i) Each project undertaken by a firm engaged in the practice of architecture in the Commonwealth of Pennsylvania must be under the personal supervision of a partner in the case of a partnership or limited liability partnership, a member of the board of governors in the case of a professional association, a shareholder in the case of a professional corporation, a director in the case of a business corporation or member or manager in the case of limited liability company who holds a certificate to engage in the practice of architecture in this Commonwealth pursuant to this act. The seal of such individuals must appear on all drawings, specifications and other design documents issued by the firm for such projects.

(j) The board shall by promulgation of rules and regulations, require any partnership, professional association, professional corporation, business corporation, limited liability company or limited liability partnership practicing architecture in this State to file with the board information concerning its officers, directors, partners, professional association board of governors, beneficial owners, members or managers and such other aspects of its organization as the board deems appropriate. In the administration of this subsection, the board may require each partnership, professional association, professional corporation, business corporation, limited liability company or limited liability partnership to pay an annual filing fee in the amount determined by the board by promulgation of rules and regulations.

(k) No corporation, professional association, partnership, limited liability company or limited liability partnership may perform any act relating to the practice of architecture which an individual architect is prohibited from doing.

(l) Nothing in this section shall be construed to prevent the practice of architecture by an individual as an employee of a person, partnership or corporation which is not an architectural firm, provided such individual holds a certificate to practice architecture in the Commonwealth in conformity with the provisions of this act and the architect's seal is affixed to all documents prepared by him or under his personal supervision for use in this Commonwealth.

(m) An architectural firm authorized to practice under subsections (a) through (l) will be allowed to offer design-build services consistent with the provisions of section 15(9).

63 Pa. Stat. § 34.15. Permitted practices.

Nothing contained in this act shall be construed to prohibit:

(1) Persons acting under the personal supervision of an architect from carrying out their normal duties in the preparation of drawings, specifications and other design and construction documents or in administering construction contracts or in performing construction management services.

(2) Engineers registered under the act of May 23, 1945 (P.L. 913, No. 367), known as the "Professional Engineers Registration Law," as amended, from performing the services for which they are duly licensed or from performing such services included in the practice of architecture as may be incidental to their engineering work.

(3) The preparation of any shop drawings or the performance of construction management services by persons customarily engaged in construction work.

* * *

(9) Design-build services strictly in accordance with the following practices: a design-build entity not authorized to practice under section 13(a) through (l) may offer design-build services if the architectural services in the design-build process are provided in accordance with the following:

(i) An architectural firm which has been authorized to practice architecture in this Commonwealth under section 13(a) through (i) shall independently contract with a design-build entity and is responsible for all material aspects of the practice of architecture as defined in section 3.

(ii) At the time a design-build entity offers a written design-build proposal for a specific project the design-build entity shall give a written disclosure to the client stating an architect will be engaged by and will be contractually responsible to the design-build entity offering design-build services and will not be responsible to the client.

(iii) The design-build entity shall agree that the architect will have direct supervision of the architectural work.

(iv) The contract between the design-build entity and the client shall set forth the name of the architectural firm which will be contractually responsible to the design-build entity for providing architectural services.

RHODE ISLAND
None.

SOUTH CAROLINA

I. Public Procurement Laws.

S.C. Code Ann. § 11-35-3245. Architect, engineer, or construction manager; performance of other work.

No architect or engineer performing design work, or construction manager performing construction management services as described in Section 11-35-2910(3), pursuant to a contract awarded under any provision of this chapter ***may perform other work on that project as a contractor*** or subcontractor either directly or through a business in which he or his architectural engineering or construction management firm has greater than a five percent interest. For purposes of this section, safety compliance and other incidental construction support activities performed by the construction manager are not considered work performed as a contractor or subcontractor. Should the construction manager perform or be responsible for safety compliance and other incidental construction support activities, and these support activities are in noncompliance with the provisions of Section 41-15-210, then the construction management firm is subject to all applicable fines and penalties.

S.C. Code Ann. § 57-3-200. Department of Transportation authorized to enter into agreements to finance construction and maintenance of highways, roads, streets, and bridges.

From the funds appropriated to the Department of Transportation and from any other sources which may be available to the Department, the Department of Transportation may expend such funds as it deems necessary to ***enter into partnership agreements with*** political subdivisions including authorized transportation authorities, and ***private entities to finance, by tolls and other financing methods, the cost of acquiring, constructing, equipping, maintaining and operating highways***, roads, streets and bridges in this State. The provisions of this Section must not be construed to confer upon the Department of Transportation or political subdivisions any power to finance by tolls or other means the acquisition, construction, equipping, maintenance or operation which the Department of Transportation or political subdivisions does not possess under other provisions of this Code.

SOUTH DAKOTA

I. Architect & Engineer Licensing Laws.

S. D. Cod. Laws § 36-18A-1. Definition of terms.
Terms used in this chapter mean:

(1) "Alteration," for the purpose of determining exemptions, any remodeling, renovation, or reconstruction to a building which changes the use, occupancy classification, or occupant load, or the exiting, structural, mechanical, or electrical systems of a building as defined by the building code;

(2) "Architect," any person licensed in good standing and legally authorized to practice architecture in this state;

* * *

(11) "Design-build," a delivery approach in which a project team of design professionals and builders perform design and construction services under contract with a client;

(12) "Engineer," a person who is qualified to practice engineering by reason of special knowledge and use of the mathematical, physical, and engineering sciences and the principles and methods of engineering analysis and design, acquired by engineering education and engineering experience;

* * *

S. D. Cod. Laws § 36-18A-11. Design-build services by contractor.
This chapter does not prohibit a contractor from offering to provide or from providing ***design-build services*** if the architectural and engineering services offered or provided in connection with ***the design-build services*** are rendered by an architect or professional engineer licensed in accordance with this chapter.

II. Licensing Board Rules.

Article 20:38:22:04. Combining architecture, engineering, and construction services.
A business entity offering a combination of architectural and engineering services together with construction services may offer to render architectural and engineering services only if:

(1) An architect and professional engineer licensed in that state participates substantially in all material aspects of the offering;

(2) There is written disclosure at the time of the offering that such architect and professional engineer is engaged by and contractually responsible to such business entity;

(3) Such business entity agrees that such architect and professional engineer will have direct supervision of the work and that such professional services will not be terminated without the consent of the person engaging the business entity; and

(4) The rendering of architectural and engineering services by such professionals will conform to the provisions of this article.

III. Public Procurement Laws.

S. D. Cod. Laws § 5-18-1. Definition of terms.

Terms used in this chapter mean:

(1) "Construction," and "constructed," in addition to their ordinary meaning, repair and alteration;

(2) "Public corporation," the state and all counties, municipalities, public school corporations and all officers, boards or commissions empowered by law to enter into contracts for the construction of public improvements;

(3) "Public improvement," is one the cost of which is payable from taxes or other funds under the control of the public corporation and also local improvements for which special assessments are to be levied;

(4) "Surety," a bond or undertaking executed by a surety company authorized to do business in the State of South Dakota and countersigned by an agent of the company resident in the State of South Dakota, but nothing in this subdivision requires countersignature of a bid bond;

(5) "Acceptance," the formal resolution of a public corporation authorizing the execution of a ***design-build*** contract;

(6) "Design-build contract," a contract between a public corporation and a design-builder to furnish the architecture, engineering, and related services as required, and the labor, materials, and other construction services for a public improvement. A design-build contract may be conditioned upon future refinements in scope and price, and may permit the public corporation to make changes in the scope of the project without invalidating the design-build contract;

(7) "Design-builder," the person that proposes to design and construct a public improvement covered by the procedures of this chapter;

(8) "Performance criteria," the requirements for the public improvement, including as appropriate, capacity, durability, production standards, ingress and egress requirements, building code requirements, or other criteria for the intended use of the public improvement, expressed in performance-oriented specifications or drawings ***suitable to allow the design-builder to make a proposal***;

(9) "Performance criteria developer," the person and the person's subcontractors retained by the public corporation to develop performance criteria;

(10) "Proposal, an offer to enter into ***a design-build contract***;

(11) "Request for proposals," the document or publication whereby a public corporation solicits proposals for ***a design-build contract.***

S. D. Cod. Laws § 5-18-15.
Public work contractor not to act as architect or engineer—Emergency projects exempt.

No person, firm, or corporation may act as architect or engineer and also contractor on any public work in this state ***where the amount to be expended upon such public work exceeds one hundred thousand dollars. Public works of an*** emergency nature which affect the public health and safety of the state and are funded through the use of an emergency appropriation or special appropriation, and full-service firms which specialize in the design, fabrication, and installation of cultural and educational exhibits are exempt from this section.

S. D. Cod. Laws § 5-18-26. Public corporations—Contracts—Conditions.

Notwithstanding any other provisions of law, ***a public corporation may enter into design-build contracts*** for public improvements, if the following conditions are met:

(1) The public corporation shall, prior to issuing any request for proposals, establish and publish procedures for the solicitation and ***award of design-build contracts***. The Bureau of Administration shall establish such procedures for the state. Such procedures shall include, at a minimum, the following:

(a) The procedure to select or designate a performance criteria developer utilizing a qualification based process and the procedure to prepare performance criteria;

(b) The procedures for the preparation and contents of a request for proposals;

(c) The procedure and standards to be used to qualify or prequalify ***design-builders***;

(d) The procedures for preparing and submitting proposals;

(e) The procedures for evaluating proposals;

(f) The procedures for negotiations between the public corporation and those submitting proposals prior to the acceptance of a proposal. Such procedure shall contain safeguards to preserve confidential information and proprietary information supplied by those submitting proposals, consistent with § 5-18-40;

(g) The procedures for awarding and executing ***design-build contracts***;

(h) The procedures for awarding ***design-build contracts*** in the event of public emergencies as defined in § 5-18-3.1; and

(i) The procedures for acting on formal protests relating to the solicitation or ***award of design-build contracts***;

(2) The public corporation shall, for each public improvement or improvements under this section, make a determination that it is in the best interest of the public to enter into ***a design-build contract*** to complete the public improvement. The determination to utilize ***design-build*** and the basis for the determination shall be recorded in the minutes of the meeting of the governing board of the public corporation. In making this determination, ***design-build projects*** should meet one or more of the following criteria:

(a) The public corporation requires a project design and construction time line that is faster than the traditional design/bid/build process would allow;

(b) The complexity of the project requires close coordination of design and construction expertise or an extreme amount of coordination;

(c) The public corporation requires early cost commitments;

(d) The project can be defined at an early stage and the public corporation is able to specify all requirements;

(3) The public corporation shall follow the procedures of other laws governing public improvement construction contracts to the extent such laws are compatible with the ***use of design build contracts.***

S. D. Cod. Laws § 5-18-28. Assignment of design responsibilities.

A ***design-builder*** may sublet responsibility for professional design services to any person licensed and registered to provide professional design services in this state. Nothing in this section limits or eliminates the responsibility or liability of any person registered pursuant to chapter 36-18A, on a ***design-build*** project to the public corporation or other third parties under existing law.

S. D. Cod. Laws § 5-18-30. Sublet of services design—Builders not qualified for.

A ***design-builder*** may contract with the public corporation to provide professional services or construction services for which the ***design-builder*** is not licensed, registered, or qualified to perform, as long as the ***design-builder*** sublets all such services required ***under the design-build contract*** to a licensed, registered, or otherwise qualified person.

S. D. Cod. Laws § 5-18-32. Contract proposal guidelines.
The performance criteria developer may not submit a proposal to enter into the ***design-build contract*** and the ***design-builder*** may not delegate or contract services under the ***design-build contract*** to the performance criteria developer.

S. D. Cod. Laws § 5-18-33. Guidelines for performance criteria developer.
The performance criteria developer shall be either an employee of the public corporation or shall be engaged in accordance with normal procedures for contracting with architects or engineers. With the approval of the public corporation, the developer may delegate or contract for the development of specific aspects of the design criteria to other consultants. The performance criteria developer may be retained at the public corporation's option through to the completion of the ***design-build contract***.

S. D. Cod. Laws § 5-18-35. Request for proposals—Requirements.
A request for proposals shall be prepared for each ***design-build contract*** containing, at a minimum, the following elements:

(1) The identity of the public corporation which will award the ***design-build contract*** and the identity of the performance criteria developer;

(2) The procedures to be followed for submitting proposals, the criteria for evaluation of a proposal and its relative weight, and the procedures for making awards;

(3) The proposed terms and conditions for ***the design-build contract***;

(4) The performance criteria;

(5) A description of the drawings, specifications, or other submittals to be submitted with the proposal, with guidance as to the form and level of completeness of the drawings, specifications, or submittals that is acceptable;

(6) A schedule for planned commencement and completion of the ***design-build*** contract;

(7) Budget limits for ***the design-build contract***, if any;

(8) Affirmative action, disadvantaged business, or set-aside goals or requirements for the ***design-build*** contract, if any;

(9) The qualifications the ***design-builder*** is required to have;

(10) Requirements for performance and payment bonds, and insurance. At a minimum, these requirements shall meet the requirements of § 5-21-1; and

(11) The compensation, if any, to be given to ***design-builders*** submitting proposals who are not awarded the project.

S. D. Cod. Laws § 5-18-37. Prequalification of design-builders—Elements required.
A public corporation may prequalify ***design-builders for design-build contracts*** by advertising requests for letters of interest in accordance with § 5-18-3. Requests for letters of interest shall contain, at a minimum, the following elements:

(1) The identity of the public corporation;

(2) A description of the proposed public improvement;

(3) Budget limits for the proposed public improvement;

(4) The requirements the ***design-builder*** will be required to have; and
(5) The criteria and their relative weight for prequalification.

S. D. Cod. Laws § 5-18-38. Requirement of deposit or bond.

Any proposal, submitted pursuant to this section, shall be accompanied by a deposit or bond meeting the requirements of § 5-18-6. The deposit or security may be forfeited if the proposal is accepted but the ***design-builder*** fails to execute ***the design-build contract.***

S. D. Cod. Laws § 5-18-39. Sealed bid guidelines.

Any proposal shall be sealed and may not be opened until expiration of the time established for making proposals as set forth in the request for proposals. To the extent required by the request for proposals, any proposal shall identify each person to whom the ***design builder*** proposes to sublet obligations under ***the design build contract***. At a minimum, any proposal shall identify each person to whom the ***design-builder*** proposes to sublet any design obligations or general construction obligations. Any person so identified may not be replaced without the approval of the public corporation. Any proposal shall establish a cost of the ***design-build contract*** that may not be exceeded if the proposal is accepted without change. The maximum cost in the proposal may be converted to fixed prices by negotiated agreement between the public corporation and the selected ***design-builder***.

S. D. Cod. Laws § 5-18-41. Review of proposals.

Once received, any proposal shall be submitted to the performance criteria developer for review. Clarifications may be required to ensure conformance of any proposal with the performance criteria. In seeking clarifications, the performance criteria developer may not reveal any aspect of any ***design-builder's proposal*** to any other ***design-builder***. No proposal may be considered until the performance criteria developer issues a written opinion that the proposal is consistent with the performance criteria. Once the performance criteria developer has issued such an opinion, the proposal shall be submitted to the governing body of the public corporation for review and evaluation. No proposal or ***design-build*** contract may be accepted unless the public corporation determines there was adequate competition for such contract.

TENNESSEE

I. Public Procurement Laws.

Tenn. Code Ann. § 12-10-124. Common accounts—Certain purchases for which competitive bidding not required—Contracting by public invitation for proposals.

(a) An authority which operates or maintains more than one (1) project may maintain a common account or accounts and buy supplies and services in common; provided, that expenditures periodically shall be allocated to each project on an appropriate basis.

(b) An authority in the operation, maintenance, and routine repairs of a project may purchase goods, supplies and services which are generally sold to the public by advertised price without the necessity of competitive bidding; provided, that no purchase shall exceed five thousand dollars ($5,000) or any larger limit as shall be allowed for such purchases under the regulations of a municipal corporation with which the authority has contracted.

(c) An authority may contract for services by public invitation for proposals. Such contracts may be for a period of three (3) years. Professional services on an "at-will" basis may be contracted by an interview process. ***Construction contracts for design-build or for remodeling and maintenance may be by public invitation for proposals as well as by competitive bid.***

Tenn. Code Ann. § 4-15-102 Powers and duties.

(a) The commission has the power and authority, except as otherwise provided herein, to approve and supervise all projects involving:

(1) Any improvement to real property funded by public or private funds or both in which the state of Tennessee or any of its departments, institutions or agencies has an interest; and

(2) The demolition of any building or structure located on real property in which the state of Tennessee or any of its departments, institutions or agencies has an interest except for the demolition of any building or structure acquired by the department of transportation for highway rights-of-way.

(b)(1) No contract for the improvement of real property in which the state of Tennessee or any of its departments, institutions or agencies has an interest, and no contract for the demolition of any building or structure on real property in which the state of Tennessee or any of its departments, institutions or agencies has an interest shall be awarded until the plans therefor have been submitted to and approved by the commission.

* * *

TEXAS

I. Public Procurement Laws.

Texas Education Code § 44.031. Purchasing Contracts.

(a) Except as provided by this subchapter, all school district contracts, except contracts for the purchase of produce or vehicle fuel, valued at $25,000 or more in the aggregate for each 12-month period shall be made by the method, of the following methods, that provides the best value for the district:

(1) competitive bidding;

(2) competitive sealed proposals;

(3) a request for proposals, for services other than construction services;

(4) a catalogue purchase as provided by Subchapter B, Chapter 2157, Government Code; [FN1]

(5) an interlocal contract;

(6) ***a design/build contract***;

(7) a contract to construct, rehabilitate, alter, or repair facilities that involves using a construction manager; or

(8) a job order contract for the minor construction, repair, rehabilitation, or alteration of a facility.

(b) Except as provided by this subchapter, in determining to whom to award a contract, the district may consider:

(1) the purchase price;

(2) the reputation of the vendor and of the vendor's goods or services;

(3) the quality of the vendor's goods or services;

(4) the extent to which the goods or services meet the district's needs;

(5) the vendor's past relationship with the district;

(6) the impact on the ability of the district to comply with laws and rules relating to historically underutilized businesses;

(7) the total long-term cost to the district to acquire the vendor's goods or services; and

(8) any other relevant factor that a private business entity would consider in selecting a vendor.

(c) The state auditor may audit purchases of goods or services by the district.

(d) The board of trustees of the district may adopt rules and procedures for the acquisition of goods or services.

(e) To the extent of any conflict, this subchapter prevails over any other law relating to the purchasing of goods and services except a law relating to contracting with historically underutilized businesses.

(f) This section does not apply to a contract for professional services rendered, including services of an architect, attorney, or fiscal agent. A school district may, at its option, contract for professional services rendered by a financial consultant or a technology consultant in the manner provided by Section 2254.003, Government Code, in lieu of the methods provided by this section.

(g) Notice of the time by when and place where the bids or proposals, or the responses to a request for qualifications, will be received and opened shall be published in the county in which the district's central administrative office is located, once a week for at least two weeks before the deadline for receiving bids, proposals, or responses to a request for qualifications. If there is not a newspaper in that county, the advertising shall be published in a newspaper in the county nearest the county seat of the county in which the district's central administrative office is located. In a two-step procurement process, the time and place where the second-step bids, proposals, or responses will be received are not required to be published separately.

* * *

Texas Education Code § 44.036. Design-Build Contracts for Facilities.

(a) In this section:

(1) ***"Design-build contract" means a single contract with a design-build firm for the design and construction of a facility.***

(2) ***"Design-build firm"*** means a partnership, corporation, or other legal entity or team that includes an engineer or architect and builder qualified to engage in building construction in Texas.

(3) "Design criteria package" means a set of documents that provides sufficient information to permit ***a design-build firm*** to prepare a response to a school district's request for qualifications and any additional information requested, including criteria for selection. The design criteria package must specify criteria the district considers necessary to describe the project and may include, as appropriate, the legal description of the site, survey information concerning the site, interior space requirements, special material requirements, material quality standards, conceptual criteria for the project, special equipment requirements, cost or budget estimates, time schedules, quality assurance and quality control requirements, site development requirements, applicable codes and ordinances, provisions for utilities, parking requirements, or any other requirement, as applicable.

(b) A school district may use the ***design-build*** method for the construction, rehabilitation, alteration, or repair of a facility. In using that method and in entering into a contract for the services of a ***design-build firm***, the contracting school district and ***the design-build firm*** shall follow the procedures provided by Subsections (c)-(j).

(c) The district may designate an engineer or architect to act as its representative. If the district's engineer or architect is not a full-time employee of the district, any engineer or architect designated shall be selected on the basis of demonstrated competence and qualifications in accordance with Subchapter A, Chapter 2254, Government Code.

(d) The district shall prepare a request for qualifications that includes general information on the project site, project scope, budget, special systems, selection criteria, and other information that may assist ***potential design-build firms*** in submitting proposals for the project. The district shall also prepare the design criteria package that includes more detailed information on the project. If the preparation of the design criteria package requires engineering or architectural services that constitute the practice of engineering within the meaning of The Texas Engineering Practice Act (Article 3271a, Vernon's Texas Civil Statutes) or the practice of architecture within the meaning of Chapter 478, Acts of the 45th Legislature, Regular Session, 1937 (Article 249a, Vernon's Texas Civil Statutes), those services shall be provided in accordance with the applicable law.

(e) The district shall evaluate statements of qualifications and select ***a design-build firm*** in two phases:

(1) In phase one, the district shall prepare a request for qualifications and evaluate each offeror's experience, technical competence, and capability to perform,

the past performance of the offeror's team and members of the team, and other appropriate factors submitted by the team or firm in response to the request for qualifications, except that cost-related or price-related evaluation factors are not permitted. Each offeror must certify to the district that each engineer or architect that is a member of its team was selected based on demonstrated competence and qualifications. The district shall qualify a maximum of five offerors to submit additional information and, if the district chooses, to interview for final selection.

(2) In phase two, the district shall evaluate the information submitted by the offerors on the basis of the selection criteria stated in the request for qualifications and the results of any interview. The district may request additional information regarding demonstrated competence and qualifications, considerations of the safety and long-term durability of the project, the feasibility of implementing the project as proposed, the ability of the offeror to meet schedules, costing methodology, or other factors as appropriate. The district may not require offerors to submit detailed engineering or architectural designs as part of the proposal. The district shall rank each proposal submitted on the basis of the criteria set forth in the request for qualifications. The district shall select ***the design-build firm*** that submits the proposal offering the best value for the district on the basis of the published selection criteria and on its ranking evaluations. The district shall first attempt to negotiate with the selected offeror a contract. If the district is unable to negotiate a satisfactory contract with the selected offeror, the district shall, formally and in writing, end negotiations with that offeror and proceed to negotiate with the next offeror in the order of the selection ranking until a contract is reached or negotiations with all ranked offerors end.

(f) Following selection of ***a design-build firm*** under Subsection (e), that firm's engineers or architects shall complete the design, submitting all design elements for review and determination of scope compliance to the district or district's engineer or architect before or concurrently with construction.

(g) An engineer shall have responsibility for compliance with the engineering design requirements and all other applicable requirements of The Texas Engineering Practice Act (Article 3271a, Vernon's Texas Civil Statutes). An architect shall have responsibility for compliance with the requirements of Chapter 478, Acts of the 45th Legislature, Regular Session, 1937 (Article 249a, Vernon's Texas Civil Statutes).

(h) The district shall provide or contract for, independently of ***the design-build firm***, the inspection services, the testing of construction materials engineering, and the verification testing services necessary for acceptance of the facility by the district. The district shall select those services for which it contracts in accordance with Section 2254.004, Government Code.

(i) ***The design-build firm*** shall supply a signed and sealed set of construction documents for the project to the district at the conclusion of construction.

(j) A payment or performance bond is not required for, and may not provide coverage for, the portion of ***a design-build contract*** under this section that includes design services only. If a fixed contract amount or guaranteed maximum price has not been determined at the time ***a design-build contract*** is awarded, the penal sums of the performance and payment bonds delivered to the district must each be in an amount equal to the project budget, as specified in the design criteria package. ***The design-build firm*** shall deliver the bonds not later than the 10th day after the date ***the design-build firm*** executes the contract unless ***the design-build firm*** furnishes a bid bond or other financial security acceptable to the district to ensure that ***the design-build firm*** will furnish the required performance and payment bonds when a guaranteed maximum price is established.

Texas Education Code § 51.780. Design-Build Contracts for Facilities.

(a) In this section:

(1) ***"Design-build contract" means a single contract with a design-build firm for the design and construction of a facility.***

(2) "***Design-build firm"*** means a partnership, corporation, or other legal entity or team that includes an engineer or architect and builder qualified to engage in building construction in Texas.

(3) "Design criteria package" means a set of documents that provides sufficient information to permit ***a design-build firm*** to prepare a response to an institution's request for qualifications and any additional information requested, including criteria for selection. The design criteria package must specify criteria the institution considers necessary to describe the project and may include, as appropriate, the legal description of the site, survey information concerning the site, interior space requirements, special material requirements, material quality standards, conceptual criteria for the project, special equipment requirements, cost or budget estimates, time schedules, quality assurance and quality control requirements, site development requirements, applicable codes and ordinances, provisions for utilities, parking requirements, or any other requirement, as applicable.

(b) An institution may use ***the design-build method*** for the construction, rehabilitation, alteration, or repair of a facility. In using that method and in entering into a contract for the services of ***a design-build firm***, the contracting institution and ***the design-build firm*** shall follow the procedures provided by Subsections (c)-(k).

(c) The board may designate an engineer or architect to act as its representative. If the board's engineer or architect is not a full-time employee of the institution, any engineer or architect designated shall be selected on the basis of demonstrated competence and qualifications in accordance with Subchapter A, Chapter 2254, Government Code.

(d) The institution shall prepare a request for qualifications that includes general information on the project site, project scope, budget, special systems, selection criteria, and other information that may assist ***potential design-build firms*** in submitting proposals for the project. The institution shall also prepare the design criteria package that includes more detailed information on the project. If the preparation of the design criteria package requires engineering or architectural services that constitute the practice of engineering within the meaning of The Texas Engineering Practice Act (Article 3271a, Vernon's Texas Civil Statutes) or the practice of architecture within the meaning of Chapter 478, Acts of the 45th Legislature, Regular Session, 1937 (Article 249a, Vernon's Texas Civil Statutes), those services shall be provided in accordance with the applicable law.

(e) The board or its representative shall publish the request for qualifications in a manner prescribed by the board.

(f) The board or its representative shall evaluate statements of qualifications and select ***a design-build firm*** in two phases:

(1) In phase one, the board or its representative shall prepare a request for qualifications and evaluate each offeror's experience, technical competence, and capability to perform, the past performance of the offeror's team and members of the team, and other appropriate factors submitted by the team or firm in response to the request for qualifications, except that cost-related or price-related evaluation factors are not permitted. Each offeror must certify to the board that each engineer or architect that is a member of its team was selected based on demonstrated competence and qualifications. The board or its representative shall qualify a maximum of five offerors to submit additional information and, if the board or its representative chooses, to interview for final selection.

(2) In phase two, the board or its representative shall evaluate the information submitted by the offerors on the basis of the selection criteria stated in the request for qualifications and the results of any interview. The board or its representative may request additional information regarding demonstrated competence and qualifications, considerations of the safety and long-term durability of the project, the feasibility of implementing the project as proposed, the ability of the offeror to meet schedules, costing methodology, or other factors as appropriate. The board or its representative may not require offerors to submit detailed engineering or architectural designs as part of the proposal. The board or its representative shall rank each proposal submitted on the basis of the criteria specified in the request for qualifications. The board or its representative shall select ***the design-build firm*** that submits the proposal offering the best value for the institution on the basis of the published selection criteria and on its ranking evaluations. The board or its representative shall first attempt to negotiate with the selected offeror a contract. If the board or its representative is unable to negotiate a satisfactory contract with the selected offeror, the institution shall, formally and in writing, end all negotiations with that offeror and proceed to negotiate with the next offeror in the order of the selection ranking until a contract is reached or negotiations with all ranked offerors end.

(g) Following selection of ***a design-build firm*** under Subsection (f), that firm's engineers or architects shall complete the design, submitting all design elements for review and determination of scope compliance by the institution's engineer or architect before or concurrently with construction.

(h) An engineer shall have responsibility for compliance with the engineering design requirements and all other applicable requirements of The Texas Engineering Practice Act (Article 3271a, Vernon's Texas Civil Statutes). An architect shall have responsibility for compliance with the requirements of Chapter 478, Acts of the 45th Legislature, Regular Session, 1937 (Article 249a, Vernon's Texas Civil Statutes).

(i) The institution shall provide or contract for, independently of ***the design-build firm***, the inspection services, the testing of construction materials engineering, and the verification testing services necessary for acceptance of the facility by the institution. The institution shall select those services for which it contracts in accordance with Section 2254.004, Government Code.

(j) ***The design-build firm*** shall supply a signed and sealed set of construction documents for the project to the institution at the conclusion of construction.

(k) A payment or performance bond is not required for, and may not provide coverage for, the portion of ***a design-build contract*** under this section that includes design services only. If a fixed contract amount or guaranteed maximum price has not been determined at the time ***a design-build contract*** is awarded, the penal sums of the performance and payment bonds delivered to the institution shall each be in an amount equal to the project budget, as specified in the design criteria package. ***The design-build firm*** shall deliver the bonds not later than the 10th day after the date ***the design-build firm*** executes the contract unless ***the design-build firm*** furnishes a bid bond or other financial security acceptable to the institution to ensure that ***the design-build firm*** will furnish the required performance and payment bonds when a guaranteed maximum price is established.

II. Attorney General Opinions.

This opinion states that a school district may not under the Professional Services Procurement Act contract for "guaranteed program services" that include both construction manager-at-risk and architect and engineering services to be rendered by a single "program manager." Attorney General Opinion JM-940 (1988) is superseded by statute. Under the "guaranteed program management" services all services would be provided by the program manager, except actual construction services, necessary for the construction, improvement, and renovation of the district's school buildings, specifically including architectural, engineering, planning, design, scheduling, procurement, risk management-controlled insurance programs, construction management, quality control, project close-out, and operation and management services for the entire building program. These services would be provided by a single corporate entity, the "program manager." The school district in turn will contract for construction services with one or more contractors and assign those contracts to the program manager. Held that the Professional Services Procurement Act prohibits a school district from selecting a provider of "professional services" or awarding a contract for professional services on the basis of competitive bidding. In addition to architect and engineering services, the program manager will provide "construction management" services. Additionally, the program manager upon or after assignment of the construction contracts will guarantee the project's schedule and budget. Architect and engineering services must be obtained separately from the construction manager-at-risk services, therefore, the school district may not under the Professional Services Procurement Act contract for "guaranteed program services" that include both construction manager-at-risk and architect and engineering services to be rendered by a single "program manager." Sections 44.031 and 44.038 of the Education Code prevail over Attorney General Opinion JM-940 to the extent the opinion is inconsistent with these statutes. Accordingly, Attorney General Opinion JM-940 (1988) is superseded by statute. Tex. Atty. Gen. Op. JC-0224; 2000 WL 659129 (Tex.A.G.)

This opinion addressed the issues of how a school district determines that a design/build contract provides the "best value" to the school district and whether it must follow competitive procurement requirements to let a design/build contract. Held that in the absence of specific procurement procedures or evaluation criteria, the district may evaluate vendors' submissions using the criteria listed in section 44.031(b), including price and qualifications. As to determining who has the best value, nothing in the statutes expressly defines or directs a school district in determining best value. Therefore, a school district should establish, by rule, its own procedure and criteria to determine the purchasing method that will provide the best value in a particular instance. Also stated that a school district selecting the design/build purchasing method must comply with section 44.036 of the Education Code, which effectively adopts a mechanism for competitive procurement—but not competitive bidding—of design/build contracts. Tex. Atty. Gen. Op. JC-0037; 1999 WL 270054 (Tex.A.G.)

This opinion addresses whether the city contract is illegal as a design/build contract, but the Attorney General declined to make that determination. However, as a general matter, the opinion states that a municipality would not have the authority to contract for construction of a project under a design/build procedure if the contract is awarded pursuant to competitive bidding and includes architectural or engineering components. Tex. Atty. Gen. Op. LO 98-060; 1998 WL 486322 (Tex.A.G.)

This opinion addressed whether a county was authorized to build a detention facility using a design/build firm to provide architectural plans and construct the facility and using an architect to oversee construction. Under the facts, the design/build firm would "provide conceptual facility design drawings certified by an architect." The Professional Services Procurement Act prohibits the county from obtaining architectural services by competitive bidding. Thus, a county may not procure by competitive bidding a design/build contract that includes architectural services. The use of an architect, retained pursuant to the Professional Services Procurement Act to oversee construction of the project, would not cure the fact that the county obtained other architectural services, i.e. the certified drawings, by competitive bidding contrary to law. Held that as a general matter, the Professional Services Procurement Act, prohibits a county from obtaining architectural services by competitive bidding. Therefore, a county may not procure by competitive bidding a design/build contract that includes architectural services. Section 351.102 of the Local Government Code authorizes a county commissioners court to enter into a design/build contract with a private vendor to design and construct a detention facility, however, provided that the county adheres to the statute's procedures to obtain the contract and that the contract meets all statutory requirements. Tex. Atty. Gen. Op. LO 96-117; 1996 WL 650176 (Tex.A.G.)

A 1990 opinion looked into whether a public body could award "design/build"' contracts for construction of public buildings on the basis of competitive bids and related questions. In particular, the commissioners court of Montgomery County was considering the construction of certain public works through the award "design/build"' contracts. The key legal issue was the Texas Professional Services Procurement Act, article 664-4, V.T.C.S., and whether the inclusion of architectural design services as a component of the design/build contract violated the prohibition on award of professional services through a competitive bidding process. The opinion state that, "The design/build concept is a significant departure from the traditional arrangement. Under the design/build concept, the owner contracts with a single party for both the design and construction of the entire project. The single party may be either the architect/engineering firm, the contractor, or both acting in tandem as a joint venture . . . The entity contracting with the owner undertakes either to design and build the entire project using the owner's financial resources and present the owner with a finished product or to present the owner with the finished product on a "turn-key"' basis." The Professional Services Procurement Act bars any state agency, political subdivision, county, municipality, district, authority or publicly-owned utility of the State of Texas from making any contract for, or engage the professional services of, any licensed architect, land surveyor, or registered engineer, or any group or association thereof, selected on the basis of competitive bids. Contracts made in violation of any of the provisions of the act are declared void. As a result, the opinion concluded that the public owner lacked authority to make a contract for the construction of public works under the "'design/build"' concept when the resulting contract is awarded pursuant to competitive bidding and includes architectural or engineering services as a component of the contract. Tex. Atty. Gen. Op. JM-1189; 1990 WL 508381 (Tex.A.G.)

UTAH

I. Architect/Engineer Licensing Laws.

Utah Code Ann. § 13-8-2 Contractual limitations of liability arising from services of design professionals prohibited.

(1) As used in this section:

(a) "Agreement" means a contract, promise, covenant, or understanding.

(b) "Contractor" means any person engaged by an owner to develop or assist in the development of the owner's land.

(c) "Design professional" means an architect, engineer, or land surveyor. It includes any other person who, for a fee or other compensation, performs services similar to the services of an architect, engineer, or land surveyor in connection with the development of land.

(d) "Development" means the construction, alteration, repair, maintenance, or improvement of land, including any related moving, demolition, or excavation.

(e) "Land" means any real property, including any building, fixture, improvement, appurtenance, structure, road, highway, or other development.

(f) "Liability" includes liability arising by contract, indemnity, contribution, tort, or otherwise.

(g) "Owner" means the holder of any legal or equitable title or interest in property.

(h) "Subcontractor" means any person engaged by a contractor to develop or assist in the development of land.

(2) An agreement between an owner and a contractor may not limit the owner's or a design professional's liability to the contractor for any claim arising from services performed by the design professional in connection with the development of land. ***This subsection does not apply if the owner and the contractor are the same person or entity or are controlled by the same person or entity.***

(3) An agreement between a contractor and a subcontractor may not limit the owner's or a design professional's liability to the subcontractor for any claim arising from services performed by the design professional in connection with the development of land.

(4) This section does not apply if the design professional is retained under a single contract to perform both the design and the construction of the project, such as in a design-build or turn-key project.

(5) This section may not be construed to affect any limitation of a design professional's liability to an owner or other design professional that may exist in an agreement between the owner and the design professional or between design professionals.

(6) This section does not affect or impair the obligations of agreements in existence as of May 1, 1988.

II. Public Procurement Laws.

Utah Code Ann. § 63-56-21. Use of competitive sealed proposals in lieu of bids — Procedure.

(1) (a) When, according to rules established by the Procurement Policy Board, the chief procurement officer, the head of a purchasing agency, or a designee of either officer above the level of procurement officer determines in writing that the use of

competitive sealed bidding is either not practicable or not advantageous to the state, ***a contract may be entered into by competitive sealed proposals.***

(b) (i) Competitive sealed proposals may be used for the procurement of services of consultants, professionals, and providers as defined by the policy board by rule, whether or not the determination described in this subsection has been made.

(ii) The policy board shall make rules establishing guidelines to assure maximum practicable competition in those procurements, including the relative importance, if any, of the fee to be charged by an offeror.

(iii) The rules may provide that it is either not practicable or not advantageous to the state to procure certain types of supplies, services, or construction by competitive sealed bidding or competitive sealed proposals.

(2) (a) Proposals shall be solicited through a request for proposals.

(b) Public notice of the request for proposals shall be given in accordance with policy board rules.

(3) (a) Proposals shall be opened so as to avoid disclosure of contents to competing offerors during the process of negotiation.

(b) A register of proposals shall be prepared in accordance with policy board rules and shall be open for public inspection after the contract is awarded.

(4) The request for proposals shall state the relative importance of price and other evaluating factors.

(5) (a) As provided in the request for proposals and under policy board rules, discussions may be conducted with responsible offerors who submit proposals for the purpose of assuring full understanding of, and responsiveness to, solicitation requirements.

(b) Offerors shall be accorded fair and equal treatment with respect to any opportunity for discussion and revision of proposals, and revisions may be permitted after submissions and before the contract is awarded for the purpose of obtaining best and final offers.

(c) In conducting discussions, there shall be no disclosure of any information derived from proposals submitted by competing offerors.

(6) (a) Award shall be made to the responsible offeror whose proposal is determined in writing to be the most advantageous to the state, taking into consideration price and the evaluation factors set forth in the request for proposals.

(b) No other factors or criteria shall be used in the evaluation.

(c) The contract file shall contain the basis on which the award is made.

Utah Code Ann. § 63-56-36. Alternative methods of construction contracting management.

(1) (a) Rules shall provide as many alternative methods of construction contracting management as determined to be feasible.

(b) These rules shall:

(i) grant to the chief procurement officer or the head of the purchasing agency responsible for carrying out the construction project the discretion to select the appropriate method of construction contracting management for a particular project; and

(ii) require the procurement officer to execute and include in the contract file a written statement setting forth the facts which led to the selection of a particular method of construction contracting management for each project.

(c) Before choosing a construction contracting management method, the chief procurement officer or the head of the purchasing agency responsible for carrying out the construction project shall consider the following factors:

(i) when the project must be ready to be occupied;

(ii) the type of project;

(iii) the extent to which the requirements of the procuring agencies and the ways in which they are to be met are known;

(iv) the location of the project;

(v) the size, scope, complexity, and economics of the project;

(vi) the source of funding and any resulting constraints necessitated by the funding source;

(vii) the availability, qualification, and experience of state personnel to be assigned to the project and how much time the state personnel can devote to the project; and

(viii) the availability, qualifications, and experience of outside consultants and contractors to complete the project under the various methods being considered.

(2) (a) Rules adopted by state public procurement units and local public procurement units to implement this section may authorize the use of a Construction Manager/General Contractor as one method of construction contracting management.

(b) Those rules shall require that:

(i) the Construction Manager/General Contractor shall be selected using one of the source selection methods provided for in Sections 63-56-20 through 63-56-35.8 of this chapter; and

(ii) when entering into any subcontract that was not specifically included in the Construction Manager/General Contractor's cost proposal submitted under the requirements of Subsection (2)(c)(ii), the Construction Manager/General Contractor shall procure that subcontractor by using one of the source selection methods provided for in Sections 63-56-20 through 63-56-35.8 of this chapter in the same manner as if the subcontract work was procured directly by the state.

(3) Procurement rules adopted by the State Building Board under Subsection (1) for state building construction projects may authorize the use of a design-build provider as one method of construction contracting management.

Utah Code Ann. § 63-56-36.1. Procurement of design-build transportation project contracts.

(1) As used in this section:

(a) "***Design-build transportation project*** contract" means ***the procurement of both the design and construction of a transportation project in a single contract*** with a company or combination of companies capable of providing the necessary engineering services and construction.

(b) "Transportation agency" means:

(i) the Department of Transportation;

(ii) a county of the first or second class, as defined in Section 17-1a-101;

(iii) a municipality of the first class, as defined in Section 10-2-301;

(iv) a public transit district that has more than 200,000 people residing within its boundaries; and

(v) a public airport authority, as created under Title 17A, Chapter 2, Part 15, Airport Authorities.

(2) Except as provided in Subsection (3), a transportation agency may award ***a design-build transportation project contract*** for any transportation project that has an estimated cost of at least $50,000,000 by following the requirements of this section.

(3) The Department of Transportation may:

(a) award a design-build transportation project contract for any transportation project by following the requirements of this section; and

(b) make rules, by following the procedures and requirements of Title 63, Chapter 46a, Utah Administrative Rulemaking Act, establishing requirements for the procurement of its design-build transportation project contracts in addition to those required by this section.

(4) (a) Before entering a ***design-build*** transportation project contract, a transportation agency may issue a request for qualifications to prequalify potential contractors.

(b) Public notice of the request for qualifications shall be given in accordance with policy board rules.

(c) A transportation agency shall require, as part of the qualifications specified in the request for qualifications, that potential contractors at least demonstrate their:

(i) construction experience;

(ii) design experience;

(iii) financial, manpower, and equipment resources available for the project; and

(iv) experience in other ***design-build transportation projects*** with attributes similar to the project being procured.

(d) The request for qualifications shall identify the number of eligible competing proposers that the transportation agency will select to submit a proposal, which must be at least two.

(5) (a) The transportation agency shall:

(i) evaluate the responses received from the request for qualifications;

(ii) select from their number those qualified to submit proposals; and

(iii) invite those respondents to submit proposals based upon the transportation agency's request for proposals.

(b) If the transportation agency fails to receive at least two qualified eligible competing proposers, the transportation agency shall readvertise the project.

(6) The transportation agency shall issue a request for proposals to those qualified respondents that:

(a) includes a scope of work statement constituting an information for proposal that may include:

(i) preliminary design concepts;

(ii) design criteria, needs, and objectives;

(iii) warranty and quality control requirements;

(iv) applicable standards;

(v) environmental documents;

(vi) constraints;

(vii) time expectations or limitations;

(viii) incentives or disincentives; and

(ix) other special considerations;

(b) requires submitters to provide:

(i) a sealed cost proposal;

(ii) a critical path matrix schedule, including cash flow requirements;

(iii) proposal security; and

(iv) other items required by the department for the project; and

(c) may include award of a stipulated fee to be paid to submitters who submit unsuccessful proposals.

(7) The transportation agency shall:

(a) evaluate the submissions received in response to the request for proposals from the prequalified proposers;

(b) comply with rules relating to discussion of proposals, best and final offers, and evaluations of the proposals submitted; and

(c) after considering price and other identified factors, award the contract to the responsible proposer whose proposal is most advantageous to the state.

Utah Code Ann. § 63-56-43.1. Selection as part of design-build or lease.
Notwithstanding any other provision of this chapter, architect-engineer services may be procured under Title 63A, Chapter 5, State Building Board—Division of Facilities Construction and Management, as part of the services obtained in ***a design-build contract*** or as part of the services obtained in a lease contract for real property, provided that the qualifications of those providing the architect-engineer services are part of the consideration in the selection process.

Utah Code Ann. § 63-56-5 Definitions.

As used in this chapter:

(1) "Architect-engineer services" are those professional services within the scope of the practice of architecture as defined in Section 58-3a- 102, or professional engineering as defined in Section 58-22-102.

(2) "Business" means any corporation, partnership, individual, sole proprietorship, joint stock company, joint venture, or any other private legal entity.

(3) "Change order" means a written order signed by the procurement officer, directing the contractor to suspend work or make changes, which the appropriate clauses of the contract authorize the procurement officer to order without the consent of the contractor or any written alteration in specifications, delivery point, rate of delivery, period of performance, price, quantity, or other provisions of any contract accomplished by mutual action of the parties to the contract.

(4) (a) "Construction" means the process of building, renovation, alteration, improvement, or repair of any public building or public work.

(b) "Construction" does not mean the routine operation, routine repair, or routine maintenance of existing structures, buildings, or real property.

(5) (a) "Construction Manager/General Contractor" means any contractor who enters into a contract for the management of a construction project when that contract allows the contractor to subcontract for additional labor and materials that were not included in the contractor's cost proposal submitted at the time of the procurement of the Construction Manager/General Contractor's services.

(b) "Construction Manager/General Contractor" does not mean a contractor whose only subcontract work not included in the contractor's cost proposal submitted as part of the procurement of construction is to meet subcontracted portions of change orders approved within the scope of the project.

(6) "Contract" means any state agreement for the procurement or disposal of supplies, services, or construction.

(7) "Cooperative purchasing" means procurement conducted by, or on behalf of, more than one public procurement unit, or by a public procurement unit with an external procurement unit.

(8) "Cost-reimbursement contract" means a contract under which a contractor is reimbursed for costs which are allowed and allocated in accordance with the contract terms and the provisions of this chapter, and a fee, if any.

(9) (a) "Design-build" means the procurement of architect-engineer services and construction by the use of a single contract with the design-build provider.

(b) This method of design and construction can include the design-build provider supplying the site as part of the contract.

* * *

VERMONT

None.

APPENDIX E

VIRGINIA

I. State Licensing Laws (Contractors and Design Professionals).

Va. Stat. § 2.1-1.6. (Effective until January 1, 2001) State boards.
A. There shall be, in addition to such others as may be established by law, the following permanent collegial bodies affiliated with a state agency within the executive branch:

* * *

Architects, Professional Engineers, Land Surveyors, Certified Interior Designers and Landscape Architects, Board for

* * *

Contractors, Board for

* * *

Design-Build/Construction Management Review Board

* * *

Va. Stat. § 2.1-1.6 (Effective January 1, 2001) State boards.
A. There shall be, in addition to such others as may be established by law,the following perma nent collegial bodies affiliated with a state agency within the executive branch:

* * *

Architects, Professional Engineers, Land Surveyors, Certified Interior Designers and Landscape Architects, Board for

* * *

Contractors, Board for

* * *

Design-Build/Construction Management Review Board

* * *

Va. Stat. § 2.1-483.1:2. Assistance to the Design-Build/Construction Management Review Board.
The Director of the Division of Engineering and Buildings or his designee shall serve as a member of the ***Design-Build/***Construction Management Review Board (the Review Board) created pursuant to § 11-41.2:3. The Division shall provide staff support to the Review Board in the conduct of its duties in accordance with § 11-41.2:4.

Va. Stat. § 9-6.25:2 Policy boards, commissions and councils.
There shall be, in addition to such others as may be designated in accordance with § 9-6.25, the following policy boards, commissions and councils:

* * *

Board for Architects, Professional Engineers, Land Surveyors, Certified Interior Designers and Landscape Architects

* * *

Board for Contractors

* * *

Design-Build/Construction Management Review Board

* * *

Va. Code Ann. § 11-41.2:3 Design-Build/Construction Management Review Board created; membership; terms; staffing; seal.

A. There is hereby created the ***Design-Build***/Construction Management Review Board, hereinafter referred to as the Review Board, which shall be composed of nine members to be appointed by the Governor as follows: the Director of the Division of Engineering and Buildings of the Department of General Services, or his designee; two Class A general contractors selected from a list recommended by the Associated General Contractors; one architect and one engineer selected from a list recommended by the Consulting Engineers Council of Virginia, the Virginia Society of the American Institute of Architects, and the Virginia Society of Professional Engineers; and four representatives of public bodies other than the Commonwealth selected from a list recommended by the Virginia Municipal League and the Virginia Association of Counties. Each such list, other than those recommended as representatives of public bodies, shall include the names of at least four persons who are experienced and actively engaged in competitive sealed bidding or competitive negotiation and in ***design-build*** or construction management procedures. The list for representatives of public bodies shall include at least four persons who are experienced in competitive sealed bidding or competitive negotiation and in ***design-build*** or construction management procedures. The Director of the Division of Engineering and Buildings or his designee shall be a nonvoting member of the Review Board, except in the event of a tie vote of the Review Board.

B. The initial terms of the Review Board shall be as follows: three members shall be appointed for two-year terms, three members shall be appointed for three-year terms and three members shall be appointed for four-year terms.

Thereafter, all appointments shall be for terms of four years, except that appointments to fill vacancies shall be for the unexpired terms. No person shall be eligible to serve for more than two successive full terms, except the Director of the Division of Engineering and Buildings, who shall serve until a successor qualifies.

C. The Review Board shall elect its chairman and vice-chairman from among its members. Members shall receive no compensation for their services as members of the Review Board, but shall receive reasonable expenses.

D. The Review Board shall meet monthly to conduct its business as required by § 11-41.2:4. However, monthly meetings may be canceled by the chairman if there is no business before the Review Board. Five members shall constitute a quorum.

E. Such staff support as is necessary for the conduct of the Review Board's business shall be furnished by the Division of Engineering and Buildings of the Department of General Services pursuant to § 2.1-483.1:2.

F. The Review Board shall adopt a seal by which it shall authenticate its proceedings.

Va. Code Ann. § 54.1–406. [Architect and Engineer Licensing Laws] License required.

A. Unless exempted by §§ 54.1-401, 54.1-402, or § 54.1-402.1, a person shall hold a valid license prior to engaging in the practice of architecture or engineering which includes design, consultation, evaluation or analysis and involves proposed or existing improvements to real property. Unless exempted by § 54.1-401 or § 54.1-402.1, a person shall hold a valid license prior to engaging in the practice of land surveying.

B. Unless exempted by § 54.1-402, any person, partnership, corporation or other entity offering to practice architecture, engineering, or land surveying without being registered or licensed in accordance with the provisions of this chapter, shall be subject to the provisions of § 54.1-111 of this title.

C. Any person, partnership, corporation or other entity which is not licensed or registered to practice in accordance with this chapter and which advertises or promotes through the use of the words "architecture," "engineering" or "land surveying" or any modification or derivative thereof in its name or description of its business activity in a manner that indicates or implies that it practices or offers to practice architecture, engineering or land surveying as defined in this chapter shall be subject to the provisions of § 54.1-111.

* * *

F. Notwithstanding the provisions of this section, a contractor who is licensed pursuant to the provisions of Chapter 11 (§ 54.1-1100 et seq.) of this title shall not be required to be licensed or registered to practice in accordance with this chapter when negotiating design-build contracts or performing services other than architectural, engineering or land surveying services under a design-build contract. The architectural, engineering or land surveying services offered or rendered in connection with such contracts shall only be offered and rendered by an architect, engineer or land surveyor licensed in accordance with this chapter.

54.1-1103. [Contractor Licensing Laws]. Necessity for license; requirements for water well drillers and landscape irrigation contractors; exemption.

A. No person shall engage in, or offer to engage in, contracting work in the Commonwealth unless he has been licensed under the provisions of this chapter. The Board may waive any provision of this chapter for Habitat for Humanity, its local affiliates or subsidiaries, and any other nonprofit organization exempt from taxation under § 501 (c)(3) of the Internal Revenue Code for the purpose of constructing single- family dwellings that will be given to or sold below the appraised value to low-income persons. Prior to a joint venture engaging in, or offering to engage in, contracting work in the Commonwealth, (i) each contracting party of the joint venture shall be licensed under the provisions of this chapter or (ii) a license shall be obtained in the name of the joint venture under the provisions of this chapter.

B. Except as provided in § 54.1-1117, the issuance of a license under the provisions of this chapter shall not entitle the holder to engage in any activity for which a special license is required by law.

* * *

D. Notwithstanding the other provisions of this section, an architect or professional engineer who is licensed pursuant to Chapter 4 (§ 54.1-400 et seq.) of this title shall not be required to be licensed or certified to engage in, or offer to engage in, contracting work or operate as an owner-developer in the Commonwealth in accordance with this chapter when bidding upon or negotiating ***design-build contracts*** or performing services other than construction services ***under a design-build contract.*** However, the construction services offered or rendered in connection with such contracts shall only be offered or rendered by a contractor licensed or certified in accordance with this chapter.

II. Public Procurement Laws.

Va. Code Ann. § 11–37. Definitions.

The words defined in this section shall have the meanings set forth below throughout this chapter.

"Best value," as predetermined in the solicitation, means the overall combination of quality, price, and various elements of required services that in total are optimal relative to a public body's needs.

"Competitive sealed bidding" is a method of contractor selection which includes the following elements:

1. Issuance of a written Invitation to Bid containing or incorporating by reference the specifications and contractual terms and conditions applicable to the procurement. Unless the public body has provided for prequalification of bidders, the Invitation to Bid shall include a statement of any requisite qualifications of potential contractors. When it is impractical to prepare initially a purchase description to support an award based on prices, an Invitation to Bid may be issued requesting the submission of unpriced offers to be followed by an Invitation to Bid limited to those bidders whose offers have been qualified under the criteria set forth in the first solicitation.

2. Public notice of the Invitation to Bid at least ten days prior to the date set for receipt of bids by posting in a designated public area, or publication in a newspaper of general circulation, or both. Public notice may also be published on the Department of General Services' central electronic procurement Web site and other appropriate Web sites. Effective July 1, 2002, posting by state agencies, departments and institutions on the public Internet procurement Web site designated by the Department of General Services shall be required. In addition, bids may be solicited directly from potential contractors. Any additional solicitations shall include businesses selected from a list made available by the Department of Minority Business Enterprise.

3. Public opening and announcement of all bids received.

4. Evaluation of bids based upon the requirements set forth in the invitation, which may include special qualifications of potential contractors, life- cycle costing, value analysis, and any other criteria such as inspection, testing, quality, workmanship, delivery, and suitability for a particular purpose, which are helpful in determining acceptability.

5. Award to the lowest responsive and responsible bidder. When the terms and conditions of multiple awards are so provided in the Invitation to Bid, awards may be made to more than one bidder.

6. Competitive sealed bidding shall not be required for procurement of professional services.

"Competitive negotiation" is a method of contractor selection which includes the following elements:

1. Issuance of a written Request for Proposal indicating in general terms that which is sought to be procured, specifying the factors which will be used in evaluating the proposal and containing or incorporating by reference the other applicable contractual terms and conditions, including any unique capabilities or qualifications which will be required of the contractor.

2. Public notice of the Request for Proposal at least ten days prior to the date set for receipt of proposals by posting in a public area normally used for posting of public notices and by publication in a newspaper or newspapers of general circulation in the area in which the contract is to be performed so as to provide reasonable notice to the maximum number of offerors that can be reasonably anticipated to submit proposals in response to the particular request. Public notice may also be published on the Department of General Services' central electronic procurement Web site and other appropriate Web sites. Effective July 1, 2002, publishing by state agencies, departments and institutions on the public Internet procurement Web site designated by the Department of General Services shall be required. In addition, proposals may be solicited directly from potential contractors.

* * *

"Construction" means building, altering, repairing, improving or demolishing any structure, building or highway, and any draining, dredging, excavation, grading or similar work upon real property.

"Construction management contract" means a contract in which a party is retained by the owner to coordinate and administer contracts for construction services for the benefit of the owner, and may also include, if provided in the contract, the furnishing of construction services to the owner.

"Design-build contract" means a contract between a public body and another party in which the party contracting with the public body agrees to both design and build the structure, roadway or other item specified in the contract. "Goods" means all material, equipment, supplies, printing, and automated data processing hardware and software.

* * *

Va. Code Ann. § 11–41. Methods of procurement.

A. All public contracts with nongovernmental contractors for the purchase or lease of goods, or for the purchase of services, insurance, or construction, shall be awarded after competitive sealed bidding, or competitive negotiation as provided in this section, unless otherwise authorized by law.

B. Professional services shall be procured by competitive negotiation.

C. 1. Upon a determination made in advance by the public body and set forth in writing that competitive sealed bidding is either not practicable or not fiscally advantageous to the public, goods, services, or insurance may be procured by competitive negotiation. The writing shall document the basis for this determination. Upon a written determination made in advance by (i) the Governor or his designee in the case of a procurement by the Commonwealth or by a department, agency or institution thereof or (ii) the local governing body in the case of a procurement by a political subdivision of the Commonwealth, that competitive negotiation is either not practicable or not fiscally advantageous, insurance may be procured through a licensed agent or broker selected in the manner provided for the procurement of things other than professional services in subdivision 3 b of the definition of "competitive negotiation" in § 11-37. The basis for this determination shall be documented in writing.

2. Construction may be procured only by competitive sealed bidding, except that ***competitive negotiation may be used in the following instances*** upon a determination made in advance by the public body and set forth in writing that competitive sealed bidding is either not practicable or not fiscally advantageous to the public, which writing shall document the basis for this determination:

a. By the Commonwealth, its departments, agencies and institutions on a fixed ***price design-build basis or construction management basis under § 11-41.2***;

b. By any public body for the alteration, repair, renovation or demolition of buildings when the contract is not expected to cost more than $500,000;

c. By any public body for the construction of highways and any draining, dredging, excavation, grading or similar work upon real property; * * *

Va. Code Ann. § 11-41.2. Design-build or construction management contracts for Commonwealth authorized.

A. Notwithstanding any other provisions of law, the Commonwealth may enter into contracts on a fixed price ***design-build basis*** or construction management basis in accordance with the provisions of this section and s 2.1-51.31. Procedures to implement this section and any changes to such procedures shall be adopted by the Secretary of Administration after a public hearing and approval by the House Appropriations and Senate Finance Committees.

B. Procurement of construction by ***the design-build method*** shall be a two-step competitive negotiation process. In the first step, offerors shall be requested to submit their qualifications. Based upon the information submitted and any other relevant information which the Commonwealth may obtain, no more than five offerors deemed most suitable for the project shall be selected by the Commonwealth and requested to submit proposals.

C. ***Design-build contracts*** may be used by the Commonwealth only for those types of construction projects designated in the procedures adopted by the Secretary of Administration to implement this section.

Va. Code Ann. § 11-41.2:02 Fixed-price or not-to-exceed-price design- build and construction management contracts for juvenile correctional facilities authorized.

Notwithstanding the provisions of § 11-41.2, but subject to the procedures adopted by the Secretary of Administration to implement the provisions of that section, the Commonwealth may enter into contracts for juvenile correctional facilities on a fixed-price or not-to-exceed-price ***design-build basis*** or construction management basis, including related leases, lease/purchase contracts, agreements relating to the sale of securities to finance such facilities, and similar financing agreements.

Va. Code Ann. § 11-41.2:2 Design-build or construction management contracts for public bodies other than the Commonwealth; eligibility requirements; award of contract; records to be kept.

A. While the competitive sealed bid process remains the preferred method of construction procurement for public bodies in the Commonwealth, any public body other than the Commonwealth may enter into a contract for construction on a fixed price or not-to- exceed price ***design-build*** or construction management basis provided the public body complies with the requirements of this section and has obtained the approval of the ***Design-Build***/Construction Management Review Board (the Review Board) pursuant to §11-41.2:5. Prior to making a determination as to the use of ***design-build*** or construction management for a specific construction project, the public body shall have in its employ or under contract a licensed architect or engineer with professional competence appropriate to the project who shall advise the public body regarding the use of ***design-build*** or construction management for that project and who shall assist the public body with the preparation of the Request for Proposal and the evaluation of such proposals. Prior to issuing a Request for Proposal for any ***design-build*** or construction management contract for a specific construction project, the public body shall:

1. Have adopted, by ordinance or resolution, written procedures governing the selection, evaluation and award of ***design-build*** and construction management contracts. Such procedures shall be consistent with those described in this chapter for the procurement of nonprofessional services through competitive negotiation. Such procedures shall also require Requests for Proposals to include and define the criteria of such construction project in areas such as site plans; floor plans; exterior elevations; basic building envelope materials; fire protection information plans; structural, mechanical (HVAC), and electrical systems; and special telecommunications; and may define such other requirements as the public body determines appropriate for that particular construction project. Except as may otherwise be approved by the Review Board, such procedures for:

a. ***Design-build construction projects*** shall include a two-step competitive negotiation process consistent with the Review Board's regulations.

b. Construction management projects shall include selection procedures and required construction management contract terms consistent with the Review Board's regulations.

2. Have documented in writing that for a specific construction project (I) a ***design-build*** or construction management contract is more advantageous than a competitive sealed bid construction contract; (ii) there is a benefit to the public body by using a ***design-build*** or construction management contract; and (iii) competitive sealed bidding is not practical or fiscally advantageous.

B. Once approved by the Review Board in accordance with § 11-41.2:5, the public body may procure a ***design-build*** or construction management contract for the specific construction project presented to the Review Board. Unless otherwise specified in the Request for Proposal, such contract shall be awarded to the fully qualified offeror who submits an acceptable proposal at the lowest cost in response to the Request for Proposal.

C. The public body shall provide information as requested by the Review Board to allow post-project evaluation by the Review Board.

Va. Code Ann. § 11-41.2:4 Duties of the Design-Build/Construction Management Review Board; transitional provisions relating to regulations.

A. The Review Board shall have the following duties:

1. Review submissions by public bodies other than the Commonwealth of draft or adopted ordinances or resolutions to determine if the process for the selection, evaluation and award of a ***design-build*** or construction management contract is in compliance with the provisions of subdivision A 1 of § 11-41.2:2;

2. Determine if the public body has complied with the provisions of § 11-41.2:2 relating to the retention of a licensed architect or engineer;

3. Review the findings and the basis of such findings submitted by the public body to determine if the public body has complied with the requirements of §11-41.2:2 and that the findings made by the public body pursuant to § 11-41.2:2 are not unreasonable;

4. Develop guidelines relating to the documents and information to be reviewed by the Review Board;

5. Make post-project evaluations of construction projects procured by ***design-build*** or construction management contracts entered into by public bodies other than the Commonwealth, including cost and time savings, effectiveness of the selection, evaluation and award of such contracts, and the benefit to the public body; and

6. Report to the General Assembly and the Governor on or before December 1, 1999, and again on or before December 31, 2002, concerning the Review Board's evaluation of and findings regarding all ***design-build*** and construction management construction undertaken by public bodies other than the Commonwealth since July 1, 1996, and any recommendations relating to future use of ***design-build*** or construction management contracts by such public bodies.

B. The Review Board shall adopt regulations, as it deems appropriate, based on the substantive requirements of Chapter IX of the Capital Outlay Manual of the Commonwealth, for a two-step competitive negotiation process which shall be applied to ***design-build*** and construction management projects undertaken by public bodies other than the Commonwealth. For construction management projects, such regulations shall also include applicable provisions of the Required Construction Management Contract Terms of the Capital Outlay Manual. Such regulations shall also allow the Review Board to approve deviations from provisions of the Capital Outlay Manual that it deems appropriate. Such regulations shall be adopted in accordance with the Administrative Process Act.

Va. Code Ann. § 11-41.2:5 Review by the Review Board for design-build or construction management approval; effect of disapproval; review of Review Board decision.

The Review Board shall conduct such inquiry it deems appropriate and may require the submission of additional documents or information by the public body, in a form prescribed by the Review Board, to determine if the public body has complied with the provisions of § 11-41.2:2.

Within sixty days of the receipt of the request for review, the Review Board shall render a decision, unless a different timetable is agreed to by the public body. If the Review Board determines that the public body has complied with the provisions of § 11-41.2:2 and the findings made by the public body pursuant to subdivision A 2 of § 11-41.2:2 are not unreasonable, the Review Board shall approve such use. If the Review Board determines that (I) the public body has not complied with the provisions of § 11-41.2:2 or (ii) the findings made by the public body pursuant to subdivision A 2 of § 11-41.2:2 are unreasonable, it shall disapprove such use, and the public body shall not use a ***design-build*** or construction management contract to procure construction for the proposed project. If no decision is made by the Review Board within the sixty-day period or as otherwise agreed to by the public body, the proposed use of a ***design-build*** or construction management contract shall be deemed approved.

Any public body other than the Commonwealth which has been aggrieved by any action of the Review Board shall be entitled to a review of such action. Appeals from such actions shall be in accordance with the provisions of the Administrative Process Act (§ 9-6.14:1 et seq.).

Va. Code Ann. § 11-46 Prequalification generally; prequalification for construction.

A. Prospective contractors may be prequalified for particular types of supplies, services, insurance or construction, and consideration of bids or proposals limited to prequalified contractors. Any prequalification procedure shall be established in writing and sufficiently in advance of its implementation to allow potential contractors a fair opportunity to complete the process.

B. Any prequalification of prospective contractors for construction by a public body subsequent to July 1, 1995, shall be pursuant to a prequalification process for construction projects adopted by the public body. Such process shall be consistent with the provisions of this subsection.

The application form used in such process shall set forth the criteria upon which the qualifications of prospective contractors will be evaluated. The application form shall request of prospective contractors only such information as is appropriate for an objective evaluation of all prospective contractors pursuant to such criteria. Such form shall allow the prospective contractor seeking prequalification to request, by checking the appropriate box, that all information voluntarily submitted by the contractor pursuant to this subsection shall be considered a trade secret or proprietary information subject to the provisions of subsection D of § 11-52.

In all instances in which the public body requires prequalification of potential contractors for construction projects, advance notice shall be given of the deadline for the submission of prequalification applications. The deadline for submission shall be sufficiently in advance of the date set for the submission of bids for such construction so as to allow the procedures set forth in this subsection to be accomplished.

At least thirty days prior to the date established for submission of bids or proposals under the procurement of the contract for which the prequalification applies, the public body shall advise in writing each contractor which submitted an application whether that contractor has been prequalified. In the event that a contractor is denied prequalification, the written notification to such contractor shall state the reasons for such denial of prequalification and the factual basis of such reasons.

A decision by a public body denying prequalification under the provisions of this subsection shall be final and conclusive unless the contractor appeals the decision as provided in § 11-63.

A public body may deny prequalification to any contractor only if the public body finds one of the following:

1. The contractor does not have sufficient financial ability to perform the contract that would result from such procurement. If a bond is required to ensure performance

of a contract, evidence that the contractor can acquire a surety bond from a corporation included on the United States Treasury list of acceptable surety corporations in the amount and type required by the public body shall be sufficient to establish the financial ability of such contractor to perform the contract resulting from such procurement;

2. The contractor does not have appropriate experience to perform the construction project in question;

3. The contractor or any officer, director or owner thereof has had judgments entered against him within the past ten years for the breach of contracts for governmental or nongovernmental construction, including, but not limited to, ***design-build*** or construction management;

* * *

Local Correctional Facilities.

Va. Code Ann. § 53.1-95.18 Design-build contracts.

An authority created pursuant to this article may enter into a contract for a jail on a fixed price or not-to-exceed price ***design-build basis*** or construction management basis in accordance with procedures consistent with those described in the Virginia Public Procurement Act (§ 11-35 et seq.) for procurement of nonprofessional services through competitive negotiation. The authority may authorize payment to no more than three responsive bidders who are not awarded ***the design-build contract*** if the authority determines that such payment is necessary to promote competition. The authority shall not be required to award ***a design-build contract*** to the lowest bidder but may consider price as one factor in evaluating the proposals received. The authority shall maintain adequate records to allow post-project evaluation by the Commonwealth.

Juvenile Corrections Private Management Act.

Va. Code Ann. § 66-25.7 Fixed-price or not-to-exceed-price design-build-operate and related contracts authorized.

Notwithstanding any other provisions of law to the contrary, but in accordance with the procedures consistent with those described in the Virginia Public Procurement Act (§ 11-35 et seq.) for procurement of nonprofessional services through competitive negotiation, the Director may enter into ***design-build-operate contracts for juvenile correctional facilities*** on a fixed-price or not-to-exceed- price basis, including related leases, lease/purchase contracts, agreements relating to the sale of securities to finance such facilities, and similar financing agreements and agreements for correctional services. For the purposes of this section, ***"design-build-operate contract"*** means a contract between the Commonwealth and another party in which the party contracting with the Commonwealth agrees to ***(i) design, build and operate the juvenile correctional facility or (ii) design and build the juvenile correctional facility*** where the facility is to be operated by a third party. The Director shall maintain adequate records to allow post-project evaluation.

WASHINGTON

I. Architect and Engineer Licensing Laws.

Wash. Rev. Code § 18.08.410. Application of chapter.

This chapter shall not affect or prevent:

(1) The practice of naval architecture, landscape architecture, engineering, space planning, interior design, or any legally recognized profession or trade by persons not registered as architects;

(2) Drafters, clerks, project managers, superintendents, and other employees of architects, engineers, naval architects, or landscape architects from acting under the instructions, control, or supervision of their employers;

(3) The construction, alteration, or supervision of construction of buildings or structures by contractors or superintendents employed by contractors or the preparation of shop drawings in connection therewith;

(4) Owners or contractors from engaging persons who are not architects to observe and supervise construction of a project;

(5) Any person from doing design work including preparing construction contract documents and administration of the construction contract for the erection, enlargement, repair, or alteration of a structure or any appurtenance to a structure, if the structure is to be used for a residential building of up to and including four dwelling units or a farm building or is a structure used in connection with or auxiliary to such residential building or farm building such as a garage, barn, shed, or shelter for animals or machinery;

(6) Any person from doing design work including preparing construction contract documents and administering the contract for construction, erection, enlargement, alteration, or repairs of or to a building of any occupancy up to four thousand square feet of construction;

(7) Design-build construction by registered general contractors if the structural design services are performed by a registered engineer;

(8) Any person from designing buildings or doing other design work for any structure prior to the time of filing for a building permit; or

(9) Any person from designing buildings or doing other design work for structures larger than those exempted under subsections (5) and (6) of this section, if the plans, which may include such design work, are stamped by a registered engineer or architect.

II. Public Procurement Laws.

Wash. Rev. Code § 39.10.020. Definitions. (Effective until July 1, 2001.)

Unless the context clearly requires otherwise, the definitions in this section apply throughout this chapter.

(1) "Alternative public works contracting procedure" means the ***design-build*** and the general contractor/construction manager contracting procedures authorized in RCW 39.10.050 and 39.10.060, respectively.

(2) "Public body" means the state department of general administration; the University of Washington; Washington State University; every city with a population greater than one hundred fifty thousand; every city authorized to use ***the design-build procedure*** for a water system demonstration project under RCW 39.10.065(3); every

county with a population greater than four hundred fifty thousand; and every port district with a population greater than five hundred thousand.

(3) "Public works project" means any work for a public body within the definition of the term public work in RCW 39.04.010.

Wash. Rev. Code § 39.10.050. Department of General Administration.

Design-build procedure—Which public bodies may use. (Effective until July 1, 2001.)

(1) Notwithstanding any other provision of law, and after complying with RCW 39.10.030, the following public bodies ***may utilize the design-build procedure*** of public works contracting for public works projects authorized under this section: The state department of general administration; the University of Washington; Washington State University; every city with a population greater than one hundred fifty thousand; every county with a population greater than four hundred fifty thousand; and every port district with a population greater than five hundred thousand. The authority granted to port districts in this section is in addition to and does not affect existing contracting authority under RCW 53.08.120 and 53.08.130. For the purposes of this section, ***"design-build procedure"*** means a contract between a public body and another party in which ***the party agrees to both design and build the facility***, portion of the facility, or other item specified in the contract.

(2) Public bodies authorized under this section may utilize the ***design-build*** procedure for public works projects valued over ten million dollars where:

(a) The construction activities or technologies to be used are highly specialized and a ***design-build approach*** is critical in developing the construction methodology or implementing the proposed technology;

(b) The project design is repetitive in nature and is an incidental part of the installation or construction; or

(c) Regular interaction with and feedback from facilities users and operators during design is not critical to an effective facility design.

(3) Public bodies authorized under this section may also use ***the design-build procedure*** for the following projects that meet the criteria in subsection (2)(b) and (c) of this section:

(a) The construction or erection of preengineered metal buildings or prefabricated modular buildings, regardless of cost; or

(b) The construction of new student housing projects valued over five million dollars.

(4) Contracts for ***design-build services*** shall be awarded through a competitive process utilizing public solicitation of proposals for ***design-build services***. The public body shall publish at least once in a legal newspaper of general circulation published in or as near as possible to that part of the county in which the public work will be done, a notice of its request for proposals for ***design-build services*** and the availability and location of the request for proposal documents. The request for proposal documents shall include:

(a) A detailed description of the project including programmatic, performance, and technical requirements and specifications, functional and operational elements, minimum and maximum net and gross areas of any building, and, at the discretion of the public body, preliminary engineering and architectural drawings;

(b) The reasons for using the ***design-build procedure***;

(c) A description of the qualifications to be required of the proposer including, but not limited to, submission of the proposer's accident prevention program;

(d) A description of the process the public body will use to evaluate qualifications and proposals, including evaluation factors and the relative weight of factors. Evaluation factors shall include, but not be limited to: Proposal price; ability of professional personnel; past performance on similar projects; ability to meet time and budget requirements; ability to provide a performance and payment bond for the project; recent, current, and projected work loads of the firm; location; and the concept of the proposal;

(e) The form of the contract to be awarded;

(f) The maximum allowable construction cost and minority and women enterprise total project goals;

(g) The amount to be paid to finalists submitting best and final proposals who are not awarded a ***design-build contract***; and

(h) Other information relevant to the project.

(5) The public body shall establish a committee to evaluate the proposals based on the factors, weighting, and process identified in the request for proposals. Based on its evaluation, the public body shall select not fewer than three nor more than five finalists to submit best and final proposals. The public body may, in its sole discretion, reject all proposals. ***Design-build contracts*** shall be awarded using the procedures in (a) or (b) of this subsection.

(a) Best and final proposals shall be evaluated and scored based on the factors, weighting, and process identified in the initial request for proposals. The public body may score the proposals using a system that measures the quality and technical merits of the proposal on a unit price basis. Final proposals may not be considered if the proposal cost is greater than the maximum allowable construction cost identified in the initial request for proposals. The public body shall initiate negotiations with the firm submitting the highest scored best and final proposal. If the public body is unable to execute a contract with the firm submitting the highest scored best and final proposal, negotiations with that firm may be suspended or terminated and the public body may proceed to negotiate with the next highest scored firm. Public bodies shall continue in accordance with this procedure until a contract agreement is reached or the selection process is terminated.

(b) If the public body determines that all finalists are capable of producing plans and specifications that adequately meet project requirements, the public body may award the contract to the firm that submits the responsive best and final proposal with the lowest price.

(6) The firm awarded the contract shall provide a performance and payment bond for the contracted amount. The public body shall provide appropriate honorarium payments to finalists submitting best and final proposals who are not awarded a ***design-build contract***. Honorarium payments shall be sufficient to generate meaningful competition among potential proposers on ***design-build projects***.

Wash. Rev. Code § 39.10.065. Demonstration projects—Contract deadline—Transfer of authority to other public body. (Effective until July 1, 2001.)

(1) In addition to the projects authorized in RCW 39.10.050 and 39.10.060, public bodies may use the general contractor/construction manager or ***design-build***

procedure for demonstration projects valued between three million dollars and ten million dollars as follows:

(a) Three demonstration projects by the department of general administration; and

(b) One demonstration project by each of the public bodies authorized in RCW 39.10.020(2) other than the department of general administration.

(2) Public bodies shall give weight to proposers' experience working on projects valued between three million dollars and ten million dollars in the evaluation process for the selection of a general contractor/construction manager or ***design-build firm*** for demonstration projects authorized in subsection (1) of this section.

(3) Cities which supply water to over three hundred fifty thousand people may use ***the design-build procedure*** for one water system demonstration project valued over ten million dollars. Use of the ***design-build procedure*** shall be deemed to effect compliance with the requirement for competitive bids under RCW 43.155.060.

(4) All contracts authorized under this section must be entered into before July 1, 1999.

(5) In the event that a public body determines not to perform a demonstration project using its authority under this section, it may transfer its authority to another public body.

Baseball Stadium Procurement.

Wash. Rev. Code § 39.10.120. Application of chapter.

(1) Except as provided in subsections (2) and (3) of this section, the alternative public works contracting procedures authorized under this chapter are limited to public works contracts signed before July 1, 2001. Methods of public works contracting authorized by RCW 39.10.050 and 39.10.060 shall remain in full force and effect until completion of contracts signed before July 1, 2001.

(2) For the purposes of a baseball stadium as defined in RCW 82.14.0485, ***the design-build contracting procedures*** under RCW 39.10.050 shall remain in full force and effect until completion of contracts signed before December 31, 1997.

(3) For the purposes of a stadium and exhibition center, as defined in RCW 36.102.010, ***the design-build contracting procedures*** under RCW 39.10.050 shall remain in full force and effect until completion of contracts signed before December 31, 2002.

Wash. Rev. Code § 36.58.090. Contracts with vendors for solid waste handling systems, plants, sites, or facilities—Requirements—Vendor selection procedures.

(1) Notwithstanding the provisions of any county charter or any law to the contrary, and in addition to any other authority provided by law, ***the legislative authority of a county may contract with one or more vendors for one or more of the design, construction, or operation of, or other service related to, the solid waste handling systems, plants, sites, or other facilities*** in accordance with the procedures set forth in this section. When a contract for design services is entered into separately from other services permitted under this section, procurement shall be in accord with chapter 39.80 RCW. For the purpose of this chapter, the term "legislative authority" shall mean the board of county commissioners or, in the case of a home rule charter county, the official, officials, or public body designated by the charter to perform the functions authorized therein.

(2) If the legislative authority of the county decides to proceed with the consideration of qualifications or proposals for services from vendors, the county shall publish

notice of its requirements and request submission of qualifications statements or proposals. The notice shall be published in the official newspaper of the county at least once a week for two weeks not less than sixty days before the final date for the submission of qualifications statements or proposals. The notice shall state in summary form (a) the general scope and nature of the design, construction, operation, or other service, (b) the name and address of a representative of the county who can provide further details, (c) the final date for the submission of qualifications statements or proposals, (d) an estimated schedule for the consideration of qualifications, the selection of vendors, and the negotiation of a contract or contracts for services, (e) the location at which a copy of any request for qualifications or request for proposals will be made available, and (f) the criteria established by the legislative authority to select a vendor or vendors, which may include but shall not be limited to the vendor's prior experience, including design, construction, or operation of other similar facilities; respondent's management capability, schedule availability and financial resources; cost of the services, nature of facility design proposed by the vendor; system reliability; performance standards required for the facilities; compatibility with existing service facilities operated by the public body or other providers of service to the public; project performance guarantees; penalty and other enforcement provisions; environmental protection measures to be used; consistency with the applicable comprehensive solid waste management plan; and allocation of project risks.

(3) If the legislative authority of the county decides to proceed with the consideration of qualifications or proposals, it may designate a representative to evaluate the vendors who submitted qualifications statements or proposals and conduct discussions regarding qualifications or proposals with one or more vendors. ***The legislative authority or representative may request submission of qualifications statements and may later request more detailed proposals from one or more vendors*** who have submitted qualifications statements, or the representative may request detailed proposals without having first received and evaluated qualifications statements. The representative shall evaluate the qualifications or proposals, as applicable. If two or more vendors submit qualifications or proposals that meet the criteria established by the legislative authority of the county, discussions and interviews shall be held with at least two vendors. Any revisions to a request for qualifications or request for proposals shall be made available to all vendors then under consideration by the city or town and shall be made available to any other person who has requested receipt of that information.

(4) Based on criteria established by the legislative authority of the county, the representative shall recommend to the legislative authority a vendor or vendors that are initially ***determined to be the best qualified to provide one or more of the design, construction, or operation of, or other service related to, the proposed project or services. The legislative authority may select one or more qualified vendors for one or more of the design, construction, or operation of, or other service related to, the proposed project or services.***

(5) The legislative authority or its representative may attempt to negotiate a contract with the vendor or ***vendors selected for one or more of the design, construction, or operation*** of, or other service related to, the proposed project or services on terms that the legislative authority determines to be fair and reasonable and in the best interest of the county. If the legislative authority or its representative is unable to negotiate such a contract with any one or more of the vendors first selected on terms that it determines to be fair and reasonable and in the best interest of the

county, negotiations with any one or more of the vendors shall be terminated or suspended and another qualified vendor or vendors may be selected in accordance with the procedures set forth in this section. If the legislative authority decides to continue the process of selection, negotiations shall continue with a qualified vendor or vendors in accordance with this section at the sole discretion of the legislative authority until an agreement is reached with one or more qualified vendors, or the process is terminated by the legislative authority. The process may be repeated until an agreement is reached.

(6) Prior to entering into a contract with a vendor, the legislative authority of the county shall make written findings, after holding a public hearing on the proposal, that it is in the public interest to enter into the contract, that the contract is financially sound, and that it is advantageous for the county to use this method for awarding contracts compared to other methods.

(7) Each contract shall include a project performance bond or bonds or other security by the vendor that in the judgment of the legislative authority of the county is sufficient to secure adequate performance by the vendor.

(8) The provisions of chapters 39.12, 39.19, and 39.25 RCW shall apply to a contract entered into under this section to the same extent as if the systems and plants were owned by a public body.

(9) The vendor selection process permitted by this section shall be supplemental to and shall not be construed as a repeal of or limitation on any other authority granted by law.

(10) The alternative selection process provided by this section may not be used in the selection of a person or entity to construct a publicly owned facility for the storage or transfer of solid waste or solid waste handling equipment unless the facility is either (a) privately operated pursuant to a contract greater than five years, or (b) an integral part of a solid waste processing facility located on the same site. Instead, the applicable provisions of RCW 36.32.250 and chapters 39.04 and 39.30 RCW shall be followed.

Wash. Rev. Code § 43.34.080. Capitol campus design advisory committee—Generally.

(1) The capitol campus design advisory committee is established as an advisory group to the capitol committee and the director of general administration to review programs, planning, design, and landscaping of state capitol facilities and grounds and to make recommendations that will contribute to the attainment of architectural, aesthetic, functional, and environmental excellence in design and maintenance of capitol facilities on campus and located in neighboring communities.

* * *

(4) The advisory committee shall review plans and designs affecting state capitol facilities as they are developed. The advisory committee's review shall include:

(a) The process of solicitation and selection of appropriate professional design services including design-build proposals;

(b) Compliance with the capitol campus master plan and design concepts as adopted by the capitol committee;

(c) The design, siting, and grouping of state capitol facilities relative to the service needs of state government and the impact upon the local community's economy, environment, traffic patterns, and other factors;

(d) The relationship of overall state capitol facility planning to the respective comprehensive plans for long-range urban development of the cities of Olympia, Lacey, and Tumwater, and Thurston county; and

(e) Landscaping plans and designs, including planting proposals, street furniture, sculpture, monuments, and access to the capitol campus and buildings.

Department of Transportation.

Wash. Rev. Code § 47.20.750. Design-build—Finding. (Expires April 30, 2001.)

The legislature finds and declares that a contracting procedure that facilitates construction of transportation facilities in a more timely manner may occasionally be needed to ensure that construction can proceed simultaneously with the design of the facility.

Wash. Rev. Code § 47.20.755. Design-build—Procedures—Definition. (Expires April 30, 2001.)

The department of transportation shall develop, by January 1, 1999, a process for awarding competitively bid highway construction contracts for projects over ten million dollars that may be constructed using a ***design-build procedure***. As used in this section, ***"design-build procedure" means a method of contracting under which the department of transportation contracts with another party for such party to both design and build the structures, facilities, and other items specified in the contract.***

The process developed by the department shall, at a minimum, include the scope of services required under the ***design-build procedure***, contractor prequalification requirements, criteria for evaluating technical information and project costs, contractor selection criteria, and an issue resolution procedure. If a request for proposal will be used, the requirements of RCW 39.10.050 (4), (5), and (6) apply.

Wash. Rev. Code § 47.20.760. Design-build—Other statutes applicable. (Expires April 30, 2001.)

RCW 39.10.080, 39.10.090, and 39.10.100 and the notice requirements of RCW 39.10.030 (2), (3), and (4) apply to chapter 195, Laws of 1998.

Wash. Rev. Code § 47.20.765. Design-build—When used. (Expires April 30, 2001.)

The department may use ***the design-build procedure*** for public works projects over ten million dollars where:

(1) The construction activities are highly specialized and ***a design-build approach***, as defined in RCW 47.20.755, is critical in developing the construction methodology; or

(2) The projects selected provide opportunity for greater innovation and efficiencies between the designer and the builder; or

(3) Significant savings in project delivery time would be realized.

Wash. Rev. Code § 47.20.770. Design-build—Release of sureties. (Expires April 30, 2001.)

Sureties are not responsible for damages, including corrective work, attributable to the design aspect of a design-build project.

Wash. Rev. Code § 47.20.775. Design-build—Demonstration projects. (Expires April 30, 2001.)

A demonstration program consisting of two projects must be implemented using ***the design-build method of contracting,*** as prescribed under a process developed by the department of transportation under RCW 47.20.755. The department shall select projects valued over ten million dollars for the demonstration program.

The department shall present progress reports on the demonstration projects to the legislative transportation committee and the alternative public works oversight committee during the course of the performance of the demonstration projects. The department shall present a final detailed report to the legislative transportation committee within one year of completion of the demonstration projects. The report must detail the advantages and disadvantages ***of the design-build construction process*** and make recommendations for possible changes in law, and in how the process may be used for future department projects.

WEST VIRGINIA

W. Va. Code § 18-5-9a. Energy Contracts for County Boards of Education. Energy-savings contracts.

(a) For the purposes of this section:

* * *

(2) "Energy-savings contract" means a contract for the evaluation and recommendation of energy operations conservation measures, and for implementation of one or more such measures. The contract shall provide that payments, except obligations upon termination of the contract before its expiration, are to be made over time. A county board of education may supplement these payments with federal, state or local funds to reduce the annual cost or to lower the initial amount to be financed.

(3) "Qualified provider" means a person, firm or corporation experienced in the design, implementation and installation of energy-conservation measures.

(b) County boards of education are hereby authorized to enter into performance-based contracts with qualified providers of energy-conservation measures for the purpose of reducing energy operating costs of school buildings.

(c) A board of education may enter into an energy-savings contract with a qualified provider to reduce energy operating costs significantly. Before entering into such a contract or before the installation of equipment, modifications or remodeling to be furnished under such a contract, the ***qualified provider shall first issue a proposal*** summarizing the scope of work to be performed. ***Such a proposal shall contain estimates of all costs of installation, modifications or remodeling, including the costs of design, engineering, installation,*** maintenance, repairs or debt service, as well as estimates of the amounts by which energy operating costs will be reduced. If the board finds, after receiving the proposal, that the proposal includes more than one energy-conservation measure designed to save energy operating costs, the board may enter into a contract with the provider pursuant to this section.

* * *

WISCONSIN

I. Public Procurement Laws.

Wis. Stat. Ann. § 59.79. Milwaukee County.
In a county with a population of 500,000 or more, the board may:

* * *

(11) Baseball. Own and operate a professional baseball team, appropriate money for the purchase of a franchise for such team, and in the discretion of the board organize and maintain a nonprofit corporation for such team ownership and operation or participate with others in the formation and operation of such corporation.

* * *

(13) **Design-build** construction process. Let a contract for the construction of a sheriff's department training academy, that is located in the county, using the **design-build** construction process.

Highway Projects.
Wis. Stat. Ann. § 84.11. Local bridge construction and reconstruction.
(1) Definitions. In this section:
(a) "Construction" includes reconstruction.
(b) "Local bridge" means a bridge which is not on the state trunk highway system or on marked routes of the state trunk highway system designated as connecting highways.
(1g) Administration. The department shall administer the local bridge construction program and shall have all powers necessary and convenient to implement this section.

* * *

(5n) ***Design-build contracts. (a) In this subsection, "design-build contract" means a contract for a project under which the engineering, design and construction services are provided by a single entity.***
(b) Notwithstanding any other provision of this section and ss. 84.01(13) and 84.06(2), the department may enter into ***a design-build contract*** for the design and construction of a bridge for which funding is provided under s. 84.11(5), 1993 stats., and for which no contract for construction is awarded before May 1, 1999. The department may enter into a contract under this paragraph only if all of the following conditions are met:
1. ***The design-build contract*** is awarded through a competitive selection process that utilizes, at a minimum, contractor qualifications, quality, completion time and cost as award criteria. In order to be eligible to participate in the selection process, the contractor must be prequalified by the department as a design consultant and as a contractor.
2. ***The design-build contract*** is approved by the secretary of the federal department of transportation under an experimental program described under section 1307(d) of P.L. 105-178 pursuant to the authority granted under section 1307(e) of P.L. 105-178.

3. ***The design-build contract is approved by the governor.***

(c) No later than 5 years after October 29, 1999, the department shall submit a report to the governor, and to the legislature under s. 13.172(2), describing the effectiveness of ***the design-build process*** contracting procedures under this subsection.

* * *

II. Attorney General Opinions.

This opinion dealt with a county housing authority utilizing the 'Turnkey' construction method without the necessity of complying with the bidding requirements of state procurement laws. The opinions defined "Turnkey" as contracts for completed housing to be produced by the developer on his own land or on land which authority may provide and upon completion of construction and the 'turning over of the keys' to the authority, payment for the project is made by the authority. The attorney general concluded that a county is required to utilize competitive bids with respect to the construction of public works, but a county housing authority is separate and apart from the county. A county housing authority would have to comply with the bidding requirements unless some other statute excepted it from that duty. Section 59.075(4) creates such an exception and, therefore, acquisition may done using the 'Turnkey' construction method without the necessity of complying with the bidding requirements if the provisions of federal law are complied with. 66 Wis. Op. Atty. Gen. 31; 1977 WL 36103 (Wis.A.G.).

WYOMING

None.

INDEX

References are to sections numbers or appendixes.

C

D

E

P

Q

R